TABLE B.3 (Continued)

The Standard Normal Distribution
Areas under the standard normal curve from 0 to z for various values of z.

	.00	.01	.02	.03	.04	.05	.06	.07	.08	.09
1.5	.4332	.4345	.4357	.4370	.4382	.4394	.4406	.4418	.4429	.4441
1.6	.4452	.4463	.4474	.4484	.4495	.4505	.4515	.4525	.4535	.4545
1.7	.4554	.4564	.4573	.4582	.4591	.4599	.4608	.4616	.4625	.4633
1.8	.4641	.4649	.4656	.4664	.4671	.4678	.4686	.4693	.4699	.4706
1.9	.4713	.4719	.4726	.4732	.4738	.4744	.4750	.4756	.4761	.4767
2.0	.4772	.4778	.4783	.4788	.4793	.4798	.4803	.4808	.4812	.4817
2.1	.4821	.4826	.4830	.4834	.4838	.4842	.4846	.4850	.4854	.4857
2.2	.4861	.4864	.4868	.4871	.4875	.4878	.4881	.4884	.4887	.4890
2.3	.4893	.4896	.4898	.4901	.4904	.4906	.4909	.4911	.4913	.4916
2.4	.4918	.4920	.4922	.4925	.4927	.4929	.4931	.4932	.4934	.4936
2.5	.4938	.4940	.4941	.4943	.4945	.4946	.4948	.4949	.4951	.4952
2.6	.4953	.4955	.4956	.4957	.4959	.4960	.4961	.4962	.4963	.4964
2.7	.4965	.4966	.4967	.4968	.4969	.4970	.4971	.4972	.4973	.4974
2.8	.4974	.4975	.4976	.4977	.4977	.4978	.4979	.4979	.4980	.4981
2.9	.4981	.4982	.4982	.4983	.4984	.4984	.4985	.4985	.4986	.4986
3.0	.4987	.4987	.4987	.4988	.4988	.4989	.4989	.4989	.4990	.4990
3.1	.4990	.4991	.4991	.4991	.4992	.4992	.4992	.4992	.4993	.4993
3.2	.4993	.4993	.4994	.4994	.4994	.4994	.4994	.4995	.4995	.4995
3.3	.4995	.4995	.4995	.4996	.4996	.4996	.4996	.4996	.4996	.4997
3.4	.4997	.4997	.4997	.4997	.4997	.4997	.4997	.4997	.4997	.4998
3.5	.4998	.4998	.4998	.4998	.4998	.4998	.4998	.4998	.4998	.4998
3.6	.4998	.4998	.4999	.4999	.4999	.4999	.4999	.4999	.4999	.4999
3.7	.4999	.4999	.4999	.4999	.4999	.4999	.4999	.4999	.4999	.4999
3.8	.4999	.4999	.4999	.4999	.4999	.4999	.4999	.4999	.4999	.4999
3.9	.49995	.49995	.49996	.49996	.49996	.49996	.49996	.49996	.49997	.49997
4.0	.49997									
4.5	.499997									
5.0	.4999997									

Adapted from *Standard Mathematical Tables*, 25th ed. (Boca Raton: Chemical Rubber Company Press, 1978), p. 524. Reprinted with permission.

GENERAL STATISTICS

Second Edition

Warren Chase

Fred Bown
Framingham State College

John Wiley & Sons, Inc.

New York　　*Chichester*　　*Brisbane*　　*Toronto*　　*Singapore*

ACQUISITIONS EDITOR / Brad Wiley II
PRODUCTION MANAGER / Linda Muriello
DESIGNER / Kevin Murphy
PRODUCTION SUPERVISOR / Micheline Frederick
MANUFACTURING MANAGER / Lorraine Fumoso
COPY EDITOR / Elizabeth Swain
COVER AND INTERIOR OPENING ART / Roy Wiemann

Recognizing the importance of preserving what has been written, it is a policy of John Wiley & Sons, Inc. to have books of enduring value published in the United States printed on acid-free paper, and we exert our best efforts to that end.

Library of Congress Cataloging in Publication Data:
ISBN 0-471-61901-9

Printed in the United States of America

10 9 8 7 6 5 4 3 2 1

To

Ginny   Becky
Gerry   Maura
Kenny   Joe
Judy
Mladen
Xiaodan

A Note from the Publisher

John Wiley & Sons, Inc., is committed to publishing textbooks of pedagogical integrity and technical accuracy. To these ends, the following measures were built into the revision program for *General Statistics*, 2e.

After both a survey of adopters and a review program, the authors rewrote many parts of the original text. After further review, the authors, publisher's copyeditors, and a professional statistics teacher each proofread the textual material, examples and solutions, student exercises, students' answers, and teacher's solutions manual for presentation consistency and mathematical and notational accuracy.

Preface

The purpose of this book is to present a first course in statistics with an emphasis on statistical inference. It is appropriate for students in a wide variety of disciplines, the only prerequisite being a knowledge of intermediate high school algebra. Great care has been taken in writing the book to make the subject matter understandable. All technical terms are defined in easy to grasp language; definitions, important formulas, and summaries of statistical tests are set off in boxes for quick reference. Concepts are introduced and reinforced with examples and exercises from a wide range of fields from sports to medicine. All chapters begin with an introduction and end with a summary of the important ideas of the chapter.

New Features of the Second Edition

- The chapters on measures of central tendency and measures of dispersion have been combined and the discussion has been streamlined.

- The treatment of Minitab has been updated and expanded. With the exception of the introduction and the chapter on probability, all chapters contain optional sections at the end showing how to apply Minitab to the concepts developed in the chapter. These sections constitute a self-contained introduction to Minitab.

- The treatment of P values has been expanded. A survey of our users of the first edition showed varying opinions on how much emphasis should be placed on P values versus the traditional approach to hypothesis testing. Although some of our exercises on hypothesis testing ask for solutions using both the traditional approach and the P value approach, most do not specify which approach to use. We give answers in terms of both approaches. The instructor can then decide which to use.

- Optional sections on sampling, box-and-whisker plots, counting techniques (combinatorics), and the design of experiments have been added.

- There has been a substantial increase in the number of exercises, and there is now a *Student Solutions Manual* containing the complete solutions to selected exercises.

- An *Instructor's Manual* contains the complete solutions to all exercises, a chapter on multiple regression that may be reproduced for class use, and a test bank. The test bank contains true–false, fill-in-the-blank, multiple-choice, and computational exercises. The test bank is also available on a floppy disk for either an IBM PC or a Macintosh hardware environment. The disk may be edited by the instructor.

- The *Minitab Supplement to Accompany General Statistics*, 2nd ed., by Anne Sevin, has been updated to the current release of Minitab. This supplement provides material for a lab course or homework assignments, giving further instruction in computation, analysis, and decision making with the computer. A data disk that includes data on 1000 randomly selected subjects from the Framingham Heart Study is available for replication to any adopter of the supplement or the main text.

- A *Study Guide* to review the course has been prepared by Professor James Curl of Modesto Junior College. The study guide also includes a review of algebra and notation.

The book is organized to give the instructor maximum flexibility in choosing the order of topics in the course. For instance, some instructors like to introduce the students to regression and correlation early in the course. Although we cover these topics in Chapter 10, the first three sections of Chapter 10 can be used to provide an early introduction to these topics right after completing Chapter 3. In Chapter 7, where estimation and decision making (hypothesis testing) are introduced, the sections are written so that these topics may be treated in any order.

Another example of flexibility is our treatment of probability. We believe that students taking a first course in statistics need to develop both an understanding of the notions of event and the probability of an event, and the ability to calculate some simple probabilities. Instructors who wish to move quickly through probability may cover these topics in Chapter 4, sections 4.1–4.3. Compound events (including the Addition Rule and Multiplication Rule) are covered in sections 4.4–4.7, but may be omitted if desired. The text is written in such a way that only sections 4.1–4.3 are required. Even the treatment of contingency tables in Chapter 12, which in many books requires a prior understanding of independent events and the Multiplication Rule, is self-contained and requires no prior exposure to these topics.

It seems to us that one of the problems encountered in teaching (and studying) statistics is that it's easy to get bogged down in details of the early topics in the course, such as descriptive statistics and probability. We favor moving rapidly through these topics so that more time can be spent on inferential statistics (studying populations using samples). At Framingham State College, topics from the first 10 chapters constitute an introductory statistics course for liberal arts majors. We move very rapidly through the first 4 chapters. About all the student needs to know from these chapters is measures of central tendency (mean, median, mode), variance, standard deviation, percentiles, some graphical techniques, and what is meant by probability of an event (the first 3 sections of Chapter 4). We spend a fair amount of time on Chapter 6 (Normal Distributions) and then cover selected topics from Chapters 7 through 10. Depending on how rapidly one moves through Chapters 1 through 4, additional material might be covered such as Analysis of Variance (Chapter 11) or Analysis of Categorical Data (Chapter 12). After Chapter 9, the remaining chapters are independent—none is a prerequisite of any of the others. See Figure 1 for a chart indicating prerequisites.

In selecting topics for a course, one thing to keep in mind is that some sections are labeled as "optional" because most instructors would agree that they indeed are optional. That doesn't mean that other sections may not be

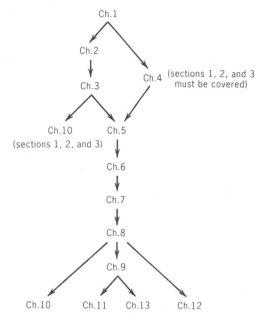

FIGURE 1
Prerequisite Chart. Chapter at Tail of Arrow Is a Prerequisite for Chapter at Point of Arrow.

legitimately viewed as optional by some instructors. Chapter 4 is a case in point. We view the last four sections as optional, although only one is labeled as such.

The theme underlying this book is that statistics is a very general and powerful discipline, and its methods are applicable in a variety of areas. We feel that a first course in statistics should stress this idea. Students who will make substantial use of statistics in their field may wish to follow up a general statistics course with a statistics course that is oriented toward their own discipline.

Acknowledgments

We wish to thank Dr. William Castelli of the Framingham Heart Study for providing us with data and reprints of papers from the heart study. We also wish to thank Robert Garrison, Michael Hartman, and Paul Sorlie of the National Institutes of Health for their role in compiling the heart study data and for providing us with computer tapes containing the data.

We want to thank the following reviewers of the second edition for their many valuable suggestions:

Geraldine C. Banda
Glassboro State College

Carole Bernett
Harper College

J. Stanley Laughlin
Idaho State University

Kenneth Schoen
Worcester State College

Louis F. Bush
San Diego City College

Nancy Schoeps
University of North Carolina

James M. Edmondson
Santa Barbara City College

Robert A. Schley
Mt. San Jacinto Community College

Janet Evert
Erie Community College

Lynn Smith
Gloucester County College

Shu-Ping Hodgson
Central Michigan University

John D. Spurrier
University of South Carolina

Special thanks are in order for James Curl of Modesto Community College whose exhaustive comparative review of our first edition and a number of other books, provided a wealth of information and ideas; and also for Mack Hill of Worcester State College who worked through all the examples and exercises and checked the manuscript of the second edition for accuracy. We also want to express our gratitude to our editor, Brad Wiley; his hard work, ethusiasm, and encouragement contributed in large part to making the second edition a reality.

W.C.
F.B.

Contents

GENERAL STATISTICS

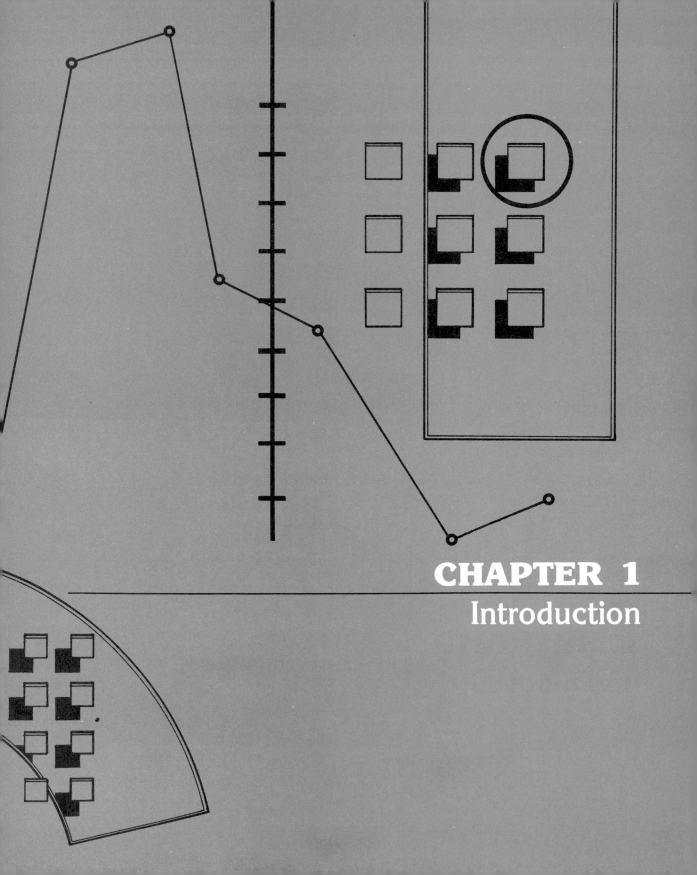

CHAPTER 1
Introduction

THE NATURE OF STATISTICS

SAMPLING (OPTIONAL)

SUMMARY

NOTES

THE NATURE OF STATISTICS

To get an idea of what statistics is all about, we consider a few concrete illustrations of the kinds of problems that involve statistics.

- A pollster interested in the outcome of an upcoming election interviews a certain number of voters and, based on the results obtained, makes a prediction as to who will probably win the election.

- The Environmental Protection Agency conducts tests on a certain number of cars of the same make and model in order to estimate the average gas mileage for all cars of that make and model.

- Each year the Federal Bureau of Investigation publishes the "Uniform Crime Reports." Among other things, this document reports the violent crime rate (the number per 100,000 people) for each of the Metropolitan Statistical Areas in the United States.
- In order to get an idea of the economic status of a town, an economist obtains the salaries of all wage earners in the town and then computes the average.
- To estimate the average age of all adult Americans, a sociologist selects 1000 adults, computes their average age, and then uses this figure as an estimate.

All of the above illustrations make use of the raw material of statistics, namely, **data** (or **data values**). Here we use these terms in a very broad sense. That is, a data value is simply a piece of information that might be numerical, such as the annual snowfall in Boston, or a person's weight or age. Or it might be nonnumerical information, such as the color of a car, a person's ethnic status, or whom you favor in the next presidential election. We give the following definition of **statistics:**

Definition *Statistics* is the science of collecting, simplifying, and describing data, as well as making inferences (drawing conclusions) based on the analysis of data.

As the definition suggests, there are two branches of statistics. The branch that deals with collecting, simplifying, and giving properties of data is called **descriptive statistics.** An important objective of descriptive statistics is to organize, summarize, or describe the data so as to make it more comprehensible. For example, suppose that we obtained a list of the salaries of all the wage earners in Boston. This list would be so long that it would be incomprehensible. But if we were to find the average of all these salaries, then we would understand something about the economic status of the residents of Boston.

The other branch of statistics, which involves drawing conclusions based on the analysis of data, is called **inferential statistics.** Since the major emphasis in this book will be on inferential statistics, let us examine this concept carefully. The pollster who predicts the outcome of an election based on a knowledge of only *some* of the votes and the sociologist who estimates the average age of *all* adult Americans based on a knowledge of the average of the ages of *some* of these adults are both using inferential statistics. Apparently, it was impractical for these researchers to obtain all the data they were interested in (i.e., all the votes or all the ages); therefore, in both cases a judgment was made about the larger body of data that was being studied by means of information obtained from only some of these data values. This motivates the following definition:

> **Definition** The entire collection of all the elements we are interested in is called a *population*. (These elements might be people, automobiles, data values, etc.) A collection of some of the elements obtained from the population is called a *sample* from the population.

In an investigation such as a voter preference study, we may think of the population as consisting of all the voters or all the votes they will cast. The votes are the **data values** of interest, and it is these values that we are really investigating. In general, we are ultimately interested in data values in statistics. For this reason, in this book we will often think of populations as consisting of data values.

Now that we have introduced the terms *population* and *sample*, we can give a more precise definition of inferential statistics.

> **Definition** *Inferential statistics* is concerned with making judgments (or inferences) about a population based on the properties of some sample obtained from the population.

We said that trying to estimate the average age of all adult Americans by using the average of the ages of some of these adults is a typical problem in inferential statistics. The average age for the population of all adult Americans is an example of what is called a **parameter.** The average of the ages for a sample of adult Americans is an example of a **statistic.**

> **Definition** A numerical property of a population is called a *parameter*. A numerical property of a sample is called a *statistic*. (By "numerical property" we mean a property that is expressible as a number.)

For another example of these terms consider the population of all voters (or votes) in the 1936 presidential election. The percentage of the voters that were for Roosevelt is a parameter. Viewing the voters in Peoria, Illinois, as a sample from this population, we see that the percentage in this sample that were for Roosevelt is a statistic.

Many problems in inferential statistics involve estimating the value of, or making some decision concerning, a parameter based on the value of a statistic.

The methods used in making inferences in statistics are probabilistic. For example, suppose that a pollster interviews 100 voters selected by chance and finds that 96 of them favor candidate A in an upcoming election. Then he would say that the evidence points to candidate A winning the election, because it would be highly unlikely or improbable that so many voters (in the sample of 100 voters) would be for A if A were not going to win the election. **Probability, which deals with the laws of chance, plays an important role in statistics.** Therefore, we will study this topic in some detail in this book.

Exercises

In Exercises 1.1 and 1.2, discuss similarities and differences between the following terms.

1.1   Population and sample.

1.2   Parameter and statistic.

In Exercises 1.3–1.5, which are true?

1.3   Generally, the value of a parameter is unknown.

1.4   The value of a parameter always remains unchanged from sample to sample.

1.5   The value of a statistic always remains unchanged from sample to sample.

In Exercises 1.6–1.10, determine whether the results given are examples of descriptive or inferential statistics.

1.6   In the 1988 presidential election, voters in Massachusetts cast 1,184,323 and 1,387,398 votes for George Bush and Michael Dukakis, respectively (*Source: The 1989 Information Please Almanac*, 1989, p. 44).

1.7   As of January 1, 1988, the Nielsen Company estimated the percentage of television sets in United States homes (*Source: World Almanac*, 1989, p. 356) as follows:

Color TV sets	96%	Black and white only	4%
Two or more sets	60%	One set	40%

1.8   The mid–1988 population of the United States was estimated to be 246.1 million, and the 1987 median age estimated to be 32.1 years (*Source: The 1989 Information Please Almanac*, 1989, p. 2). Note: Estimating a median age of 32.1 years is saying that 50% of the population is estimated to be younger than 32.1 years of age.

1.9   In the 1988 presidential election, George Bush and Michael Dukakis received 426 and 46 electoral votes, respectively (*Source: The 1989 Information Please Almanac*, 1989, p. 44).

1.10   The population of Worcester, Massachusetts in 1980 was 161,799, a decrease of 8.4% from the 1970 population of 176,572 (*Source: World Almanac*, 1988, p. 539).

1.11   The Bureau of the Census estimates that 5% of the black population was missed in the 1980 population census (*Source:* Hansen, M. H., and B. A. Bailar, "How to Count Better," *Statistics: A Guide to the Unknown*, 3rd ed., p. 209). Is 5% the value of a parameter or a statistic?

1.12   A researcher is looking into aspects of life at a large eastern university. The researcher wishes to estimate the proportion of students at the university who are commuters. The researcher does not know that 4/10 of the students are commuters. Is 4/10 the value of a parameter or a statistic?

1.13 A linguist was interested in the population of words in James Joyce's *Ulysses*. The word *the* occurs 14,887 times (*Source:* Simon, H. A., "The Sizes of Things," *Statistics: A Guide to the Unknown*, 3rd ed., p. 143). Is 14,887 the value of a parameter or a statistic?

1.14 A sample of 50 federal employees in a large midwestern city showed an average age of 38.2 years. Is 38.2 the value of a parameter or a statistic?

1.15 Consider the problem of estimating the average grade-point average (GPA) of the 750 seniors at a college. (The average, which is unknown, is the sum of 750 GPAs divided by 750.)

(a) What is the population? How many data values are in the population?

(b) What is the parameter of interest?

(c) Suppose that a sample of 10 seniors is selected and their GPAs are 2.72, 2.81, 2.65, 2.69, 3.17, 2.74, 2.57, 2.17, 3.48, 3.10. Calculate a statistic that you would use to estimate the parameter.

(d) Suppose that another sample of 10 seniors was selected. Would it be likely that the value of the statistic is the same as in part (c)? Why or why not? Would the value of the parameter remain the same?

1.16 Unaware that 35% of the 10,000 voters in his district still support him, a politician decides to estimate his political strength. A sample of 200 voters shows 40% support him.

(a) What is the population?

(b) What is the value of the parameter of interest?

(c) What is the value of the statistic of interest?

(d) Compare your answers in (b) and (c). Is it surprising they are different? If the politician were to sample another 200 voters, which of the two numbers would most likely change? Explain.

1.17 A sociologist was interested in estimating some aspects of family life in a town. Information about the entire town (unknown to the sociologist) and the results of a sample obtained by the sociologist follow:

Number of Children per Family	Number of Families in Town	Number of Families in Sample
0	120	6
1	180	10
2	270	12
3	300	8
4	80	6
5	50	8

(a) Identify the elements of the population and give the population size.

(b) Identify the elements of the sample and give the sample size.

(c)   Suppose that the sociologist were interested in the *number* of families with more than two children.

  (i)   Calculate a statistic to estimate the parameter of interest. (*Hint:* The population is 20 times the size of the sample.)

  (ii)   What is the value of the parameter?

(d)   Suppose that the sociologist were interested in the *proportion* of families with less than four children.

  (i)   What is the value of the statistic?

  (ii)   What is the value of the parameter?

1.18   In the Massachusetts State Lottery Megabucks game, six numbers are selected from the set of 36 numbers 1, 2, 3, 4, ... , 33, 34, 35, 36. A player felt that since the inception of the lottery, too high a percentage of single-digit numbers were being selected. He obtained a partial list of past drawings and observed that 28.6% of numbers were single digit (*Source:* Massachusetts State Lottery Commission, Public Relations Dept., 15 Rockdale St., Braintree, MA).

(a)   If the numbers were generated randomly, what would the value of the parameter of interest be?

(b)   What is the value of the statistic?

1.2   SAMPLING (OPTIONAL)

We said that in inferential statistics we use samples to make judgments about populations. Hopefully the samples we obtain will be **representative** of the population, that is, will resemble the population. There are many ways of obtaining samples; we mention only a few here.

Random Samples

Random sampling is one of the most important types of sampling in statistics.

> **Definition**   A *random sample* is a sample obtained from the population in such a manner that all samples of the same size have equal likelihood of being selected. Any method of obtaining random samples is called *random sampling*.

For example, one method of random sampling is the **lottery method.** With this method, elements in the population are identified by a name or number written on a tag. The tags are placed in a container and are then well stirred up. A tag is then drawn by chance from the container, and this process is repeated until the desired number of tags is obtained.

When the elements in the population are identified by numbers (such as employee identification numbers for employees of a large corporation), we may use the **random number method** to obtain random samples. Random numbers are found in Appendix Table B.1. Instructions on how to use the table are found in Appendix A. Suppose there are 9000 employees with ID numbers 0001, 0002, 0003, and so on, up to 9000. We can use the random number table to obtain a random sample of, say, 5 employees, by choosing a sequence of 5 ID numbers that occurred in a purely random manner. When the sampling is done in such a way that there are no repetitions in the sample (for example, by ignoring ID numbers that repeat) we call this **simple random sampling.**

Stratified Samples

Another type of sample frequently encountered is a **stratified sample.**

> **Definition**   If the population is divided into subpopulations, called *strata*, and we take a random sample from each stratum, the resulting sample is a *stratified sample.*

For example, the students at a college could be divided into strata according to class: freshmen, sophomores, juniors, seniors. To obtain a stratified sample, we could take a random sample from each class. In stratified sampling, the size of the sample from each stratum is often proportional to the size of the stratum in the population. For example, suppose 40% of the student body are freshmen, 25% sophomores, 20% juniors, and 15% seniors. To obtain a stratified sample of size 100, we could sample 40 freshmen, 25 sophomores, 20 juniors, and 15 seniors. The result would be a **proportional stratified sample.**

Stratified samples are often easier to obtain than true random samples. For example, if we wanted a sample of Massachusetts voters, it would be inconvenient to compile a list or computer file of all voters from which to select a random sample. However, we could sample voters from each voting precinct, thereby obtaining a stratified sample.

Cluster Samples

Even though a population can be divided into strata, it may not be convenient to sample from each stratum. For example, instead of sampling from each voting precinct in the state, we could randomly select some of the precincts and sample from each of these. Thus if we wanted a sample of 1000 voters, we might select 20 precincts and randomly sample 50 voters from each precinct. This kind of sample is called a **cluster sample.**

> **Definition**   A *cluster sample* is obtained by selecting some of the strata and then sampling from each of these.

Systematic Samples

If we have a list of the elements in the population, an easy way to obtain a sample is by **systematic sampling.** For example, from a list of all employees of a corporation, we could choose by chance some starting point on the list and then select perhaps every 5th name on the list. The starting point could be chosen by selecting an ID number from the random number table.

> **Definition**   From a list of members of a population, choose a starting point by chance and then select every nth element on the list (for some appropriate value of n). The result is a *systematic sample*.

For example, suppose you have a list of 100 ID numbers, and you want a sample of size 12. Notice that $100/12 \doteq 8.33$ (where $\doteq$ means "is approximately equal to"). Round *up,* getting 9. Randomly select a starting point. Then select every 9th ID number until you have 12 numbers. You could use the random number table to select a starting ID number. If the starting point was 004, the next selected would be 013.

The samples discussed thus far involve a chance process to obtain the sample and fall under the general heading of **probability samples.** However, occasionally we encounter samples that are not probability samples. An example of this is a **sample of convenience.**

Samples of Convenience

A statistics instructor who was teaching two sections of statistics wanted to compare two books, so one book was used in one section and the other book was used in the other section. The two statistics sections are examples of **samples of convenience.**

> **Definition**   A *sample of convenience* is a sample that already exists and is available for study; the elements in the sample are not chosen by a chance process. By contrast, a *probability sample* is obtained by a chance process; each element in the population has a certain probability of being selected.

A sample of convenience can be useful provided we can be fairly confident that the sample is not biased for some reason. For example, we should not choose a statistics section as representative of statistics students if it is a section consisting of honors students.

Sometimes common sense is the most important element of a sampling strategy. In the 1936 presidential campaign between Franklin Roosevelt and Republican opponent Alf Landon, the publishers of *Literary Digest* magazine conducted a telephone survey of some 2.5 million people—an enormous sample by today's standards. On the basis of the survey, the magazine predicted a landslide win for Alf Landon and, of course, the opposite occurred. The flaw in the sampling scheme was that in 1936 only the relatively well-to-do (many of whom were Republicans) possessed telephones. In other words, the sample was biased.

The A. C. Nielsen Company conducts surveys to determine the percentage (share) of T.V. viewers watching various shows. In the 1960s some 1600 families were used in a sample selected as representative of T.V.-viewing families in America. A device called an audometer was attached to their T.V. sets to record the shows being viewed. The audometer was connected by telephone lines to Nielsen computers. A Texas viewer who happened to be in the sample had a strong aversion to then president of the United States—Lyndon Johnson. One night he became enraged at something Johnson said on the news, and fired his shotgun into the T.V., destroying the T.V. and the audometer. He bought a new T.V. the next day and called the Nielsen Company to have them install a new audometer. A few weeks later the same thing happened. After this occurred a number of times, the Nielsen Company became tired of installing new audometers, and officials decided to remove this individual from their sample. However, they were overruled by Mr. Nielsen, the president of the company. His reasoning was that in order for the sample to be representative, the lunatic fringe must also be included.

In Part III of this book (Inferential Statistics) we use samples to study populations. The methods we use there are valid for random samples. However, many statisticians would use these techniques on samples of convenience as well.

Exercises

1.19   List all simple random samples of size 3 from the population {a, b, c, d}.

1.20   For the population {0, 1, 2, 3, 5, 7},

    (a)   Give an example of a proportional stratified sample of size 3 where the strata are {0, 1, 2, 3}, {5, 7}.

    (b)   Give a cluster sample of size 3 where the clusters are {0, 1, 2}, {3, 5}.

    (c)   Give a systematic sample of size 3 where the elements are listed in order: 0, 1, 2, 3, 5, 7.

1.21 A business has 100 employees, and each employee is identified with a 2-digit ID number. A complete list of the identification numbers is 00, 01, 02, 03, ... , 99. Assume there are 4 departments A, B, C, and D within the business consisting of the following employees:

A: 00, 01, ... , 09   B: 10, 11, ... , 29

C: 30, 31, ... , 59   D: 60, 61, ... , 99

(a) Give a systematic sample of 10 ID numbers.

(b) Give a sample of 10 ID numbers using Appendix Table B.1. Start with the first column of numbers 10, 37, 08, ... List the elements sampled for each of the following cases:

(i) Simple random sample.

(ii) Proportional stratified sample.

(iii) Cluster sample, assuming department B is randomly selected.

1.3 SUMMARY

Statistics is the science of collecting, simplifying, and describing data as well as making inferences based on the analysis of data.

The collection of all the elements (often data values) we are interested in is called a **population**. A collection of some of the elements obtained from the population is called a **sample** from the population.

Descriptive statistics involves collecting, simplifying, and giving the properties of data. **Inferential statistics** is concerned with making judgments about a population based on properties of some sample obtained from the population.

A number representing a numerical property of a population is called a **parameter**. A number representing a numerical property of a sample is called a **statistic**.

A **random sample** is a sample obtained from the population in such a manner that all samples of the same size have equal likelihood of being selected. A **stratified sample** is obtained by dividing the population into strata or subpopulations and sampling from each stratum. A **cluster sample** is obtained by selecting some of the strata and sampling from these. A **systematic sample** can be chosen from a list of the members of the population by selecting a starting point on the list by chance, and then choosing every nth element on the list for some appropriate n. A **sample of convenience** is a sample that already exists and is available for study; the elements of the sample are not selected by any chance process.

Notes

Information Please Almanac (Boston: Houghton Mifflin Company, 1989).

Massachusetts State Lottery Commission, Braintree, MA.

Tanur, J. M., F. Mosteller, W. H. Kruskal, E. L. Lehmann, R. F. Link, R. S. Pieters, G. R. Rising, eds., *Statistics: A Guide to the Unknown*, 3rd ed. (Pacific Grove, CA: Wadsworth & Brooks/Cole, 1989).

The World Almanac and Book of Facts (New York: Newspaper Enterprise Association, 1988).

The World Almanac and Book of Facts (New York: Newspaper Enterprise Association, 1989).

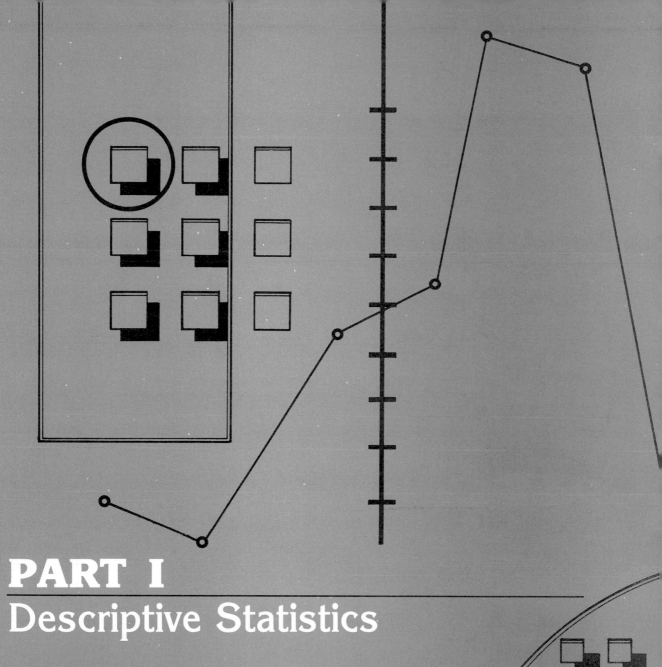

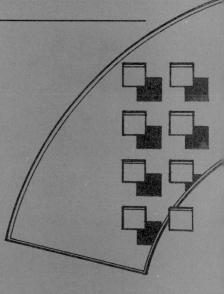

PART I
Descriptive Statistics

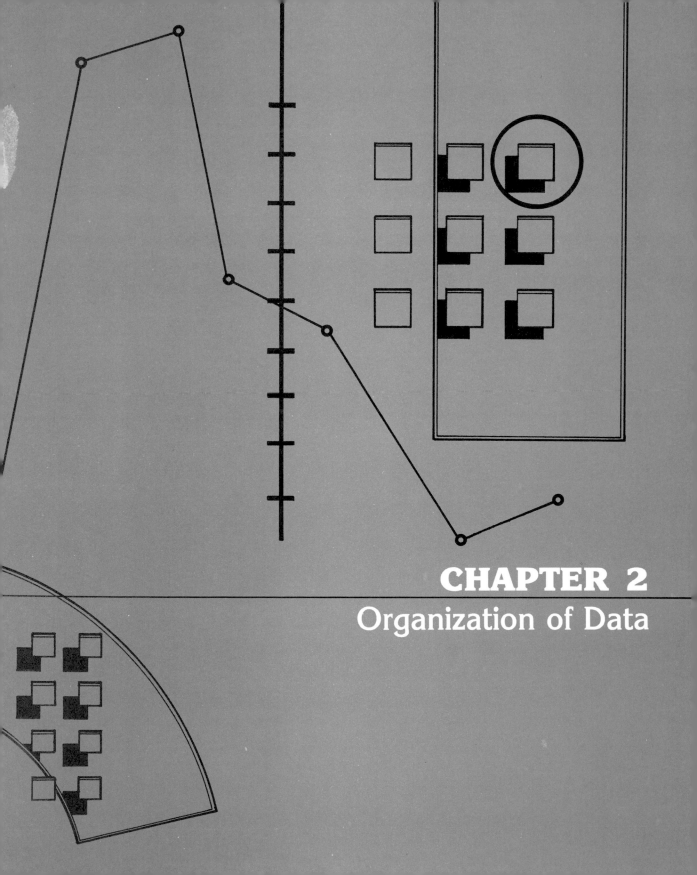

CHAPTER 2
Organization of Data

2.1 INTRODUCTION

In the last chapter we said that descriptive statistics involves organizing and describing data in such a way as to make it more comprehensible. In this chapter we concentrate on how to organize or summarize the data. This will involve presenting the data in a form that is easy to comprehend. It will also involve representing the data graphically. In the next chapter we will continue our study of descriptive statistics by discussing how to describe data by means of certain numerical characteristics (such as the average of the data values).

Occasionally, the data values we study constitute all the data values we happen to be interested in; that is, they may constitute all the data values from a population. But more often in practice we can only obtain a sample from the population, in which case we would hope that the sample was *representative* of the population, that is, resembled the population. In this case, the essential information we obtain from the sample would also be true of the population as a whole.

Notation

It may be helpful for the reader if we point out here a certain use of symbols that will be encountered throughout this book. If we have a collection of data values, we often represent an arbitrary or unspecified data value from this collection by a lower-case letter from near the end of the alphabet, such as x or y or z.

SUMMARIZING DATA

Frequency Distributions

Rather than write out all the data values in a collection of data, including repetitions, it is sometimes more convenient to simply list the distinct values in the collection together with the number of times each value occurs.

Definition   Given a collection of data values, the specification of all the distinct values in the collection together with the number of times each of these values occurs in the collection is called a *frequency distribution*. The number of times a value occurs is called its *frequency*. The frequency of a data value x is denoted by f_x (or just f).

Example 2.1

Construct a frequency distribution for the following data, which represent the number of breakdowns for each truck in a fleet of 15 delivery trucks in 1 year:

$$6 \quad 11 \quad 5 \quad 1 \quad 6 \quad 6 \quad 7 \quad 5 \quad 7 \quad 6 \quad 1 \quad 7 \quad 5 \quad 3 \quad 6$$

Solution

The distinct values are 1, 3, 5, 6, 7, and 11, which occur 2, 1, 3, 5, 3, and 1 times, respectively. This is an adequate description of the frequency distribution. However, it is also common to describe the frequency distribution by means of a table. The first column of the table contains the distinct values. In the second column next to each distinct value, we record its frequency. So the frequency distribution is represented as

x	f
1	2
3	1
5	3
6	5
7	3
11	1

Example 2.2

Frequency distributions may also be used to summarize nonnumerical data. The Federal Bureau of Investigation gathered information on all murders committed in 1987. The data value associated with each murder was the type of weapon used. Table 2.1 gives the number of murders (f) for each weapon type (x). For convenience we have added a column giving the percentage for each weapon type.

TABLE 2.1
Murders by Type of Weapon

Type of Weapon (x)	Frequency (f)	Percentage (approximate)
Handgun	7,807	44
Rifle	772	4
Shotgun	1,095	6
Knife	3,619	20
Personal weapon (hands, fists, etc.)	1,162	7
Other weapon[1] (club, poison, etc.)	3,404	19
Total	17,859	100

[1]Includes weapons not specified.
Source: Data based on *Crime in the United States—Uniform Crime Reports* (Washington, D.C.: Federal Bureau of Investigation, 1988, p. 10).

Grouped Frequency Distributions

The frequency distributions just discussed can be helpful in simplifying data if the number of distinct data values is not too large. But if the number of distinct data values is large, which is often the case with numerical data, the frequency distribution will not be of much value in simplifying the data.

Given a large collection of numerical data values, we can simplify these data and make them more comprehensible as follows: Divide an interval containing these data into a small number of line segments (or subintervals), usually of equal width. These segments are called **classes.** We then tell how many data values fall into each class. The result is called a **grouped frequency distribution.**

For example, a department store described the ages of its 100 employees with the following grouped frequency distribution:

Age	Number of Employees
20–29	30
30–39	35
40–49	20
50–59	10
60–69	5

Classes { ... } Class frequencies

The values 20, 30, 40, 50, and 60 are called **lower class limits.** The values 29, 39, 49, 59, and 69 are called **upper class limits.** The distance from one lower class limit to the next lower class limit is called the **class width.** For the above grouped frequency distribution, the class width is 10. (*Note:* This is the distance from lower class limit to lower class limit, not from lower class limit to upper class limit.)

We illustrate the procedure for constructing a grouped frequency distribution for the data in Table 2.2, which represent the scores of 40 high school students on a scientific achievement test.

Notice that the data in Table 2.2 are ranked, that is, arranged in the order of increasing magnitude. When the data values are ranked, it makes it easier to count the number of values in each class. However, it is not essential to rank the data in order to construct a grouped frequency distribution. But if we are working with unranked data, we must be very careful when counting the number of data values in each class.

TABLE 2.2
Scores on a Scientific Achievement Test

46	58	65	70	76
49	59	66	71	78
50	59	66	71	79
53	60	66	72	80
54	62	66	73	82
55	63	68	73	83
55	64	68	73	84
57	65	69	74	88

Procedure for Constructing a Grouped Frequency Distribution

1. Decide on the starting point, that is, the lower class limit of the first class. The lowest data value in Table 2.2 is 46. The lower class limit should be 46 or a number somewhat less than this. The number 45 seems convenient, so we will use this.

2. Choose the number of classes and the class width W. Each class will have the same width. The distance from our starting point (45) to the highest data value (88) is 43 units. We can cover the distance 43 with 9 classes having width 5. Notice that we chose the number of classes and the class width so that their product (45) is somewhat larger than the distance we wish to cover (43).

3. We then describe the classes by giving the class limits. The lower and upper class limits for a class will be the smallest and largest numbers, respectively, that could be data values for the class. We have said that the lower class limit for the first class will be 45. The lower class limit for the second class is obtained by adding the class width ($W = 5$) to this, giving 50. The upper class limit for the first class is the largest possible value that could conceivably be a data value in that class. Since we are dealing with integers, that value would be 49. Proceeding in this manner, we obtain the following classes:

    45–49
    50–54
    55–59
    60–64
    65–69
    70–74
    75–79
    80–84
    85–89

4. Now, to complete our grouped frequency distribution, we list each class with its class limits. To the right of each class we record the **frequency of the class, f.** This is the number of data values in the class. In Table 2.3 we display the grouped frequency distribution for the data in Table 2.2. (Keeping tallies is optional but is especially recommended when the data are unranked.)

In some situations we may have some predetermined value in mind for the number of classes (or class width). For example, suppose we had decided beforehand that we wanted nine classes for the data in Table 2.2. To find an appropriate class width, divide the distance we wish to cover (43) by the number of classes: $\frac{43}{9} = 4.78$. We then choose a convenient number somewhat larger than this to be sure we cover the entire distance with our classes. Since the data values are whole numbers, we would choose 5 as the class width.

TABLE 2.3
**Grouped Frequency Distribution for the Data in
Table 2.2**

Class	Class Limits	Tally	Class Frequency (f)
1	45–49	||	2
2	50–54	|||	3
3	55–59	ⅣⅠ |	6
4	60–64	||||	4
5	65–69	ⅣⅠ ||||	9
6	70–74	ⅣⅠ |||	8
7	75–79	|||	3
8	80–84	||||	4
9	85–89	|	1

Suppose H is the highest data value in a data set, and we wish to have k classes with L as the lower class limit of the first class. The *class width W* should be a convenient number somewhat larger than

$$\frac{H - L}{k}$$

The *lower class limits* are: $L, L + W, L + 2W, \ldots$. (There will be k of these.) The *upper class limits* are the largest conceivable data values for each class.

Sometimes it is of interest to have a **grouped relative frequency distribution.** This is similar to a grouped frequency distribution except that instead of the frequency f of each class, we list the relative frequency f/n of each class (where n = the total number of data values). For example, in Table 2.3, which describes the scores of 40 high school students on a scientific achievement test, the relative frequency of the fifth class with limits 65–69 is

$$\frac{f}{n} = \frac{9}{40} = .225$$

The relative frequency can be important if we wish to generalize the properties of our sample. For instance, since the fraction of students in our sample receiving scores between 65 and 69 was $\frac{9}{40}$ = .225, we would expect a similar property to hold for the population of scores of all high school students on this test, provided that our sample was representative of the population. In other words, we would expect about 22.5% of all high school students to have scores between 65 and 69.

Another concept encountered in statistics is the **grouped cumulative frequency distribution**. Here again we list the classes, but this time, next to each class we list the total number of data values that are less than or equal to the upper class limit for that class. Finally, there is the **grouped cumulative relative frequency distribution**. Here, next to each class we list the proportion of the data values that are less than or equal to the upper class limit of the class. Sometimes all these ideas are combined in one table as in Table 2.4. (This table refers to the data in Table 2.2.)

Sometimes it is useful to describe the classes in terms of **class boundaries.** For example, for the grouped distribution discussed above, the upper class boundary for the first class is halfway between the upper class limit of the first class (49) and the lower class limit of the second class (50). So the upper class boundary of the first class is 49.5. This is also the lower class boundary of the second class. The distance between the upper and lower class boundaries of a class is equal to the class width 5. Therefore, the lower class boundary of the first class is 44.5. The classes described in terms of the class boundaries are

$$
\begin{array}{c}
44.5-49.5 \\
49.5-54.5 \\
54.5-59.5 \\
59.5-64.5 \\
64.5-69.5 \\
69.5-74.5 \\
74.5-79.5 \\
79.5-84.5 \\
84.5-89.5
\end{array}
$$

Note that since these class boundaries involve the decimal part .5, none of the data values in Table 2.2 will fall exactly on a class boundary

Another useful device is the **class mark** of a class. This is the number halfway between the lower and upper class limits of a class. For example, the class mark of the first class in Table 2.4 is the number halfway between 45 and 49, namely, 47. The class marks of the remaining classes are 52, 57, 62, 67, 72, 77, 82, and 87. (Notice that these class marks are whole numbers. When the class width is an odd number, the class marks will be whole numbers, which is convenient, but not necessary.) Class marks can be useful when performing computations involving the data. For example, we can simplify computations by approximating each data value in a class by the class mark. Also, this may be necessary if the original data are unavailable.

Sometimes the original data will involve decimals. For example, suppose

TABLE 2.4

Class	Class Limits	f	Relative f	Cumulative f	Cumulative Relative f
1	45–49	2	.050	2	.050
2	50–54	3	.075	5	.125
3	55–59	6	.150	11	.275
4	60–64	4	.100	15	.375
5	65–69	9	.225	24	.600
6	70–74	8	.200	32	.800
7	75–79	3	.075	35	.875
8	80–84	4	.100	39	.975
9	85–89	1	.025	40	1.000

we had data with one decimal place of accuracy, and the data values ranged from a low of 3.5 to a high of 15.1. Further, suppose we want to have five classes with 3.5 as the lower class limit of the first class. To determine the class width we first calculate

$$\frac{15.1 - 3.5}{5} = \frac{11.6}{5} = 2.32$$

Now we choose a convenient value somewhat larger than 2.32 for the class width. Since our data are assumed to have one decimal place, we will choose a class width of W = 2.4. The class limits and class boundaries are then described as follows:

Class Limits	Class Boundaries
3.5–5.8	3.45–5.85
5.9–8.2	5.85–8.25
8.3–10.6	8.25–10.65
10.7–13.0	10.65–13.05
13.1–15.4	13.05–15.45

We should keep in mind that the number of classes should not be too large; otherwise not much simplification of the data will result. On the other hand, if the number of classes is too small, too much simplification may result and we would lose too much information. It seems reasonable that the number of classes should be between 5 and 15.

Finally, we should point out that there is a trade-off in the use of any grouped distribution. We gain simplicity but lose certain information. For example, a grouped frequency distribution tells us how many data values are in each class but not what the data values are.

Exercises

2.1 **(a)** A new business recorded the number of incoming telephone calls over the first 20 days of business. The numbers of calls received were 4, 4, 1, 10, 12, 6, 4, 6, 9, 12, 12, 1, 1, 1, 12, 10, 4, 6, 4, and 1. Construct the frequency distribution.

(b) In part (a), note how the frequency distribution simplified the data. Now give an example of 20 data values such that a frequency distribution does not simplify the data very much.

2.2 Over a 20-day period, the number of employee absences from work recorded by the owner of a landscaping business was

$$1 \quad 2 \quad 0 \quad 0 \quad 1 \quad 2 \quad 2 \quad 1 \quad 0 \quad 0$$
$$4 \quad 0 \quad 1 \quad 1 \quad 3 \quad 2 \quad 1 \quad 3 \quad 0 \quad 1$$

(a) Construct the frequency distribution.

(b) What percentage of days were at least three employees not at work?

2.3 The following data represent maturity (in days) of selected tax-free money funds (*Source: The Boston Sunday Globe*, April 30, 1989, p. A7).

$$20 \quad 22 \quad 22 \quad 22 \quad 24 \quad 25 \quad 25 \quad 26 \quad 28 \quad 30 \quad 31 \quad 32 \quad 33$$
$$34 \quad 36 \quad 39 \quad 39 \quad 40 \quad 41 \quad 42 \quad 42 \quad 48 \quad 50 \quad 53 \quad 53$$

(a) Construct the frequency distribution.

(b) What percentage of these funds have a maturity of less than 30 days?

2.4 The following data are the age at inauguration of United States presidents (*Source: The 1989 Information Please Almanac*, 1989, p. 609).

42 43 46 47 48 49 49 50 50 51 51 51 51 52
52 54 54 54 54 55 55 55 55 56 56 56 57 57
57 57 58 60 61 61 61 62 64 64 65 68 69

(a) Construct the frequency distribution.

(b) Using 40 as the starting point, construct a grouped frequency distribution with six classes.

(c) To describe the distribution of data values to a friend, would you use the frequency distribution in part (a) or the grouped frequency distribution in part (b)? Give reasons for your choice.

2.5 The data below are the number of home runs hit by American League home run leaders in the years 1949–1988 (*Source: The 1989 Information Please Almanac*, 1989, p. 946).

43 37 33 32 43 32 37 52 42 42 42 40 61 48
45 49 32 49 44 44 49 44 33 37 32 32 36
32 39 46 45 41 22 39 39 43 40 40 49 42

(a) Construct the frequency distribution.

(b) Using 22 as the starting point, construct a grouped frequency distribution with eight classes.

(c) To describe the distribution of data values to a friend, would you use the frequency distribution in part (a) or the grouped frequency distribution in part (b)? Give reasons for your choice.

2.6 The data below are the number of home runs hit by National League home run leaders in the years 1949–1988 (*Source: The 1989 Information Please Almanac*, 1989, pp. 946–947).

54 47 42 37 47 49 51 43 44 47 46 41 46 49
44 47 52 44 39 36 45 45 48 40 44 36 38
38 52 40 48 48 31 37 40 36 37 37 49 39

Using 31 as the starting point with a class width of 3, construct a grouped frequency distribution.

2.7 Refer to the data in Exercise 2.3.

(a) Using 20 as the starting value, and a class width of 7, construct a grouped frequency distribution.

(b) What is the class mark for the third class?

(c) What is the upper class boundary of the second class?

(d) What is the relative frequency of the fourth class?

2.8 The data below are the number of English language Sunday newspapers (per state) in the United States as of February 1, 1988 (*Source: The 1989 Information Please Almanac*, 1989, p. 300).

3	6	6	7	7	8	8	9	9	10	10	11	12
12	14	17	19	19	19	20	23	23	23	24	25	25
26	26	27	28	28	32	35	35	36	37	45	46	
47	48	51	52	54	70	72	73	87	93	108	118	

(a) Using 1 as the starting value, construct a grouped frequency distribution with 7 classes.

(b) What is the class width?

(c) What is the class mark for the fifth class?

(d) What is the lower class boundary of the first class?

(e) What is the relative frequency of the third class?

2.9 A state policeman issued 200 speeding tickets over a 1-year period. The speeds ranged from 61 to 84 miles per hour. Suppose you were to construct a grouped frequency distribution with 8 classes. Give upper and lower class limits, upper and lower class boundaries, and the class mark for each class. Use 61 as the starting point.

2.10 The 1985 birth rates (per 1000 population) for 33 selected countries range from 9.6 to 26.6 (*Source: The 1988 Information Please Almanac*, 1988, p. 139). Suppose you were to construct a grouped frequency distribution with a starting point of 9.6 and class width of 2.9. Give upper and lower class limits, upper and lower class boundaries, and the class mark for each class. (*Note:* Since the data values are decimal, it is reasonable to have a class width with the same number of decimal places.)

2.11 The following data represent weights (in pounds) of meat obtained from 30 beef cattle:

204.0	205.1	214.9	222.6	222.8	198.4	222.2	230.9
220.0	222.4	215.9	207.6	208.2	228.0	208.4	219.5
194.2	192.9	196.4	202.1	212.9	203.8	208.9	
206.3	210.6	195.9	235.9	228.5	216.9	189.8	

(a) Complete the following table.

Class	Class Boundaries	Frequency	Cumulative Frequency
1	189.75–196.75		

(b) What is the class mark for the third class?

(c) What is the class width?

2.12 The following data give the 1985 death rates (per 1000 population) for selected countries (*Source: The 1988 Information Please Almanac*, 1988, p. 139).

Hong Kong	4.6	United States	8.7	Belgium	11.2
Singapore	5.2	Ireland	9.0	Sweden	11.3
Japan	6.2	Yugoslavia	9.1	Denmark	11.4
Cuba	6.4	Switzerland	9.2	W. Germany	11.5
Israel	6.6	Greece	9.4	Czechoslovakia	11.8
Mauritius	6.8	Italy	9.5	Luxembourg	11.8
Canada	7.2	Portugal	9.6	United Kingdom	11.8
Malta	7.4	Finland	9.8	Austria	11.9
Australia	7.5	France	10.1	E. Germany	13.5
New Zealand	8.4	Poland	10.3	Hungary	13.9
Netherlands	8.5	Norway	10.7		

(a) Complete the following table:

Class	Class Limits	Frequency	Cumulative Frequency	Class Mark
3	8.3–10.1			

(b) What is the class width?

(c) What is the relative frequency of the third class?

2.13 The following data represent cholesterol readings (mg/dl) for randomly selected women from the Framingham Heart Study:

287	242	200	260	298	278	195	265	230	300	215	224
228	291	236	244	234	278	302	244	281	217	221	156
198	267	198	204	280	182	185	204	256	234	172	

(a) Complete the following grouped frequency distribution.

Class	Class Boundaries	Frequency	Cumulative Relative Frequency
2	171.5–190.5		

(b) What is the class mark for the fourth class?

(c) What is the class width?

(d) What is the relative frequency of the third class?

(e) Use the cumulative relative frequency column to find the percentage of readings less than 248.

2.14 The owner of a small business wants to analyze profits over the past 30 years. Profits (in thousands of dollars) were obtained for the 30-year period. The ranked data values are

$$
\begin{array}{cccccccccc}
15 & 17 & 18 & 19 & 20 & 20 & 20 & 21 & 23 & 23 \\
24 & 24 & 24 & 24 & 24 & 25 & 25 & 25 & 25 & 25 \\
26 & 26 & 27 & 27 & 28 & 29 & 30 & 30 & 31 & 32
\end{array}
$$

(a) Complete the following grouped frequency distribution.

Class	Class Limits	Frequency	Cumulative Relative Frequency
2	18–20		

(b) What is the class width?

(c) What is the lower class boundary of the first class?

(d) What is the relative frequency of the second class?

(e) Use the cumulative relative frequency column to find the percentage of profits less than 30.

2.3 GRAPHIC REPRESENTATIONS

Graphic representations are, in a sense, pictures of data and as such sometimes make the data easier to comprehend.

One way to graphically represent a frequency distribution is by means of a **dot diagram.** We construct a dot diagram as follows: We use a horizontal axis to represent the data values. Above each distinct data value on the horizontal axis we place dots, the number of dots being equal to the frequency of the data value.

Example 2.3
Construct a dot diagram for the data in Example 2.1.

Solution
Recall that the frequency distribution is

x	f
1	2
3	1
5	3
6	5
7	3
11	1

The dot diagram is shown in Figure 2.1.

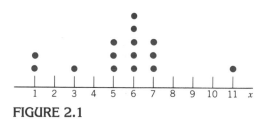

FIGURE 2.1

A **bar graph** or a **pie chart** can be used to depict a frequency distribution, especially when the data values are nonnumerical. These are illustrated in the following example.

Example 2.4
In this example we construct a bar graph and pie chart for the frequency distribution of Example 2.2 (murders by type of weapon for 1987). We construct a bar graph as follows: Along the horizontal axis we indicate the types of weapon. (These are the distinct data values.) The vertical axis measures the frequency with which each type of weapon was used. Above each type of weapon we construct a bar having a height equal to the frequency for that type of weapon. See Figure 2.2*a*. Notice that the bars are separated (*noncontiguous* is the term often used). This is always the case when we are dealing with nonnumerical categories such as the type of weapon.

A pie chart is often used to represent the frequency of each data value as a percentage of the total number of data values. We construct a circle, then divide the circle into sectors, one for each distinct data value (type of weapon). The size of a given sector is proportional to the percentage of murders committed with that type of weapon. See Figure 2.2*b* and refer to Example 2.2.

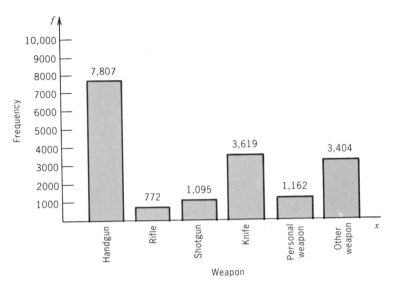

FIGURE 2.2*a*
A Bar Graph (Murders by Type of Weapon)

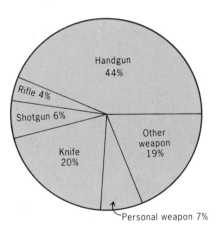

FIGURE 2.2*b*
A Pie Chart (Percentage of Murders by Type of Weapon)

Notice how the sectors of the circle were constructed. For example, handguns accounted for 44% of all 1987 murders; therefore the degree measure of the sector is

$$(.44) \times (360°) = 158.4°$$

A graphic representation of a grouped frequency distribution that is often used is a type of bar graph called a **frequency histogram.** Here again the horizontal or x axis represents the data and the vertical or f axis represents the frequency. Along the x axis we display the classes by labeling the class boundaries (not the class limits). Above each class we draw a bar having a width equal to the class width and a height equal to the class frequency. This idea is illustrated in the following example.

Example 2.5
Construct a frequency histogram for the grouped frequency distribution of Table 2.3 that summarized 40 test scores on a scientific achievement test.

Solution
Recall that the class boundaries are 44.5, 49.5, 54.5, and so on. The histogram is displayed in Figure 2.3.

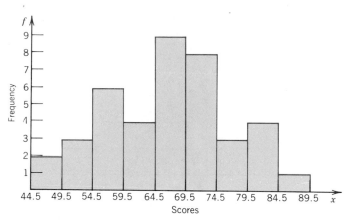

FIGURE 2.3
Frequency Histogram for the Data in Table 2.3

Remark   If, in a histogram, we make the heights of the bars equal to the relative frequencies of the classes rather than the frequencies, we have a **relative frequency histogram.** If the heights of the bars are the cumulative frequencies of the classes, we have a **cumulative frequency histogram.** Finally, if the heights of the bars are the cumulative relative frequencies of the classes, we have what is called a **cumulative relative frequency histogram.**

Another way of graphically displaying a grouped frequency distribution is by means of a **frequency polygon.** (The frequency polygon is especially useful in conveying the shape of the distribution.) The following example illustrates how a frequency polygon is constructed for the test score data.

Example 2.6

We will construct a frequency polygon using the grouped frequency distribution in Table 2.3. This grouped frequency distribution is summarized in the accompanying Table 2.5 along with the class marks. First draw a horizontal x axis and a vertical frequency axis. Label the class marks on the x axis; above each class mark place a dot at a height equal to the frequency of the class. The dots are then connected by line segments (see Figure 2.4). The first and last dots are connected to two points on the x axis, one of them being one class width to the left of the first class mark; the other, one class width to the right of the last class mark. These two points are usually not labeled. The class width in this case is 5; therefore we connect the first and last dots to the points 42 and 92, respectively.

TABLE 2.5

Class	Class Limits	Frequency (f)	Class Mark
1	45–49	2	47
2	50–54	3	52
3	55–59	6	57
4	60–64	4	62
5	65–69	9	67
6	70–74	8	72
7	75–79	3	77
8	80–84	4	82
9	85–89	1	87

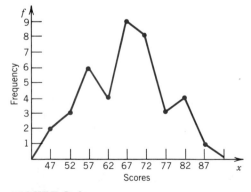

FIGURE 2.4
Frequency Polygon for Table 2.5

A **relative frequency polygon** is constructed in a manner similar to a frequency polygon with the exception that the dots are placed at a height equal to the relative frequency of the class, f/n, rather than the frequency of the class.

Another graph that can be helpful in picturing a grouped frequency distribution is the **cumulative frequency polygon.**

Example 2.7

We will construct a cumulative frequency polygon for the grouped frequency distribution in Table 2.3. The cumulative frequencies are given in Table 2.4. Draw the horizontal and vertical axes as usual. On the horizontal or x axis label the class boundaries. Above the upper class boundary of each class place a dot at a height equal to the cumulative frequency of that class. These dots are then connected by drawing line segments from one dot to the next. The cumulative frequency polygon is shown in Figure 2.5a. Note that a line segment is drawn from the lower class boundary of the first class to the first dot.

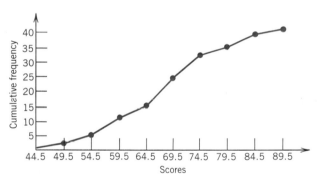

FIGURE 2.5a
Cumulative Frequency Polygon

The cumulative relative frequency polygon is constructed in a similar manner, the difference being that the height of the dots is the cumulative relative frequency. See Figure 2.5b.

Using this graph we can get some quick information. Suppose you were interested in how good a score of 66 is relative to other scores. Locate 66 on the horizontal axis, read up to the graph and across to the vertical axis. We see that 66 corresponds to a cumulative relative frequency of .50, meaning that about 50% of the scores were below 66. So 66 divides the bottom 50% from the top 50%. This is called the *50th percentile* or *median*. We will discuss these concepts

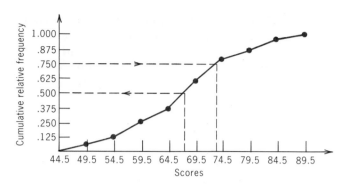

FIGURE 2.5*b*
Cumulative Relative Frequency Polygon

in the next chapter. If we want the score that separates the top 25% from the bottom 75%, read across from .75 on the vertical axis and down to the horizontal, getting (approximately) 73.

Exercises

2.15   For each of the following frequency distributions, construct a dot diagram:

(a)

x	f
0	3
1	6
2	5
3	2
4	2

(b)

x	f
1	2
2	2
3	2
4	2
5	2
6	2

2.16   In the Massachusetts State Lottery Numbers game, four numbers are selected randomly from the set of 10 numbers 0, 1, 2, 3, 4, 5, 6, 7, 8, 9. The dot diagram below represents the first 5 drawings (5 games, 20 digits in all) in the month of March 1989 (*Source:* Massachusetts State Lottery Commission, Public Relations Dept., 15 Rockdale St., Braintree, MA). Construct a frequency distribution for the data set.

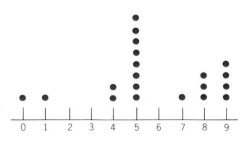

For each of Exercises 2.17–2.20, construct (a) a frequency histogram, (b) a frequency polygon, and (c) a cumulative frequency polygon. (d) Find the class width for each grouped frequency distribution.

2.17 The following grouped frequency distribution represents the time (in minutes) for students in an economic class to complete a quiz.

Class	Class Limits	Frequency
1	10–14	1
2	15–19	5
3	20–24	10
4	25–29	12
5	30–34	4

2.18 The following grouped frequency distribution represents the distance (in feet) from the pin of the shots of 32 golfers in a hole-in-one contest.

Class	Class Limits	Frequency
1	10–14	4
2	15–19	12
3	20–24	10
4	25–29	5
5	30–34	1

2.19 The following grouped frequency distribution represents the age (in years) of 59 patients of a psychiatric counseling center.

Class	Class Limits	Frequency
1	21–27	3
2	28–34	7
3	35–41	12
4	42–48	15
5	49–55	12
6	56–62	7
7	63–69	3

2.20 The grouped distribution below represents the weight (in pounds) of 30 elementary school children selected to participate in a physical fitness study.

Class	Class Limits	Frequency
1	46–48	6
2	49–51	6
3	52–54	6
4	55–57	6
5	58–60	6

2.21 Using Exercise 2.11, construct

(a) A frequency histogram.

(b) A cumulative relative frequency histogram.

(c) A cumulative relative frequency polygon.

2.22 Using Exercise 2.13, construct

(a) A frequency histogram.

(b) A frequency polygon.

(c) A cumulative frequency histogram.

(d) A cumulative frequency polygon.

2.23 The following cumulative frequency polygon represents a collection of integer data values.

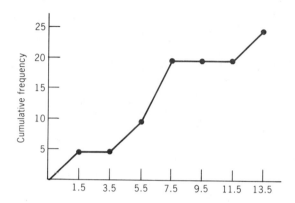

(a) How many data values are there?

(b) What are the class limits for the second class?

(c) Which class contains the most data values?

(d) What is the frequency for the third class?

(e) What is the frequency for the fifth class?

2.24 The following cumulative relative frequency polygon is for the age at inauguration of United States presidents discussed in Exercise 2.4.

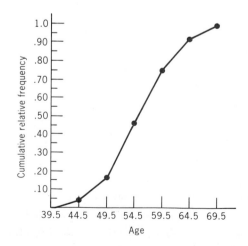

    (a)   Use the graph to estimate the age that separates the bottom 30% from the top 70% of ages. (This is called the 30th percentile. Percentiles are discussed more fully in the next chapter.)

    (b)   Use the graph to estimate the percentage of presidents younger than 60 at inauguration. (This is called the percentile rank of 60.)

2.25   The following data are the number of hazardous waste sites in the eastern north–central United States (1987) (*Source: World Almanac and Book of Facts*, 1989, p. 255):

Indiana	29	Michigan	69
Illinois	27	Wisconsin	33
Ohio	30		

    (a)   Construct a bar graph.   **(b)**   Construct a pie chart.

2.26   The following is the world distribution of nuclear reactors in operation (*Source: World Almanac and Book of Facts*, 1989, p. 171):

United States	106	U.S.S.R.	56
France	53	United Kingdom	38
Japan	36	W. Germany	21
Other countries	107		

    (a)   Construct a bar graph.   **(b)**   Construct a pie chart.

2.4 THE SHAPE OF A DISTRIBUTION

The shapes of frequency polygons (or frequency histograms) often resemble certain commonly observed types, which we display in Figure 2.6.

    In Figure 2.6 (*a*), (*b*), (*e*), and (*g*), the distributions are **symmetric** because if we draw a vertical line through the center of the graph, we see that the portion of the graph on one side of the line is a mirror image of the portion of the graph on the other side. It may be useful to know if a distribution is symmetric, as we will see in Chapter 6.

    The distributions in Figure 2.6 (*a*), (*b*), (*c*), (*d*), and (*e*) are often encountered when the data or x values represent test scores and f, of course, represents frequency. Perhaps Figure 2.6(*a*) is the most commonly encountered shape in this context. However, many college professors claim that Figure 2.6(*b*) is more and more often being encountered. This situation can occur in a class where there is a group of well-prepared students, another group of students with weak backgrounds, and not many in between.

    Figure 2.6(*d*) is also encountered when x represents heights of people, including children and adults. (The children's heights account for the skewing to the left.) Figure 2.6(*f*) might represent the number of riders or fares per day f on public transportation at various values of the fare x. Figure 2.6(*g*) might represent the number of phone calls f received by a switchboard at a department store at various times x of the business day. (The idea conveyed is that the calls come in at approximately the same rate all day.)

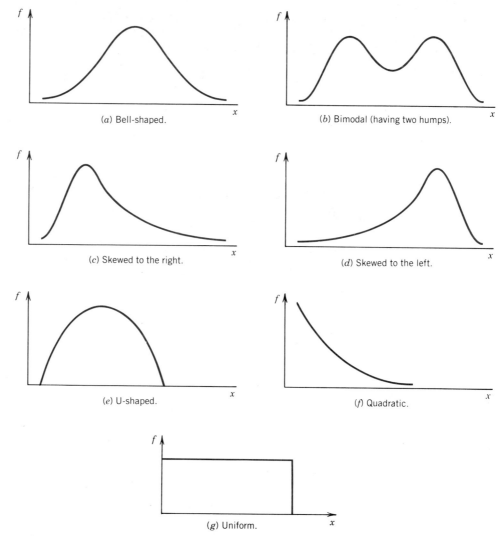

(a) Bell-shaped.

(b) Bimodal (having two humps).

(c) Skewed to the right.

(d) Skewed to the left.

(e) U-shaped.

(f) Quadratic.

(g) Uniform.

FIGURE 2.6

Frequency polygons for data tend to assume certain shapes, not by accident, but because of the nature of the population from which the data were obtained. For example, in Chapter 6 we will study a class of populations called **normal populations.** A large collection of data from a normal population will usually give rise to a frequency polygon that resembles approximately Figure 2.6(a). It may be important to know if a population is normal, as we will see later. By examining the frequency polygon for a collection of data, we may be able to get a clue as to whether or not a population is approximately normal.

Exercises

2.27 Characterize the distribution shapes in each of Exercises 2.17–2.20 by selecting the name of the distribution that most closely approximates the data.

2.5

STEM-AND-LEAF PLOTS (OPTIONAL)

Stem-and-leaf plots are a way of organizing data in such a way that the data values themselves are used to construct the bars of a figure that resembles a histogram. Thus the stem-and-leaf plot portrays the shape of a distribution and at the same time gives the individual data values. The idea of a stem-and-leaf plot was developed by Professor John Tukey of Princeton University.

To see how a stem-and-leaf plot works, consider the following data, which represent annual salary (in thousands of dollars) for 20 employees of a small business:

$$27 \quad 32 \quad 47 \quad 54 \quad 36 \quad 44 \quad 56$$
$$49 \quad 53 \quad 51 \quad 35 \quad 29 \quad 62 \quad 19$$
$$65 \quad 37 \quad 48 \quad 33 \quad 41 \quad 45$$

The data range from 19 to 65. We could group the data into categories according to the tens digit. For example, the data values between 40 and 49 inclusive have 4 as a tens digit, and so on. To construct a stem-and-leaf plot, list the tens digits: 1, 2, 3, 4, 5, 6 in a vertical column. These are called **stems.** Just to the right of this draw a vertical line. Going through the data, record the last digit (units digit) of each data value to the right of the appropriate stem. These are called **leaves.** For example, there are 2 data values with the first digit 6: 62 and 65. Therefore, next to 6 we would write a 2 and a 5.

$$6 \mid 2 \quad 5$$

The entire stem-and-leaf plot is shown in Figure 2.7. Notice that the stems are equivalent to classes. For example, the data values with stem 3 fall in the class 30–39. The number of leaves is the frequency of the class. The display of the leaves resembles the bars of a histogram.

```
1 | 9
2 | 7 9
3 | 2 6 5 7 3
4 | 7 4 9 8 1 5
5 | 4 6 3 1
6 | 2 5
```

FIGURE 2.7
Stem-and-Leaf Plot for Twenty Salaries

If we rotated Figure 2.7 counterclockwise by 90°, it would resemble a histogram. This distribution of data looks bell shaped.

Example 2.8

Construct a stem-and-leaf plot for the data in Table 2.6, which represent the number of farms (in thousands) for each state.

TABLE 2.6
Number of Farms (in thousands)

Alabama	52	Montana	24
Alaska	1	Nebraska	57
Arizona	9	Nevada	2
Arkansas	50	New Hampshire	3
California	79	New Jersey	8
Colorado	27	New Mexico	14
Connecticut	4	New York	42
Delaware	3	North Carolina	73
Florida	39	North Dakota	33
Georgia	49	Ohio	88
Hawaii	4	Oklahoma	71
Idaho	24	Oregon	37
Illinois	87	Pennsylvania	57
Indiana	78	Rhode Island	1
Iowa	109	South Carolina	28
Kansas	70	South Dakota	36
Kentucky	99	Tennessee	96
Louisiana	36	Texas	160
Maine	8	Utah	14
Maryland	17	Vermont	7
Massachusetts	6	Virginia	50
Michigan	61	Washington	38
Minnesota	93	West Virginia	21
Mississippi	46	Wisconsin	82
Missouri	115	Wyoming	9

Source: Statistical Abstract of the United States (Washington, D. C.: U.S. Bureau of the Census, 1987, p. 621).

Solution

Notice that the data values range from 1 to 160. We could use the tens and hundreds digits for the stems and the units digits for the leaves. For example, the value 87 could be represented as 8 | 7 . The value 115 could be represented as 11 | 5 . The stem-and-leaf plot is given in Figure 2.8. Notice that the data is skewed positively (i.e., to the right).

```
 0 | 1 9 4 3 4 8 6 2 3 8 1 7 9
 1 | 7 4 4
 2 | 7 4 4 8 1
 3 | 9 6 3 7 6 8
 4 | 9 6 2
 5 | 2 0 7 7 0
 6 | 1
 7 | 9 8 0 3 1
 8 | 7 8 2
 9 | 9 3 6
10 | 9
11 | 5
12 |
13 |
14 |
15 |
16 | 0
```

FIGURE 2.8
Stem-and-Leaf Plot for the Data in
Table 2.6

Choosing the Stems and Leaves

The choice of stems and leaves will depend on the nature of the data set. For example, suppose the data ranged from 115 to 982. We could use the first digit as the stem and the second two as the leaf. So the stems would be: 1, 2, 3, ..., 9. Data values such as 115 and 132 would be represented

$$1 \mid 15 \quad 32$$

On the other hand if the data range from, say, 115 to 198, we would not want to use the first digit as the stem—we would only have one stem. We could use the first two digits as stems. The stems would be 11, 12, 13, ..., 19. Data values such as 123 and 127 would be represented

$$12 \mid 3 \quad 7$$

The presence of a decimal point in the data does not change the above considerations. For example, if the data ranged from 11.5 to 98.2, we could use the first digit as the stem. So data values 11.5 and 13.2 would be represented

$$1 \mid 15 \quad 32$$

Similarly, if the data ranged from 11.5 to 19.8 we would use the first two digits for the stem. Data values 12.3 and 12.7 would be represented

$$12 \mid 3 \quad 7$$

The following thought may have occurred to the reader: if a stem-and-leaf plot displays the data value 12 | 3 , how is this to be interpreted? Is it 123 or 12.3 or 1.23? The answer is that it is up to the maker of the stem-and-leaf plot to explain the plot. For example, a note could be included, such as: 12 | 3 represents 12.3.

Exercises

2.28   The following data (discussed in Exercise 2.13) give the cholesterol readings of randomly selected women from the Framingham Heart Study. Construct a stem-and-leaf plot.

287	242	200	260	298	278	195	265	230	300	215	224
228	291	236	244	234	278	302	244	281	217	221	156
198	267	198	204	280	182	185	204	256	234	172

2.29   The following data (which we discussed in Exercise 2.8) represent the number of English language Sunday newspapers in the 50 states. Construct a stem-and-leaf plot. Describe the shape of the distribution.

3	6	6	7	7	8	8	9	9	10	10	11	12
12	14	17	19	19	19	20	23	23	23	24	25	25
26	26	27	28	28	32	35	35	36	37	45	46
47	48	51	52	54	70	72	73	87	93	108	118

2.30   The following data (discussed in Exercise 2.12) represent death rates (per 1000 population) for selected countries. Construct a stem-and-leaf plot.

4.6	5.2	6.2	6.4	6.6	6.8	7.2	7.4	7.5	8.4	8.5
8.7	9.0	9.1	9.2	9.4	9.5	9.6	9.8	10.1	10.3	10.7
11.2	11.3	11.4	11.5	11.8	11.8	11.8	11.9	13.5	13.9

2.31   The following data (from Exercise 2.11) represent the weights (in pounds) of meat obtained from 30 beef cattle. Construct a stem-and-leaf plot. Describe the shape of the distribution.

204.0	205.1	214.9	222.6	222.8	198.4
222.2	230.9	220.0	222.4	215.9	207.6
208.2	228.0	208.4	219.5	194.2	192.9
196.4	202.1	212.9	203.8	208.9	206.3
210.6	195.9	235.9	228.5	216.9	189.8

(*Hint:* For example, use 21 | 49 to represent the value 214.9.)

2.6 MISLEADING GRAPHS (OPTIONAL)

There are many abuses of statistics. We cite some of the more common abuses of graphical methods in the following examples.

Example 2.9

The treasurer of a company was instructed to prepare a report to the board of directors describing the company's profits as a percent of sales for a 5-year period. The data are

Year	Profit (% of sales)
1986	6.1
1987	6.7
1988	7.3
1989	7.5
1990	8.0

This kind of data is often described by a type of bar graph shown in Figure 2.9 (in an acceptable form).

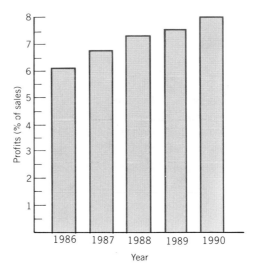

FIGURE 2.9
Bar Graph

Figure 2.9 shows a modest growth in profits over the years. However, the treasurer was overanxious to impress the board, and therefore constructed instead a **truncated bar graph,** shown in Figure 2.10. Notice how this graph exaggerates the growth in profits. The word *truncated* refers to the fact that the graph is cut off at the bottom and starts at 5.5. This sort of thing is generally frowned upon.

Bar Graph

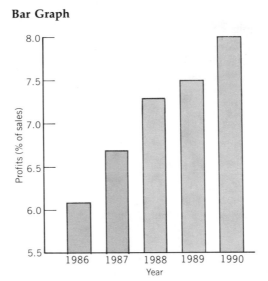

FIGURE 2.10
Truncated Bar Graph

Example 2.10
A railroad prepared some advertising material in which it portrayed a doubling of its yearly freight (measured in carload lots) between 1985 and 1990. In 1985 it shipped 4000 carload lots and in 1990 it shipped 8000 carload lots. This growth was represented graphically in Figure 2.11 by means of two freight cars, one for 1985 and the other for 1990.

FIGURE 2.11
Misleading Pictogram

The freight car for 1990 in Figure 2.11 is twice as high as the one for 1985, which the railroad management believes is justified because its freight doubled. But it is also twice as long. Therefore, its area is four times the area of the freight car for 1985, which might suggest a quadrupling of business to the unwary reader. Figure 2.12 shows a more reasonable way to represent the growth. This is called a **pictogram.**

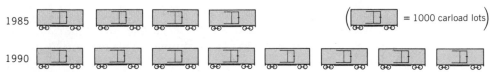

FIGURE 2.12
Pictogram Showing Growth in Freight from 1985 to 1990

2.7 USING MINITAB (OPTIONAL)

Most chapters in this textbook have an optional section that shows the reader how to use the Minitab computer package to solve statistical problems. To read this material, no prior knowledge of computers is necessary. Consult your computing center for directions on how to log on the computer and call up the Minitab package.

Entering Data

The Minitab **worksheet** consists of a collection of columns internal to the computer. On some computers the number of columns is 50. These columns will contain your data and the results of your computations. The columns are labeled C1 and C2 for columns 1 and 2, and so forth. Data and commands can be entered at a computer terminal keyboard. The keyboard may be connected to what is essentially a television screen, which will display your work, or it may be part of a printer that will display your work on paper (much like a typewriter). This paper is called a "hard-copy" printout.

The easiest way to enter a collection of data into one of the columns of the Minitab worksheet is to use the SET command. If you type

        SET THE FOLLOWING DATA IN C1
        1   4   3   6
        END

this will place the four data values in column 1 inside the computer:

C1
1
4
3
6

With the SET command you may type data on more than one line if you wish. (You may also type data in a column.) Make sure your data values are separated by spaces or commas. Don't use commas in numbers. The number one thousand five hundred should not be written 1,500. Minitab will interpret this as two numbers. Instead, type 1500. After you have typed your data, you should type END to instruct Minitab that you are finished entering data.

When typing Minitab commands, you need only type the first four letters of the command name (or three letters with a command like SET). You must also specify columns where data are located and occasionally constants or expressions that may be needed. Any additional text after the command name is optional and may be included for clarity when reading printouts. It will be ignored by the Minitab package. Thus, you may type

SET C1

This would achieve the same result as

SET THE FOLLOWING DATA IN C1

Be careful, however, not to include any unnecessary numbers in the additional text. Don't type, for example,

SET THE FOLLOWING 30 DATA VALUES IN C1

(Certain other symbols should be avoided in additional text because they have special meaning in Minitab. These are the symbols + ; - * &.)

When Minitab is expecting a command, it "prompts" you with the symbol

MTB>

You then type the command. When data are expected, Minitab types

DATA>

after which you enter your data. After typing a line, depress the carriage return. Minitab will respond with the appropriate prompt, or it will print the desired output. When reading Minitab printouts in this text, keep in mind that those lines preceded by prompts (such as MTB> or DATA>) are typed by the user. All other lines are output from the computer.

Histograms

A psychologist developed a test of aggressiveness with scores on a scale of 0 to 200. The test was given to 30 male students at an eastern university. In the computer printout below, we place the scores in column 1 (C1) and then tell the computer to construct a histogram for the data in C1. (We entered the first six lines of this printout after the prompts; the computer printed the rest.)

```
MTB > SET THE FOLLOWING DATA IN C1
DATA> 136 133 122 105 141 118 148 113 128 125 120 146
DATA> 137 135 98 126 108 93 140 110 134 131 102 115
```

```
DATA> 106 110 118 147 136 146
DATA> END
MTB > HISTOGRAM FOR C1

Histogram of C1  N = 30

Midpoint  Count
    95      1   *
   100      2   **
   105      2   **
   110      3   ***
   115      2   **
   120      4   ****
   125      2   **
   130      2   **
   135      6   ******
   140      2   **
   145      3   ***
   150      1   *
```

Notice that for each interval (class), the midpoint is given. The number of observations (frequency) for each class is also given, along with asterisks representing the frequencies. These asterisks constitute the bars for the histogram. In fact, if you tip the printout on its side, it will very much resemble the histograms discussed in this chapter.

In the above printout the distance between the midpoints is 5, so this is the class width. The intervals are defined in terms of the midpoints. Since the midpoint of the first interval is 95, this interval contains any data ≥ 92.5 and < 97.5. If the value 97.5 had occurred in the data, it would be counted in the next interval.

In the above printout, the computer selected a convenient class width and the midpoints, since it was not told to do otherwise. It selected a first midpoint of 95 and a class width of 5. If we wanted to select the first midpoint and the class width, we could do so by means of **subcommands.** If you wish to use subcommands, type a semicolon (;) at the end of the HISTOGRAM command and then hit the return key. Minitab then expects a subcommand and prompts you with the symbol

SUBC>

Sometimes subcommands are indented to help distinguish them from main commands, but this is optional. To specify the class width we use the INCREMENT subcommand, and to specify the first midpoint we use the START subcommand. Each subcommand is followed by a semicolon except the last one, which is followed by a period. The following program gives us a histogram with class width 10 and 90 as the midpoint of the first class:

```
MTB > HISTOGRAM FOR C1;
SUBC>    INCREMENT = 10;
SUBC>    START AT 90.
```

```
Histogram of C1 N = 30

Midpoint  Count
    90.0    1  *
   100.0    2  **
   110.0    6  ******
   120.0    5  *****
   130.0    6  ******
   140.0    6  ******
   150.0    4  ****
```

The HISTOGRAM command, like many Minitab commands, can be used on more than one column at a time. For example,

HISTOGRAM FOR C1 C3 C4

would give three histograms, one each for C1, C3, and C4.

Suppose we decide that we want to have a histogram with 10 classes and the smallest data value as the lower class limit for the first class. We can use the commands

MAXIMUM FOR C1

MINIMUM FOR C1

to find the largest and smallest data values.*

```
MTB > MAXIMUM FOR C1
    MAXIMUM =    148.00
MTB > MINIMUM FOR C1
    MINIMUM =     93.000
```

Therefore, we must cover the distance from 93 to 148 with 10 classes. Now

$$\frac{148 - 93}{10} = \frac{55}{10} = 5.5$$

Choosing a convenient number somewhat larger than 5.5, we can use 6 as a class width. To construct a histogram we can use the following procedure.

If the lower class limit of the first class is to be 93, then we can obtain the

---

*The DESCRIBE command (discussed in Section 3.4) also gives the maximum and minimum values (as well as other information).

first midpoint by adding half the class width to get 96. This histogram is given below.*

```
MTB > HISTOGRAM FOR C1;
SUBC>   INCREMENT = 6;
SUBC>   START AT 96.

Histogram of C1  N = 30

Midpoint  Count
    96.00    2    **
   102.00    1    *
   108.00    5    *****
   114.00    2    **
   120.00    4    ****
   126.00    3    ***
   132.00    3    ***
   138.00    5    *****
   144.00    3    ***
   150.00    2    **
```

Dotplots

A histogram groups the data into relatively few intervals. A **dotplot** is a variation on the histogram that does as little grouping as possible. Thus the data are grouped into many intervals of small width. A dot is given for each data value. Hence a dotplot resembles a dot diagram and is useful for small data sets, whereas a histogram is useful for large data sets. Below is a dotplot for the data discussed above.

```
             MTB > DOTPLOT FOR C1
```

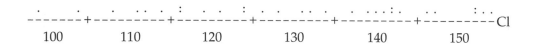

```
  .      .    .   . . .  :   . .   :  .  .  .. .  . ...:  ..      :..
-------+---------+---------+---------+---------+---------+---------C1
     100       110       120       130       140       150
```

---

*The midpoints printed here are not the class marks. Since we add 3 to the lower class limit of 93 to get the first midpoint, this value is halfway between the lower class limit of the first class and the lower class limit of the second class. The class marks are halfway between the class boundaries. If we want to use the class marks, we add 3 to the lower class boundary to get the first class mark: 92.5 + 3 = 95.5. If we use 95.5 as the first midpoint, the class marks would be printed as the midpoints. In either case we get essentially the same histogram.

Stem-and-Leaf Plots

Below is a stem-and-leaf plot for the same 30 data values.

```
MTB > STEM-AND-LEAF PLOT FOR C1

Stem-and-leaf of C1     N = 30
Leaf Unit = 1.0
   1     9  3
   2     9  8
   3    10  2
   6    10  568
   9    11  003
  12    11  588
  14    12  02
  (3)   12  568
  13    13  134
  10    13  5667
   6    14  01
   4    14  6678
```

Notice that Minitab tells you the leaf digit unit is 1. This means it is the ones digit. Thus for example, 9 3 represents the number 93. If the leaf unit were 10, then 9 3 would mean 930. With a leaf unit of .01, we would read 9 3 as .93. The first column gives the number of data values on that line plus the number of values in the lines toward the outer tail of the distribution (that is, the outer tail nearest the line). For example, the 6 on the fourth line means there are 6 data values in the nearest tail: lines 1 through 4. The exception to this rule is the line containing the median (a kind of central value, to be discussed in Chapter 3). Next to the line containing the median, the number of data values on *that* line is given in parentheses. We will see that the median for these data is 125.5. This is halfway between the values 125 and 126, which appear on the eighth line.

Sometimes the leaves for each stem are stretched over more than one line. The number of lines is usually 1, 2, or 5. In the printout above there are two lines per stem. Thus we see the values 93 and 98 as

9   3

9   8

instead of

9   38

The values 90–94 go on the first line and 95–99 on the second line. Note the distance between the lowest possible value on the first line (90) and the lowest possible value on the second line (95) is 5. This is called the increment. We can control the increment with a subcommand. If we wanted all the values with the same tens digits on the same line, then the distance (increment) between lines would be 10. We would then type

STEM-AND-LEAF FOR C1;

INCREMENT = 10.

The increment must always be a 1, 2, or 5 with perhaps some leading or trailing zeros, such as 10, .1, .01, 2, .2, .02, 200, etc.

Stored Constants

While data sets are stored in columns, single numbers may also be stored, and as such are referred to as **stored constants.** Locations of stored constants are labeled K1, K2, etc. We saw that for the data discussed above, the maximum value was 148. We could have stored this for future reference with the following variation of the MAXIMUM command given previously:

MAXIMUM FOR C1, STORE IN K1

The value of 148 would be stored in K1. Later if we wanted to see the maximum value, we could type

PRINT K1

Minitab would print the value of K1. The PRINT command can be used to print columns and/or constants. For example,

PRINT C1, C3–C5, K1

would print columns C1, C3, C4, C5, and constant K1.

Annotating a Program

If you wish to type any explanatory comment on your output, you can use the symbol # followed by your comment. Minitab ignores everything on the line following #. For example,

HISTOGRAM FOR C1 # SCORES FOR 30 MALES

Naming Columns

We may use the NAME command to name columns. Minitab will then use the names when printing information about these columns. Always enclose the name in single quotes. For example,

NAME FOR C1 IS 'SCORES'

As usual, we can abbreviate the above command:

NAME C1 'SCORES'

If you wish, you may use this name in subsequent commands. For example,

HISTOGRAM FOR 'SCORES'

We can name more than one column at a time:

NAME FOR C2 IS 'PRICE', FOR C3 IS 'QUANTITY'

A name may be up to 8 characters long, but may not begin or end with a blank, or contain the symbol #, or a single quote. You may use the NAME command to change a column name. To erase a column name, but not the column contents, give it a blank name:

NAME FOR C1 IS ' '

When you are through using Minitab, type STOP (and hit the carriage return).

MTB >STOP

Practice Quiz (Answers can be found after the chapter review exercises)

Find the errors (if any) in the following.

1. SET THE FOLLOWING DATA IN C1: 1   4   5
   END

2. SET THE FOLLOWING DATA IN C1
   913   847   312   2,104
   END

3. SET THE FOLLOWING DATA IN COLUMN 1
   5   3   7
   END

4. MAXIMUM FOR C1
   MINIUM FOR C1

5. SETTHE FOLLOWING DATA IN C1
   1   2   3   4
   END

6. PLEASE SET THE FOLLOWING DATA IN C1
   1   2   3   4
   END

7. SET IN C1
   2   7   32
           12
   END

8. SET THE DATA FOR BRAND 1 in C1
   11   8   2   3   9
   END

9. MAXIMUM FOR C1, MINIMUM FOR C1

10. STEM-AND-LEAF FOR C5;
    INCREMENT = 13.

Exercises

Suggested exercises for use with Minitab are 2.21 (a), 2.22 (a), 2.32 (b), 2.33 (c), 2.38 (a), 2.40 (a), and 2.45 (a). Some of these exercises ask you to construct a frequency histogram based on grouped frequency distributions found in previous exercises. The histogram command in Minitab does not require that you first construct a grouped frequency distribution; it does everything at once. Use the previous exercises only to determine necessary information about the grouped distribution, such as the starting point for the first class.

SUMMARY

Various ways of summarizing and simplifying data have been discussed in this chapter. A **frequency distribution** gives us the distinct data values in a collection of data together with the number of times each occurs. A **grouped frequency distribution** is obtained by giving classes or intervals together with the number of data values in each class. (There are certain variations on this idea that are often useful, such as the **cumulative grouped frequency distribution.**)

Describing data graphically can also be useful. We saw that **dot diagrams, bar graphs,** and **pie charts** are used to represent frequency distributions. A grouped frequency distribution can be represented by means of a **frequency histogram** or a **frequency polygon.**

Stem-and-leaf plots are also useful for displaying data.

Review Exercises

2.32   The following data are the number of deaths by horsekicks in the Prussian Army from 1875 to 1894 for 14 corps (*Source:* Andrews, D. F., and A. M. Herzberg, "The Number of Deaths by Horsekicks in the Prussian Army," in *Data,* p. 18. New York: Springer-Verlag):

3	4	5	5	6	6	7	8	9	9
10	11	11	11	12	14	15	15	17	18

(a)   Construct the frequency distribution.

(b)   Construct a dot diagram.

2.33   The data below are magnitudes on the Richter scale of 20 earthquakes selected from 1000 seismic events occuring near the Fiji Islands between 1964 and 1974 (*Source:* extracted from the World Wide Seismic Network Tape by Professor John Woodhouse of Harvard University. Data from Donoho, A., D. Donoho, and M. Gasko, *MacSpin Graphical Data Analysis Software,* p. 115). The data are

4.7	4.3	4.5	4.8	5.1	5.5	4.7	4.4	4.7	4.6
4.5	5.7	4.9	4.5	4.3	4.1	4.3	4.6	5.1	4.7

(a)   Construct the frequency distribution.

(b)   Give the percentage of earthquakes for which the magnitude was less than 5.0.

(c)   Construct a dot diagram.

In Exercises 2.34–2.36, use the following data. An interviewer was interested in the TV habits of people 18 years or older in a typical large city. She selected 20 homes and obtained the information in Table 2.7 from 1 adult in each home.

2.34   Construct a frequency distribution for the number of children.

2.35   Would it be worthwhile constructing a frequency distribution for "Age"? For "Number of Hours Watching TV per Week"? Give your reasons.

2.36   What proportion of those sampled are female? Does this proportion seem representative of the proportion of people in a typical city who are 18 years or older? What might account for such a large proportion of females?

In Exercises 2.37–2.40, use the information in Table 2.7 but assume that the interviewer categorized the number of hours of watching TV as follows: seldom (0–6 hours inclusive), occasional (7–13 hours inclusive), frequent (14–20 hours inclusive), very frequent (21–27 hours inclusive), and excessive (28 or more hours). (For consistency you may regard this last category as 28–34 hours.)

2.37   Construct a grouped frequency distribution for "Number of Hours" with classes based on the categories defined above.

2.38   Using Exercise 2.37, construct

(a)   A frequency histogram.   **(b)**   A frequency polygon.

TABLE 2.7

Age	Sex	Number of Hours Watching TV per Week	Number of Children
71	F	16	3
30	M	23	2
63	F	19	4
66	F	31	5
47	M	28	1
29	F	4	2
26	F	11	2
47	F	21	3
31	F	17	4
82	M	13	2
74	F	29	5
53	F	7	3
32	F	3	3
40	M	18	0
30	F	21	1
19	F	28	0
68	F	12	2
55	F	3	3
24	F	24	2
38	F	27	3

2.39 Construct a grouped frequency distribution for "Age." Use a starting point of 19 and a class width of 11.

2.40 Using Exercise 2.39, construct

(a) A frequency histogram. **(b)** A frequency polygon.

2.41 The finance committee of a town was interested in the economic makeup of families in that town. A sample of 200 families was taken to determine their income. Assume, for example, that 34 is taken to be $34,000. Refer to Table 2.8, and assume that the income was recorded to the nearest thousand.

(a) Complete Table 2.8.

(b) Construct a cumulative frequency histogram.

(c) If you felt that the sample was representative of the economic status of families in this town, what would you estimate the proportion of families to be that

(i) Earn more than $35,500?

(ii) Earn between $32,500 and $35,500?

TABLE 2.8

Class	Class Boundaries	Frequency	Relative Frequency	Cumulative Frequency
1			.05	
2		30		
3		12		
4	32.5–35.5	47		
5				149
6		28		
7		5		
8				
Total		200		

2.42 The following frequency distribution for magnitudes of 1000 earthquakes is the full set of data referred to in Exercise 2.33.

x	4.0	4.1	4.2	4.3	4.4	4.5	4.6	4.7	4.8	4.9	5.0	5.1	5.2
f	46	55	90	85	101	107	101	98	65	54	47	43	29

x	5.3	5.4	5.5	5.6	5.7	5.8	5.9	6.0	6.1	6.2	6.3	6.4
f	21	20	14	9	8	0	2	3	1	0	0	1

(a) Construct a grouped frequency distribution with 5 classes using 4.0 as a starting point.

(b) What is the class width?

(c) Find the percentage of magnitudes 5.0 or larger.

(d) Construct a frequency histogram.

(e) Characterize the distribution shape by selecting the name of the distribution that most closely approximates the data.

2.43 Select 60 digits from the random number table (Appendix Table B.1) as follows: Start at the beginning of some column in the table. Record the first 60 digits by moving down this column to the next column, and so on.

(a) Construct a dot diagram.

(b) If the entire random number table generates numbers in a chance manner, approximately what proportion of times should each integer occur? Do your data reflect this? What should be the approximate shape of the distribution?

In Exercises 2.44–2.48, use the following data. A psychologist observed reaction times to a stimulus and recorded 50 reaction times, in seconds:

2.7	3.1	4.0	3.3	3.6	2.1	1.3	1.8	3.7	3.7
3.7	3.4	6.4	3.4	3.3	2.9	1.6	3.3	4.5	3.1
3.5	2.6	3.0	3.3	1.6	0.9	1.5	2.7	2.7	2.6
2.3	1.3	1.7	2.2	3.2	2.8	3.1	3.4	2.6	1.7
2.5	3.8	2.0	3.8	1.4	2.8	2.4	4.5	3.2	2.3

2.44 Refer to the above data. Using 8 classes, construct a table with columns for class limits, class boundaries, class marks, frequencies, relative frequencies, cumulative frequencies, and cumulative relative frequencies. What is the class width? (Use the minimum data value as a starting point, i.e., lower class limit of first class.)

2.45 Using Exercise 2.44, construct

(a) A frequency histogram. **(b)** A cumulative frequency histogram.

2.46 Refer to the above data. Construct a table using 12 classes, giving frequencies and cumulative frequencies. (Use the minimum data value as a starting point.)

2.47 Using Exercise 2.46, construct

(a) A frequency histogram. **(b)** A cumulative frequency histogram.

2.48 Refer to the above data.

(a) Construct a frequency distribution.

(b) Compare part (a) with the grouped frequency distribution in Exercise 2.44. Was it worthwhile grouping the data? Give your reasons.

2.49 The following data are observations of time in days from remission induction to relapse for 51 patients with acute nonlymphoblastic leukemia (*Source:* Matthews, D. T., and V. T. Farewell, "On Testing for a Constant Hazard against a Change-point Alternative," *Biometrics* 38: 463–468, 1982. Data from Chambers, J., W. Cleveland, B. Kleiner, and P. Tukey, *Graphical Methods for Data Analysis.* Boston: Duxbury Press, 1983, p 382).

24	46	57	57	64	65	82	89	90	90	111
117	128	143	148	152	166	171	186	191	197	
209	223	230	247	249	254	258	264	269	270	
273	284	294	304	304	332	341	393	395	487	
510	516	518	518	534	608	642	697	955	1160	

(a) Construct a grouped frequency distribution with 6 classes. Use 24 as a starting point. In your table, include a cumulative relative frequency column.

(b) Construct a frequency polygon.

(c) Characterize the distribution shape by selecting the name of the distribution that most closely approximates the data.

(d) Construct a cumulative frequency polygon.

(e) What is the percentage of observations for which the time to relapse lasted more than 403 days?

2.50 Consider the following grouped frequency distribution:

Class	Class Limits	Frequency
1	60–61	13
2	62–63	27
3	64–65	45
4	66–67	43
5	68–69	32
6	70–71	38
7	72–73	21
8	74–75	14
9	76–77	2

(a) Find the class width.

(b) Give the class boundaries for the third class.

(c) Find the class mark of the fifth class.

(d) Construct a frequency histogram.

(e) Construct a cumulative frequency histogram.

2.51 The cumulative frequency polygon and cumulative relative frequency polygon below represents heights (to the nearest inch) of tenors in the New York Choral Society in 1979 (*Source:* Chambers, J., W. Cleveland, B. Kleiner, and P. Tukey, *Graphical Methods for Data Analysis.* Boston: Duxbury Press, 1983, p. 350).

(a) How many classes are there?

(b) Find the class width.

(c) How many data values are there?

(d) What are the class limits for the second class?

(e) Find the frequency of the second class.

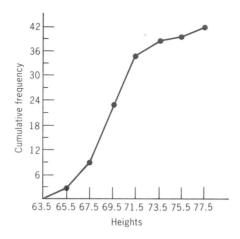
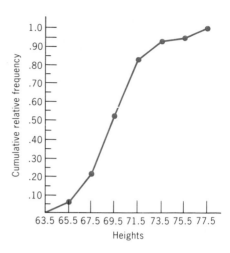

(f) Use the cumulative relative frequency polygon to estimate the data value that separates the lower 50% of the data from the upper 50% of the data. (This is called the 50th percentile.)

(g) Use the cumulative relative frequency polygon to estimate the percentage of heights less than 71.5.

For Exercises 2.52–2.58, use the following data. A Florida city hospital and a Massachusetts city hospital maintained records in a certain year indicating the age at admittance for those patients who spent at least 3 days at the hospital. The data are given in Table 2.9. These incidence rates were based upon the previous year's population census of the respective cities. (By incidence rate we mean the number of admittances per 100,000 of the city population with ages in this category.)

TABLE 2.9

Age at Admittance	Number at Florida Hospital	Number at Massachusetts Hospital	Incidence Rate (per 100,000)	
			Florida	Massachusetts
0–9	27	41	.3	.4
10–19	19	32	.7	.6
20–29	31	60	1.6	1.9
30–39	47	72	6.2	5.8
40–49	42	70	11.4	12.1
50–59	68	40	13.6	13.8
60–69	87	37	17.2	17.8
70–79	145	33	28.3	26.4
80–89	34	15	44.2	49.4
Total	500	400		

2.52 Use the columns for number of admittances.

   (a) Construct a frequency polygon for the Florida hospital.

   (b) Construct a frequency polygon for the Massachusetts hospital.

   (c) Comment on differences between the polygons.

2.53 Consider the incidence rate per 100,000.

   (a) Construct an incidence polygon* for the Florida hospital.

   (b) Construct an incidence polygon* for the Massachusetts hospital.

   (c) Comment on any similarities and differences between the polygons.

2.54 Consider the number of admittances.

   (a) Construct a cumulative frequency polygon for the Florida hospital.

   (b) Construct a cumulative frequency polygon for the Massachusetts hospital.

2.55 Compare the frequency polygon and the cumulative frequency polygon for the Massachusetts hospital. [See Exercises 2.52(b) and 2.54(b).] Do you see any connection between the two? Could you have drawn the cumulative frequency polygon (roughly) from the frequency polygon if no numbers had been displayed?

2.56 If you have not already done so, construct a column of cumulative frequencies for the Florida hospital regarding the number of admittances. Using this column, find the following:

   (a) The number admitted who were 40 years or older.

   (b) The number admitted who were 69 years or younger.

   (c) The number admitted who were more than 29 but less than 80 years of age.

2.57 Repeat Exercise 2.56 but use the frequency column.

2.58 Repeat Exercise 2.56 but replace the "number" by the "proportion," that is, the relative frequency.

2.59 There were 6000 full-time state jobs created between fiscal 1984 and fiscal 1989 in the Commonwealth of Massachusetts (*Source: The Boston Globe,* February 22, 1989, p. 14). The number of jobs (rounded off) within governmental departments were: Human Services 3000; Higher Education 1200; Public Safety 500; Administration and Finance 900; Environmental 400.

   (a) Construct a bar graph.   **(b)** Construct a pie chart.

2.60 The following data give the number of audits per 10,000 individual tax returns for various regions of the United States in 1987 (*Source: Changing Times,* Sept. 1988, p. 40):
   West 164; Southwest 156; North-Atlantic 99; Midwest 89; Central 85; Southeast 84; Mid-Atlantic 75.

   (a) Construct a bar graph.   **(b)** Construct a pie chart.

2.61 **(a)** Use the data for "Age" in Table 2.7 to construct a stem-and-leaf plot. For stems, use 1, 2, 3, 4, 5, 6, 7, 8, that is, 10–19, 20–29, 30–39, etc.

   (b) Construct a frequency histogram with classes defined by the stems.

---

*By "incidence polygon" we mean a graph similar to a frequency polygon, except that frequencies are replaced by incidence rates. The height of a dot is the incidence rate rather than frequency.

(c) Rotate your stem-and-leaf plot 90° counterclockwise and compare its appearance with the frequency histogram in part (b). Could the stem-and-leaf plot be used as a frequency histogram?

2.62 The data below are the number of English language morning newspapers (per state) in the United States as of February 1, 1988 (*Source: The 1989 Information Please Almanac*, 1989, p. 300).

15	3	6	9	47	9	11	2	29	11	2	4	16
12	11	7	5	14	5	8	7	11	11	6	10	5
4	4	1	11	3	25	11	5	11	9	5	34	
1	9	5	9	35	1	4	16	8	9	7	6	

(a) Construct a stem-and-leaf plot.

(b) Rotate your stem-and-leaf plot 90° counterclockwise. Characterize the distribution shape by selecting the name of the distribution that most closely approximates the data.

Exercises 2.63 and 2.64 are designed to show the relationship between frequency and cumulative frequency. The grouped frequency distributions are depicted by frequency histograms, with frequencies displayed above each bar.

2.63 Consider the following frequency histogram:

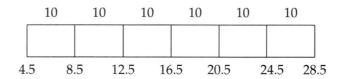

Construct a cumulative frequency polygon. What do you notice about the sequence of line segments?

2.64 Consider the following frequency histogram:

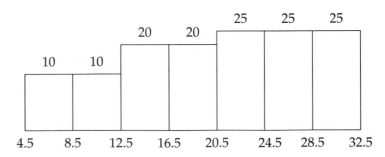

Construct a cumulative frequency polygon. What do you notice about the sequence of line segments?

2.65 For each of the following sketches of the shapes of frequency distributions, sketch a corresponding cumulative frequency polygon.

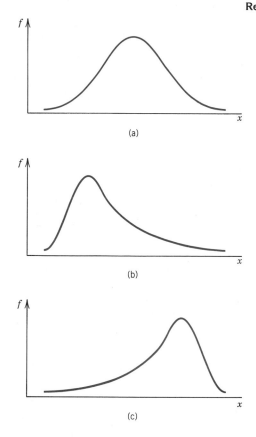

(a)

(b)

(c)

2.66   The following cumulative frequency polygon depicts the cumulative frequency of base hits for a baseball player over a period of years. In what years did he obtain more base hits than the previous year? Less? The same? (Think of each hit as having a date x associated with it. In the following figure, the class boundaries are December 31 of the year given.)

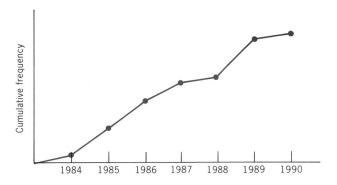

2.67   This problem deals with frequency histograms for grouped frequency distributions in which not all classes are of equal width. It illustrates that the area of each rectangle should be proportional to the frequency for that class. Consider the following data indicating profit in millions of dollars for a large corporation over the indicated time period.

Time Period	Profit
1975–1976	4
1977–1980	6
1981–1986	6
1987–1988	1
1989	2
1990	3

(a) Construct a frequency histogram (using profit for frequency). Does the histogram accurately reflect the fact that 1990 was the best year for profits for the company? Why?

(b) To adjust for the different lengths of times recorded, consider the amount of profit per year's time over a specified interval. For example, the 2-year period of 1975–1976 indicates a profit of $4 \div 2 = \$2$ million per year. Construct a frequency histogram using the amount of profit per year's time as the height of the bars. Is this histogram quite different from that of part (a)? Which gives more accurate information?

Answers to Practice Quiz

1. The data must be on a separate line (or lines) from the command line.

2. This program probably does not accomplish what the user intended. The 2,104 will be interpreted by Minitab as two numbers: 2 and 104. Type 2104 instead.

3. Don't type COLUMN 1. Type C1.

4. This program is all right. The user won't get high marks for the spelling of minimum, but Minitab is friendly and doesn't care about this as long as the first four characters of the command can be found in its dictionary of commands.

5. Minitab is friendly, but not this friendly. It looks at the first four characters of the first word of a command (including blanks). There is no SETT command in its dictionary.

6. The user who wrote this program forgot that Minitab expects the command to be the first word. If you insist on being this polite, put the PLEASE somewhere after the first word.

7. Nothing wrong.

8. The 1 in BRAND 1 is an unnecessary constant, but Minitab is expecting a column. Minitab can ignore letters in explanatory text, but not numbers. The following error message will be returned:

   ERROR ARGUMENT IS A CONSTANT OR MATRIX BUT A COLUMN WAS EXPECTED

   You can avoid this difficulty by typing BRAND ONE.

9. You should put only one command to a line.

10. The value for INCREMENT must be 1, 2, or 5 times a power of 10.

Notes

Andrews, D. F., and A. M. Herzberg, "The Number of Deaths by Horsekicks in the Prussian Army" *Data* (New York: Springer-Verlag, 1985).

Changing Times (Washington D.C.: The Kiplinger Washington Editors Inc., Sept. 1988).

Crime in the United States—Uniform Crime Reports (Washington, D.C.: Federal Bureau of Investigation, 1988).

Information Please Almanac (Boston: Houghton Mifflin Company, 1988, 1989).

Massachusetts State Lottery Commission, Braintree, MA.

Matthews, D. E., and V. T. Farewell (1982), "On Testing for a Constant Hazard Against a Change-Point Alternative," *Biometrics* 38, 463–468. Data in Chambers, J. M., W. S. Cleveland, B. Kleiner, and P. A. Tukey, *Graphical Methods for Data Analysis* (Boston: Duxbury Press, 1983).

Statistical Abstract of the United States (Washington, D. C.: U.S. Bureau of the Census, 1987).

The Boston Globe (Boston: Globe Newspaper Company, February 22, 1989).

The Boston Sunday Globe (Boston: Globe Newspaper Company, April 30, 1989).

The World Almanac and Book of Facts (New York: Newspaper Enterprise Association, 1988).

The World Almanac and Book of Facts (New York: Newspaper Enterprise Associarion, 1989).

Woodhouse, J., "World Wide Seismic Network Tape,". Data in Donoho, A. W., D. L. Donoho, and M. Gasko, *MacSpin User Manual* (Austin, TX: D^2 Software Inc.).

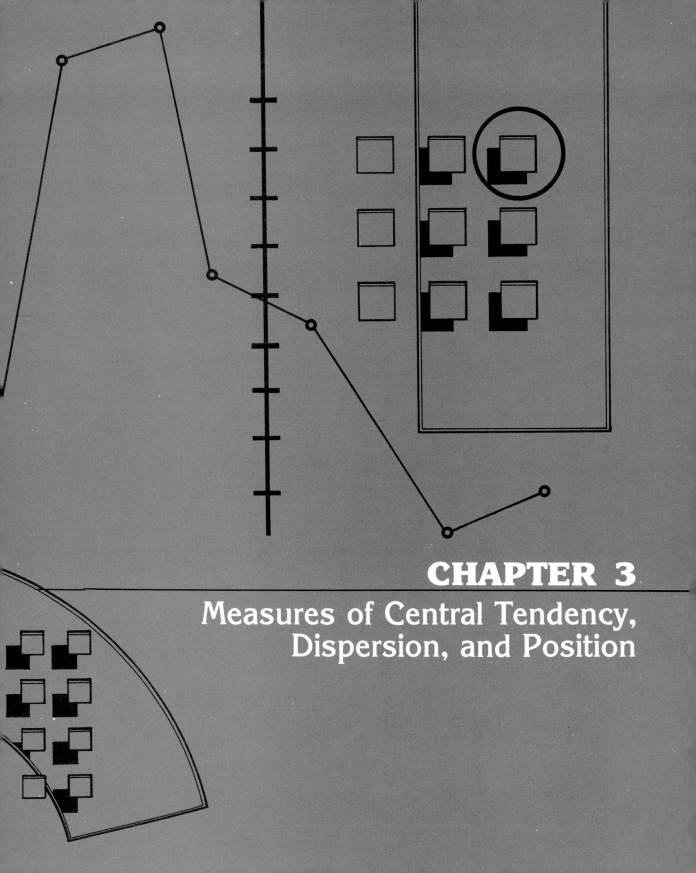

CHAPTER 3
Measures of Central Tendency, Dispersion, and Position

3.1 INTRODUCTION

In this chapter we continue our study of certain topics in descriptive statistics. We will investigate ways of describing numerical data by giving certain properties of the data. Specifically, we will deal here with the concept of a **measure of central tendency.** A measure of central tendency for a collection of data values is a number that is meant to convey the idea of a typical or representative value for the data. We will study three measures of central tendency: the **mean,** the **median,** and the **mode.** However, the one we will use most often in later chapters is the mean.

We will also study **measures of dispersion,** which reflect the amount of spread or variability in a collection of data. **Variance** and **standard deviation** are such measures. Finally we discuss **measures of position,** which describe how a data value relates to the other data in a collection of data values. For example, if your score on a test was at the 99th percentile, this would be a measure of position because it means that 99% of the scores were lower than yours.

In this and future chapters, we will need a rule for rounding off decimal answers. In statistics it is common to retain one or two more decimal places than were present in the original data. In this book we favor retaining two more decimal places than were present in the original data. Only the final answer should be rounded; do not round off during intermediate steps. When we round off a number, our answer will be approximate; this is indicated by using the symbol $\doteq$, which means "is approximately equal to."

3.2 THREE MEASURES OF CENTRAL TENDENCY

The Mean

> **Definition**   Given a collection of n data values, the *mean* of these data is simply the average of these data values. Therefore, it is the sum of the data values divided by n.
>
> $$\text{mean} = \frac{\text{sum of the data values}}{n}$$

The collection of data we are working with may constitute a population or a sample from a population. We use the Greek letter μ to represent a population mean and the symbol $\bar{x}$ to represent a sample mean (if x represents an arbitrary data value from the sample). In many cases we don't know enough about the population to find the value of μ exactly. In such cases we can often estimate μ by the sample mean $\bar{x}$ for some appropriate sample from the population.

Many of the formulas involving the mean are equally valid whether we are talking about the mean μ of a population or the mean $\bar{x}$ of a sample. When this is the case, we represent the mean simply by writing the word *mean*.

Example 3.1
Find the mean for the sample data that represent the number of times five members of a family caught colds in 1 year: 4, 2, 0, 1, and 0.

Solution
The number of data values is $n = 5$. Thus

$$\overline{x} = \frac{4 + 2 + 0 + 1 + 0}{5} = \frac{7}{5} = 1.4$$

Remark A graphic interpretation of the mean is that it is the balancing point for the frequency distribution. That is, if equal weights are placed along the number axis, one for each data value, the number axis will balance if a fulcrum is placed at the mean. Figure 3.1 illustrates this idea for the data in Example 3.1.

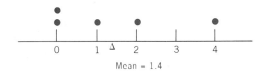

Mean = 1.4

FIGURE 3.1
A Frequency Distribution Balances at the Mean

Summation Convention

We said that we can use a symbol such as x to represent any data value from some collection of data. We use the symbol Σx to represent the sum of these values. The Greek letter sigma, Σ, is used to represent summation of the values of whatever symbol follows. So Σx means sum up the values of x.

For instance, consider the data values: 5, 10, 6, 9, 4, 3, and 5. If we use the symbol x to represent any of these values, then

$$\Sigma x = 5 + 10 + 6 + 9 + 4 + 3 + 5 = 42$$

Suppose that we wished to sum up the squares of some data values. We would use the symbol Σx^2 for this. Thus, for the data above,

$$\Sigma x^2 = 5^2 + 10^2 + 6^2 + 9^2 + 4^2 + 3^2 + 5^2 = 292$$

Notice that $(\Sigma x)^2 = (42)^2 = 1764$. So $(\Sigma x)^2$ is not the same as Σx^2.

Using our summation convention, we can write the formula for the mean of a collection of n data values as

$$mean = \frac{\Sigma x}{n}$$

For the data values 5, 10, 6, 9, 4, 3, and 5, we have $n = 7$. Therefore

$$\text{mean} = \frac{\Sigma x}{n} = \frac{42}{7} = 6$$

Weighted Mean

When we find the mean of a collection of data values, we must add up each data value in the collection, including repetitions. This can be inefficient if there are many repetitions. What we can do is take each *distinct* data value, multiply it by its frequency, add the results, and then divide by the total number of data values. For example, the following frequency distribution represents the ages of 20 students in a college history class.

x	f
16	1
18	4
19	9
20	3
21	2
30	1
	20

Notice that $\Sigma f = 20$. The mean of these ages is

$$\text{mean} = \frac{(16)(1) + (18)(4) + (19)(9) + (20)(3) + (21)(2) + (30)(1)}{20}$$

$$= \frac{391}{20} = 19.55$$

In general, if we have a collection of n data values described in terms of a frequency distribution, the mean is given by

$$\text{mean} = \frac{\Sigma xf}{n} \text{ where } n = \Sigma f$$

Notice, however, that x in this formula represents each *distinct* data value and f its frequency. A mean calculated this way is sometimes referred to as a **weighted mean,** but it is really just the mean of all the data values in the collection (including repetitions) obtained in an efficient manner.

A variation of the above formula may be used to estimate the mean of a collection of data from a grouped frequency distribution for the data (even if the original data are unknown). Assume that X represents an arbitrary class mark of a grouped frequency distribution. (The class mark is halfway between the lower and upper class limits of the class.) We can approximate each data value in the class by the class mark. Therefore, if f is the frequency of the class, we can view the value X as having frequency f. We can then approximate the mean of the original collection of n data values by

$$\frac{\Sigma Xf}{n}$$

Example 3.2
Estimate the mean for the sample data described by the grouped frequency distribution in Table 2.5 (which is reproduced as part of Table 3.1 below).

TABLE 3.1

Class	Class Limits	Class Frequency (f)	Class Mark (X)	Xf
1	45–49	2	47	94
2	50–54	3	52	156
3	55–59	6	57	342
4	60–64	4	62	248
5	65–69	9	67	603
6	70–74	8	72	576
7	75–79	3	77	231
8	80–84	4	82	328
9	85–89	1	87	87
		Sum = 40		Sum = 2665
		n _____↑		ΣXf _____↑

From the above we see that

$$\bar{x} \doteq \frac{\Sigma Xf}{n} = \frac{2665}{40} \doteq 66.63$$

The exact value of $\bar{x}$ for the original data (in Table 2.2) is 66.75.

Remark   A histogram balances at the mean, as shown in Figure 3.2. This is the histogram for the data in the above example. It was constructed in Example 2.5.

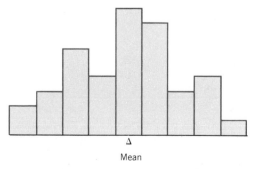

Mean

FIGURE 3.2
A Histogram Balances at the Mean (66.63)

The Median

> **Definition**   The *median* for a collection of data values is the number that is exactly in the middle position of the list when the data are ranked (i.e., arranged in increasing order of magnitude). If we use the symbol x to represent an arbitrary data value in the collection, then the median is represented by the symbol $\tilde{x}$ (read "x tilde").

Assume that we have an odd number of data values. For example, consider the ranked list of five data values: 5, 6, 9, 11, and 15. Clearly, the middle number is the third data value; that is, the location of the median is 3. The median itself is 9.

Now suppose that we have an even number of data values. For example, consider the ranked list of four data values: 1, 4, 7, and 9. Here there is no middle data value. We will agree that the median in a case like this is the number that is halfway between the two data values closest to the middle. So the median here is the number that is halfway between the second and third data value in the ranked list. Notice that this is also the average of the second and third data values; that is, the median is

$$\frac{4 + 7}{2} = 5.5$$

In a case like this we say that the location of the median is 2.5, since the median is that number which is halfway between the second and third data values (in the ranked list).

To find the location of the median in a ranked list of n data values, we can use the formula

$$\text{location of } \tilde{x} = \frac{n + 1}{2}$$

Notice that this formula works for both of the data sets discussed above. For the first set, $n = 5$; so the location is $\frac{6}{2} = 3$. For the second set, $n = 4$; so the location is $\frac{5}{2} = 2.5$. When the location involves .5, as in 2.5, the median is halfway between the two data values nearest to the location. (When these two have the same value, this value is the median.) *Note:* Do not confuse the *location* of $\tilde{x}$ with the *value* of $\tilde{x}$.

Example 3.3
Find the median for the data in Example 3.1, namely, 4, 2, 0, 1, and 0.

Solution
First rank the data

$$0 \quad 0 \quad 1 \quad 2 \quad 4$$

The number of data values is $n = 5$.

$$\text{location of } \tilde{x} = \frac{n + 1}{2} = \frac{5 + 1}{2} = 3$$

So $\tilde{x}$ is the third data value in the ranked list. Therefore,

$$\tilde{x} = 1$$

Example 3.4
Find the median for the following data, which are eight systolic blood pressure readings from the Framingham Heart Study: 126, 110, 122, 98, 116, 124, 128, 144. (Systolic pressure is arterial pressure when the heart pumps. It is measured in millimeters of mercury.)

Solution
First rank the data

$$98 \quad 110 \quad 116 \quad 122 \quad 124 \quad 126 \quad 128 \quad 144$$

There are $n = 8$ data values. Thus

$$\text{location of } \tilde{x} = \frac{n + 1}{2} = \frac{8 + 1}{2} = \frac{9}{2} = 4.5$$

This means that $\tilde{x}$ is halfway between the fourth and fifth data values in the ranked list. Thus

$$\tilde{x} = 123$$

The above two examples were simple enough so that they could have been done by inspection, without first using a formula to find the location of $\tilde{x}$. But when the number of data values is large, the formula for the location of $\tilde{x}$ comes in handy.

The Mode

> **Definition** The *mode* for a collection of data values is the data value that occurs most frequently (if there is one).

Example 3.5
Find the mode for the data in Example 3.1, namely, 4, 2, 0, 1, and 0.

Solution

The data value that occurs most frequently is 0. So the mode = 0.

Example 3.6

Find the mode for the data in Example 3.4, namely, 126, 110, 122, 98, 116, 124, 128, and 144.

Solution

No one data value occurs most frequently. So there is no mode.

Exercises

In Exercises 3.1–3.7, find the (a) mean, (b) location of the median, (c) median, and (d) mode (if it exists).

3.1   The number of copies of a book sold by a bookstore over a 6-day period was

$$7 \quad 12 \quad 6 \quad 12 \quad 2 \quad 9$$

3.2   The diastolic blood pressure readings of 12 randomly selected males aged 45–49 from the Framingham Heart Study were

$$94 \quad 84 \quad 74 \quad 90 \quad 98 \quad 92 \quad 74 \quad 90 \quad 80 \quad 98 \quad 78 \quad 80$$

3.3   The number of students advised over a 5-day period by a professor was

$$0 \quad 4 \quad 5 \quad 0 \quad 7$$

3.4   The number of traffic tickets issued by a police department over a 7-day period was

$$19 \quad 17 \quad 14 \quad 21 \quad 19 \quad 16 \quad 34$$

3.5   The weights (in pounds) of 10 new-born babies at a local hospital were recorded as

$$9.0 \quad 7.4 \quad 7.6 \quad 9.4 \quad 7.6 \quad 6.4 \quad 7.6 \quad 8.3 \quad 8.5 \quad 7.1$$

3.6   The number of patients serviced by a dentist over a 5-day period was

$$10 \quad 8 \quad 13 \quad 12 \quad 7$$

3.7   Six cars of the same make and model were tested; the gas mileage of each was recorded with the following results (in miles per gallon):

$$24.0 \quad 25.7 \quad 23.1 \quad 26.3 \quad 29.7 \quad 24.2$$

3.8   In each part below, use the grouped frequency distribution to estimate $\bar{x}$ for the original data.

(a)

Class	Class Limits	Frequency
1	9–17	14
2	18–26	9
3	27–35	7
4	36–44	4
5	45–53	3
6	54–62	1

(b)

Class	Class Limits	Frequency
1	9–17	1
2	18–26	3
3	27–35	4
4	36–44	7
5	45–53	9
6	54–62	14

In doing Exercises 3.9 and 3.10, using the fact that a distribution balances at its mean, estimate $\bar{x}$ without calculating.

3.9   **(a)**

Class	Class Limits	Frequency
1	3–7	2
2	8–12	3
3	13–17	6
4	18–22	3
5	23–27	2

(b)

Class	Class Limits	Frequency
1	3–7	2
2	8–12	3
3	13–17	6
4	18–22	6
5	23–27	3
6	28–32	2

3.10   **(a)**   2   5   6   7   10

(b)   Consider the dot diagram

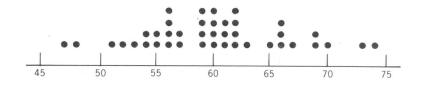

(c)   Below is a frequency histogram representing the distribution of 1000 randomly selected observations.

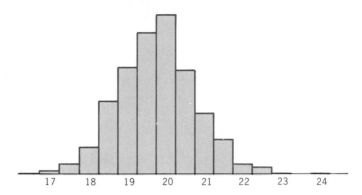

3.11   A statistics student claimed that the median of the data values 9, 8, 12, 4, and 6 must be 12. The student reasoned that the location of the median is 3 and the third value in the list is 12. Is there anything wrong with the student's reasoning?

3.12   The median family income in the United States for a year was reported to be $29,744. Interpret this statement (*Source: The 1989 Information Please Almanac, 1989, p. 54*).

3.13   The following data are the percentage of annual salary typically awarded employees who are fired or laid off. Figures are the percentages of total annual salary plus bonuses, if any, typically paid to anyone whose employment is involuntarily terminated for reasons other than misconduct (*Source: U.S. News & World Report, December 19, 1988, p. 70*).

Country	Manufacturing Workers	Chief Executives
Japan	102	261
Netherlands	25	228
Germany	100	200
Spain	183	185
Britain	22	156
Italy	46	154
Korea	125	125
Canada	23	115
Mexico	110	146
France	54	83
United States	29	79
Sweden	42	42
Brazil	19	27
Argentina	24	131
Hong Kong	18	77

(a)   Find the median percentage for manufacturing workers. Would your answer change if the five largest percentages were increased by 25 percentage points?

(b)   Find the median percentage for chief executives. Would your answer change if the five smallest percentages were decreased by 25 percentage points?

3.14 The cumulative relative frequency polygon below represents the distribution of observations of time in days from remission induction to relapse for 51 patients with acute nonlymphoblastic leukemia (*Source:* Matthews, D. E., and V. T. Farewell, "On Testing for a Constant Hazard against a Change-point Alternative," *Biometrics* 38: 463–468, 1982.) Estimate the median time to relapse.

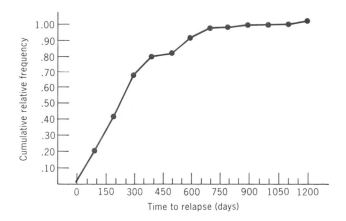

3.15 An experiment was conducted to determine whether increased hibernation results in increased life span for hamsters [*Source:* Lyman et al. (1981)]. The cumulative relative frequency polygon below represents the distribution of age at death in days for 144 hamsters. Estimate the median life span for these hamsters.

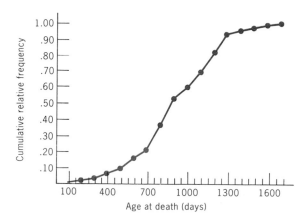

3.16 For data values 2, 5, 2, 6, 0, calculate each of the following:

(a) Σx (b) $\Sigma(x - \bar{x})$

(c) $\Sigma(x - \bar{x})^2$ (d) Σx^2

(e) $(\Sigma x)^2$ (f) $\Sigma 7x$ [Compare your answer with $7(\Sigma x)$.]

[Note that the answers for (d) and (e) are not the same.]

COMPARISONS OF THE MEAN, MEDIAN, AND MODE

All three measures of central tendency (the mean, median, and mode) may vary in value from one sample to the next from the same population. But for some of the most commonly encountered populations, the measure that varies the least is the mean. This gives the mean certain advantages in the area of statistical inference. For example, if we wish to estimate a population parameter (such as a measure of central tendency for a population) by its corresponding sample statistic, we would want the values of the statistic to vary from the parameter by as little as possible.

Also, there is a simple formula for the mean in terms of all the data values in the collection. This makes it easy to perform algebraic manipulations on the mean. There are no corresponding simple formulas for the median or mode in terms of the data values.

Another important property of the mean is that it depends on the value of every piece of data in the collection. Therefore, the mean uses all the information contained in the data, which is usually desirable. The median and mode do not depend on all the data values in quite the same way as the mean. The mode depends on only one data value—the most frequent one. Of course, in determining the mode we must also take into consideration the frequencies of the various data values; but other than this, no other information is needed except the value of the most frequent piece of data. Similarly, the median depends only on the value in the middle position in a ranked list of the data values. The rest of the data values only affect the median insofar as they are used to determine which data value is in the middle position.

On the other hand, there are some situations where it may be desirable to ignore or downplay certain atypical data values such as those we encounter when a distribution is skewed, and this the mean cannot do. For example, consider the following data values, which represent the annual salaries of 11 people:

$$20,000 \quad 22,000 \quad 23,000 \quad 23,500 \quad 24,000 \quad 24,000$$
$$24,500 \quad 25,000 \quad 26,000 \quad 26,000 \quad 70,000$$

Now the mean salary is $28,000, which does not seem typical. After all, 10 of the 11 people earn less than this figure. The median, however, is $24,000, which seems more typical.

The mean and median will always exist for a collection of data, but the mode may not, which is a disadvantage. However, one advantage of the mode is that when it does exist it will be a data value. For example, the mean number of children in American families is 2.5 children, which seems slightly awkward. However, the most frequent number of children in a family (the mode) is 2 children, which seems more satisfactory as a representation of the typical family.

Sometimes two researchers looking at the same data will be interested in different measures of central tendency. For instance, in many towns the school system is supported by a tax based on the assessed value of each home. For example, the tax might be $50 per $1000 of assessed value. An owner of a house assessed at $10,000 would have a tax bill of $500. An educator interested in a town's ability to provide quality education for all its students might look at the mean of the assessed values of all the homes.

However, suppose that a sociologist is interested in a measure of the social or economic status of a town. The mean (assessed) home value might not be appropriate. To understand this, consider the following example: There are two towns, town A and town B. Now assume that there are 100 homes in each town. In town A, 99 homes are assessed at $30,000 each and one home is assessed at $7,030,000. In town B there are 100 homes, each assessed at $100,000.

Therefore, the mean (assessed) home value in town A is

$$\text{mean} = \frac{(30,000)(99) + 7,030,000}{100}$$

$$= \frac{10,000,000}{100}$$

$$= \$100,000$$

It should be clear that the mean assessed home value in town B is also $100,000. For the educator, the mean is a satisfactory indication of what each town can provide because both towns can provide the same funding per student. (We are assuming roughly the same number of students in each town.)

However, the sociologist would point out that the data for town A are skewed to the right, and we have seen that in a situation like this the mean may not be a satisfactory measure of typicalness (at least for some purposes). The fact is that town A has mostly very modest homes except for one mansion. Town A does not seem to be in the same social class as town B. As we have seen, the median might be a better measure in this case. The median home value for town A is $30,000 (halfway between the 50th and 51st data value in the ranked list). The median home value in town B is $100,000. These seem to give a much better picture of the social status of the two towns.

This discussion points out a relationship between the mean and median that will be helpful to keep in mind when choosing a measure of central tendency. When a distribution is skewed, this pulls the mean away from the median. If the distribution is skewed to the right, the mean is pulled to the right (in a positive direction). If the distribution is skewed to the left, the mean will be pulled to the left. When a distribution is symmetric (the right half is a mirror image of the left half), the mean and median will be equal. These ideas are shown graphically in Figure 3.3, where we display the (smoothed-out) shapes of some frequency polygons.

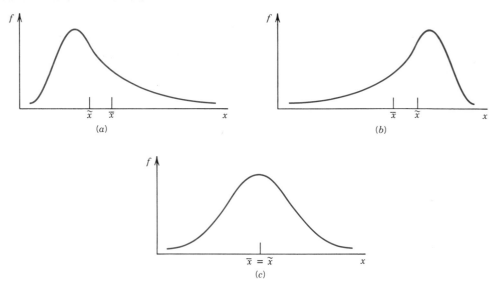

FIGURE 3.3
Relationships Between the Mean and the Median. (*a*) Skewed to the right. (*b*)
Skewed to the left. (*c*) Symmetric.

Exercises

3.17 Answer each of the following either True or False.
   (a) For a symmetric distribution, the mean and the median have the same value.
   (b) For a distribution skewed to the left, the value of the median is generally smaller than the value of the mean.
   (c) The value of the mean for a finite collection of data values is necessarily one of the data values.
   (d) The value of the median for a finite collection of data values is necessarily one of the data values.

3.18 In Table 2.7 we considered "Number of Hours Watching TV per Week for 20 Adults." The data are

   16 23 19 31 28   4 11 21 17 13
   29   7   3 18 21 28 12   3 24 27

   (a) Find the mean.
   (b) Find the median.
   (c) Find the mode (if it exists).
   (d) Describe the shape of the distribution as approximately symmetric, skewed to the right, or skewed to the left.
   (e) Which measure, if any, appears to best summarize the data in terms of typical value?

3.19 In Table 2.7 we considered "Number of Children for 20 Families." The data are

> 3 2 4 5 1 2 2 3 4 2 5 3 3 0 1 0 2 3 2 3

(a) Find the mean.

(b) Find the median.

(c) Find the mode (if it exists).

(d) Describe the shape of the distribution as symmetric, skewed to the right, or skewed to the left.

(e) Which measure, if any, appears to best summarize the data in terms of typical value?

3.20 The data below are the number of English language Sunday newspapers (per state) in the United States as of February 1, 1988 (*Source: The 1989 Information Please Almanac*, 1989, p. 300).

> | 3 | 6 | 6 | 7 | 7 | 8 | 8 | 9 | 9 | 10 | 10 | 11 | 12 |
> |----|----|----|----|----|----|----|----|----|----|-----|-----|----|
> | 12 | 14 | 17 | 19 | 19 | 19 | 20 | 23 | 23 | 23 | 24 | 25 | 25 |
> | 26 | 26 | 27 | 28 | 28 | 32 | 35 | 35 | 36 | 37 | 45 | 46 | |
> | 47 | 48 | 51 | 52 | 54 | 70 | 72 | 73 | 87 | 93 | 108 | 118 | |

(a) Find the mean.

(b) Find the median.

(c) Describe the shape of the distribution as approximately symmetric, skewed to the right, or skewed to the left.

(d) Of the two measures, which appears to best summarize the data in terms of typical value?

3.21 The dot diagram below represents the distribution of 25 data values.

(a) Without calculating, which is larger, the mean or median value?

(b) Which measure would best summarize the data in terms of a typical value?

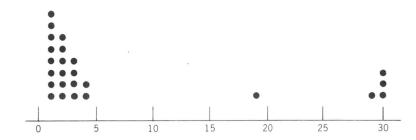

3.4 MEASURES OF DISPERSION

In Section 3.2 we studied the notion of a measure of central tendency, but measures of central tendency may not be completely adequate in describing certain properties of data. For example, consider the following two collections of data: 99, 100, 100, 101; and 50, 100, 100, 150. Both have the same mean, median, and mode (namely, 100), and yet the second set of data values possesses much more spread or variability than the first. Apparently, a measure of central tendency is incapable of detecting differences in the spread or variability in a collection of data values.

Yet it may be of great importance to know how much variability there is. For instance, a manufacturer of steel girders advertises the strength of the girders as being 10,000 pounds per square inch on the average. Consider the collection of data values, which consists of the actual strengths of the various girders to be used in the construction of a bridge. This manufacturer would want the variability of these data values from the mean of 10,000 pounds to be at a minimum.

In contrast, if an educator develops an examination to be used by a college to screen applicants for admission, it would be desirable for the scores on this exam to have as much variability as possible so that the college could more easily distinguish between applicants.

For these reasons, statisticians have developed ways of measuring variability—measures of dispersion. The measure of dispersion that is perhaps the simplest and easiest to calculate is the **range.**

Definition   The *range R* for a collection of data values is the difference between the highest (H) and lowest (L) data values.

$$R = H - L$$

Example 3.7
The following is a list of the number of books read by the 10 students in a history class while doing research for a term paper: 6, 11, 5, 1, 6, 6, 7, 5, 7, and 6. Find the range of the data.

Solution
For these data $L = 1$, $H = 11$. Thus

$$R = H - L = 11 - 1 = 10$$

The range is a rather rough measure of dispersion: it depends only on the highest and lowest data values and ignores the ones in between. Hence it may not give us all the information we need about the variability of all the data values. To get a better understanding of this, we construct a dot diagram for the data in Example 3.7. (See Figure 3.4*a*.) Now compare this with a dot diagram for the following data: 10, 3, 9, 2, 2, 11, 2, 1, 9, and 11. (See Figure 3.4*b*.) These

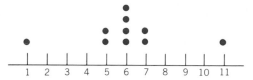

FIGURE 3.4a
Number of Books Read by History Students

FIGURE 3.4b
Number of Books Read by Sociology Students

data represent the number of books read by 10 students in a sociology class preparing for a term paper.

Both sets of data have the same range (10) and the same mean (6), but somehow there appears to be less variability among the data in Figure 3.4a. This is because more of the data values appear to cluster near the middle (in fact, the mean 6), while there are fewer out toward the extreme values of 1 and 11. Apparently, the range is incapable of detecting these differences in the two sets of data. We would like to have a measure of dispersion that is more sensitive to such differences. More precisely, we would like to have a measure of dispersion that would somehow indicate that there is less variability for the data of Figure 3.4a than the data of Figure 3.4b. We will discuss such a measure next.

We will find it convenient to measure variability in a collection by examining the degree to which the data values cluster about (or deviate from) their mean. For simplicity, we consider the following data: 1, 5, and 6. The mean of these data values is 4. To study variability from the mean, perhaps we should find the difference between each data value and the mean, and then average these differences. These differences, called **deviations,** are

$$1 - 4 = -3$$
$$5 - 4 = 1$$
$$6 - 4 = 2$$

The average of these deviations is

$$\frac{-3 + 1 + 2}{3} = 0$$

It turns out that no matter what data values we use, the average of the differences between the data values and their mean is always 0. Therefore, such an average will not be a satisfactory measure of dispersion. The average of these deviations is 0, because their sum is 0—the negative values cancel the positive values. We can get around this difficulty by squaring the deviations. This eliminates the negative signs.* The squares of the deviations are

$$(1 - 4)^2 = (-3)^2 = 9$$
$$(5 - 4)^2 = 1^2 = 1$$
$$(6 - 4)^2 = 2^2 = 4$$

The average of these is

$$\frac{9 + 1 + 4}{3} = \frac{14}{3} \doteq 4.67$$

Note that to calculate this, we added each value of $(x - \text{mean})^2$ and divided by the number of data values. Using the summation convention, we can write this average symbolically as

$$\frac{\Sigma(x - \text{mean})^2}{n}$$

(where n is the number of data values).

Although averaging squares of numbers may not be a familiar part of our experience, this average does lead to a valid measure of dispersion in the following sense: if we have two collections of data, the one for which this average is larger is the one whose values tend to deviate more from their mean.

The reader might get the feeling that the above average would be a satisfactory measure of dispersion for either a population or a sample. However, most statisticians prefer to use it only for a population. (They prefer to use a slightly different expression for samples, which we will discuss shortly.) When the collection of data values we are dealing with is a population and not a sample, we often use N to represent the number of data values rather than n. We are now in a position to define the term **population variance.**

---

*If we used the absolute value of each deviation, this would also eliminate the negative signs. But it turns out that for technical reasons, it is mathematically more convenient to use the squares of the deviations.

> **Definition**   Given a population of N data values, we define the *population variance*, represented by the symbol σ^2 (where σ is the Greek lowercase "sigma"), as
>
> $$\sigma^2 = \frac{\Sigma(x - \mu)^2}{N}$$
>
> where μ is the population mean $\Sigma x/N$.

Example 3.8

In Example 3.7 we discussed the number of books read by each student preparing for a term paper in a history class of 10 students: 6, 11, 5, 1, 6, 6, 7, 5, 7, and 6. Viewing this collection of data values as a population, find the population variance, σ^2.

Solution

First find the population mean, μ. Now $N = 10$ so

$$\mu = \frac{\Sigma x}{N}$$

$$= \frac{6 + 11 + 5 + 1 + 6 + 6 + 7 + 5 + 7 + 6}{10}$$

$$= \frac{60}{10} = 6$$

Some find it convenient to use a table such as Table 3.2 to calculate variance.

TABLE 3.2

x	$x - \mu$	$(x - \mu)^2$
6	0	0
11	5	25
5	-1	1
1	-5	25
6	0	0
6	0	0
7	1	1
5	-1	1
7	1	1
6	0	0
	Sum =	54
		↑
		$\Sigma(x - \mu)^2$

Hence

$$\sigma^2 = \frac{\Sigma(x - \mu)^2}{N}$$

$$= \frac{54}{10} = 5.4$$

If you were to calculate the population variance for the number of books read by the sociology class (Figure 3.4b), you would find it to be 16.6. Comparing this with the variance of 5.4 for the history class confirms our feeling that the sociology class shows more variability.

Many results in statistics involve the square root of the variance. We call the square root of the population variance the **population standard deviation.** Some statisticians like to think of standard deviation as a sort of typical distance from the mean for the data values. The units of standard deviation will be the same as the units of the data set—if your data is given in pounds, the standard deviation will be in pounds.

> **Definition**   Given a population of N data values, we define the *population standard deviation* as the square root of the population variance, $\sqrt{\sigma^2} = \sigma$. Therefore,
>
> $$\sigma = \sqrt{\frac{\Sigma(x - \mu)^2}{N}}$$

For example, for the population of Example 3.8, the population standard deviation $\sigma = \sqrt{5.4} \doteq 2.32$.

Variance and Standard Deviation for a Sample

In practice, the populations we study are usually enormous and we rarely know enough about them to find σ^2 exactly. What we usually do is obtain a sample from the population and then try to estimate σ^2 using information from the sample. Suppose that there are n data values in a sample obtained from our population of interest. Our first impulse might be to try to estimate σ^2 as follows: assuming that we do not know μ, we can estimate μ by $\bar{x}$, and then estimate σ^2 by

$$\frac{\Sigma(x - \bar{x})^2}{n}$$

This does provide an estimate for σ^2 and, in fact, is used as such by some statisticians. However, it can be shown using more advanced techniques than we have at our disposal that the above expression tends (on the average) to underestimate σ^2. Replacing the n in the denominator of the above expression by $n - 1$ gives us a slightly larger ratio; the resulting expression provides an estimate for σ^2 that many statisticians feel is more desirable. With this in mind we give the following definition.

Definition Assume that we have a sample of n data values from some population. We define the *sample variance, s^2*, by the equation

$$s^2 = \frac{\Sigma(x - \bar{x})^2}{n - 1}$$

The square root of the sample variance is the *sample standard deviation, s.*

$$s = \sqrt{\frac{\Sigma(x - \bar{x})^2}{n - 1}}$$

In the future, when it is impossible or impractical to calculate the exact value of σ^2 for a population, we can think of s^2 for some appropriate sample from the population as an estimate for σ^2. Of course, s^2 may also be viewed as a measure of variability of the data values in the sample from their mean $\bar{x}$.

Example 3.9
Find the sample variance and sample standard deviation for the following data, which represent the number of fire alarms answered by Boston Engine Company 10 over a 5-day period: 4, 3, 7, 4, 2.

Solution
The sample mean is

$$\bar{x} = \frac{4 + 3 + 7 + 4 + 2}{5} = \frac{20}{5} = 4$$

TABLE 3.3

x	$x - \bar{x}$	$(x - \bar{x})^2$
4	0	0
3	−1	1
7	3	9
4	0	0
2	−2	4
		Sum = 14
		↑
		$\Sigma(x - \bar{x})^2$

Table 3.3 is helpful in calculating the variance:

$$s^2 = \frac{\Sigma(x - \bar{x})^2}{n - 1} = \frac{14}{4} = 3.5$$

$$s = \sqrt{3.5} \doteq 1.87$$

Alternative Formulas for Finding Variance

There is a way of computing variance that is often more convenient because it does not require finding the mean first. Even if the mean is known, this method can result in less error due to round off when the original data are whole numbers and the mean involves a decimal. This method consists of using the following formulas to find σ^2 and s^2 that are equivalent to the original formulas given for σ^2 and s^2. That is, they will give the same result.

Alternative Formulas for Computing Variance   For a population of N data values

$$\sigma^2 = \frac{N(\Sigma x^2) - (\Sigma x)^2}{N^2}$$

For a sample of n data values

$$s^2 = \frac{n(\Sigma x^2) - (\Sigma x)^2}{n(n - 1)}$$

The reader will usually find that **these formulas are more convenient than the original formulas unless the mean is a known whole number.** For example, consider the sample data 1, 3, and 4. The sample mean is

$$\bar{x} = \frac{\Sigma x}{n} = \frac{1 + 3 + 4}{3} = \frac{8}{3} \doteq 2.67$$

TABLE 3.4

x	x^2
1	1
3	9
4	16
8	26
↑	↑
Σx	Σx^2

To find s^2 using the original formula would be slightly unpleasant. But observe how the alternate formula avoids working with such decimals:

$$s^2 = \frac{n(\Sigma x^2) - (\Sigma x)^2}{n(n-1)} = \frac{3(26) - 8^2}{(3)(2)}$$

$$= \frac{78 - 64}{6} = \frac{14}{6} \doteq 2.33$$

Example 3.10

The actual voltages of 6-volt batteries will vary somewhat, but a battery manufacturer wanted to keep this variability to a minimum. A quality control procedure was put in place whereby, periodically, a sample of a half-dozen batteries were measured for voltage. If the standard deviation exceeds .3 volts, the production process is checked. The following voltage readings were obtained:

$$6.1 \quad 5.7 \quad 5.8 \quad 6.0 \quad 5.8 \quad 6.3$$

Using the alternate formula, find the sample variance and the sample standard deviation.

Solution
See Table 3.5.

TABLE 3.5

x	x^2
6.1	37.21
5.7	32.49
5.8	33.64
6.0	36.00
5.8	33.64
6.3	39.69
Sums 35.7	212.67 n = 6

Therefore, $\Sigma x = 35.7$ and $\Sigma x^2 = 212.67$. Thus

$$s^2 = \frac{n(\Sigma x^2) - (\Sigma x)^2}{n(n-1)} = \frac{6(212.67) - (35.7)^2}{(6)(5)}$$

$$= \frac{1276.02 - 1274.49}{30} = \frac{1.53}{30} = .051$$

$$s = \sqrt{.051} \doteq .226$$

Since $s < .3$ the process was judged to be under control.

We have seen that data are sometimes described by a frequency distribution that gives each distinct data value x together with the number of times it occurs, f. When data are presented this way, the above formulas should be adjusted to reflect this fact. In Section 3.2 we discussed how the formula for the mean should be adjusted in a case like this (weighted mean). Similar considerations would lead to the following adjusted formula for sample variance:

$$s^2 = \frac{n(\Sigma x^2 f) - (\Sigma x f)^2}{n(n-1)}$$

where x represents the distinct data values, f their frequencies, and n the total number of data values ($n = \Sigma f$).

The above formula may be used to estimate the sample variance for data described by a grouped frequency distribution. We would use the class marks (X) as the distinct data values and f would be the class frequencies. This amounts to approximating each data value in a class by its class mark.

Example 3.11
Estimate the sample standard deviation for the test scores described by the grouped frequency distribution in Table 2.5. (For convenience we reproduce this table along with other pertinent information in Table 3.6.)

TABLE 3.6

Class Limits	Frequency (f)	Class Mark (X)	Xf	$X^2f = X(Xf)$
45–49	2	47	94	4,418
50–54	3	52	156	8,112
55–59	6	57	342	19,494
60–64	4	62	248	15,376
65–69	9	67	603	40,401
70–74	8	72	576	41,472
75–79	3	77	231	17,787
80–84	4	82	328	26,896
85–89	1	87	87	7,569
Sums	40		2665	181,525

Solution

$$n = 40$$
$$\Sigma Xf = 2665$$
$$\Sigma X^2 f = 181,525$$

Therefore,

$$s \doteq \sqrt{\frac{n(\Sigma X^2 f) - (\Sigma Xf)^2}{n(n-1)}}$$

$$\doteq \sqrt{\frac{(40)(181,525) - (2665)^2}{(40)(39)}} = \sqrt{\frac{158,775}{1560}} = \sqrt{101.779}$$

$$\doteq 10.09$$

The exact value of the sample standard deviation for the original data (in Table 2.2) is 10.23.

The Empirical Rule

As an example of how standard deviation may be used to convey information about variability in a single collection of data without comparing it to another collection of data, we consider the case of a **normal population.** Recall that a normal population of data values will have a bell-shaped histogram. It can be shown that for such a population we will find about 68% of the data within one standard deviation of the mean, about 95% within two standard deviations of the mean, and about 99.7% within three standard deviations of the mean. See Figure 3.5.

FIGURE 3.5
Distribution of Data in a Normal Population

Suppose we have a fairly large sample from a normal population and we don't know μ and σ. We would expect the sample to have properties somewhat similar to the population. Thinking of the sample mean $\bar{x}$ as a rough estimate for μ, and the sample standard deviation s as a rough estimate for σ, we expect:

Empirical
Rule

Between	Percentage of Data (approximate)
$\bar{x} - s$ and $\bar{x} + s$	68
$\bar{x} - 2s$ and $\bar{x} + 2s$	95
$\bar{x} - 3s$ and $\bar{x} + 3s$	99.7

This is known as the **Empirical Rule.** See Figure 3.6.

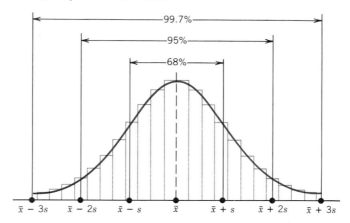

FIGURE 3.6
The Empirical Rule

Sometimes it's important to know if our sample came from a normal population. If we have a large sample obtained from a normal population, we would expect it to satisfy the following conditions:

1. A histogram for the sample should be roughly bell shaped.
2. The Empirical Rule should be satisfied.

If these two conditions are satisfied, statisticians may feel justified in assuming that the sample came from a normal population. This is only a very rough check of normality. More sophisticated ways of checking for normality are discussed later in the book.

Example 3.12
Consider the test score data in Table 2.2 (this data is also displayed in Table 3.7 at the beginning of the next section). Find the percentage of data values within one, two, and three standard deviations (S.D.) of the mean. Compare with the Empirical Rule.

Solution
We have seen in Examples 3.2 and 3.11 that $\bar{x} \doteq 67$ and $s \doteq 10$. Examining the data we see

Within	Number of Data	Percent	Empirical Rule %
1 S.D. (57–77)	26	65	68
2 S.D.'s (47–87)	38	95	95
3 S.D.'s (37–97)	40	100	99.7

These percentages are not out of line with the Empirical Rule. Furthermore, a histogram for the data is roughly bell shaped (see Example 2.5). These results are consistent with the sample coming from a normal population.

The Empirical Rule applies to data sets from normal populations, but many statisticians also use it if the distribution is roughly mound shaped. However, it would not apply to highly skewed data. But there is a result that applies to any set of data, no matter what the shape of the distribution. This is called **Chebyshev's Theorem,** which is discussed in Exercises 3.86–3.88.

Exercises

In Exercises 3.22–3.26, find the (a) range, (b) sample variance, and (c) sample standard deviation.

3.22 A public library employee in a small town reported the number of books signed out each day over a 10-day period as

27 14 12 16 23 9 7 16 6 10

3.23 A grocery store owner recorded the number of cases of a soft drink sold each month over a 6-month period. The numbers recorded were

24 16 27 34 36 19

3.24 The following data represent systolic blood pressures of randomly selected females aged 30–34 from the Framingham Heart Study:

138 124 116 128 140 136

3.25 A high school guidance counselor sampled 10 college-bound seniors and recorded their mathematics S.A.T. scores. They were

623 504 519 473 629 705 513 510 630 594

3.26 The daily number of absentees at a large plant over a 10-day period was

30 17 18 18 29 32 14 10 8 33

3.27 Which of the following measures of dispersion use all data values in calculating its value? Assume that there are more than two distinct data values.

(a) The range.   (b) The standard deviation.

3.28 For the data set 5, 5, 5, 5, 5, and 5, compute the sample standard deviation. Does your answer reflect the fact that there is no spread in the data?

3.29 Consider the two data sets

(i)   1 2 3 7 8 12 13 14
(ii)   1 7 7 7 8 8 8 14

(a) Construct a dot diagram for each set. Locate the mean of each data set.

(b) Which of the two sets appears to have the least spread about the mean?

(c) Compute the sample standard deviations for both sets. Do these values agree with your conclusion in part (b)?

3.30 Suppose that you are undecided between two cans of tennis balls (three balls per can) for use in a match. Upon testing, you find that both cans of balls bounce the same height on the average. How do you choose which can of balls to use? (*Hint:* Is the average bounce your only consideration?)

3.31   Find the *population* variance σ^2 for the following data.

    (a)   14   14   14   14   0   0   0   0   **(b)**   1   2   3   4   5   6   7

    (c)   5   4   4   6   0   1

In Exercises 3.32–3.35, estimate the sample mean $\bar{x}$ and the sample standard deviation s for each of the grouped frequency distributions.

3.32

Class	Class Limits	Frequency
1	3–7	2
2	8–12	3
3	13–17	6
4	18–22	3
5	23–27	2

3.33

Class	Class Limits	Frequency
1	3–7	2
2	8–12	3
3	13–17	6
4	18–22	6
5	23–27	3
6	28–32	2

3.34   The grouped frequency distribution below represents the number of hours of racquetball played each week at the Sloaner Fitness Club.

Class	Class Limits	Frequency
1	0–1	2
2	2–3	17
3	4–5	8
4	6–7	13
5	8–9	10

3.35   The grouped frequency distribution below is the cost (in cents) of a can of soup at various stores.

Class	Class Limits	Frequency
1	63–67	3
2	68–72	3
3	73–77	8
4	78–82	11
5	83–87	0
6	88–92	1

3.36   The mean and standard deviation of a sample of 50 randomly generated data values are 73.38 and 4.54, respectively. The data and frequency histogram are given below.

    (a)   Find the percentage of data values within one, two, and three standard deviations of the mean. Compare with the Empirical Rule.

(b)   Are the results of part (a) and the histogram consistent with sampling from a normal population?

63	65	66	66	67
67	70	70	70	70
70	71	71	71	71
71	71	72	72	72
72	72	72	72	72
72	73	73	74	74
75	75	75	75	75
76	76	77	77	77
78	78	78	78	79
80	80	81	83	84

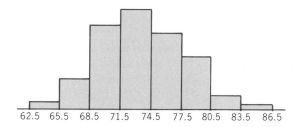

62.5   65.5   68.5   71.5   74.5   77.5   80.5   83.5   86.5

3.37   The data below are the number of home runs hit by American League home run leaders in the years 1949–1988 (*Source: The 1989 Information Please Almanac*, 1989, p. 946).

22	32	32	32	32	32	32	33	33	36	37	37	37	39
39	39	40	40	40	41	42	42	42	42	43	43	43	
44	44	44	45	45	46	48	49	49	49	49	52	61	

The mean and sample standard deviation of the data are 40.67 and 7.13, respectively. Find the percentage of data values within one, two, and three standard deviations of the mean. Compare with the Empirical Rule.

3.5   MEASURES OF POSITION

Measures of position are used to describe the standing or place occupied by a data value relative to the rest of the data. They are most commonly used with large collections of data. For example, suppose you are told that you received a raw score of 733 on a standard test. This does not give you much of an idea of how well you did. But if you are told that relative to the other scores on the test, your score was the 99th percentile, then this means that 99% of the people taking the test had scores that were between the lowest score and your score, while 1% of the people had scores between your score and the highest score. This gives you a better idea of how well you did. In general, the ***m*th percentile**

is the number that separates the bottom $m\%$ of the data from the top $(100 - m)\%$ of the data. It is denoted by the symbol P_m. (We have already discussed a percentile in Section 3.2, because the median is the 50th percentile.)

To find percentiles, we first rank the data, that is, list the data values in ascending order of magnitude. For example, Table 3.7 contains the ranked scores of 40 high school seniors on a scientific achievement test. (These were discussed in Chapter 2.)

Suppose that we want to find the 80th percentile, P_{80}. We first find the location of P_{80} in the ranked list. Since about 80% of the data values will be less than P_{80}, the location of P_{80} should be about 80% of the way down the ranked list of data. There are $n = 40$ data values and 80% of 40 is $(.80)(40) = 32$. Notice that 32 is a whole number. In this case we will agree that the location of P_{80} is 32.5. This means that P_{80} is halfway between the 32nd and 33rd data values. In other words, it is the average of the 32nd and 33rd data values (74 and 76). So $P_{80} = 75$.

Now suppose that we want to find the 83rd percentile, P_{83}. Eighty-three percent of 40 is $(.83)(40) = 33.2$. When we get a decimal, we agree that the location is the next higher whole number. So the location of P_{83} is 34. From Table 3.7 we see that the 34th data value is 78. So $P_{83} = 78$.

Procedure for Finding Percentiles To find the mth percentile, P_m, we first find its location in the ranked list of n data values. Evaluate $(m/100)n$.

(a) If $(m/100)n$ is a whole number, then

$$\text{location of } P_m = \left(\frac{m}{100}\right)n + .5$$

So P_m is halfway between the data value in position $(m/100)n$ and the data value in the next position; that is, it is the average of these two data values.

(b) If $(m/100)n$ is not a whole number, then

$$\text{location of } P_m = \text{next higher whole number}$$

The percentile P_m is the data value in this location.

TABLE 3.7
Scores on a Scientific Achievement Test

46	58	65	70	76
49	59	66	71	78
50	59	66	71	79
53	60	66	72	80
54	62	66	73	82
55	63	68	73	83
55	64	68	73	84
57	65	69	74	88

Example 3.13

For the data in Table 3.7 find

(a) P_{10}.   **(b)** P_{66}.

Solution

(a) In this case $m = 10$.

$$\left(\frac{m}{100}\right) n = \left(\frac{10}{100}\right)(40) = (.10)(40) = 4$$

Thus

$$\text{location of } P_{10} = 4 + .5 = 4.5$$

This means that P_{10} is the average of the 4th and 5th data values. Therefore, from Table 3.7, $P_{10} = 53.5$.

(b) In this case $m = 66$.

$$\left(\frac{m}{100}\right) n = \left(\frac{66}{100}\right)(40) = (.66)(40) = 26.4$$

Thus

$$\text{location of } P_{66} = 27$$

Since the 27th data value in Table 3.7 is 71, $P_{66} = 71$. *Note:* Do not confuse the location of a percentile with the value of the percentile itself.

A term closely related to the idea of percentile is the **percentile rank** of a data value. Earlier we discussed a standard test for which the score 733 was the 99th percentile. We then say that 99 is the percentile rank of the score 733.

> **Definition** If a data value x is the mth percentile, that is, if $x = P_m$, then we say that m is the *percentile rank* of x.

We can find the approximate percentile rank of a data value as follows: Look at the data value 62 in Table 3.7. There are 12 out of 40 data values that precede 62 in the ranked list. Since

$$\tfrac{12}{40} = .30$$

30% of the scores are less than 62. So the percentile rank of 62 is 30.

> **Formula for Percentile Rank of a Data Value x**   The percentile rank m is (approximately)
>
> $$m = \left(\frac{\text{number of data values less than } x}{\text{total number of data values}} \right) \cdot 100$$

Example 3.14
Find the percentile rank of the data value 58 in Table 3.7.

Solution
There are 8 data values less than 58 in the list of 40 data values. So the percentile rank is

$$m = \left(\tfrac{8}{40} \right) \cdot 100 = 20$$

One often sees the term **quartiles** in statistics. These are defined as follows:

> **Definition**   The *first quartile* Q_1, the *second quartile* Q_2, and the *third quartile* Q_3 are defined as
>
> $$Q_1 = P_{25}, \qquad Q_2 = P_{50}, \qquad Q_3 = P_{75}$$

Observe that the quartiles divide the data into quarters. Also note that the second quartile is the 50th percentile, which is also the median.

Example 3.15
Find the quartiles for the data in Table 3.7.

Solution
According to the above definition, we must find the percentiles P_{25}, P_{50}, and P_{75}.

(a)   Find P_{25}: Now $(.25)(40) = 10$. Thus

$$\text{location of } P_{25} = 10.5$$

This means that P_{25} is the average of the 10th and 11th data values. Thus

$$Q_1 = P_{25} = 59$$

(b)   Find P_{50}: Now $(.50)(40) = 20$. Thus

$$\text{location of } P_{50} = 20.5$$

This means that P_{50} is the average of the 20th and 21st data values. Thus

$$Q_2 = P_{50} = 66$$

(c)   Find P_{75}: Now $(.75)(40) = 30$. Thus

$$\text{location of } P_{75} = 30.5$$

This means that P_{75} is the average of the 30th and 31st data values so that

$$Q_3 = P_{75} = 73$$

The Standard Score

Shirley received a 60 on a mathematics examination and Bob received an 80 on a history examination. How can we find out which student did better relative to his or her classmates? We could answer this question by finding out the percentile ranks of each score. But there is another device for comparing the scores, which we will examine. It is called the **standard score** and has certain technical advantages over percentiles.

The distributions of scores on the two exams are given in Figure 3.7 along with the means on both exams. Notice that the scores of both students are 10 points above their means. However, the mathematics scores appear to be much less variable than the history scores. Therefore, the history scores tend to spread farther from their mean. It would appear that Shirley did better relative to her classmates than Bob did relative to his classmates. Only 1 student (out of 16) did better than Shirley, but 4 students (out of 16) did better than Bob. To determine the standing of a particular score relative to the other scores of the class, we should consider not only where the score is in relation to the mean, but also the amount of variability in the scores.

Consider the following information:

	Mean	Standard Deviation
Mathematics scores	50	5
History scores	70	10

The fact that Shirley did better is reflected by the fact that her score was two standard deviations above her class mean, whereas Bob's score was only one

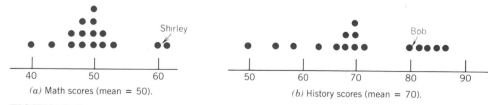

(a) Math scores (mean = 50).   *(b)* History scores (mean = 70).

FIGURE 3.7

standard deviation above his class mean. The number of standard deviations a data value is above or below the mean is called its **standard score** or **z score**. To obtain a formula for the z score (z), observe that Shirley's z score was calculated as follows:

$$\frac{60 - 50}{5} = 2$$

Apparently, to calculate the z score for a data value x, we would use the following formula:

> Formula for the z score: $z = \dfrac{x - \text{mean}}{\text{standard deviation}}$

When dealing with a population, we would write

$$z = \frac{x - \mu}{\sigma}$$

When dealing with a sample, we would write

$$z = \frac{x - \bar{x}}{s}$$

Essentially, calculating a z score amounts to changing our units of measurement to standard deviations—much like changing from inches to centimeters. Note the two-sided ruler in Figure 3.8 for the math scores. Observe that a test score below the mean has a negative z score: $(45 - 50)/5 = -1$, meaning that 45 is one standard deviation *below* the mean.

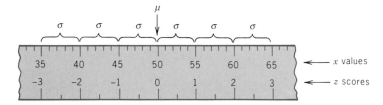

FIGURE 3.8
z scores Measure the Number of Standard Deviations Above or Below the Mean

---

Example 3.16
The (raw) scores on a psychology examination had a mean of 40 and a standard deviation of 6 points. The (raw) scores on a sociology examination had a mean of 60 and a standard deviation of 20 points.

(a) Bill received a raw score of 58 on the psychology examination and Jack received a raw score of 90 on the sociology examination. Which student did better relative to his class?

(b) Eileen had a standard score of $z = -1.2$ on the sociology exam. Find her raw score (i.e., x value).

Solution

(a) Bill's z score is obtained from the formula

$$z = \frac{x - 40}{6} = \frac{58 - 40}{6} = 3$$

Jack's z score is

$$z = \frac{x - 60}{20} = \frac{90 - 60}{20} = 1.5$$

So Bill did better relative to his class.

(b) Eileen's z score is $z = -1.2$. So x is 1.2 standard deviations below the mean. The equation relating the two is

$$z = \frac{x - 60}{20}$$

We substitute $z = -1.2$ and solve for x:

$$-1.2 = \frac{x - 60}{20}$$

$$(-1.2)(20) = x - 60$$

$$x = 60 - 24 = 36$$

Remark   In contrast to percentiles, standard scores are easy to manipulate arithmetically and are of great use in simplifying many problems (as we shall see in Chapter 6). For this reason they are much more commonly used in statistics. However, when using standard scores to compare data values from different distributions, ideally the distributions should have roughly similar shapes.

Exercises

3.38   The following ranked data represent the miles driven each day by a salesman over a 30-day period:

33   37   43   44   44   55   58   65   65   66   71   74   75   75   78
81   81   81   82   84   86   86   87   89   89   92   92   93   93   95

Find

(a) P_{33}.   (b) P_{87}.   (c) Q_1.   (d) Q_3.

(e)   The percentile rank of 74.

(f)   The nine deciles are defined as follows: The first decile D_1 is the 10th percentile P_{10}; the second decile D_2 is the 20th percentile P_{20}; ... ; and the 9th decile D_9 is the 90th percentile P_{90}. Find D_3.

3.39   Consider the following ranked data:

.02	.09	.14	.25	.37	.55	.55	.56	.60
.77	.86	.93	1.15	1.34	1.41	1.75	2.01	
2.16	2.23	3.69	3.90	4.50	4.88	7.79	9.56	

Find

(a) P_{20}.   **(b)** P_{95}.   **(c)** Q_2.   **(d)**   The percentile rank of 3.69.

3.40   Consider the 50 data values that represent the nearest hour of births in a city hospital in a 6-day period. Note that 0 represents a birth between 11:31 P.M. and 12:30 A.M. inclusive, while 23 represents a birth between 10:31 P.M. and 11:30 P.M. inclusive.

19	13	6	1	5	17	19	22	16	14
5	17	11	18	2	7	11	19	21	19
15	22	12	20	23	9	17	12	19	9
5	21	11	6	6	17	5	6	16	7
22	21	5	8	22	12	6	9	6	9

Find

(a) P_{20}.   **(b)** P_{80}.   **(c)** Q_1.   **(d)**   The percentile rank of 8.

3.41   The following data (discussed in Exercise 3.14) give time (in days) from remission to relapse for 51 patients with acute nonlymphoblastic leukemia.

24	46	57	57	64	65	82	89	90	90	111
117	128	143	148	152	166	171	186	191	197	
209	223	230	247	249	254	258	264	269	270	
273	284	294	304	304	332	341	393	395	487	
510	516	518	518	534	608	642	697	955	1160	

Find

(a)   The first and third quartiles.   **(b)**   The median.

(c)   The 95th percentile.   **(d)**   The percentile rank of 111.

3.42   The following frequency distribution represents the magnitude on the Richter scale of 1000 seismic events occurring near the Fiji Islands during 1964–1974 (*Source:* Extracted from the World Wide Seismic Network Tape by Professor John Woodhouse of Harvard University. Data from Donoho, A., D. Donoho, and M. Gasko, *MacSpin Graphical Data Analysis Software*, p. 115).

x	4.0	4.1	4.2	4.3	4.4	4.5	4.6	4.7	4.8
f	46	55	90	85	101	107	101	98	65

x	4.9	5.0	5.1	5.2	5.3	5.4	5.5	5.6
f	54	47	43	29	21	20	14	9

x	5.7	5.8	5.9	6.0	6.1	6.2	6.3	6.4
f	8	0	2	3	1	0	0	1

(a) Find the quartiles.   (b) Find the 90th percentile.

(c) Find the percentile rank of 5.7.

3.43 The data below represents the age at inauguration for United States presidents (*Source: The 1989 Information Please Almanac*, 1989, p. 607).

$$
\begin{array}{ccccccccccccccc}
42 & 43 & 46 & 47 & 48 & 49 & 49 & 50 & 50 & 51 & 51 & 51 & 51 & 52 \\
52 & 54 & 54 & 54 & 54 & 55 & 55 & 55 & 55 & 56 & 56 & 56 & 57 & 57 \\
57 & 57 & 58 & 60 & 61 & 61 & 61 & 62 & 64 & 64 & 65 & 68 & 69 \\
\end{array}
$$

(a) Find the quartiles.   (b) Find the 80th percentile.

(c) Find the percentile rank of 60.

3.44 A collection of data values has mean 30 and standard deviation 10.

(a) Find the z score for each of the following data values:

     (i) $x = 30$.   (ii) $x = 50$.

     (iii) $x = 10$.   (iv) $x = 65$.

(b) Find the data value corresponding to each of the following z scores:

     (i) $z = -2$.   (ii) $z = 1$.   (iii) $z = 2.5$.

     (iv) $z = 1.5$.   (v) $z = 0$.

3.45 Repeat Exercise 3.44 for a distribution with mean 40 and standard deviation 15.

3.46 By inspecting the dot diagram below,

(a) Give the z value for each data value.

(b) Find the distance from each data value x to $\bar{x}$, assuming $s = 6$.

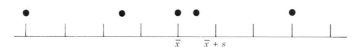

3.47 On the diagram below, locate the data values x that satisfy the following conditions:

(a) x is 3.5 standard deviations above the mean.

(b) x is 2 standard deviations below the mean.

(c) x is zero standard deviations from the mean.

3.48 The (raw) scores on a mathematics examination had a mean of 65 and a standard deviation of 5. The (raw) scores on an economics examination had a mean of 74 and a standard deviation of 10. You scored 70 on the mathematics exam and 79 on the economics exam. In which examination was your score better relative to the class?

3.49 The (raw) scores on a statistics examination had a mean of 78 and a standard deviation of 6. The (raw) scores on a biology examination had a mean of 60 and a standard deviation of 15. Your biology professor told you that you received a raw score of 80. Your statistics professor told you that your raw score was 1.5 standard deviations above the mean.

(a) In which examination did you do better relative to the class?

(b) What was your raw score on the statistics examination?

3.50 Suppose that 5% of all students have a higher grade-point average than yours. What is your percentile rank?

BOX-AND-WHISKER PLOTS (OPTIONAL)

A **box-and-whisker plot** (or boxplot) can be used to display graphic and numerical features of collections of data. We will illustrate the idea using the exam score data in Table 3.7. (See Section 3.5.)

- First we compute five numbers: the lowest data value (L), the highest value (H), and the three quartiles, Q_1, Q_2, Q_3. Examining the ranked data in Table 3.7, we see that

$$L = 46 \text{ and } H = 88$$

Furthermore, we have already computed the three quartiles in Example 3.15.

$$Q_1 = 59 \qquad Q_2 = 66 \qquad Q_3 = 73$$

- Next we draw a horizontal axis with the five numbers marked on the axis. Above the axis construct a **box** with the two vertical sides above Q_1 and Q_3 on the axis (see Figure 3.9). These sides are called the **hinges.** Also construct a vertical line in the box above Q_2.

- Finally draw vertical lines above L and H and connect these to the hinges using horizontal lines. These horizontal lines are called **whiskers.** Note that each portion of the box and each whisker contains about one-fourth of the data.

An examination of Figure 3.9 suggests that the data between Q_1 and Q_3 are quite symmetrically distributed since Q_2 (the median) is exactly halfway between Q_1 and Q_3. The right whisker is only slightly longer than the left whisker,

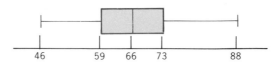

FIGURE 3.9
A Box-and-Whisker Plot for the Exam Scores in Table 3.7

indicating only a very slight skewing to the right. Overall, however, the data appear to be rather symmetrically distributed.

Example 3.17

The following data represent the length of long-distance phone calls (in minutes) made by a business office in 1 day:

$$3 \quad 7 \quad 2 \quad 14 \quad 4 \quad 29 \quad 3 \quad 9 \quad 1 \quad 20 \quad 10 \quad 7 \quad 2 \quad 42 \quad 3 \quad 5$$

Construct a box-and-whisker plot for the data. What does the box-and-whisker plot tell us about the way the data are distributed?

Solution

Since the data are unranked, we must first rank them:

$$1 \quad 2 \quad 2 \quad 3 \quad 3 \quad 3 \quad 4 \quad 5 \quad 7 \quad 7 \quad 9 \quad 10 \quad 14 \quad 20 \quad 29 \quad 42$$

Now, $L = 1$, $H = 42$. Find Q_1 (which is the 25th percentile P_{25}): There are $n = 16$ data values: $(.25)(16) = 4$. So

$$\text{location of } Q_1 = 4.5$$

Thus Q_1 is the average of the 4th and 5th data values in the ranked list. Therefore,

$$Q_1 = 3$$

Find Q_2 (the 50th percentile): Now $(.50)(16) = 8$. So

$$\text{location of } Q_2 = 8.5$$

Hence Q_2 is the average of the 8th and 9th data values.

$$Q_2 = 6$$

Find Q_3 (the 75th percentile): Now $(.75)(16) = 12$. So

$$\text{location of } Q_3 = 12.5$$

and

$$Q_3 = 12$$

The box-and-whisker plot is given in Figure 3.10.

FIGURE 3.10
A Box-and-Whisker Plot for the Telephone Data

Note that in Figure 3.10 the right portion of the box is longer than the left portion, and the right whisker is considerably longer than the left. This indicates that the data are skewed to the right.

The box-and-whisker plots discussed in this section are sometimes called **quick boxplots.** Another version of the box-and-whisker plot is discussed in the Minitab section at the end of this chapter. No knowledge of Minitab is necessary to understand the discussion.

Exercises

3.51   Construct a box-and-whisker plot for the data in Exercise 3.41. Describe the shape of the distribution.

3.52   Construct a box-and-whisker plot for the data in Exercise 3.42. Describe the shape of the distribution.

3.53   Construct a box-and-whisker plot for the data in Exercise 3.43. Describe the shape of the distribution.

3.7 USING MINITAB (OPTIONAL)

READ and LET commands

When you have more than one collection of data you can use the SET command repeatedly to enter the data. However, when each collection of data has the same number of data values, it is sometimes more convenient to use the READ command. With this command, each collection of data is entered in a column.

The printout below deals with math competency scores of 10 students before and after taking a course to increase basic math skills. We read the "before" and "after" scores into columns 1 and 2 for each student, one row at a time. The difference (after score minus before score) for each student is calculated and stored in column 3 by the command

$$LET \ C3 = C2 - C1$$

This gives the increase for each student. Finally, we print the increases for the 10 students.

```
MTB > READ THE FOLLOWING DATA INTO C1, C2
DATA> 38 58
DATA> 36 55
DATA> 40 65
DATA> 49 70
DATA> 56 67
DATA> 35 40
```

```
DATA> 45 60
DATA> 65 83
DATA> 57 68
DATA> 42 61
DATA> END
        10 ROWS READ
MTB > LET C3 = C2 - C1
MTB > PRINT C3

C3
    20  19  25  21  11  5  15  18  11  19
```

Note: A list of consecutive columns may be abbreviated with a dash. For example, we could type

$$READ\ INTO\ C1 - C3$$

instead of

$$READ\ INTO\ C1,\ C2,\ C3$$

It should be noted that if a collection of data is already in a column, and you SET or READ a new collection of data into the same column, the old data is automatically erased.

Arithmetic

In the above program we used the LET expression to compute a column of differences. Minitab performs arithmetic operations on columns term by term. The operations of addition, subtraction, multiplication, and division are denoted $+$, $-$, $*$, and $/$, respectively. Exponentiation (raising to a power) is denoted by $**$. For example, suppose we have the columns

C1	C2
9	3
4	2
8	2

Then

$$C1 + C2 = \begin{array}{c} 12 \\ 6 \\ 10 \end{array} \qquad C1 - C2 = \begin{array}{c} 6 \\ 2 \\ 6 \end{array} \qquad C1 + 5 = \begin{array}{c} 14 \\ 9 \\ 13 \end{array}$$

$$C1 * C2 = \begin{array}{c} 27 \\ 8 \\ 16 \end{array} \qquad C1/C2 = \begin{array}{c} 3 \\ 2 \\ 4 \end{array}$$

The square of column 1 is

$$C1 ** 2 = \boxed{\begin{array}{c} 81 \\ 16 \\ 64 \end{array}}$$

Using the LET command we can form new columns by means of expressions involving various combinations of the above operations. For example, suppose we typed

$$\text{LET } C4 = (C1 - C2) ** 2/C2$$

Minitab would first compute the difference $C1 - C2$, then square it, and then divide by C2, getting

C4
12
2
18

Minitab evaluates expressions as follows: expressions are repeatedly scanned from left to right, evaluating expressions within parentheses first, then exponentiation ** is performed, then * and /, then + and −. Note the differences in the following:

$$\text{Let } C5 = 2 * (C1 + C2) \rightarrow \quad \begin{array}{|c|} \hline C5 \\ \hline 24 \\ 12 \\ 20 \\ \hline \end{array}$$

$$\text{Let } C6 = 2 * C1 + C2 \rightarrow \quad \begin{array}{|c|} \hline C6 \\ \hline 21 \\ 10 \\ 18 \\ \hline \end{array}$$

When in doubt about an expression, you can usually use parentheses to get what you want. We can use LET to do some editing. For example, in the second row of C6, suppose we wanted a 5 instead of a 10. Type

$$\text{LET } C6(2) = 5$$

Describing Data

In Section 2.7 we presented the scores on an aggressiveness test given to 30 males. These scores were entered into column 1 of the Minitab worksheet as follows:

```
MTB > SET THE FOLLOWING DATA IN C1
DATA> 136 133 122 105 141 118 148 113 128 125 120 146
DATA> 137 135 98 126 108 93 140 110 134 131 102 115
DATA> 106 110 118 147 136 146
DATA> END
```

For the remainder of this section, we assume this data is in column 1. We can find the mean and median for the data with the commands

MEAN FOR C1

MEDIAN FOR C1

After each command, the computer prints out the answer.

```
MTB > MEAN FOR C1
    MEAN    =     124.23
MTB > MEDIAN FOR C1
    MEDIAN  =     125.50
```

(*Note:* Currently there is no MODE command in Minitab.) We could also store these values. For example, to store the mean in K1, type

MEAN FOR C1, STORE IN K1

To compute the standard deviation, we can use the STANDARD DEVIA-TION command (a synonym is STDEV).

```
MTB > STANDARD DEVIATION FOR C1
    ST.DEV. =     15.750
```

(As with the mean and median, we could also store the standard deviation.)

The DESCRIBE command gives the number of data values, the mean, median, standard deviation, and other information.

```
MTB > DESCRIBE C1
```

	N	MEAN	MEDIAN	TRMEAN	STDEV	SEMEAN
C1	30	124.23	125.50	124.65	15.75	2.88

	MIN	MAX	Q1	Q3
C1	93.00	148.00	110.00	136.25

The TRMEAN given above is the **trimmed mean.** This can be useful when there are atypical extreme data values, because it is obtained by discarding the top 5% and bottom 5% of the data values and computing the mean of the remaining 90%. The SEMEAN is the **standard error of the mean.** This is the standard deviation divided by the square root of the sample size. (We will see how this is used in Chapter 6.) Note that the first quartile Q_1 and the third quartile Q_3 are also given. (The median is the second quartile and also the 50th percentile.)

Standard Scores

Suppose we want the standard score for the data value 146. Even if we do not know the mean and standard deviation, we can calculate this by telling the computer to compute and print the value of

$$(146 - \text{MEAN(C1))/STDEV(C1)}$$

This is accomplished in the following printout:

```
MTB > LET K1 = (146 - MEAN(C1)) / STDEV(C1)
MTB > PRINT K1
K1     1.38205
```

(When using another command name, such as MEAN, in a LET command, we must place parentheses around the column.) To obtain the standard score for every value in C1 we type

> CENTER C1 PUT IN C2

Minitab will compute the standard score for each value in column 1 and store the results in column 2.

Percentiles

Suppose we want to find some percentiles (other than those given by DE-SCRIBE). We first rank the data and store it in, say, column 2. This is accomplished as follows:

```
MTB > SORT C1 PUT INTO C2
```

Suppose we want the 35th percentile of the 30 data values in column 2. Since $(.35)(30) = 10.5$, the location of P_{35} is 11. The eleventh data value in column 2 is denoted by C2(11). The printout of this is obtained as follows:

```
MTB > LET K1 = C2(11)
MTB > PRINT K1
K1     118.000
```

Suppose we want the 60th percentile. Since $(.60)(30) = 18$, the location of P_{60} is 18.5. So P_{60} is the average of the 18th and 19th data values in column 2.

```
MTB > LET K2 = (C2(18) + C2(19)) / 2
MTB > PRINT K2
K2     132.000
```

COUNT and COPY Commands

The COUNT command gives us the number of data values in a column. If we type

COUNT C1

the computer will print the number of data values in column 1. If we wish to store the number of data values (as a stored constant), we can type

Let K1 = COUNT(C1)

The COUNT command can be used along with the COPY command to find the percentile rank of a data value. Suppose we want to find the percentile rank of the data value 115 in column 1. This would be the percentage of data values that are less than 115. The next lower possible value is 114 (since we are dealing with whole numbers), and 0 is the smallest possible value for these data. The following COPY command, along with the USE subcommand, will take the values in column 1 between 0 and 114 (inclusive) and place them in column 2. (Any values previously in column 2 will automatically be erased.)

COPY C1 INTO C2;
USE ROWS WHERE C1 = 0:114.

Note that 0:114 means 0 to 114 inclusive. If we had said 0,114 or 0 to 114, only the values 0 and 114 would be selected.*

The percentage of data values less than 115 is then obtained by multiplying 100 times the number of data values in C2 divided by the number of data values in C1.

100 * COUNT(C2)/COUNT(C1)

The following printout shows that the percentile rank is 30.

```
MTB > COPY C1 INTO C2;
SUBC>   USE ROWS WHERE C1 = 0:114.
MTB > LET K1 = 100 * COUNT(C2) / COUNT(C1)
MTB > PRINT K1
K1     30.0000
```

---

*With the USE subcommand, we can also specify *rows* to be used by giving row numbers.

USE ROWS 3 5 7:9.

will select rows 3, 5, 7, 8, 9. That is, the 3rd, 5th, 7th, 8th, and 9th values in column 1 would be selected.

Box-and-Whisker Plots

A box-and-whisker plot can be obtained using the BOXPLOT command. The following is a box-and-whisker plot for the 30 data values in column 1:

MTB > BOXPLOT FOR C1

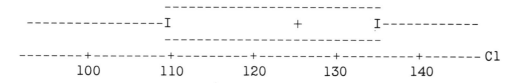

```
                            --------------------------
      ----------------I                 +          I-----------
                            --------------------------
      --------+---------+---------+---------+---------+------- C1
            100       110       120       130       140
```

Minitab does a box-and-whisker plot somewhat differently than we did in Section 3.6. The box is essentially the same. The median is in the center and the lower and upper hinges of the box are essentially the first and third quartiles. But the whiskers are a little different. Instead of drawing the whiskers out to the lowest and highest values, Minitab draws the whiskers to the *lower adjacent value* and *upper adjacent value*. To understand these, we must first understand the concept of *inner fences, outer fences,* and *H-spread.*

H-spread = distance between hinges = $Q_3 - Q_1$

Inner fences are at $Q_1 - 1.5 \cdot$ (H-spread) and $Q_3 + 1.5 \cdot$ (H-spread)

Outer fences are at $Q_1 - 3 \cdot$ (H-spread) and $Q_3 + 3 \cdot$ (H-spread)

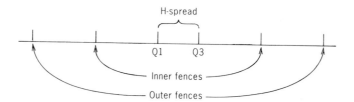

The *lower* and *upper adjacent values* are the smallest and largest values (respectively) that are still within the inner fences. The whiskers are drawn from the hinges to these values. Values between the inner and outer fences are *possible outliers,* denoted by an asterisk (*). Values beyond the outer fences are *probable outliers,* denoted by a zero, 0. The term *outlier* refers to a data value that is in some sense atypical. For example, former basketball player Wilt Chamberlain's height (7 ft 2 in.) is atypical. An outlier can sometimes be a recording mistake or keypunch error—a data transcriber types 143 instead of 1.43.

Practice Quiz (Answers can be found after the chapter review exercises.)

1. Show how one READ command can be used to accomplish what the following SET commands do.

```
SET C1
10   11   15   3
END
SET C2
4   8   1   5
END
```

2. Describe the status of the Minitab worksheet after running the following program:

```
READ INTO C1 C2
2    4
1    2
3    3
END
LET C3 = 4 * (C1 + C2)
LET C4 = 4 * C1 + C2
LET C5 = 12/C1 + C2
LET C6 = 12/(C1 + C2)
LET C7 = C1 ** 2 + C2
```

3. The COPY command can be used to copy rows from more than one column. What will columns 3 and 4 look like after the following program?

```
READ INTO C1 C2
1    5
4    7
2    8
3    6
END
COPY C1  C2 INTO C3  C4;
   USE ROWS WHERE C1 = 2    4.
```

In Questions 4 through 10, determine what is wrong (if anything).

4. READ INTO C1, C2
```
    4    6
    8    1
    3    9
    7
    END
```

5. SET C1
```
    11   14   21   18   9
    END
    MEAN FOR C1
    MEDIUM FOR C1
```

6.   SET C1, C2
     4   8
     7   2
     3   9
     END

7.   READ C1, C2
     3   8
     7   4
     1   9
     END
     LET K1 = C1 + C2

8.   DESCRIBE THE 3 DATA VALUES IN C1

9.   LET K1 = MEAN C1

10. PRINT K1/K2

Exercises

Suggested exercises for use with Minitab are 3.38 (a, c, e), 3.44 (a), 3.54 (a, b, f, g), 3.60, 3.72, 3.73, and 3.76 (a, b, c, d).

3.8 SUMMARY

In this chapter we studied the notion of a **measure of central tendency,** which is a number that is a typical or representative value for a collection of data. We investigated three measures of central tendency: the **mean** (the average value), the **median** (the middle value in a ranked list), and the **mode** (the most frequent value).

In this book the measure we will use the most is the mean. The mean of a population is denoted by the Greek letter μ; while we use the symbol $\bar{x}$ to represent the mean of a sample (when x is used to represent an arbitrary data value). One important point to keep in mind is this: in practice we rarely have enough information about a population to calculate μ. In this case we might estimate the value of μ by $\bar{x}$.

A **measure of dispersion** is a number that conveys an idea of how much spread or variability there is in a collection of data values. In this chapter we studied the following measures of dispersion:

1.   The **range** R is the difference between the highest (H) and the lowest (L) data values.

$$R = H - L$$

2.   The **population variance** σ^2 for a population of N data values is the average of the squares of the deviations of the data values from the population mean μ.

$$\sigma^2 = \frac{\Sigma(x - \mu)^2}{N} = \frac{N(\Sigma x^2) - (\Sigma x)^2}{N^2}$$

3. The **population standard deviation** σ is the square root of the population variance.

4. The **sample variance** s^2 for a sample of n data values is a "modified average" of the squares of the deviations of the data values from their (sample) mean $\bar{x}$.

$$s^2 = \frac{\Sigma(x - \bar{x})^2}{n - 1} = \frac{n(\Sigma x^2) - (\Sigma x)^2}{n(n - 1)}$$

5. The **sample standard deviation** s is the square root of the sample variance.

Given a large sample from a normal population, the **Empirical Rule** says that

Between	Approximate Percent of Data
$\bar{x} - s$ and $\bar{x} + s$	68
$\bar{x} - 2s$ and $\bar{x} + 2s$	95
$\bar{x} - 3s$ and $\bar{x} + 3s$	99.7

Another concept we studied in this chapter was the idea of a **measure of position,** which describes the location of a data value relative to the other data values in the collection. The measures of position we studied were the **percentile** and the **standard score.** If a data value is, for example, the 95th percentile, then this data value divides the bottom 95% of the data values from the top 5% of the data values. The standard score z of a data value x is the number of standard deviations that data value is above or below the mean.

$$z = \frac{x - \text{mean}}{\text{standard deviation}}$$

Because it is mathematically more convenient to use, the standard score will be more often used in this book than the percentile.

Box-and-whisker plots can be useful in portraying how data are distributed.

Review Exercises

3.54 Consider the following data:

$$29 \quad 25 \quad 20 \quad 21 \quad 30 \quad 25 \quad 27 \quad 31 \quad 32 \quad 20$$

Find each of the following:

(a) Mean. (b) Median. (c) Mode (if it exists).
(d) Range. (e) Sample variance. (f) Sample standard deviation.

(g) P_{85}. (h) Percentile rank of 21. (i) z score for 29.
(j) Data value corresponding to $z = 1.099$.

3.55 Consider the following ranked data:

$$2 \quad 5 \quad 6 \quad 6 \quad 7 \quad 8 \quad 11 \quad 11 \quad 11 \quad 12 \quad 14 \quad 15$$

Find each of the following:

(a) Mean. **(b)** Median. **(c)** Mode (if it exists).

(d) Range. **(e)** Sample variance. **(f)** Sample standard deviation.

(g) Q_3. **(h)** Percentile rank of 8. **(i)** z score for $x = 7$.

(j) Data value corresponding to $z = -1.78$.

3.56 Consider the following data:

$$12 \quad 14 \quad 12 \quad 16 \quad 17 \quad 19 \quad 12 \quad 13 \quad 20 \quad 175$$

(a) Find the mean. **(b)** Find the median.

(c) Find the mode (if it exists). **(d)** Which measure best represents the data?

3.57 Consider the following data:

$$19 \quad 20 \quad 23 \quad 27 \quad 1 \quad 22$$

(a) Find the mean. **(b)** Find the median.

(c) Find the mode (if it exists). **(d)** Which measure best represents the data?

3.58 The data below are the world's 20 most populous countries and their 1988 estimated populations, in millions (*Source: The 1989 Information Please Almanac,* 1989, p. 139).

Country	Population	Country	Population
China	1087	Mexico	83.5
India	816.8	Vietnam	65.2
USSR	286	Philippines	63.2
United States	246.1	West Germany	61.2
Indonesia	177.4	Italy	57.3
Brazil	144.4	United Kingdom	57.1
Japan	122.7	France	55.9
Nigeria	111.9	Thailand	54.7
Bangladesh	109.5	Egypt	53.3
Pakistan	107.5	Turkey	52.9

(a) Characterize the distribution of data values as skewed to the left, symmetric, or skewed to the right. Which measure, mean or median, best represents the data? Which is most likely larger?

(b) Find the mean. (*Note:* The sum of the 20 data values is 3813.6.)

(c) Exclude China and India and find the mean for the remaining 18 countries. Compare your answer with that obtained in part (b). Did deleting China and India make a big difference in the mean population?

(d) Find the median.

(e) Exclude China and India and find the median for the remaining 18 countries. Compare your answer with that obtained in part (d). Did deleting China and India make a big difference in the median population?

3.59 The data below, in thousands of barrels, are the production of crude petroleum for countries in the western hemisphere in 1987 (*Source: The 1989 Information Please Almanac,* 1989, p. 377). (*Note:* In the table below, T & T represents Trinidad and Tobago.)

Country	Production	Country	Production
United States	3,041,910	Ecuador	98,915
Mexico	914,690	Peru	63,875
Venezuela	547,500	T & T	60,225
Canada	538,375	Chile	10,950
Brazil	208,415	Bolivia	6,570
Argentina	141,255	Guatemala	1,460
Colombia	132,130		

(a) Characterize the distribution of data values as skewed to the left, symmetric, or skewed to the right. Which measure, mean or median, best represents the data? Which measure is most likely larger?

(b) Find the mean production per country. (*Note:* The sum of the 13 data values is 5,766,270.)

(c) Exclude the United States and find the mean production per country of the remaining 12 countries. Compare your answer with that obtained in part (b). Did deleting the United States make a big difference in the mean production?

(d) Find the median production per country.

(e) Exclude the United States and find the median production per country of the remaining 12 countries. Compare your answer with that obtained in part (d). Did deleting the United States make a big difference in the median production?

3.60 A meteorologist was asked to compare average wind speeds for "The Windy City," Chicago, and Boston. Table 3.8 gives average wind speed per month based on records kept for a period of 20 years.

TABLE 3.8

Month	Boston	Chicago
January	14.2	11.5
February	14.1	11.6
March	13.9	11.9
April	13.3	12.1
May	12.2	10.6
June	11.4	9.1
July	10.9	8.1
August	10.7	8.1
September	11.3	8.7
October	12.0	9.8
November	12.9	10.9
December	13.7	10.9

(a) Find the means of the monthly wind speed for each city.

(b) Find the medians of the monthly wind speeds for each city.

(c) Does it seem to make much difference for either city whether the mean or median is used to typify the wind speed?

3.61 This exercise demonstrates that the mean can be viewed in terms of proportions. Consider the data values 2, 3, 3, 3, 3, 3, 3, 5, 5, and 10, which have a mean of 4.

(a) Write an expression that gives the value of $\bar{x}$ in terms of the distinct data values and their proportions (i.e., relative frequencies).

(b) What would the mean be of a collection of data values consisting of 1000 2's, 6000 3's, 2000 5's, and 1000 10's?

Exercises 3.62 and 3.63 are to be done as a unit.

3.62 Over a five-game period a basketball player scored 9, 12, 8, 5, and 6 points. Therefore the mean number of points scored per game was $\bar{x} = 8$.

(a) Suppose the player had scored four more points in each game. What would be the mean number of points per game? Compare your answer with $\bar{x} = 8$.

(b) Suppose the player had scored c additional points per game, where c is a constant value. What would be the mean number of points scored per game? Express your answer in terms of $\bar{x}$ and k.

(c) Suppose the player had doubled the number of points scored in each game. What would be the mean number of points scored per game? Compare your answer with $\bar{x} = 8$.

(d) Suppose the player had scored k times as many points in each game, where k is a constant value. What would be the mean number of points scored per game? Express your answer in terms of $\bar{x}$ and k.

3.63 Consider the data values 0, 2, 5, 8, 9, and 12. A new collection of data values is obtained in the following way. Each of the above data values is multiplied by four, and the resulting product increased by seven. Use the results of Exercise 3.62 (b) and (d) to find the mean of the new collection of data values. Note that $\bar{x} = 6$ for the original set of data (that is, 0, 2, 5, 8, 9, 12).

3.64 In Section 3.2 we mentioned that the mean is the balancing point for a set of data. This is reflected by the fact that $\Sigma(x - \bar{x}) = 0$ [e.g., see Exercise 3.16 (b).]

(a) Consider the data values 0, 2, 5, 8, 9, and 12. Verify the fact given in the previous sentence for these data.

(b) Given the six data values above, suppose that we supplement the data with a 4 (giving the data 0, 2, 5, 8, 9, 12, and 4). If we were to supplement another value, what must it be so that the mean again is 6? (*Hint:* We do not want to disturb the balancing point.)

3.65 The following distribution (discussed in Exercise 3.15) gives life span (in days) for 144 hamsters:

x	116	264	314	331	364	397	430	446	496	512	545
f	1	2	1	1	2	1	1	2	2	1	3

x	562	579	612	645	678	694	711	727	744	760	777
f	2	3	1	1	4	1	4	4	2	10	1

x	793	810	826	843	860	876	884	893	909	942	959
f	1	5	6	3	4	3	1	3	3	1	1

x	975	992	1008	1025	1041	1058	1074	1091	1107	1124	1132
f	4	1	3	1	2	2	4	1	6	4	1

x	1140	1157	1174	1190	1207	1223	1256	1273	1289	1306	1355
f	3	1	3	1	3	2	3	3	5	1	1

x	1372	1388	1421	1438	1504	1587	1620
f	1	1	1	1	1	2	1

Find

(a) The quartiles.

(b) The 10th and 90th percentiles.

(c) The percentile rank of each of the following:
   (i) 884   (ii) 1091

3.66 Use the grouped frequency distribution below to estimate the sample mean $\bar{x}$ and the sample standard deviation s.

Class	Class Limits	Frequency
1	9–17	14
2	18–26	9
3	27–35	7
4	36–44	4
5	45–53	3
6	54–62	1

3.67 Use the grouped frequency distribution below to estimate the sample mean $\bar{x}$ and the sample standard deviation s.

Class	Class Limits	Frequency
1	9–17	1
2	18–26	3
3	27–35	4
4	36–44	7
5	45–53	9
6	54–62	14

3.68 The data below together with a grouped frequency distribution for the data are the number of home runs hit by National League home run leaders for the years 1949–1988 (*Source: The 1989 Information Please Almanac*, 1989, pp. 946–947).

54	47	42	37	47	49	51	43
44	47	46	41	46	49	44	47
52	44	39	36	45	45	48	40
44	36	38	38	52	40	48	48
31	37	40	36	37	37	49	39

Class	Class Limits	Frequency
1	31–33	1
2	34–36	3
3	37–39	8
4	40–42	5
5	43–45	7
6	46–48	9
7	49–51	4
8	52–54	3

(a) Using the grouped frequency distribution, estimate $\bar{x}$ and s.

(b) Now find $\bar{x}$ and s from the data, and compare with your answer in (a).
(*Note:* $\Sigma x = 1733$ and $\Sigma x^2 = 76{,}251$.)

3.69 A collection of data values has mean 87 and variance 16.

(a) Find the z score for each of the following data values:

(i) $x = 95$. **(ii)** $x = 77$. **(iii)** $x = 90$. **(iv)** $x = 83$.

(b) Find the data value x corresponding to each of the following z scores:

(i) $z = 1$. **(ii)** $z = 0$. **(iii)** $z = -1.25$. **(iv)** $z = 1.5$.

3.70 Consider the data 7, 8, 8, 16, 19, 20. Find the distance, expressed in standard deviations, between each data value and the mean.

A businessman was trying to decide whether to accept a new position in Albany, New York; or Reno, Nevada. Being an avid golfer, he wanted to compare temperatures in the two cities. For Exercises 3.71–3.75, consider the following data, which represent the average monthly temperatures in Fahrenheit degrees for Albany and Reno based on a standard 30-year period (*Source: Statistical Abstract of the United States*, 1984, p. 217).

Albany

January	21.1	July	71.4
February	23.4	August	69.2
March	33.6	September	61.2
April	46.6	October	50.5
May	57.5	November	39.7
June	66.7	December	26.5

Reno

January	32.2	July	69.5
February	37.4	August	66.9
March	40.6	September	60.2
April	46.4	October	50.3
May	54.6	November	39.7
June	62.4	December	32.5

3.71 Compare the ranges of temperatures for the above cities.

3.72 Compare the means $\bar{x}$ for the above cities.

3.73 Compare the sample standard deviations for the above cities.

3.74 Construct dot diagrams for both cities.

3.75 Based on the results of Exercises 3.71–3.74, which city should the businessman choose? Why?

The data below are to be used in Exercises 3.76–3.79. A midwestern company that uses the railways to transport its goods wants to open a factory in another city. A major consideration is the railroad distance between that city and other cities. One of the cities being considered is Boston. The following data represent railroad distances (in hundreds of miles) between Boston and various cities (*Source: Hammond Almanac*, 1983, p. 222).

City	Miles to Boston ($\times 100$)	City	Miles to Boston ($\times 100$)
New York City	2.3	St. Paul	14.1
Philadelphia	3.2	Mobile	14.4
Baltimore	4.2	Kansas City	15.7
Washington, D.C.	4.5	New Orleans	15.7
Buffalo	4.9	Miami	15.8
Pittsburgh	6.7	Oklahoma City	17.4
Cleveland	6.8	Dallas	18.6
Detroit	7.5	Houston	19.3
Cincinnati	9.4	Denver	20.4
Indianapolis	9.6	Albuquerque	23.6
Chicago	10.2	El Paso	24.1
Atlanta	10.9	Salt Lake City	25.3
St. Louis	12.0	Seattle	31.6
Jacksonville	12.1	Portland	32.2
Birmingham	12.2	Los Angeles	32.4
Des Moines	13.8	San Francisco	32.8
Memphis	13.8		

3.76 Find each of the following for the above data.
- (a) The mean.
- (b) The sample standard deviation s.
- (c) The median.
- (d) P_{95}.
- (e) P_{10}.

3.77 For the above data, find the z score for
- (a) San Francisco.
- (b) New York City.

3.78 Which of the above cities is approximately
- (a) .862 standard deviations below the mean?
- (b) 1.027 standard deviations above the mean?

3.79 Find the city for which the corresponding data value has approximate percentile rank:
- (a) 12
- (b) 91.

3.80 Consider the following grouped frequency distribution. The data represent length (in miles) of principal rivers in the United States (*Source: Hammond Almanac*, 1983, p. 301).

Class Limits	Frequency	Class Mark	Cumulative Frequency
426–758	37	592	
759–1091	10	925	
1092–1424	5	1258	
1425–1757	2	1591	
1758–2090	3	1924	
2091–2423	2	2257	
2424–2756	1	2590	
2757–3089	0	2923	
3090–3422	0	3256	
3423–3755	1	3589	

(a) Complete the cumulative frequency column.

(b) Construct a frequency histogram.

(c) Estimate the mean using the grouped frequency distribution. Compare with the mean 914.15 of the original data.

(d) Estimate the sample standard deviation s using the grouped frequency distribution. Compare with the sample standard deviation 630.08 of the original data.

3.81 The following data and graph refer to the number of newspapers in each state (discussed in Exercise 3.20).

3	6	6	7	7
8	8	9	9	10
10	11	12	12	14
17	19	19	19	20
23	23	23	24	25
25	26	26	27	28
28	32	35	35	36
37	45	46	47	48
51	52	54	70	72
73	87	93	108	118

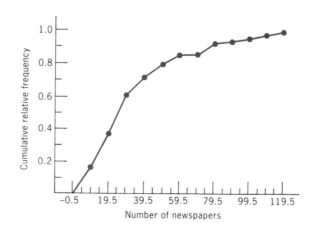

(a) Use the graph to estimate the median.

(b) Use the graph to estimate the 90th percentile.

Now use the data to find

(c) The median; compare your answer with part (a).

(d) The 90th percentile; compare your answer with part (b).

Exercises 3.82 and 3.83 are to be done as a unit.

3.82 Each week a newspaper carrier notes the number of new subscribers. Over a 6-week period the carrier recorded 0, 2, 5, 8, 9, and 12 new subscribers. Note the sample variance is $s^2 = 20.4$.

(a) Suppose 3 were added to each of the above data values. What would the new sample variance be? Compare your answer with $s^2 = 20.4$.

(b) Suppose the carrier added a constant c to each data value. What would the new sample variance be?

(c) Suppose the number of subscribers per week doubled. What would the new sample variance be? Compare your answer with $s^2 = 20.4$.

(d) Suppose the carrier obtained k times as many new subscribers per week, where k is a constant value. What would the new sample variance be? Express your answer in terms of s^2 and k.

3.83 Consider the data values 1, 1, 4, 4, and 5, which have sample variance 3.50. A new collection of data values is obtained in the following way. Each of the above data values is multiplied by three, and the resulting product is increased by six. Use the results of Exercise 3.82 (b) and (d) to find the sample variance of the new collection of data values.

3.84 The mean and sample standard deviation of the data below are 30.05 and 6.23, respectively.

(a) Find the percentage of data values within one, two, and three standard deviations of the mean. Compare with the Empirical Rule.

(b) Are the results of part (a) and the histogram consistent with sampling from a normal population?

17	18	21	22	23	24	24
25	25	26	26	26	26	27
27	29	29	29	29	30	30
30	30	31	31	31	32	32
32	32	33	35	36	36	36
37	39	39	40	43	44	

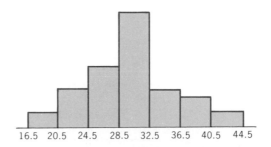

16.5 20.5 24.5 28.5 32.5 36.5 40.5 44.5

3.85   The mean and standard deviation of the sample of 50 randomly generated data values below are 99.68 and 10.37, respectively. Find the percentage of data values within one, two, and three standard deviations of the mean. Compare with the Empirical Rule. The data are

76	82	85	86	87	87	88	90	90
90	91	91	92	93	94	94	95	95
95	96	96	97	98	98	98	99	
100	100	101	102	102	103	104	105	
105	106	106	107	108	108	108	109	
109	109	113	115	118	119	120	124	

In Exercises 3.86–3.88, use Chebyshev's Theorem, which states that given a collection of data values, then for any number $k > 1$, the proportion of these data values that fall within k standard deviations of the mean is at least $1 - 1/k^2$.

3.86   Suppose that a collection of data values has mean 150 and standard deviation 15. Applying Chebyshev's Theorem, at least what proportion of data values lie between

(a)   120 and 180?   (b)   132 and 168?   (c)   90 and 210?

3.87   Assume that a collection of 1000 data values has mean 400 and standard deviation 25.

(a)   At least how many data values lie between 325 and 475?

(b)   At least how many data values lie between 300 and 500?

(c)   At *most*, how many scores are smaller than 350 or larger than 450?

3.88   A collection of data values has mean 150 and standard deviation 15. Use Chebyshev's Theorem in each of the following:

(a)   Find the minimum proportion of data values that lie between 127.5 and 172.5.

(b)   Find the maximum proportion of data values that are smaller than 112.5 or larger than 187.5.

Answers to Practice Quiz

1.   READ IN C1, C2
     10   4
     11   8
     15   1
     3   5

2.

C3	C4	C5	C6	C7
24	12	10	2	8
12	6	14	4	3
24	15	7	2	12

3.

C3	C4
4	7
2	8

4. Columns are not of equal length, which is a requirement for use of the READ command.

5. Nothing is wrong. Median is misspelled, but Minitab looks only at the first four letters.

6. SET command can be used only to enter one collection of data at a time. Use READ command instead.

7. K1 is a constant, but C1 + C2 will be a column (the column of sums). Type

LET C3 = C1 + C2

8. Here 3 is an unnecessary constant. Type

DESCRIBE C1

9. Since MEAN is used in the LET command, we must have parentheses around C1. Type

LET K1 = MEAN (C1)

10. PRINT is used only to print stored constants or columns. Type

LET K3 = K1/K2
PRINT K3

Notes

Information Please Almanac (Boston: Houghton Mifflin Company, 1989).

Lyman, C. P., R. C. O'Brien, G. C. Greene, and E. D. Papafrangos (1981), "Hibernation and Longevity in the Turkish Hamster *Mesocricetus brandti*," *Science* 212, 668–670. Data in Chambers, J. M., W. S. Cleveland, B. Kleiner, and P. A. Tukey, *Graphical Methods for Data Analysis* (Boston: Duxbury Press, 1983).

Matthews, D. E., and V. T. Farewell (1982), "On Testing for a Constant Hazard Against a Change-point Alternative," *Biometrics* 38, 463–468. Data in Chambers, J. M., W. S. Cleveland, B. Kleiner and P. A. Tukey, *Graphical Methods for Data Analysis* (Boston: Duxbury Press, 1983).

Statistical Abstract of the United States (Washington, D. C.: U.S. Bureau of the Census, 1984).

The Hammond Almanac (Maplewood, NJ: Hammond Almanac, Inc., 1983)

U.S. News & World Report (Washington D. C.: U.S. News & World Report, Inc., Dec. 19, 1988).

Woodhouse, J., "World Wide Seismic Network Tape". Data in Donoho, A. W., D. L. Donoho, and M. Gasko, *MacSpin User Manual* (Austin, TX: D^2 Software Inc.).

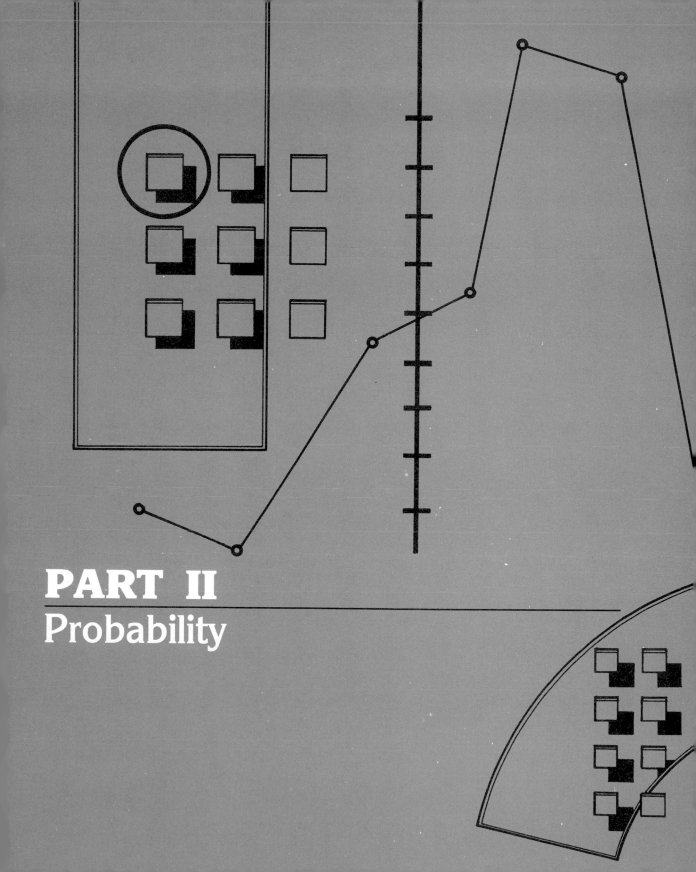

PART II
Probability

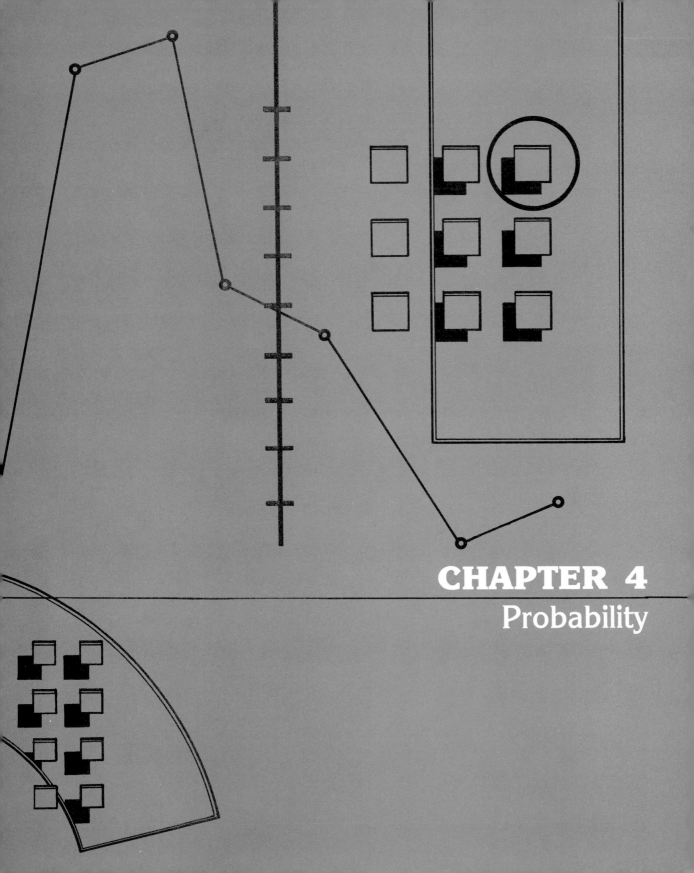

CHAPTER 4
Probability

4.1 INTRODUCTION

In Chapter 1 we pointed out that probability, which deals with the laws of chance, plays an important role in statistical inference. We discussed a hypothetical situation in which a pollster interviews 100 voters selected by chance and finds that 96 of them plan to vote for a certain political candidate, candidate A, in an upcoming election. Based on this information, the pollster would predict that A would win the election. The reason for this is that it seems highly unlikely or improbable that so many voters in the sample would be for A if A were not going to win the election.

On the basis of the sample of 100 voters, the pollster made a decision about the population of all voters. The decision that candidate A would win the election was equivalent to saying that the percentage of all voters in the population that are (or will be) for candidate A is greater than 50%. (We are assuming a two-candidate race.)

This was not a difficult decision to make given the results of the survey. It seems intuitively clear that the chances or probability of so many voters being for A would be very small if A were not going to win. But what if only 66 of the 100 voters were for candidate A? Can it also be said that this many voters for A would be very unlikely unless A were going to win? The answer to this is not so obvious.*

In later chapters dealing with statistical inference we will need to be able to deal with questions such as this. Therefore, in this chapter we will develop the fundamentals of probability theory. We will talk about what is meant by the probability or likelihood associated with various occurrences or events and also show how to calculate the probability of certain types of events.

4.2 SAMPLE SPACES AND EVENTS

The first technical term we will study is **experiment.** We will see that most of the other terms in probability follow quite naturally from this term.

> **Definition**   Any activity that yields a result or an outcome is called an *experiment*.

Normally, there are a variety of possible outcomes of an experiment and the one that occurs when the experiment is performed is a matter of chance.

Example 4.1

Consider the experiment of tossing a coin. There are two possible outcomes: heads (*H*) or tails (*T*).

---

*Actually, it can be shown that if candidate A were not going to win the election, the odds against A getting more than 65 votes out of the 100 are about 1000 to 1 (if the voters are selected by a truly random process, which is not always so easy).

Example 4.2

Consider the experiment of interviewing a voter to determine whether he or she favors nuclear power. Here the possible outcomes might be: yes or no or undecided.

Example 4.3

Consider the experiment of randomly selecting an adult American. This means selecting an adult in such a way that each adult has an equal likelihood of being selected.* The adult selected is the outcome of this experiment. Since any adult could be selected, the collection of all adult Americans constitutes the possible outcomes.

We will be interested in the collection of all possible outcomes of an experiment. An important thing to keep in mind about this collection is that not only will it include all possibilities, but also no two outcomes in this collection can occur at the same time. For instance, in Example 4.1, when we toss a coin we must get either H or T, but we cannot get both of these at the same time.

> **Definition**   The collection (or set) of all possible distinct outcomes that can occur when an experiment is performed is called the *sample space* for the experiment. This collection of outcomes must have the property that when the experiment is performed, one and only one of these outcomes must occur.

We often use a symbol such as S to represent a sample space, and we sometimes describe this set by specifying the outcomes inside braces. So, for Example 4.1, we could describe S as

$$S = \{H, T\}$$

(The outcomes may be listed in any order inside the braces.)

The sample spaces associated with the experiments discussed so far are finite, but sample spaces associated with some experiments are infinite, as the following example shows.

Example 4.4

Consider the following experiment: a coin is flipped until a tail occurs. We could get a tail on the first flip, represented symbolically by T; or we could get a head

---

*In practice this may be very difficult to do, but we need not worry about this for now.

on the first flip and a tail on the second flip, represented by *HT*; or we could get two heads followed by a tail represented by *HHT*, and so on. Notice that the number of possible outcomes is infinite. We could represent the sample space as

$$S = \{T, HT, HHT, HHHT, \ldots\}$$

Example 4.5

Microscopic particles known as "genes" are present in the cells of all living organisms and determine the inherited characteristics of each organism. Most genes occur in pairs. In humans a single gene pair determines whether the body will have normal pigmentation (coloring) or complete lack of pigmentation (albinism). We will use the symbol *C* to represent the gene that calls for normal pigmentation and the symbol *c* will represent the gene for albinism. A person might have each of these in a gene pair. We could represent this pair by the symbol *Cc*, called a "genotype." A human with this genotype will have normal pigmentation because the gene *C* is "dominant," while the gene *c* is "recessive."

In producing an offspring, each parent contributes one gene. For example, suppose that two people with the same genotype, *Cc*, produce an offspring. The gene *c* from the mother and the gene *C* from the father might go into the new cell, resulting in genotype *cC* for the offspring. (Note that when representing the genotype of an offspring, we will find it convenient to let the first symbol represent the gene from the mother and the second symbol represent the gene from the father.)

The particular combination of genes (genotype) that occurs is a matter of chance. Observing the pigmentation genotype of an offspring is an experiment. Let us find the sample space of the experiment. The following diagram gives the possible genotypes that can occur.

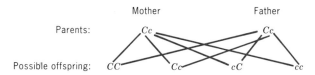

The sample space for the experiment is

$$S = \{CC, Cc, cC, cc\}$$

Note that since *C* is a dominant gene, each of the above genotypes will result in a child with normal pigmentation except *cc*, which will result in albinism. Also note that both parents have normal pigmentation.

Example 4.6

Sickle-cell anemia is a disorder of the blood characterized by red blood cells that are crescent shaped. People with sickle-cell anemia can exhibit severe anemia, headaches, nausea, and other symptoms. Until recently, few lived beyond age 40. Sickle-cell anemia is found only in blacks. This disease is inherited and is determined by a single gene pair that affects the hemoglobin in the blood. We will call

$$H = \text{normal hemoglobin gene}$$
$$h = \text{altered hemoglobin gene for sickle-cell anemia}$$

Each person has a pair of hemoglobin genes (a genotype), one gene inherited from the mother and one from the father. A person with genotype *HH* will have normal blood cells. Someone with genotype *hh* will have sickle-cell anemia. A person with mixed genotype *Hh* is called a "carrier" of the sickle-cell trait. The gene *h* is not completely recessive. That is, someone with genotype *Hh* will not actually have sickle-cell anemia but will exhibit episodes of the sickle-cell trait that can be brought on by stress. As in Example 4.5, observing the hemoglobin genotype of an offspring is an experiment in which the possible hemoglobin genotypes are the outcomes. If a normal mother and a father who is a carrier produce an offspring, find the sample space and use this to show that this couple cannot produce a child with sickle-cell anemia.

Solution

We use the same procedure as was used in Example 4.5.

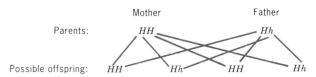

The sample space consists of the possible distinct outcomes. Since there are only two different outcomes listed above,

$$S = \{HH, Hh\}$$

It is clear from this that these parents cannot produce a child with sickle-cell anemia (but they could produce a carrier).

Events

Consider the experiment of flipping two coins, a nickel and a dime. If we get a head on the nickel and a tail on the dime, we will represent this by the symbol *HT*. This is one possible outcome of this experiment. The others are tail on the nickel, head on the dime, *TH*; two heads, *HH*; and two tails, *TT*. Therefore,

$$S = \{HT, TH, HH, TT\}$$

(It is worth noting that we would have the same sample space if the experiment consisted of tossing a single coin twice.)

Now we may be interested in whether or not a match occurs when we perform this experiment. This can occur in two ways, namely, if either of the outcomes *HH* or *TT* occurs. These two outcomes define what we mean by a match; either of these outcomes constitutes a match. So we can describe what is meant by a match by listing the outcomes that constitute the match inside of braces, as follows:

$$\{HH, TT\}$$

It is important to stress that in order for the match to occur, just one of these outcomes need occur when we perform this experiment (indeed, when we perform an experiment only one outcome *can* occur at one time).

Since a match can be completely specified by listing those outcomes that describe the different ways the match can occur, we may as well simply view the match as this list of outcomes. Thus we will view a match as the collection of outcomes.

$$\{HH, TT\}$$

As we said, the match occurs if either of these constituent outcomes occurs. A match is an example of what is called an **event,** which is defined as follows.

> **Definition** An *event* is a collection (or set) of some of the possible outcomes from the sample space. In other words, an event is a subset of the sample space. We say that the event *occurs* if, when we perform the experiment, one of its constituent outcomes occurs.

The outcomes that define an event are usually listed inside of braces. We often use capital letters, especially from the beginning of the alphabet, to represent events. If A is some finite event, we will use the symbol $n(A)$ to represent the number of distinct outcomes in A.

Notice that an event can consist of a single outcome. Therefore, each possible outcome in the sample space is an event, sometimes called a **simple event.** Also, an event could consist of all the possible outcomes in the sample space; that is, S is an event. This event must occur each time the experiment is performed.

Example 4.7
For the experiment of tossing a nickel and dime, let the event E be "a head on the nickel." Describe E as a subset of the sample space.

Solution
A head on the nickel will occur if we get either HT or HH (since the first symbol refers to the nickel). Therefore,

$$E = \{HT, HH\}$$

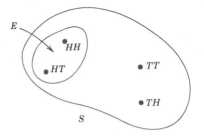

FIGURE 4.1
A Sample Space and an Event *E*
(see Example 4.7)

Figure 4.1 depicts *E* as a subset of the sample space using a sketch called a Venn diagram.

Example 4.8
A randomly selected citizen is interviewed and the following information is recorded: sex and ethnic status. The symbols for ethnic status will be A = Asian, B = Black, H = Hispanic, I = American Indian, W = White, O = Other. If the person interviewed was a female Hispanic, we could represent this outcome as *FH*. Complete the description of the sample space *S*. Let *E* be the event that the person interviewed was female and nonwhite. Describe *E* as a subset of the sample space.

Solution
The sample space is

$$S = \{FA, MA, FB, MB, FH, MH, FI, MI, FW, MW, FO, MO\}$$
$$E = \{FA, FB, FH, FI, FO\}$$

Example 4.9
Consider the experiment and sample space of Example 4.5. Let *E* be the event that occurs if the offspring has normal pigmentation. Describe this as a subset of the sample space.

Solution
Remember that the gene for normal pigmentation *C* is dominant, so any one of the following genotypes will result in normal pigmentation: *CC, Cc, cC*. Therefore,

$$E = \{CC, Cc, cC\}$$

Exercises

4.1   People's blood types are classified as O, A, B, and AB.

   (a)   Suppose that one person is to be selected and the blood type observed. List the elements of the sample space.

   (b)   Suppose that a husband and wife are selected and both of their blood types are observed. List the elements of the sample space.

4.2   A coin is tossed three times and an H or T (H = head, T = tail) is recorded each time. List the elements of the sample space S and list the elements of the event consisting of

   (a)   All heads.   **(b)**   A head on the second toss.   **(c)**   Two tails.

4.3   Suppose that a lottery number is determined by selecting four digits in succession and recording each digit when selected. Repeated digits are allowed. List the elements of the event where

   (a)   The digits 426 are the first three digits in this order.

   (b)   The four digits consist of 4, 2, and 6 in the first three positions (not necessarily in that order) and 8 in the fourth position.

4.4   A randomly selected citizen is interviewed and the following information is recorded: employment status and level of education. The symbols for employment status are Y = employed and N = unemployed, and the symbols for level of education are 1 = did not complete high school, 2 = completed high school but did not complete college, and 3 = completed college. List the elements of the sample space, and list the elements of the following events:

   (a)   Did not complete high school.   **(b)**   Is unemployed.

4.5   In genetics there are two sex genes: X and Y. Each individual has a pair of these (a sex genotype). A female has the pair XX, a male the pair XY. In producing an offspring, each parent contributes one gene. Some inherited characteristics are carried by, or linked to, the X gene. They are called "sex-linked inherited characteristics." For example, colorblindness is one such characteristic. When this gene carries the colorblind trait, we use the lowercase x. A female with the pair xx will be colorblind. A female with sex genotype xX will not be colorblind but will be a carrier of the colorblind gene. A male with sex genotype xY will be colorblind. For a carrier mother and a normal father, find the sample space of sex genotypes for the experiment of observing the sex genotype of an offspring.

4.6   Repeat Exercise 4.5 with a normal mother and colorblind father. Use the sample space to show that a child of this couple cannot inherit colorblindness.

4.7   A hospital administrator records a 0 if a patient has no medical insurance, a 1 if the patient does have medical insurance. Also recorded is an A, B, C, D, or E representing good, fair, poor, serious, or critical condition, respectively. List the elements of the sample space S and list the elements of the event consisting of a selected patient

   (a)   With no medical insurance and in serious or critical condition.

   (b)   With medical insurance and not in critical condition.

    (c)   In good or fair condition.

    (d)   With medical insurance.

4.8   Two cards are selected in succession from an ordinary deck of cards. The suit of the first card is recorded, and then the suit of the second card is recorded. For the suits, let C = clubs, D = diamonds, H = hearts, and S = spades. List the elements of the events consisting of

    (a)   Both spades.   **(b)**   Either both spades or both clubs.

    (c)   At least one heart.   **(d)**   A diamond as the second card.

4.9   A study is to be made in a large university to try to determine a relationship, if any, between the sex of a faculty member and his or her salary. Faculty are to be interviewed and classified according to sex and salary category. Suppose that M = male, F = female, 1 = less than \$30,000, 2 = less than \$35,000 but greater than or equal to \$30,000, 3 = less than \$40,000 but greater than or equal to \$35,000, 4 = less than \$45,000 but greater than or equal to \$40,000, 5 = less than \$50,000 but greater than or equal to \$45,000, and 6 = greater than or equal to \$50,000. List the elements of the event consisting of a selected faculty member

    (a)   With a salary less than \$40,000.

    (b)   Who is female or has a salary greater than, or equal to, \$40,000.

    (c)   Who is a male with a salary greater than, or equal to, \$50,000.

    (d)   Who is a male with a salary less than \$40,000 and greater than, or equal to, \$35,000.

4.3 THE PROBABILITY OF AN EVENT

In this section we arc interested in the chances that a particular event will occur when an experiment is performed.

    Suppose that we look at the experiment of tossing a (fair) coin. We saw that the sample space is

$$S = \{H, T\}$$

Consider the event A, which consists of getting a head. Written in set language,

$$A = \{H\}$$

(*Note:* This event consists of a single outcome, which is perfectly legitimate.) If we perform this experiment over and over many times, say, 1000 times, we would expect to see about 500 heads and 500 tails, since we have a fair coin. So we expect the number of occurrences (or frequency) of heads to be about 500.

The number of occurrences of heads divided by the total number of times the experiment is performed is called the "proportion of occurrences of heads" or "the proportion of times heads occurred." In this case, the expected proportion of occurrences of heads is

$$\frac{500}{1000} = \tfrac{1}{2}$$

This expected proportion is also called the "probability of getting heads" or the "probability of event A," where $A = \{H\}$. Symbolically, we write

$$P(A) = \tfrac{1}{2} \qquad \text{or} \qquad P(H) = \tfrac{1}{2}$$

Note: The symbol $P(A)$ is read "P of A" or the "probability of A."

Let us generalize the foregoing ideas. First we will agree that if A is some event, then the number of times A occurs divided by the total number of times the experiment is performed is called the *proportion of occurrences of A* or the *relative frequency of A*. From the preceding discussion we see that the **probability of A** is the proportion of occurrences of A we expect in the long run, that is, if the experiment is performed over and over many times.

Definition   Suppose that A is some event. The *probability of A*, denoted by $P(A)$, is the expected proportion of occurrences of A if the experiment were to be repeated many times.

If we perform the experiment many times, $P(A)$ is the proportion or fraction of the times we expect to see A occur. Therefore, $P(A)$ is a proportion that will be at least 0 and at most 1. So $0 \leq P(A) \leq 1$.

If we have no idea of the value of $P(A)$ for some particular event, we could use an empirical approach to approximate the probability. For example, suppose that you are given a coin and told the coin is *not* fair. If you were asked to find the probability of observing a head when the coin is tossed, $P(H)$, you might conduct a little experiment. You could toss the coin, say, 100 times and count the number of heads. Suppose that 87 heads occurred. You might estimate $P(H)$ to be

$$\frac{87}{100} = .87$$

This is called **empirical probability.** It is only an estimate of the (true) probability (as explained in the above definition). To distinguish between empirical probability and true probability as given in the definition, some statisticians refer to the latter as **theoretical probability.** If the number of times we toss the coin becomes larger and larger, it seems likely that the empirical probability would get closer and closer to the true probability.

Sometimes we hear a statement such as "There's a 90% chance that I will pass this course." This guess (at a probability of .90) is sometimes referred to as **subjective probability.** The word *subjective* is meant to imply a guess at the true probability.

In this chapter we will be interested in finding the (true) probability of events as it is defined in the above definition. In later chapters there will be instances where we estimate probability using empirical probability.

Example 4.10

We said that any sample space S is an event that must occur each time the experiment is performed, so the proportion of times S will occur is 1. Therefore, $P(S) = 1$.

Example 4.11

Consider the experiment of tossing a fair nickel and a fair dime. Recall from Section 4.2 that the sample space is

$$S = \{HT, TH, HH, TT\}$$

(a) Find the probability associated with each outcome.

(b) Find the probability that a match will occur.

Solution

(a) It seems intuitively clear that each outcome in the sample space has an equal likelihood of occurring. So, if we performed this experiment over and over, we would expect each outcome to occur about one-fourth of the time. That is, the expected proportion of occurrences for each outcome is $\frac{1}{4}$. Therefore, each outcome has the probability $\frac{1}{4}$.

(b) Let A represent the event consisting of a match. As a set,

$$A = \{HH, TT\}$$

Since each outcome that constitutes a match can be expected to occur one-fourth of the time, and there are two such outcomes, we can expect a match to occur one-half of the time. That is,

$$P(A) = \tfrac{1}{4} + \tfrac{1}{4} = \tfrac{1}{2}$$

Example 4.12

A die is painted green on one face, red on one face, yellow on two faces, and blue on two faces. Consider the experiment of rolling the die and observing the color on the top face. Give the sample space S and find the probability associated with each outcome. Also find the probability of the event A, consisting of a red or green face showing.

Solution

The outcomes of interest are green (G), red (R), yellow (Y), and blue (B).

$$S = \{G, R, Y, B\}$$

Now there are six faces on the die and, since one is painted green, we should expect to observe green one-sixth of the time. So $P(G) = \frac{1}{6}$. Similarly,

$$P(R) = \tfrac{1}{6}, \qquad P(Y) = \tfrac{2}{6} = \tfrac{1}{3}, \qquad P(B) = \tfrac{2}{6} = \tfrac{1}{3}$$

The event A written in set language is

$$A = \{R, G\}$$

Since we expect R one-sixth of the time and G one-sixth of the time, we can expect the event A to occur $\frac{1}{6} + \frac{1}{6} = \frac{1}{3}$ of the time. Note that

$$P(A) = P(R) + P(G) = \tfrac{1}{6} + \tfrac{1}{6} = \tfrac{1}{3}$$

In the above example we saw that the probability of the event A was the sum of the probabilities of the distinct outcomes that constitute A. This result generalizes as follows:

> **Rule 4.1** Suppose that the event A consists of the n distinct outcomes o_1, $o_2, \ldots, o_n$, that is,
>
> $$A = \{o_1, o_2, \ldots, o_n\}$$
>
> then the probability of A is
>
> $$P(A) = P(o_1) + P(o_2) + \cdots + P(o_n)$$

In a finite sample space S in which each outcome is equally likely, a useful result can be obtained as a special case of Rule 4.1. The sample space of Example 4.11 was just such a sample space. There were four distinct outcomes in this sample space; therefore, each had probability $\frac{1}{4}$. The event A (a match) had two distinct outcomes. Hence its probability was

$$P(A) = \tfrac{1}{4} + \tfrac{1}{4} = \tfrac{2}{4} = \tfrac{1}{2}$$

Note that

$$P(A) = \frac{\text{number of outcomes in } A}{\text{number of outcomes in } S}$$

We can generalize this result in the form of a rule that is really a corollary of Rule 4.1.

> **Rule 4.2**   Suppose that A is an event in a finite sample space S and each outcome in S is equally likely. Let $n(A)$ represent the number of distinct outcomes in A and $n(S)$ represent the total number of distinct outcomes in S. Then
>
> $$P(A) = \frac{n(A)}{n(S)}$$

Note: Be sure not to apply Rule 4.2 unless all the outcomes in S are equally likely. Observe that Rule 4.2 does *not* apply to the event A of Example 4.12.

Example 4.13

Let C = the collection of all passenger cars produced by U.S. plants in 1987. Details of the production of 1987 cars are given in Table 4.1.

From a list of serial numbers of the cars produced in 1987, suppose that one of these cars is selected at random. Find the probability that it will be a Ford.

Solution

The set C can be viewed as the sample space of the experiment of selecting one of the cars at random (the cars are the possible outcomes). Let the event F = a Ford is selected. As a set, F is the collection of all Fords produced in 1987. Now C is a finite sample space in which each outcome (each car) is equally likely. By Rule 4.2,

$$P(F) = \frac{n(F)}{n(C)} = \frac{1,830,376}{7,098,910} \doteq .2578$$

TABLE 4.1
Passenger Cars Produced by U.S. Plants in 1987

Company	Number of Cars
Chrysler	1,109,421
Ford	1,830,376
General Motors	3,603,074
Honda	324,065
Mazda	4,200
Nissan	117,334
Toyota	43,744
VW of America	66,696
Total	7,098,910

Source: The World Almanac and Book of Facts (New York: Newspaper Enterprise Association, 1989), p. 189.

Example 4.14

Consider the experiment of rolling two ordinary dice. Find the probability that a sum of 7 will occur.

Solution

It will help us to understand this situation, without altering our results, if we imagine one die as painted red and the other as painted green. We could get a 4 on the red die and 3 on the green die. This outcome could be represented as (4,3). Note that (3,4) is a different outcome.* The sample space S will have 36 possible outcomes:

$$S = \{(1,1),\ (1,2),\ (1,3),\ (1,4),\ (1,5),\ (1,6)$$
$$(2,1),\ (2,2),\ (2,3),\ (2,4),\ (2,5),\ (2,6)$$
$$(3,1),\ (3,2),\ (3,3),\ (3,4),\ (3,5),\ (3,6)$$
$$(4,1),\ (4,2),\ (4,3),\ (4,4),\ (4,5),\ (4,6)$$
$$(5,1),\ (5,2),\ (5,3),\ (5,4),\ (5,5),\ (5,6)$$
$$(6,1),\ (6,2),\ (6,3),\ (6,4),\ (6,5),\ (6,6)\}$$

Each outcome is equally likely. Now let A be the event consisting of a sum of 7. Describing A as a set, we see that

$$A = \{(6,1),\ (5,2),\ (4,3),\ (3,4),\ (2,5),\ (1,6)\}$$

that is, A will occur if any one of these outcomes occurs. Applying Rule 4.2, we see that

$$P(A) = \frac{n(A)}{n(S)} = \frac{6}{36} = \frac{1}{6}$$

Example 4.15

For the sample space of Example 4.5, assign probabilities to each outcome. Also, find the probability that an offspring will have normal pigmentation.

Solution

Each possible pairing of genes from the parents is equally likely and there are four of them. So the probability associated with each is $\frac{1}{4}$. Let E be the event that an offspring has normal pigmentation. We pointed out in Example 4.9 that since C is a dominant gene, CC, Cc, and cC will all result in normal pigmentation. Therefore, as a set

$$E = \{CC,\ Cc,\ cC\}$$

---

*In mathematics, parentheses ( ) are often used to enclose elements when order is important. When order is not important, braces { } are usually used.

Since each outcome in the sample space is equally likely, we can use Rule 4.2 to obtain

$$P(E) = \frac{n(E)}{n(S)} = \frac{3}{4}$$

Example 4.16

Note that this example depends on an understanding of Examples 4.5 and 4.6. Suppose that parents who are both carriers of the sickle-cell trait produce an offspring. Find the probability that this offspring will also be a carrier.

Solution

A carrier has a normal hemoglobin gene H and a sickle-cell gene h.

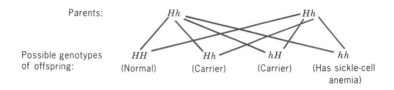

Parents: Hh Hh

Possible genotypes of offspring: HH (Normal) Hh (Carrier) hH (Carrier) hh (Has sickle-cell anemia)

Thus the sample space is

$$S = \{HH, Hh, hH, hh\}$$

Let B = the event that occurs if the offspring is a carrier. This will occur if either of the genotypes Hh or hH occur; therefore

$$B = \{Hh, hH\}$$

Each outcome is equally likely; so by Rule 4.2

$$P(B) = \frac{n(B)}{n(S)} = \frac{2}{4} = \frac{1}{2}$$

Complementary Events

Before closing this section, we should mention the notion of the **complement of an event.** To understand this idea, imagine an experiment in which a family with three children is randomly selected and we record in order from the oldest to the youngest child whether the child is a boy (B) or girl (G). The sample space could be represented as

$$S = \{GGG, BGG, GBG, GGB, GBB, BGB, BBG, BBB\}$$

The event A consisting of at least one boy is

$$A = \{BGG, GBG, GGB, GBB, BGB, BBG, BBB\}$$

The **complement of A,** denoted by $\overline{A}$, is the opposite of A, that is, the event consisting of no boys:

$$\overline{A} = \{GGG\}$$

We will assume that for a birth the probability of a boy is $\frac{1}{2}$. Since for a given birth a boy or girl is equally likely, the outcomes in the sample space are equally likely. Therefore by Rule 4.2,

$$P(A) = \frac{n(A)}{n(S)} = \frac{7}{8} \quad \text{and} \quad P(\overline{A}) = \frac{n(\overline{A})}{n(S)} = \frac{1}{8}$$

Note that

$$\boxed{P(A) + P(\overline{A}) = 1}$$

This is always true of complementary events. Such a result can be useful, because if we know the probability of one of these events, we can quickly solve for the other. We will have more to say about complementary events in Sections 4.4 and 4.6.

Exercises

4.10 A statistics class contains 14 males and 20 females. A student is to be selected by chance and the sex of the student recorded.

(a) Give a sample space S for the experiment.

(b) What is the probability that the selected student is female?

4.11 In Example 4.14, we considered the experiment of tossing two fair dice. Find the probability of each of the following events:

(a) The sum of the two dice is less than 5.

(b) At least one 6.

(c) Neither is a 6.

(d) The sum is either 7 or 11.

4.12 A health clinic in the city of Worcester employs 50 people. Four live in the town of Shrewsbury, 10 in the town of Auburn, 12 in the town of Grafton, and the rest live in Worcester. One employee is needed to participate in a clinic project. If one employee is randomly selected, find the probability that

(a) The person lives in Shrewsbury.

(b) The person lives in Worcester.

4.13 A digit is to be selected by chance.

    (a) Give a sample space S.

    In parts (b) and (c), list the elements of the event. Then find the probability that the event will occur.

    (b) The event consisting of an odd digit.

    (c) The event consisting of a number larger than 6.

4.14 An experiment consists of selecting by chance four digits and recording each in succession as E (even) or O (odd).

    (a) List the elements of a sample space S.

    In parts (b)–(d), find the probability of each event.

    (b) The event consisting of all evens.

    (c) The event consisting of at least one even.

    (d) The event consisting of at least two evens occurring in succession.

4.15 Assume a 50–50 chance that a newborn child will be a boy. In a family of four children, find the probability that

    (a) There are exactly two boys.   **(b)** The oldest two children are boys.

    (c) All are girls.   **(d)** Not all are girls.

4.16 Megabucks, a lottery game conducted by the Massachusetts State Lottery Commission, consists of selecting 6 numbers from the 36 numbers, {1, 2, 3, 4, ..., 35, 36} (no repetitions and order doesn't matter). The commission selects by chance the 6 winning numbers. You pay \$1 to play. You get a free ticket if 3 of your numbers match 3 of the winning numbers, \$40 for matching 4 numbers, \$400 for matching 5 numbers, and a large prize if you match all 6 numbers. The sample space S consist of all possible 6-number combinations that can be selected. It can be shown that $n(S) = 1,947,792$. The number of elements in S that match 0, 1, 2, 3, 4, 5, or 6 winning numbers are as follows:

Number of Matches	Number of Possibilities
0	593,775
1	855,036
2	411,075
3	81,200
4	6,525
5	180
6	1

Find the probability of

    (a) Not winning anything.   **(b)** At least one match.

    (c) Winning a free ticket.   **(d)** Winning either \$40 or \$400.

    (e) Winning Megabucks.

4.17 The following data represent the number of earned degrees conferred in the United States and Puerto Rico for the years 1960, 1970, and 1980 (*Source: Statistical Abstract of the United States*, 1984, p. 168). The data values are in thousands.

Year	Bachelor's	Master's
1960	395	75
1970	833	209
1980	999	298

A degree recipient is selected by chance (assume one degree per recipient). Find the probability that the recipient

(a) Received a master's degree.   **(b)** Received a degree in 1970.

4.4 COMPOUND EVENTS. THE ADDITION RULE

We will use the term **compound event** to mean an event that is expressed in terms of, or as a combination of, other events. It is often convenient to view an event as a compound event because there are rules for finding the probabilities of such events.

Suppose A, B are events from some sample space. We will be interested, in this and the next two sections, in properties of the following three events:

$$A \text{ or } B \qquad A \text{ and } B \qquad \text{complement of } A$$

We begin by giving concrete examples of these events followed by a formal definition of each.

Example 4.17
Suppose that we consider the experiment of randomly selecting a voter from a particular town. Let the event A correspond to selecting a voter who favors wage/price controls. Let the event B correspond to selecting a voter who is a member of a union. Now (A or B) is the event that occurs if the person selected either favors wage/price controls *or* is a member of a union or both. The event (A and B) occurs if the voter selected favors wage/price controls *and* at the same time is a union member. The (complement of A) is the event that occurs if the voter selected did not favor wage/price controls.

> **Definitions**   Let A, B be events from a sample space S.
>
> 1. *A or B* is the event that will occur if either A occurs or B occurs or both occur. As a set, the event (A or B) consists of those outcomes in A together with those outcomes in B. (This set is sometimes called the union of A with B, denoted by $A \cup B$.)
> 2. *A and B* is the event that will occur if both A occurs and B occurs at the same time. As a set, the event (A and B) consists of those outcomes that are both in A and at the same time in B; in other words, those outcomes common to both events. (This set is sometimes called the intersection of A and B, denoted by $A \cap B$.)
> 3. The **complement of A,** denoted by $\overline{A}$ or $S - A$, is the event that consists of those outcomes in S that are not in A. (Therefore, $\overline{A}$ will occur whenever A does not. In this sense, $\overline{A}$ is the opposite of A.)

Figure 4.2 is a graphic representation of the events just described using **Venn diagrams.**

The Addition Rule

We wish to develop formulas for finding the probabilities of various compound events. To this end we continue the discussion started in Example 4.17. The

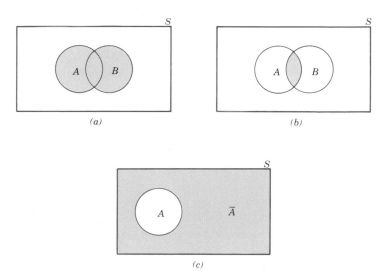

FIGURE 4.2
Compound Events. (*a*) A or B (shaded). (*b*) A and B (shaded). (*c*) $\overline{A}$ (shaded).

TABLE 4.2
Voter Preference on Wage/Price Controls versus Union Membership in a Town with 1000 Registered Voters

	A (favor W/P controls)	$\overline{A}$ (oppose W/P controls)	Row totals
B (Union)	100	300	$400 \leftarrow n(B)$
$\overline{B}$ (Nonunion)	400	200	$600 \leftarrow n(\overline{B})$
Column totals	500	500	1000 (Grand Total)
	$\uparrow$	$\uparrow$	
	$n(A)$	$n(\overline{A})$	$n(S)$

experiment consists of randomly selecting a voter from a particular town; the possible outcomes are the voters in the town. Therefore, the sample space is

$$S = \{\text{all voters in the town}\}$$

The event A occurs if the voter selected favors wage/price controls. As a set of outcomes,

$$A = \{\text{all voters in town favoring wage/price controls}\}$$

The event B corresponds to selecting a union member. As a set,

$$B = \{\text{all voters in town belonging to a union}\}$$

Now suppose that there are 1000 voters in the town. Then $n(S) = 1000$. Also, suppose $n(A) = 500$ and $n(B) = 400$. Further breakdown of the voters is given in Table 4.2.

This table is not difficult to interpret. For example, the number in the intersection of column A and row B is 100, meaning 100 voters favor wage/price controls and are also members of a union. Thus, $n(A \text{ and } B) = 100$.

Now each outcome (voter) is equally likely. Therefore let's use Rule 4.2 to evaluate some probabilities.

$$P(A) = \frac{n(A)}{n(S)} = \frac{500}{1000} = .5, \qquad P(\overline{A}) = \frac{n(\overline{A})}{n(S)} = \frac{500}{1000} = .5$$

$$P(B) = \frac{n(B)}{n(S)} = \frac{400}{1000} = .4, \qquad P(\overline{B}) = \frac{n(\overline{B})}{n(S)} = \frac{600}{1000} = .6$$

$$P(A \text{ and } B) = \frac{n(A \text{ and } B)}{n(S)} = \frac{100}{1000} = .1$$

$$P(A \text{ or } B) = \frac{n(A \text{ or } B)}{n(S)} = \frac{?}{1000}$$

We can find $n(A \text{ or } B)$, the number of voters who favor wage/price controls or are members of a union, from the table. We want the number of *distinct* voters in $(A \text{ or } B)$; that is, we do not want to count anyone twice. Now there are 500 voters in A and 400 in B. But 100 voters are in both A and B [$n(A \text{ and } B) = 100$]. If we tried to count the number of voters, $n(A \text{ or } B)$, by adding $500 + 400$, we would be counting 100 of them twice. What we should do is subtract 100 from this sum:

$$n(A \text{ or } B) = n(A) + n(B) - n(A \text{ and } B)$$
$$= 500 + 400 - 100 = 800$$

Therefore,

$$P(A \text{ or } B) = \frac{n(A \text{ or } B)}{n(S)} = \frac{800}{1000} = .8$$

Note that

$$P(A \text{ or } B) = \frac{n(A \text{ or } B)}{n(S)} = \frac{n(A) + n(B) - n(A \text{ and } B)}{n(S)}$$
$$= \frac{n(A)}{n(S)} + \frac{n(B)}{n(S)} - \frac{n(A \text{ and } B)}{n(S)}$$
$$= P(A) + P(B) - P(A \text{ and } B) = .5 + .4 - .1 = .8$$

Thus

$$P(A \text{ or } B) = P(A) + P(B) - P(A \text{ and } B)$$

This relationship is known as the **Addition Rule.** The technique we have used could be used to show that the Addition Rule is valid for any events A, B in a finite sample space in which each outcome is equally likely. (Remember, we have used Rule 4.2.) However, the Addition Rule can be shown to be valid in general (even in sample spaces where not all outcomes are equally likely). This is an important result because it enables us to find the probability on the left if we know the three on the right. In fact, if we know any of the three probabilities in the equation, we can solve for the fourth.

Rule 4.3   Addition Rule   If A, B are events from some sample space, then
$$P(A \text{ or } B) = P(A) + P(B) - P(A \text{ and } B)$$

Example 4.18

A die is rolled. Find the probability that the number on the top face is even or greater than 4 using the Addition Rule.

Solution

There are six possible outcomes: 1, 2, 3, 4, 5, and 6. As sets

$$\text{even} = \{2, 4, 6\} \qquad \text{greater than } 4 = \{5, 6\}$$

so that

$$P(\text{even}) = \tfrac{3}{6} = \tfrac{1}{2}, \qquad P(\text{greater than } 4) = \tfrac{2}{6} = \tfrac{1}{3}$$

Therefore,

$$
\begin{aligned}
P&(\text{even or greater than } 4) \\
&= P(\text{even}) + P(\text{greater than } 4) - P(\text{even and greater than } 4) \\
&= \tfrac{1}{2} + \tfrac{1}{3} - P(6) \\
&= \tfrac{1}{2} + \tfrac{1}{3} - \tfrac{1}{6} = \tfrac{2}{3}
\end{aligned}
$$

Some events, by their very nature, cannot occur at the same time. For example, for the experiment of rolling a die, let the event A correspond to an odd number showing and the event B correspond to an even number. Clearly, these events cannot occur at the same time; they have no outcomes in common. Hence $P(A \text{ and } B) = 0$.

> **Definition**   Two events A, B are said to be *mutually exclusive* if they cannot both occur together when an experiment is performed (i.e., if they have no outcomes in common).

A Venn diagram depicting mutually exclusive events is shown in Figure 4.3.

When two events A, B are mutually exclusive, $P(A \text{ and } B) = 0$. This results in a simplification of the Addition Rule because in this case

$$P(A \text{ or } B) = P(A) + P(B)$$

This result generalizes to any number of mutually exclusive events. For example, assume that events A, B, and C are mutually exclusive. (This means that no two of them can occur at the same time.) Then

$$P(A \text{ or } B \text{ or } C) = P(A) + P(B) + P(C)$$

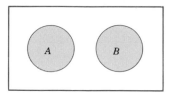

FIGURE 4.3
Mutually Exclusive Events

Example 4.19

At a meeting of the student government council of a college, there are 50 students present, of which 20 are freshmen, 15 are sophomores, 10 are juniors, and 5 are seniors. One of these students is to be selected at random to deliver a petition to the administration. Find the probability that the student selected is a freshman or sophomore.

Solution

Let A be the event that occurs if the student selected is a freshman and B the event of selecting a sophomore. Now these events are mutually exclusive. Hence

$$P(A \text{ or } B) = P(A) + P(B) = \tfrac{20}{50} + \tfrac{15}{50} = \tfrac{35}{50} = \tfrac{7}{10}$$

Exercises

4.18 In Example 4.14 we considered the tossing of two fair dice. Consider the following events: A = sum is more than 7, B = sum is even, C = sum is odd, and D = sum is less than 11.

 (a) Which pairs of events, if any, are mutually exclusive?

 (b) Use the Addition Rule to find each of the following:

 (i) $P(A \text{ or } B)$.   **(ii)** $P(B \text{ or } C)$.   **(iii)** $P(A \text{ or } C)$.

 (c) Let E − sum is less than 4, and $F = \{(3,3)\}$. Find $P(A \text{ or } E \text{ or } F)$. (*Hint:* A, E, and F are mutually exclusive.)

4.19 A card is to be randomly selected from an ordinary deck of 52 cards. Consider the following events: A = ace, B = face card, and C = club.

 (a) Which pairs of events, if any, are mutually exclusive?

 (b) Use the Addition Rule to find

 (i) $P(A \text{ or } B)$.   **(ii)** $P(A \text{ or } C)$.   **(iii)** $P(B \text{ or } C)$.

4.20 Refer to the parents of Exercise 4.5 (i.e., the carrier mother and normal father) and assume that when a parent contributes a gene from a gene pair, either gene is equally likely to be contributed.

 (a) Assign probabilities to each outcome.

 (b) Find the probability that an offspring will either carry the colorblind gene or be colorblind.

4.21 In a small college 1500 of the 4000 students are male. Also, 1200 of the 3600 students under the age of 25 are male. What proportion of the student body is either male or under the age of 25?

4.22 The data below are characteristics of the voting-age population regarding the November 1988 election in the United States. (*Source:* Data estimated from *The World Almanac and Book of Facts*, 1990, p. 339.)

	Voted	Did Not Vote
Males	47,675,000	36,856,000
Females	54,550,000	39,018,000

For a randomly selected person from the population, let A be the event that the person selected voted, and B be the event that the person selected is a male. Find each of the following:

(a) $P(B)$.   (b) $P(\overline{A})$.   (c) $P(\overline{A} \text{ and } \overline{B})$.

(d) $P(A \text{ or } B)$.   (e) $P(\overline{A} \text{ or } B)$.

4.23   There are 2000 voters in a town. Consider the experiment of randomly selecting a voter to be interviewed. (The voters in the town are the possible outcomes of the experiment.) The event A consists of being in favor of more stringent building codes; the event B consists of having lived in the town less than 10 years. The following table gives the numbers of voters falling into various categories.

	A Favor more stringent codes	$\overline{A}$ Do not favor more stringent codes
B Less than 10 years	100	700
$\overline{B}$ At least 10 years	1000	200

(a) Find $P(A)$.   (b) Find $P(\overline{B})$.   (c) Find $P(A \text{ and } B)$.

(d) Find $P(A \text{ or } B)$.   (e) Find $P(A \text{ or } \overline{B})$.

4.5 THE MULTIPLICATION RULE

Use of the Addition Rule formula to find $P(A \text{ or } B)$ will involve the probability $P(A \text{ and } B)$, unless A and B are mutually exclusive. Sometimes $P(A \text{ and } B)$ can be found easily from information given in the problem; at other times, it is not so easy. In this section we will discuss a formula for finding $P(A \text{ and } B)$.

Now we return to our discussion of the data in Table 4.2. The experiment is randomly selecting a voter from the town of 1000 voters. First, consider the probability of event B occurring, given that A has definitely occurred. In other words, we want the probability that the voter selected is a union member if we know he favors wage/price controls. The probability is called the *conditional probability of B given that A has occurred,* denoted by $P(B|A)$. Now 500 voters favor wage/price controls [$n(A) = 500$] and 100 of those are also union members [$n(A$ and $B) = 100$]. Thus the probability that the person selected is a union member given that this person favors wage/price control is $\frac{100}{500} = .2$. Therefore,

$$P(B|A) = \frac{n(A \text{ and } B)}{n(A)} = \frac{100}{500} = .2$$

By performing a simple algebraic operation on the above equation, we can discover a useful relationship. We can divide both the numerator and denominator of the fraction in the above equation by the same number, $n(S) = 1000$, without affecting the equality. Dividing by $n(S)$ and using Rule 4.2, we obtain

$$P(B|A) = \frac{n(A \text{ and } B)/n(S)}{n(A)/n(S)} \leftarrow \left(\frac{100/1000}{500/1000} = \frac{.1}{.5} = .2 \right)$$

So we see that

$$P(B|A) = \frac{P(A \text{ and } B)}{P(A)}$$

The above relationship is usually written in the following form:

$$P(A \text{ and } B) = P(A) \cdot P(B|A)$$

This result is known as the **Multiplication Rule.** Although the above discussion involved events in a finite sample space where each outcome is equally likely, nevertheless, this result is valid for events in any sample space.

Rule 4.4   Multiplication Rule   If A, B are events in some sample space, then

$$P(A \text{ and } B) = P(A) \cdot P(B|A)$$

where $P(B|A)$ is the (conditional) probability that B will occur, given that A has occurred.

Example 4.20
Two cards are dealt from a deck. Find the probability that the first is an ace and the second is a king.

Solution
Let A be the event consisting of an ace on the first card and B the event of a king on the second card. We want to find $P(A \text{ and } B)$. By Rule 4.4

$$P(A \text{ and } B) = P(A) \cdot P(B|A)$$

Four cards in the deck of 52 are aces. Thus

$$P(A) = \tfrac{4}{52}$$

Now suppose the first card was an ace. Then there are 51 cards left and 4 of them are kings. Hence, the probability of a king on the second card, given the first was an ace, is

$$P(B|A) = \tfrac{4}{51}$$

Therefore,

$$P(A \text{ and } B) = P(A) \cdot P(B|A) = \tfrac{4}{52} \cdot \tfrac{4}{51} = \tfrac{16}{2652} \doteq .006$$

Example 4.21

At a large bank 6% of the employees are computer programmers, 50% of the employees are female, and 2% of the employees are female computer programmers. If an employee is selected by chance, what is the probability that

(a)   The employee is a computer programmer, given that the employee is female.

(b)   The employee is female, given that the employee is a computer programmer.

Solution

Let F = employee is female and C = employee is a computer programmer. We are given

$$P(F) = .5 \qquad P(C) = .06$$
$$P(F \text{ and } C) = .02$$

(a)   We want $P(C|F)$. Now from the Multiplication Rule

$$P(F \text{ and } C) = P(F) \cdot P(C|F)$$

Therefore

$$P(C|F) = \frac{P(F \text{ and } C)}{P(F)} = \frac{.02}{.50} = \frac{2}{50} = \frac{1}{25} = .04$$

(b)   This asks for $P(F|C)$. We can also write (by the Multiplication Rule)

$$P(C \text{ and } F) = P(C) \cdot P(F|C)$$

Therefore

$$P(F|C) = \frac{P(C \text{ and } F)}{P(C)} = \frac{.02}{.06} = \frac{1}{3}$$

Example 4.22

A box contains seven red poker chips and three white ones. Two chips are randomly selected from the box. Find the probability that we get a red chip on the first draw and a white one on the second draw if

(a) The chips are selected without replacement (i.e., the first is not returned to the box before the second is selected).

(b) The chips are selected with replacement (first chip is returned before selecting the second).

Solution

Let $R1$ = red on first draw, $W2$ = white on second.

(a)
$$P(R1 \text{ and } W2) = P(R1) \cdot P(W2|R1)$$

Now

$$P(R1) = \tfrac{7}{10}$$

If we draw a red on the first selection, and if we do not replace the chip, there are 9 left, of which 3 are white. Hence

$$P(W2|R1) = \tfrac{3}{9} = \tfrac{1}{3}$$

and so

$$P(R1 \text{ and } W2) = \tfrac{7}{10} \cdot \tfrac{1}{3} = \tfrac{7}{30}$$

(b) To find the probability when the chips are selected with replacement, we can use the same formula as in part (a). The only difference occurs with the term $P(W2|R1)$. Since the first chip is replaced, the probability of a white on the second draw is unaffected by whether we get a red on the first or not. In other words,

$$P(W2|R1) = P(W2) = \tfrac{3}{10}$$

Therefore

$$P(R1 \text{ and } W2) = P(R1) \cdot P(W2)$$
$$= \tfrac{7}{10} \cdot \tfrac{3}{10} = \tfrac{21}{100}$$

Sometimes a *tree diagram* is helpful in analyzing probabilities of the type found in the above example. Conditional probabilities may be written on the branches of the tree. Multiplying the probabilities along the branches leading to an outcome at the end of a tree gives the probability of the outcome. See Figure 4.4 for a representation of drawing two chips without replacement as in Example 4.22(a).

In Example 4.22, part (b), we saw that when the two poker chips were selected with replacement, the probability of a white chip on the second draw

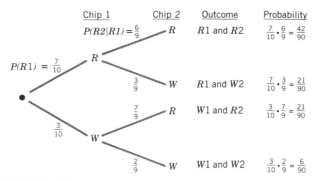

FIGURE 4.4
Drawing Two Poker Chips without Replacement. See Example 4.22 (a)

(*W2*) had nothing to do with whether or not we got a red chip on the first draw (*R1*). In this case, we call these two events **independent**.

> **Definition**   Two events *A*, *B* are called *independent* if the occurrence of one has no effect on the probability of the occurrence of the other.

If events *A*, *B* are independent, then

$$P(B|A) = P(B)$$

This results in a simplification of the Multiplication Rule, because in this case

$$P(A \text{ and } B) = P(A) \cdot P(B)$$

This result generalizes to any number of events if they are all independent (i.e., no two are dependent). For example, if *A*, *B*, *C* are independent events, then

$$P(A \text{ and } B \text{ and } C) = P(A) \cdot P(B) \cdot P(C)$$

Example 4.23
Suppose that for a certain couple, the probability of producing a child with brown eyes is $\frac{3}{4}$ and the probability of producing a child with blue eyes is $\frac{1}{4}$.

(a)   If the couple has two children, what is the probability that the first will have blue eyes and the second brown eyes?

(b)   If the couple has two children, find the probability that one has blue eyes and the other brown eyes.

(c)   If the couple has three children, find the probability that all will have brown eyes.

Solution

(a) The tree diagram in Figure 4.5 can be helpful in computing probabilities. We use the symbol A to represent a child having blue eyes and B a child having brown eyes. $A1$ will mean the first child has blue eyes; $B2$ will mean the second child has brown eyes, and so on. We wish to find $P(A1$ and $B2)$. Intuitively, it seems clear that the events $A1$ and $B2$ are independent. Therefore,

$$P(A1 \text{ and } B2) = P(A1) \cdot P(B2) = \tfrac{1}{4} \cdot \tfrac{3}{4} = \tfrac{3}{16}$$

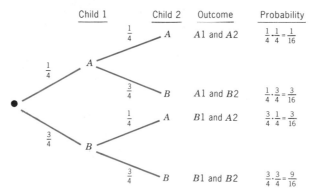

FIGURE 4.5
Tree Diagram for Eye Color

(b) One child with blue eyes and one with brown eyes can occur in two ways. The first child could have blue eyes and the second child brown eyes, or the first could have brown eyes and the second blue eyes. We can express this as the compound event

$$(A1 \text{ and } B2) \text{ or } (B1 \text{ and } A2)$$

To find the probability of the above event, we use the Addition Rule and the Multiplication Rule. First observe that the event ($A1$ and $B2$) and the event ($B1$ and $A2$) are mutually exclusive; therefore

$$P[(A1 \text{ and } B2) \text{ or } (B1 \text{ and } A2)] = P(A1 \text{ and } B2) + P(B1 \text{ and } A2)$$

Now we will use the Multiplication Rule to evaluate the probabilities on the right-hand side of the equation. Remember that eye color of one child has no bearing on eye color of the other child. So $A1$, $B2$ are independent events. Therefore,

$$P(A1 \text{ and } B2) = P(A1) \cdot P(B2) = \frac{1}{4} \cdot \frac{3}{4} = \frac{3}{16}$$

Similarly,

$$P(B1 \text{ and } A2) = P(B1) \cdot P(A2) = \frac{3}{4} \cdot \frac{1}{4} = \frac{3}{16}$$

Hence

$$P(\text{one child with blue eyes and one with brown eyes})$$
$$= P[(A1 \text{ and } B2) \text{ or } (B1 \text{ and } A2)]$$
$$= \frac{3}{16} + \frac{3}{16} = \frac{6}{16} = \frac{3}{8}$$

(c)   The event of three children having brown eyes is $(B1 \text{ and } B2 \text{ and } B3)$. Now the events $B1$, $B2$, $B3$ are independent, so that

$$P(B1 \text{ and } B2 \text{ and } B3) = P(B1) \cdot P(B2) \cdot P(B3)$$
$$= \frac{3}{4} \cdot \frac{3}{4} \cdot \frac{3}{4} = \frac{27}{64}$$

Example 4.24

Ralph and Carolyn Cummins of Clintwood, Virginia, have five children all with the same birthday, February 20, in different years (*Source: Guinness Book of World Records*, 1977, p. 36).

(a)   Find the probability that all five children in a family will have February 20 for a birthday.

(b)   Find the probability that all five children in a family will have the same birthday.

(Assume that there are no multiple births involved.)

Solution

(a)   Assuming 365 days in the year, the probability that a person will be born on February 20 is $\frac{1}{365}$. Concerning the five children, we are interested in the event A consisting of the following birthdays:

$$A = (\text{February 20 for the first child}) \text{ and}$$
$$(\text{February 20 for the second}) \text{ and } \ldots \text{ and}$$
$$(\text{February 20 for the fifth})$$

We assume the events on the right are independent. So by the Multiplication Rule for independent events

$$P(A) = [P(\text{February 20})]^5 = \left(\frac{1}{365}\right)^5 = \frac{1}{6,478,348,728,125}$$

(b)   Let B be the event that all five children have the same birthday. This could occur if they are all born on January 1 or January 2 or . . . or December 31 (365 dates). Thus

$$B = (\text{all on January 1}) \text{ or}$$
$$(\text{all on January 2}) \text{ or } \ldots$$
$$\text{or (all on December 31)}$$

The events on the right are all mutually exclusive and have the same probability that we have calculated in part (a) to be $(\frac{1}{365})^5$. So, by the Addition Rule for mutually exclusive events,

$$
\begin{aligned}
P(B) &= P(\text{all on January 1}) + P(\text{all on January 2}) \\
&\quad + \cdots + P(\text{all on December 31}) \\
&= \left(\frac{1}{365}\right)^5 + \left(\frac{1}{365}\right)^5 + \cdots + \left(\frac{1}{365}\right)^5 \leftarrow 365 \text{ terms} \\
&= (365)\left(\frac{1}{365}\right)^5 = \frac{1}{(365)^4} = \frac{1}{17{,}748{,}900{,}625}
\end{aligned}
$$

This result can also be obtained directly from the Multiplication Rule. Note that

$$
\begin{aligned}
B ={}& (\text{second child has same birthday as first}) \\
& \text{and (third has same birthday as second)} \\
& \text{and (fourth has same birthday as third)} \\
& \text{and (fifth has same birthday as fourth)}
\end{aligned}
$$

The four events on the right are independent and have the same probability, namely, $\frac{1}{365}$. So, by the Multiplication Rule,

$$
P(B) = \left(\frac{1}{365}\right)\left(\frac{1}{365}\right)\left(\frac{1}{365}\right)\left(\frac{1}{365}\right) = \frac{1}{365^4}
$$

Exercises

4.24 Two cards are to be selected from an ordinary deck of 52 cards. Assume that the first card is not replaced before the second one is drawn. Consider the following events: A = the first card is an ace; B = the second card is an ace.

(a) Find each of the following:

    (i) $P(A \text{ and } B)$.

    (ii) $P(B)$. [*Hint:* Let $\overline{A}$ = the first card is not an ace. Now $B = (A \text{ and } B)$ or $(\overline{A} \text{ and } B)$.]

    (iii) $P(A \text{ or } B)$.

(b) Are the events A, B independent?

4.25 **(a)** Redo Exercise 4.24, but assume the first card is replaced before the second card is selected.

(b) Now suppose a third card is selected after replacing the first and second cards. (Let C = third card is not an ace.) Find $P(A \text{ and } B \text{ and } C)$.

4.26 In Example 4.14 we considered the experiment of tossing two fair dice. Consider the following events: A = the sum is even; B = doubles (both numbers are the same); C = the sum is larger than five; and D = the sum is odd.

    (a) Use the Multiplication Rule to find each of the following:

        (i) $P(A|C)$.   **(ii)** $P(C|A)$.   **(iii)** $P(B|C)$.

    (b) Find

        (i) $P(A|B)$.   **(ii)** $P(D|B)$.

    (c) Which pairs of events are independent?

4.27 When all 6 ports (connections) to a mainframe computer are not in use, a computer user is randomly assigned to a port. Thus a sample space for this process is $S = \{1, 2, 3, 4, 5, 6\}$. Consider events $A = \{1, 3, 5\}$ and $B = \{1, 2\}$. Are the events A, B independent?

4.28 Refer to the parents of Exercise 4.5 (i.e., the carrier mother and normal father).

    (a) Find the probability that a child will be colorblind if it is male.

    (b) Find the probability that a child will be colorblind if it is female.

4.29 A committee of seven consists of two males and five females. Two members are to be chosen randomly to look into a specific problem. What is the probability that both males will be chosen? (*Hint:* Imagine the selection as a two-stage process—select one member, then another, without replacement.)

4.30 A sporting goods store has a large batch of cans of tennis balls on hand. Ten percent of the cans are unacceptable (that is, contain at least one defective ball).

    (a) A customer decides to purchase one can. What is the probability that the customer will be satisfied?

    (b) A customer is to purchase two cans. Find the probability that

        (i) Both cans will be satisfactory.

        (ii) Exactly one can will be satisfactory.

        (iii) At least one can will be satisfactory.

4.31 A business employs 600 men and 400 women. Five percent of the men and 10% of the women have been working there for more than 20 years. If an employee is selected by chance, what is the probability that the employee is male, given that the length of employment is more than 20 years?

4.32 In a town, 70% of the men are employed. The probability that a man will commit a crime is .10, and the probability that a man is employed, given he will commit a crime, is .05. A man is selected by chance and is employed. What is the probability that he will commit a crime?

4.33 At a college a change was proposed in the mathematics curriculum. The mathematics majors were asked whether they approved of the proposed change. The result of the survey was

	Approved	No Opinion	Did Not Approve
Male	21	6	12
Female	14	10	7

Suppose that a mathematics major is selected by chance. Find the probability that

(a) The student is female, given no opinion.

(b) The student approves of the proposed change, given the student is male.

(c) The student is male, given the student does not approve of the proposed change.

(d) The student is male and approves of the proposed change (use the Multiplication Rule).

4.34 In Exercise 4.22 we discussed the events A (person randomly selected voted in the Nov., 1988 elections) and B (person was a male). The following table gives the number of people in various categories:

	A Voted	$\overline{A}$ Did Not Vote
B Males	47,675,000	36,856,000
$\overline{B}$ Females	54,550,000	39,018,000

In parts (a) and (b), describe the probability and find the value of each:

(a) $P(B|A)$.   **(b)** $P(\overline{A}|\overline{B})$.

4.35 In a city there are 1000 married couples with both husband and wife working. Each person was asked if their salary exceeded $30,000. The following information was obtained:

		Husband	
		Less than $30,000	More than $30,000
Wife	Less than $30,000	430	410
	More than $30,000	60	100

Note that with 410 of the 1000 couples, the husband had a salary of more than $30,000 and the wife less than $30,000. If one of the couples is selected by chance,

(a) Find the conditional probability that a husband earns less than $30,000, given the wife earns less than $30,000.

(b) Find the conditional probability that a wife earns more than $30,000, given that the husband earns more than $30,000.

(c) Are the salaries of husband and wife statistically independent?

4.36   Suppose that 2% of the people in a town have a particular type of cancer. A test designed to detect the cancer has the following properties: If a person has the cancer, the test will detect it with probability .99. If the person does not have cancer, the test will indicate cancer with probability .005. If the test indicates a person has cancer, what is the probability that the person actually has cancer? Assume the probability that the test indicates cancer is .0247.

4.37   A study is being conducted on families with two children. Consider the following events: A = the older child is a boy, and B = the younger child is a boy. A family with two children is selected by chance. Find each of the following probabilities (assume a probability of $\frac{1}{2}$ that a particular child is a boy).

   (a) $P[(A \text{ and } B)|A]$ **(b)** $P[(A \text{ and } B)|(A \text{ or } B)]$

4.6 MORE ON COMPLEMENTARY EVENTS

In Section 4.3 we briefly discussed complementary events. We considered the event A, consisting of observing at least one boy in a family with 3 children, and its complement $\overline{A}$, consisting of no boys. We noted that

$$P(A) + P(\overline{A}) = \tfrac{7}{8} + \tfrac{1}{8} = 1$$

It is not hard to see that for *any* event A, $P(A) + P(\overline{A}) = 1$. First, observe that either A or $\overline{A}$ must always occur when an experiment is performed. Thus

$$P(A \text{ or } \overline{A}) = 1$$

Furthermore, the event A and event $\overline{A}$ are mutually exclusive. Therefore, using the Addition Rule

$$P(A \text{ or } \overline{A}) = P(A) + P(\overline{A}) = 1$$

Therefore, we have the following rule:

Rule 4.5   For an event A,

$$P(A) + P(\overline{A}) = 1$$

As we pointed out in Section 4.3, this rule can be useful because if we know either of the probabilities $P(A)$ or $P(\overline{A})$, we can solve for the other.

Example 4.25

In Example 4.23 we discussed a couple that had a $\frac{3}{4}$ chance of producing a child with brown eyes and a $\frac{1}{4}$ chance of producing a child with blue eyes. If this couple has three children, find the probability that not all three will have brown eyes.

Solution
Let

$$A = \text{all three have brown eyes}$$

Then

$$\overline{A} = \text{not all three have brown eyes}$$

By Rule 4.5

$$P(A) + P(\overline{A}) = 1$$

so that

$$P(\overline{A}) = 1 - P(A)$$

We have seen in Example 4.23 that $P(A) = \frac{27}{64}$. Therefore,

$$P(\overline{A}) = 1 - P(A) = 1 - \tfrac{27}{64} = \tfrac{37}{64}$$

Notice how much simpler this approach is than if we had tried to find $P(\overline{A})$ directly.

Example 4.26

A six-person jury for a civil trial was selected by chance. The defense attorney (whose client was female) was surprised that the jury did not have at least one female. Assuming the population is half female, find the probability that such a jury will have at least one female.

Solution
"At least one" is the opposite of "none." So if we let

$$A = \text{no females}$$

then

$$\overline{A} = \text{at least one female}$$

We can think of the jury as being composed of 6 males selected sequentially.

$$A = \text{all males} = MMMMMM$$

We assume the population is large and one-half male; hence the probability of any selection being a male is $\frac{1}{2}$. Thus we assume the probability of a male on one selection is independent of any other selection. Using the Multiplication Rule for independent events we get

$$P(A) = P(MMMMMM) = P(M) \cdot P(M) \cdot P(M) \cdot P(M) \cdot P(M) \cdot P(M)$$
$$= (\tfrac{1}{2})^6 = \tfrac{1}{64}$$

Using Rule 4.5, $P(A) + P(\overline{A}) = 1$. Hence

$$P(\overline{A}) = 1 - P(A) = 1 - \tfrac{1}{64} = \tfrac{63}{64} \doteq .98$$

So the chance of at least one female on the jury is about 98%.

___

Exercises

4.38   A subcommittee of two is to be selected by chance from a committee consisting of eight men and two women. What is the probability that at least one of the two women will be selected?

4.39   A fair die is to be tossed three times. Find the probability of tossing at least one 6.

4.40   Suppose that the probability that a child produced by a couple will have a particular disease is $\frac{1}{10}$. If they plan to have four children, what is the probability that at least one child will have the disease?

4.41   A large shipment of items contains 2% defective items. Five items are to be selected. What is the probability of getting at least one defective item?

4.42   The principal of each elementary school in a town with four such schools is asked to randomly select one of her grades (from grade 1 through 8) to participate in a national survey. What is the probability that at least two of the principals select the same grade level? [*Hint:* Find the probability that no two select the same grade and subtract this from 1. Assume that the principals are numbered. No two will select the same grade if the second selects a grade different from the first (probability $\frac{7}{8}$) *and* the third selects a grade different from the first two (probability $\frac{6}{8}$) *and* the fourth selects a grade different from the first three (probability $\frac{5}{8}$). Now use the Multiplication Rule to find the probability that no two select the same grade.]

4.7 COMBINATORICS (OPTIONAL)

Combinatorics is the study of counting techniques. In probability it can be important to count the number of elements in a set (see Rule 4.2).

Suppose a fast-food restaurant sells ice cream cones in two sizes (regular

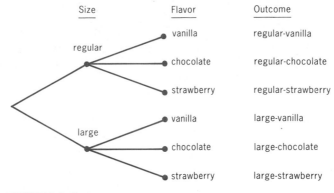

Size	Flavor	Outcome

FIGURE 4.6

and large) and three flavors (vanilla, chocolate, and strawberry). The task of ordering an ice cream cone can be broken down into two tasks:

Task 1 = specify the size
Task 2 = specify the flavor

We illustrate the process with a tree diagram (Figure 4.6).

Notice that there are 6 possible ice cream cones that could be ordered. Task 1 can be done 2 ways and Task 2 can be done 3 ways, and $2 \times 3 = 6$.

Task: Order ice cream cone = specify size and specify flavor
Number of ways: $6 = 2 \times 3$

Rule 4.6   (Fundamental Principle of Counting)   If Task 1 can be done in n_1 ways and Task 2 can be done in n_2 ways, then Task 1 *and* Task 2 performed together can be done in $n_1 \cdot n_2$ ways.

Suppose an ice cream cone is randomly ordered. Let A = event of getting a vanilla cone. Note that A can occur 2 ways and there are 6 possible cones. Thus

$$P(A) = \frac{n(A)}{n(S)} = \tfrac{2}{6} = \tfrac{1}{3}$$

Using the Fundamental Principle of Counting, this type of probability could be calculated without listing all the outcomes, which can be important when there are a large number of outcomes.

Example 4.27
The faculty in the math department are going to play the faculty in the psychology department in singles tennis. If there are 10 math professors and 15 psychology professors, how many different matches are possible?

Solution
Determining a match consists of selecting a math professor, which can be done 10 ways, and selecting a psychology professor, which can be done 15 ways. Thus the number of matches is $(10)(15) = 150$.

The Fundamental Principle of Counting can be extended to more than two tasks, as the following example shows.

Example 4.28
A lottery game consists of selecting a sequence of 3 digits by filling in the digits in 3 boxes on a lottery ticket. (*Note:* Order is important.) How many choices are there if

(a) Repeated digits are permitted (such as 292)?

(b) No repetitions are permitted?

Solution
(a) Let

$$\text{Task 1} = \text{select the first digit}$$
$$\text{Task 2} = \text{select the second digit}$$
$$\text{Task 3} = \text{select the third digit}$$

Now there are 10 digits to choose from for each task: 0, 1, 2, 3, 4, 5, 6, 7, 8, 9. So each task can be done 10 ways. Hence, the 3 digits can be selected

$$10 \cdot 10 \cdot 10 = 1000 \text{ ways}$$

Note: The possible choices are 000, 001, 002, ... , 999

(b) Suppose no repetitions are permitted. There are still 10 choices for Task 1. But once you choose a digit for Task 1, you can't use it again. So there are only 9 digits available for Task 2 and thus Task 2 can be done 9 ways. And for Task 3 there are only 8 possibilities left. So the three digits can be selected

$$10 \cdot 9 \cdot 8 = 720 \text{ ways}$$

It is worth observing that part (b) of the above example can be viewed as follows: Imagine the 10 digits 0, 1, 2, 3, 4, 5, 6, 7, 8, 9 written on 10 tags. We then choose 3 tags and arrange them in the desired order, such as

5		2		8

(Note that there won't be any repetitions.) Such an arrangement is called a **permutation,** more precisely, a **permutation of 10 objects taken 3 at a time.**

In order to be able to conveniently count the number of possible permutations, it is helpful to introduce the concept of **factorials.** If n is a positive whole number, n **factorial,** denoted by $n!$, is defined as the product

$$n! = 1 \cdot 2 \cdot 3 \cdots n$$

Thus

$$4! = 1 \cdot 2 \cdot 3 \cdot 4 = 24$$
$$3! = 1 \cdot 2 \cdot 3 = 6$$
$$2! = 1 \cdot 2 = 2$$
$$1! = 1$$

By definition $0! = 1$.

Factorials can get large very quickly. For example,

$$7! = 5,040$$
$$10! = 3,628,800$$

Many calculators have factorial keys, and also there are mathematical tables of factorials.

We saw in part (b) of Example 4.28 that there are 720 permutations of the 10 digits taken 3 at a time. Notice that

$$720 = \frac{3,628,800}{5040} = \frac{10!}{7!} = \frac{10!}{(10 - 3)!}$$

So the number of permutations of 10 objects taken 3 at a time is

$$\frac{10!}{(10 - 3)!} = 720$$

We generalize these ideas as follows:

> **Permutations**   An ordered arrangement of r objects selected from a set of n objects is a *permutation of n objects taken r at a time.* The total number of such permutations, denoted by $_nP_r$, is
>
> $$_nP_r = \frac{n!}{(n - r)!}$$

It may seem that the use of factorials is a more complicated way to compute the number of permutations. But remember, there are tables of factorials, and factorials can be computed with calculators. So factorials can shorten our work.

Example 4.29
A T.V. news department decided to interview 4 celebrities for the evening news. If the news department employs 6 reporters, how many different ways are there to assign the interviewers? (Assume no reporter will interview more than one celebrity.)

Solution
Note that the assignment of reporters to celebrities 1, 2, 3, and 4 can be thought of as an arrangement or permutation of the reporters. So we want the number of permutations of six reporters taken 4 at a time.

$$_6P_4 = \frac{6!}{(6-4)!} = \frac{6!}{2!} = \frac{720}{2} = 360$$

Suppose three friends wish to play each other at chess. How many different matches are possible? Suppose we call the players A, B, and C. We are asking how many ways are there of selecting 2 players. But notice that order doesn't matter here. A list of elements in which order doesn't matter is just a set. The sets consisting of 2 elements are

$$\{A, B\}, \{A, C\}, \{B, C\}$$

Such sets are also called **combinations**, more precisely, **combinations of 3 things taken 2 at a time.** Note there are 3 of them.

Observe that

$$\frac{3!}{2!(3-2)!} = \frac{6}{(2)(1)} = 3$$

This generalizes to a computational formula for combinations.

Combinations   A collection of r objects selected from a set of n objects is called a *combination of n objects taken r at a time*. The total number of such combinations, denoted by $_nC_r$, is

$$_nC_r = \frac{n!}{r!(n-r)!}$$

This is also denoted by $\binom{n}{r}$.

Note: **When order matters, use the permutation formula; when order doesn't matter, use the combination formula.**

Example 4.30

To complete the requirements for a major, a student must select 3 courses from a list of 7. How many different 3-course combinations are there?

Solution

We want the number of combinations of 7 courses taken 3 at a time.

$$_7C_3 = \frac{7!}{3!(7-3)!} = \frac{7!}{3!4!} = \frac{1 \cdot 2 \cdot 3 \cdot 4 \cdot 5 \cdot 6 \cdot 7}{(1 \cdot 2 \cdot 3)(1 \cdot 2 \cdot 3 \cdot 4)}$$

$$= \frac{5 \cdot 6 \cdot 7}{1 \cdot 2 \cdot 3} = \frac{5 \cdot 6 \cdot 7}{6} = 35$$

Example 4.31

The Massachusetts Megabucks lottery consists of selecting 6 numbers from the numbers: 1, 2, ... , 36. Order doesn't matter.

(a) If you play the game once, what is the probability that your 6 numbers are the same as the winning 6 numbers?

(b) How many 6-number combinations match exactly 2 of the 6 winning numbers?

(c) What is the probability that you will match exactly 2 of the 6 winning numbers?

Solution

(a) First we must calculate the number of possible 6-number combinations. We want the number of combinations of 36 things taken 6 at a time.

$$_{36}C_6 = \frac{36!}{6!(36-6)!} = \frac{36!}{6!30!}$$

$$= \frac{1 \cdot 2 \cdot 3 \cdot 4 \cdots 29 \cdot 30 \cdot 31 \cdot 32 \cdot 33 \cdot 34 \cdot 35 \cdot 36}{6! \cdot 1 \cdot 2 \cdot 3 \cdot 4 \cdots 29 \cdot 30}$$

$$= \frac{31 \cdot 32 \cdot 33 \cdot 34 \cdot 35 \cdot 36}{1 \cdot 2 \cdot 3 \cdot 4 \cdot 5 \cdot 6} = \frac{31 \cdot 32 \cdot 33 \cdot 34 \cdot 35 \cdot 36}{8 \cdot 3 \cdot 5 \cdot 6}$$

$$= (31)(4)(11)(34)(7)(6) = 1,947,792 \text{ combinations.}$$

Each combination is equally likely. Therefore the probability of selecting all 6 winning numbers is

$$\frac{1}{1,947,792} \doteq .0000005$$

So your chance of winning is about 1 in 2 million.

(b) To count the number of possibilities that match the winning combination in exactly 2 numbers, we can imagine counting the ways we can perform two tasks sequentially:

> Task 1 = specify which 2 of the 6 numbers in the winning combination are to be matched
>
> Task 2 = specify the remaining 4 numbers

Task 1 consists of selecting a subset of 2 elements from the winning set of 6 elements. We have seen that the number of ways of doing this is

$$_6C_2 = \frac{6!}{2!(6-2)!} = \frac{6!}{2!4!} = \frac{\cancel{1} \cdot 2 \cdot \cancel{3} \cdot \cancel{4} \cdot 5 \cdot 6}{(1 \cdot 2)(\cancel{1} \cdot 2 \cdot 3 \cdot 4)} = \frac{(5)(6)}{2} = 15$$

To specify the remaining 4 numbers, we must be sure not to use any of the remaining 4 numbers of the winning combination (otherwise we would have more than 2 matching numbers). Also, keep in mind that we have used up 2 numbers for the matches. Thus we have 30 numbers left from which to choose 4 numbers. Therefore Task 2 can be done $_{30}C_4$ ways.

$$_{30}C_4 = \frac{30!}{4!(30-4)!} = \frac{30!}{4!26!} = \frac{1 \cdot 2 \cdot 3 \cdots 26 \cdot 27 \cdot 28 \cdot 29 \cdot 30}{(1 \cdot 2 \cdot 3 \cdot 4)(1 \cdot 2 \cdot 3 \cdots 26)}$$

$$= \frac{27 \cdot 28 \cdot 29 \cdot 30}{1 \cdot 2 \cdot 3 \cdot 4} = 27,405$$

Using the Fundamental Principle of Counting, Task 1 and Task 2 taken together can be done

$$(15)(27,405) = 411,075$$

ways. This is the number of combinations that match exactly 2 of the 6 winning numbers.

(c) The probability of matching exactly 2 of the 6 winning numbers is

$$P(matching\ 2) = \frac{\text{Number of combinations that match 2 winning numbers}}{\text{Total number of possible combinations}}$$

$$= \frac{411,075}{1,947,792} \doteq .2110$$

Remark Although order doesn't matter in Megabucks, one player thought it did. He selected 6 numbers for his first ticket by checking off the numbers on a card. He then bought another ticket and carefully selected the same numbers, but checked them off in a *different* order. Of course the computer that reads your numbers has no idea of the order in which you checked them off. It just reads the numbers and prints them on your ticket-in ascending order of magnitude for convenience. So this brilliant gambler bought two tickets with the same combination of numbers on each. Therefore he had only a 1 in 2 million chance of winning instead of 2 in 2 million. But, you guessed it. His number won. There were 3 winning tickets and he held 2 of them. So he got 2/3 of the prize money.

Exercises

4.43 Assume there are 4 roads between towns A and B, 5 roads between towns B and C, and 8 roads between towns C and D.

(a) How many different routes are possible in traveling from A to B to C?

(b) How many different routes are possible in traveling from A to B to C to D?

4.44 Evaluate each of the following:

(a) $6!$. **(b)** $_{12}P_3$. **(c)** $_5P_5$. **(d)** $_{12}C_3$. **(e)** $_{12}C_9$. **(f)** $_5C_5$.

4.45 Consider the set $S = \{a, b, c, d\}$.

(a) Find $_4C_2$ and list the combinations.

(b) Find $_4P_2$ and list the permutations.

4.46 A newspaper editor is going to assign two reporters to cover a political convention. The assignment will be made from a pool of six women and four men. How many groups of two reporters can be assigned if

(a) Both are to be women?

(b) Both are to be men?

(c) There is to be one of each sex?

4.47 A license plate consists of 3 letters followed by 3 digits. How many license plates can be issued? (*Note:* Repetitions are allowed.)

4.48 Some computers use the binary coded decimal (BCD) system to represent characters such as A, B, 3, 4, etc. This is a system that uses a sequence of 6 bits to represent a character, where each bit is a 0 or 1. How many characters can be represented?

4.49 Assume a telephone number consists of 10 digits, and the first and fourth digits cannot be 0.

(a) How many telephone numbers are possible?

(b) How many telephone numbers are possible within the 617 area code?

(c) How many telephone numbers are possible within the 617–661 telephone exchange?

4.50   A group of six women and four men have volunteered to participate in an experiment. Three of them will be randomly selected to participate in one phase of the experiment.

    (a)   Find the probability that all are men.

    (b)   Find the probability that 2 are women.

4.51   Suppose 4 digits are to be randomly selected (repetitions allowed). [*Note:* The set of digits is {0, 1, 2, 3, 4, 5, 6, 7, 8, 9}.] Find the probability that

    (a)   5562 is selected.

    (b)   0000 is selected.

    (c)   2 is the first digit selected.

    (d)   1, 2, and 3 are the first 3 digits selected, not necessarily in the order 123.

4.52   The letters A, B, C, D, E, and F are to be randomly arranged in a row. Find the probability that

    (a)   The arrangement is ABCDEF.

    (b)   The first 3 letters, in order, are ABC.

    (c)   The letters A and B appear next to each other.

4.53   Megabucks, a lottery game conducted by the Massachusetts State Lottery Commission, consists of selecting 6 numbers from the 36 numbers, {1, 2, 3, 4, ..., 35, 36} (no repetitions and order doesn't matter). The commission selects by chance the 6 winning numbers. You pay $1 to play. You get a free ticket if 3 of your numbers match 3 of the winning numbers, $40 for matching 4 numbers, $400 for matching 5 numbers, and a large prize if you match all 6 numbers. The sample space *S* consist of all possible 6 number combinations that can be selected. Without using Exercise 4.16, find the probability of

    (a)   Winning a free ticket.   **(b)**   Winning $40.

4.54   In a 13-card bridge hand, find each of the following:

    (a)   The probability of getting 4 aces.

    (b)   The probability of getting 7 diamonds and 6 hearts.

4.8 SUMMARY

In this chapter we discussed the notion of an **experiment,** which is any activity yielding an outcome. The collection of all distinct possible outcomes is the **sample space** of the experiment. An **event** is a subset of the sample space. A major objective of the chapter is to be able to find the chances or the probability that an event will occur. We developed some techniques in the form of rules for finding such probabilities.

Rule 4.1

If A is an event consisting of the n distinct outcomes $o_1, o_2, \ldots, o_n$, then

$$P(A) = P(o_1) + P(o_2) + \cdots + P(o_n)$$

Rule 4.2

If A is an event in a finite sample space in which each outcome is equally likely, then

$$P(A) = \frac{n(A)}{n(S)}$$

where $n(A)$ = the number of distinct outcomes in A (i.e., the number of different ways A can occur) and $n(S)$ is the total number of possible distinct outcomes in S.

Rule 4.3 (Addition Rule)

For events A, B

$$P(A \text{ or } B) = P(A) + P(B) - P(A \text{ and } B)$$

If A, B are *mutually exclusive* (cannot occur at the same time), then $P(A$ and $B) = 0$. Therefore,

$$P(A \text{ or } B) = P(A) + P(B)$$

Rule 4.4 (Multiplication Rule)

For events A, B

$$P(A \text{ and } B) = P(A) \cdot P(B|A)$$

where $P(B|A)$ = probability of B occurring if A has occurred. When events A, B are *independent*, A has no effect on the probability of B. So $P(B|A) = P(B)$. In this case,

$$P(A \text{ and } B) = P(A) \cdot P(B)$$

Rule 4.5

For an event A, its *complement* $\overline{A}$ is the event that occurs, provided that A does not ($\overline{A}$ is the opposite of A).

$$P(A) + P(\overline{A}) = 1$$

Rule 4.6 (Fundamental Principle of Counting)

If Task 1 can be done in n_1 ways and Task 2 can be done in n_2 ways, then Task 1 *and* Task 2 performed together can be done in $n_1 \cdot n_2$ ways.

A **permutation** is an arrangement of objects in a particular order. The number of possible permutations of n objects taken r at a time is denoted by $_nP_r$.

$$_nP_r = \frac{n!}{(n-r)!}$$

A **combination** is a set of objects (order doesn't matter). The number of possible combinations of n objects taken r at a time is denoted by $_nC_r$.

$$_nC_r = \frac{n!}{r!(n-r)!}$$

Review Exercises

4.55   A lottery ticket has a color, R = red, B = blue, G = green followed by either 1, 2, 3, or 4. One ticket is to be selected and the combination observed. List the elements in the sample space (use the set notation $S = \{\ \}$).

4.56   A study of 1000 couples is designed to determine the relationship, if any, between educational backgrounds of husbands and wives. It has been decided to categorize the educational levels in terms of the highest degree attained. The labels are 1 = Doctorate, 2 = Master's degree, 3 = Bachelor's degree, and 4 = others. Note that both husband and wife are asked their educational status. List the elements in the sample space (use the set notation $S = \{\ \}$).

4.57   In Example 4.5 a sample space S was given as $S = \{CC, Cc, cC, cc\}$. List the elements of the following events: (a) albinism, and (b) offspring is a carrier of albinism (but has normal pigmentation).

4.58   A man buys 10 chances for a large raffle. Two tickets are to be drawn in the raffle. (Thus the man could possibly win two prizes.) For each ticket drawn, it is noted whether the man wins or loses.

(a)   Give a sample space.

(b) List the elements of the following events:

  (i) The man wins two prizes.

  (ii) The man wins exactly one prize.

4.59 A large city hospital is doing a study on the safety of certain anesthetics and is collecting data on the type used, the physical status of the patient at the time of surgery, and whether the patient survived or did not survive the surgery. For the type of anesthetic used, let 1 = cyclopropane, 2 = ether, 3 = halothane, 4 = pentothal, and 5 = other. Also, let a = good physical status, b = poor physical status, x = survived surgery, and y = did not survive surgery. List the elements of the following events:

  (a) Survived surgery.

  (b) Either received cyclopropane or halothane, and was in good physical shape.

  (c) Was in poor physical shape and survived surgery.

4.60 An urn contains 365 chips representing the days of the year. One chip is to be selected by chance. Find the probability that

  (a) Your birth date is selected.

  (b) A day in the month of July is selected.

  (c) The first day of any month is selected.

  (d) A day in a month beginning with a J is selected.

4.61 The voting list in a small town contains the names of 417 Democrats, 335 Republicans, and 248 Independents. One name is to be selected by chance. Find the probability that the name selected is

  (a) A Democrat.

  (b) A Democrat or an Independent.

  (c) Not an Independent.

4.62 Suppose two fair dice are tossed. If the sum is smaller than 10, then a fair coin will be tossed. [*Note:* 8T and 11 are elements of S (where T = tails).]

  (a) Give a sample space for the experiment.

  (b) Find the probability of obtaining

  (i) A head. **(ii)** A head or tail. **(iii)** A sum of 8 and a tail.

4.63 The Mass Millions lottery game involves selecting 6 numbers (no repetitions) from a collection of 46 numbers, {1, 2, 3, 4, ..., 44, 45, 46}. Let the sample space S consist of all possible 6-number combinations that can be selected. It can be shown that $n(S)$ = 9,366,819. Suppose you select 6 numbers. The number of elements in S that match 0, 1, 2, 3, 4, 5, or 6 numbers of the winning combination are

Number of Matches	Number of Possibilities
0	3,838,380
1	3,948,048
2	1,370,850
3	197,600
4	11,700
5	240
6	1

(a) Find the probability of matching 3 numbers. Compare with Exercise 4.16, part (b).

(b) Find the probability of matching 0 or 1 numbers.

(c) On average, approximately how many times per 10,000 would you expect to match 4, 5, or 6 numbers?

4.64 A person purchases a package of 10 ball-point pens that, unknown to the purchaser, contains 3 defective pens. Of the first 2 pens to be used by the purchaser, find the probability that

(a) Both are defective.

(b) At least 1 is defective.

(c) Exactly 1 is defective.

4.65 Refer to the parents of Exercise 4.5 (i.e., the carrier mother and normal father). Assume that these parents have two children (not twins).

(a) Find the probability that the oldest is colorblind and the other is not.

(b) Find the probability that exactly one is colorblind.

(c) Find the probability that both are colorblind.

(d) Find the probability that at least one is colorblind.

4.66 It is claimed that 70% of the residents in a large town approve of the current zoning law. Four residents are to be randomly selected. Find the probability that

(a) None of the four approves.

(b) Exactly two approve.

(c) At least one approves.

4.67 A family owns two cars. The probability that cars S and T will fail to start on a cold morning is $\frac{2}{10}$ and $\frac{3}{10}$, respectively. Assuming that the failure of one car to start is independent of the starting of the second car, find the probability that on a cold morning

(a) Both cars will fail to start.

(b) At least one of the cars will fail to start.

(c) Exactly one of the cars fails to start.

4.68 Teams S and T are in the World Series, and the first team to win four games is the series winner. Assume that the probability that team S wins any particular game is $\frac{2}{3}$. Find the probability that

(a) S wins the series in four games.

(b) The series lasts exactly five games.

4.69 Two people S and T agree to meet at a restaurant. Each person will arrive, independently of the other, at either 1:00, 1:30, or 2:00 P.M. Assume that each person has the probability $\frac{1}{3}$ of arriving at any of the three times. Find the probability that

(a) They both arrive at 1:30.

(b) They both arrive at the same time.

(c) S arrives 30 minutes before T.

4.70 A state lottery selects a four-digit number as follows: four digits are selected one at a time (at random) from {0, 1, 2, 3, 4, 5, 6, 7, 8, 9}. Assume sampling with replacement so that repetitions are allowed. (*Note:* 0939 is a possible outcome.) Suppose that you purchase a lottery ticket. Find the probability that your number is selected.

4.71 Suppose the chances that a particular husband and wife live for 25 more years are $\frac{7}{10}$ and $\frac{9}{10}$, respectively. Assume that survival of one is independent of the other. Find the probability that

(a) The husband does not live for 25 more years.

(b) Both husband and wife live for 25 more years.

(c) At least one of the two lives for 25 more years.

4.72 At one time it was thought that an enlarged thymus in infants increased the risk of sudden infant death syndrome. Subsequently, this was found to be incorrect, but at the time many such infants were treated with radiation to shrink the thymus. Over the next 40 years about 25% of those treated developed thyroid tumors and about 6% of those treated developed cancerous thyroid tumors. (There is some debate about the latter figure, but let us assume that it is correct.) Suppose that during a medical checkup one of those treated is found to have a thyroid tumor. What is the probability that it is malignant?

4.73 One thousand young adults (18–25 years) from a large city were asked whether or not they had used certain drugs at least once within a month prior to the study. The results were as follows:

Type of Drug	Number
Only marijuana	400
Only cocaine	190
Only heroin	35
Marijuana and cocaine	75
Marijuana and heroin	15
Cocaine and heroin	10
All three drugs	5
No drugs	270

Find the proportion of the young adults who used

(a) Marijuana.

(b) Either cocaine or heroin.

(c) Either marijuana or cocaine.

4.74 At a liberal arts college 90% of the freshman are enrolled in English I, 80% are enrolled in Mathematics I, and 75% are enrolled in both courses. A freshman is to be randomly selected. Find the probability that the student is

(a) Not enrolled in English I.

(b) Enrolled in either English I or Mathematics I.

(c) Enrolled in English I, given that the student is enrolled in Mathematics I.

4.75 A survey was taken of 150 residents of a small resort town to determine attitudes of the residents toward a proposed hotel development. The occupations of the residents were determined along with whether or not they approved of the hotel being built. The results follow.

	Building and Trade	Businessmen	Other
Approve	40	40	5
Disapprove	10	20	35

In parts (a)–(e) find the proportion of the 150 residents that

(a) Are businessmen.

(b) Approve of the hotel being built.

(c) Approve of the hotel being built, given that they are in building and trade.

(d) Are businessmen, given that they approve of the hotel being built.

(e) Are businessmen or approve of the hotel being built.

(f) Are events "businessmen" and "approve" independent? Explain. Are they mutually exclusive? Explain.

4.76 The data below represent characteristics of the voting-age population in the November 1988 election in the United States (*Source: The World Almanac and Book of Facts,* 1990, p. 339).

	A Registered	$\overline{A}$ Not Registered
B **Males**	55,114,000	29,417,000
$\overline{B}$ **Females**	63,439,000	30,129,000

In parts (a), (b), and (c), evaluate the following probabilities and describe the proportions they represent:

(a) $P[(\overline{B} \text{ and } A) \text{ or } (\overline{B} \text{ and } \overline{A})]$. **(b)** $P(B|A)$. **(c)** $P(\overline{B}|\overline{A})$.

4.77 Suppose Urn I contains two red balls and four green balls, while Urn II contains three red balls and five green balls. A ball is to be randomly selected from each urn and the colors recorded.

(a)   Find the probability that both balls will be green.

(b)   Find the probability that at least one ball will be green.

4.78   Refer to Exercise 4.77. An urn is to be selected by chance and then a ball selected from that urn.

(a)   Give a sample space for the experiment.

(b)   Find the probability that a green ball will be selected. [*Hint:* A green ball corresponds to (Urn I and green ball) or (Urn II and green ball).]

(c)   Find the probability that Urn I was selected, given that a green ball has been selected.

4.79   Refer to Exercise 4.77. A ball is to be chosen from Urn I and placed in Urn II. A ball is then selected from Urn II. Find the probability that the ball eventually selected from Urn II is green. [*Hint:* Note that $B = (A$ and $B)$ or $(\overline{A}$ and $B)$, where A represents a green ball from Urn I and B represents a green ball from Urn II.]

4.80   Evaluate each of the following:

(a) $12!$ **(b)** $_{20}P_3$.   **(c)** $_{20}C_3$.   **(d)** $100!/98!$.

4.81   A man has 4 pairs of trousers, 6 shirts, and 3 sweaters. How many different outfits can he choose to wear?

4.82   Consider the set $S = \{A, B, C, D, E\}$.

(a)   Find $_5C_2$ and list the combinations.

(b)   Find $_5P_2$ and list the permutations.

4.83   A hand in cribbage consists of 6 cards dealt from an ordinary deck of 52 cards. How many possible cribbage hands are there?

4.84   A basketball team has 12 players, only 5 of whom can be on the court at once. How many different sets of 5 players are there?

4.85   The letters A, B, C, D, E, F, and G are to be randomly arranged in a row. Find the probability that

(a)   The arrangement is ABCDEFG.

(b)   The first 3 letters, in order, are ABC.

(c)   The letters ABC appear together in that order.

(d)   The letters A, B, and C appear together in any order.

4.86   A hand in 5-card draw poker consists of 5 cards dealt from an ordinary deck of 52 cards. Find the probability of being dealt

(a)   4 aces.   **(b)**   5 hearts.   **(c)**   3 aces and 2 kings.

Notes

McWhirter, N., and R. McWhirter, *Guinness Book of World Records* (New York: Bantam, 1977).

Statistical Abstract of the United States (Washington, D. C.: U.S. Bureau of the Census, 1984).

The World Almanac and Book of Facts (New York: Newspaper Enterprise Association, 1989).

The World Almanac and Book of Facts (New York: Newspaper Enterprise Association, 1990).

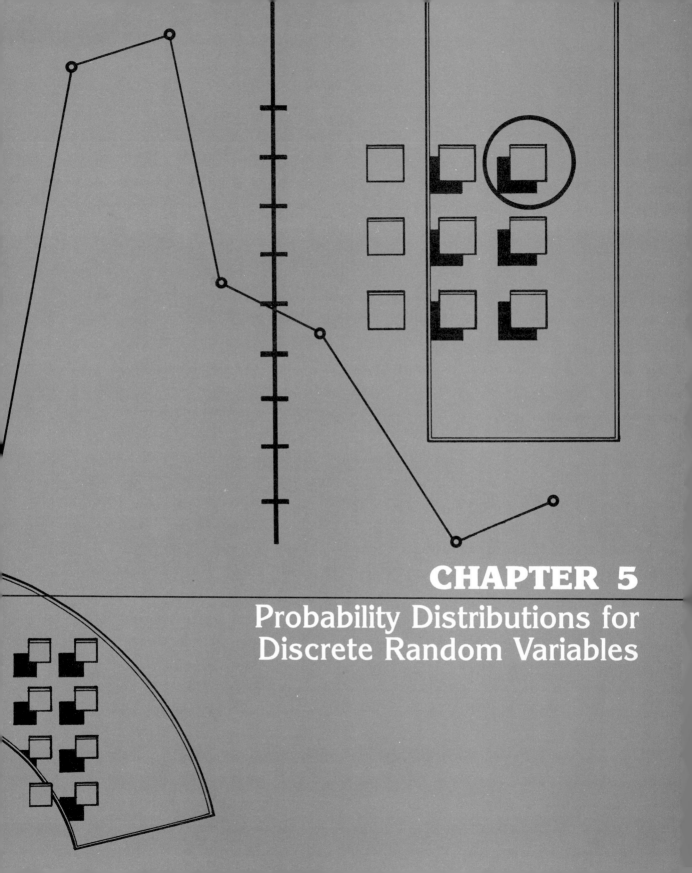

CHAPTER 5
Probability Distributions for Discrete Random Variables

5.1 INTRODUCTION

5.2 RANDOM VARIABLES

5.3 DISCRETE PROBABILITY DISTRIBUTIONS

5.4 MEAN AND VARIANCE

5.5 THE BINOMIAL PROBABILITY DISTRIBUTION

5.6 USING MINITAB (OPTIONAL)

5.7 SUMMARY
REVIEW EXERCISES
NOTES

5.1 INTRODUCTION

In Chapters 2 and 3 we were concerned with the analysis of numerical data. In Chapter 4 we studied such concepts as experiment, outcome, sample space, and probability. In this chapter we will bring these ideas together by introducing **random variables,** which describe numerical properties of the elements of a sample space.

We should also keep in focus the fact that in this book we will ultimately be interested in the study of populations. We will see that **probability distributions,** which are studied in this and the next chapter, can be used to summarize the essential features of a population of data values.

5.2 RANDOM VARIABLES

In any scientific inquiry it is helpful to translate that which is being studied into numbers. For instance, if we are interested in the fuel efficiency of various kinds of automobiles, we would look at the mileage rating (number of miles per gallon of fuel) for each kind of car. The process of assigning numbers to the objects being studied seems to be a natural one.

Given a sample space, we often study some numerical property of the various outcomes in the sample space. For example, consider the collection of adult Americans. (This is the sample space associated with the experiment of selecting an adult American at random.) A possible numerical property of interest might be the age of each person. Once we have specified the numerical property of interest, we have in essence given a rule for assigning a certain number to each element of the sample space. (Next to the name of each adult American, we assign the number that represents that person's age.) A rule for assigning such numbers is called a **random variable.**

> **Definition**   A rule that enables us to assign a number to each outcome of a sample space is called a *random variable* on the sample space. The actual number associated with a particular outcome is called the value (or data value) of the random variable associated with this outcome. When the experiment is performed, the value associated with the outcome is said to have occurred or been observed.

Example 5.1
In the 1988 presidential election a pollster was assigned to interview the oldest male and female in various households. Each was asked if he or she intended to vote for George Bush. If the male said "Yes" and the female said "No," the pollster recorded *YN*. The interview of one household can be viewed as an experiment with sample space

$$S = \{YN, NY, YY, NN\}$$

Consider the random variable "the number of yes responses" from the two voters. The possible values are 0, 1, or 2, as shown in the following table:

Outcome	*YN*	*NY*	*YY*	*NN*
Number of Yes Responses	1	1	2	0

Example 5.2
Consider the collection (sample space) of all working people in the United States. There are a number of random variables we could consider here, such as weight, height, age, salary, and so on.

We often use a symbol such as x or y or z to represent an arbitrary or unspecified value of a random variable. Thus in Example 5.1, we could have used x to represent any one of the values 0, 1, or 2. We could also use the symbol more abstractly to represent the random variable itself. Thus we could say: consider the random variable x, where $x =$ the number of yes responses.

Suppose we let $x =$ the water temperature at some randomly selected point in Lake Erie next September 1, then x could potentially be any value between freezing (32°F) and boiling (212°F). (A spot containing industrial waste might account for the possibility of extremely high temperatures.)

$$32°F < x < 212°F$$

This inequality says that x must be greater than 32°F and less than 212°F. The numbers that satisfy this inequality constitute what is called an **interval** or **line segment.** The random variable discussed here (water temperature) is called a **continuous** random variable; it can potentially assume any value on an interval (see Figure 5.1a).

On the other hand, suppose that we select three light bulbs from a production line. Let y be the random variable that counts the number of defectives. It can potentially assume any one of the values 0, 1, 2, or 3. We call y a **discrete** random variable; the values it can potentially assume constitute separated or isolated points on the real number axis. (We could get 1 or 2 defectives but not 1.5 defectives.) (See Figure 5.1b.)

> **Definitions** A random variable is *continuous* if it potentially can take on any value on some line segment or interval (there are no "breaks" between possible values). A random variable is *discrete* if the values it can potentially assume constitute a sequence of isolated or separated points on the number axis.

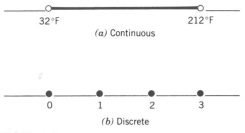

(a) Continuous

(b) Discrete

FIGURE 5.1

Continuous random variables usually measure the amount of something, whereas discrete random variables usually count something.

Here are some additional examples of continuous random variables: a person's height, the length of time to run a marathon, the mass in kilograms of a celestial object (planet, star, meteor, piece of space dust, etc.). Note in this last example that the mass of such an object could theoretically be any value greater than 0, showing that sometimes the interval of possible values of a random variable could be infinite in length.

Some additional examples of discrete random variables are the number of children in a family, the number of times a person catches a cold in a given year, and the number of tosses of a coin before a tail appears. This last example shows that sometimes the sequence of potential values of a discrete random variable can be infinite because the sequence in this example would be 1, 2, 3, ...

We will ultimately be interested in probabilities associated with various values of a random variable. A formula or table that enables us to find such probabilities is called a **probability distribution** for the random variable. We will investigate probability distributions for discrete random variables in this chapter. In Chapter 6 we will study probability distributions for continuous random variables.

Exercises

5.1   Classify the following random variables as discrete or continuous:
   (a)   The number of automobile fatalities over a 72-hour period.
   (b)   The length of time to run a 100-yard dash.
   (c)   The diameter of pebbles from a stream.
   (d)   The number of incoming phone calls to a switchboard at a city hospital.
   (e)   The barometric pressure at noontime at a specific location.

5.2   A word is randomly selected from the following sentence: "The grass needs cutting and should be cut by tomorrow afternoon." Let x be the number of letters in the selected word.
   (a)   What values can x assume?
   (b)   Is x a discrete or continuous random variable?

5.3   A green, a red, and a blue die are tossed. In each of the following, find the values that x can assume:
   (a)   Let x = the sum of the three dice.
   (b)   Let x = the minimum number of dots showing on any die.
   (c)   Let x = two times the number of dots on the green die.

5.4   One hundred voters in a suburban town are to be selected and asked whether they approve of a civic arena being built. Let x = the number of voters in the sample who approve.
   (a)   What values can x assume?
   (b)   Is x a discrete or continuous random variable?

5.5   A manufacturer of light bulbs is interested in various properties about the length of life of his product. Let x = length of life of a light bulb (measured in hours).

    (a)   What values can x assume?

    (b)   Is x a discrete or continuous random variable?

5.6   An engineer is testing samples of steel cables to be used in the construction of a suspension bridge.

    (a)   Describe a possible random variable of interest.

    (b)   Is the random variable discrete or continuous?

5.7   A health official wanted to investigate the relationship between females with no children and the incidence of breast cancer. Five hundred women, 40–45 years of age, with no children and no prior breast cancer, are selected. Let x = the number of women who develop breast cancer the following 5 years.

    (a)   What values can x assume?

    (b)   Is x a discrete or continuous random variable?

5.8   A known health effect associated with exposure to elevated levels of radon is an increased risk of developing lung cancer. The alpha track detector is a device for measuring the amount of radon in the air. The amount is measured in picocuries of radon per liter of air (pCi/l). The E.P.A. action level for radon in air is 4.0 pCi/l. An alpha track detector is placed in the cellars of 50 randomly selected homes. In each part below, classify the random variable as continuous or discrete.

    (a)   Let x = the mean pCi/l level of the 50 measurements.

    (b)   Let x = the number of homes that exceed the E.P.A. action level for radon in air.

5.3 DISCRETE PROBABILITY DISTRIBUTIONS

We said that a discrete random variable is one with the property that its possible distinct values constitute a sequence of isolated or separated points on the number axis. Consider the random variable x = the number of heads showing when a fair coin is tossed. The possible distinct values of x are 0 (if a tail) and 1 (if a head) so that x is clearly discrete. The probability that we will observe a 0 (a tail) is $\frac{1}{2}$ and the probability of observing a 1 (a head) is $\frac{1}{2}$. We sometimes write

$$P(0) = \tfrac{1}{2}, \qquad P(1) = \tfrac{1}{2}$$

The specification of the probabilities associated with the distinct values of this random variable is called its **probability distribution.** We generalize this idea with the following definition.

> **Definition**   The specification of the probabilities associated with the various distinct values of a discrete random variable is called a *discrete probability distribution*. The probability associated with the value x is denoted by the symbol $P(x)$.

If a fair coin were tossed many times, and the number of heads (0 or 1) were recorded for each toss, a population of data values would be generated. The probability distribution for x (the number of heads on a toss of the coin) provides a theoretical description of what this population should look like: Since the probability of observing a 0 is $\frac{1}{2}$ and the probability of observing a 1 is $\frac{1}{2}$, the 0's and 1's should occur in equal proportions (in the long run).

It is often convenient to think of a conceptual population of data values associated with a random variable x. This is the collection of values of x that would result if the experiment were performed many times. Sometimes we imagine the experiment being performed repeatedly for an unlimited number of times. In this case the population is viewed as infinite, because there is no reason to impose a specific limit on the number of times the experiment is performed.

Example 5.3

A student is to take a short true–false quiz consisting of two questions. Since this student has not studied for the quiz, he decides to guess at each question without even reading it. Let x = the number of correct guesses. Find the probability distribution for x.

Solution

The student's guess might be correct on the first question and wrong on the second (represented by CW), wrong on the first and correct on the second (WC), and so on. The sample space associated with the experiment of taking the quiz by guessing is

$$S = \{CW, WC, WW, CC\}$$

Since there are two choices for each question and the student guesses each time, the probability of guessing right on any question is the same as the probability of guessing wrong. It seems clear that each outcome in S is equally likely. Therefore, each has the probability $\frac{1}{4}$. The values of the random variable x are given by the following table:

Outcome	x
CW	1
WC	1
WW	0
CC	2

The value 0 will occur if *WW* occurs. Thus

$$P(0) = P(WW) = \tfrac{1}{4}$$

The value 1 will occur if either *CW* or *WC* occurs; that is, if the event {*CW*, *WC*} occurs:

$$P(1) = P(\{CW, WC\}) = \tfrac{1}{4} + \tfrac{1}{4} = \tfrac{1}{2}$$

The value 2 will occur if *CC* occurs. Therefore

$$P(2) = P(CC) = \tfrac{1}{4}$$

The probability distribution is summarized in the following table:

x	$P(x)$
0	$\tfrac{1}{4}$
1	$\tfrac{1}{2}$
2	$\tfrac{1}{4}$

where x is the number of correct guesses.

Example 5.4

A college statistics class has 20 students. The ages of these students are as follows: one student is 16 years old, four are 18, nine are 19, three are 20, two are 21, and one is 30. Let $x =$ the age of any student (randomly selected). Find the probability distribution for x.

Solution

Since each student has an equal likelihood of being selected, the probability of selecting a particular student is $\tfrac{1}{20}$. The probability of selecting a student that is, say, 19 years old is $P(19) = \tfrac{9}{20}$, since there are 9 students of that age. The probability distribution is summarized in the following table:

x	$P(x)$
16	$\tfrac{1}{20}$
18	$\tfrac{4}{20}$
19	$\tfrac{9}{20}$
20	$\tfrac{3}{20}$
21	$\tfrac{2}{20}$
30	$\tfrac{1}{20}$

In reading the preceding examples, the reader may have noted the following properties, which generalize to all discrete probability distributions.

1. Since $P(x)$ is a probability, it will always be a number between 0 and 1 inclusive.

    $$0 \leq P(x) \leq 1$$

2. The sum of the values of $P(x)$ for each distinct value of x is 1.

    $$\sum P(x) = 1$$

Another point worth noting is this: Suppose in Example 5.3 that we were interested in the probability that the number of correct guesses on the two question quiz is either 0 or 1. We write this probability: $P(x = 0 \text{ or } 1)$. Now x will be 0 for two wrong guesses, WW, and x will be 1 with the guesses CW or WC. So $x = 0$ or 1 corresponds to the event $\{WW, CW, WC\}$. Notice that

$$P(x = 0 \text{ or } 1) = P(\{WW, CW, WC\}) = \tfrac{3}{4} = \tfrac{1}{4} + \tfrac{1}{2}$$
$$= P(0) + P(1)$$

This result generalizes to any discrete distribution.*

$$P(x = a \text{ or } b) = P(a) + P(b)$$

Every discrete probability distribution must satisfy properties 1 and 2 above. Conversely, if we are given values for x and $P(x)$ such that properties 1 and 2 are satisfied, then these values define a discrete probability distribution for some random variable x. The following example illustrates this point.

Example 5.5
Consider the expression $P(x)$ defined by the equation

$$P(x) = \frac{x}{6} \qquad \text{for } x = 1, 2, 3$$

(a)   Show that $P(x)$ defines a probability distribution.

(b)   Give an example of an experiment, a sample space, and a random variable on the sample space for which this $P(x)$ is the probability distribution.

Solution
(a)   To show that $P(x)$ is a probability distribution, we need only show that the two basic properties of a distribution [(1) and (2) above] are satisfied.

---

*For those who have read Section 4.4, it should be clear that this generalization follows from the Addition Rule for mutually exclusive events.

(1) Clearly, $0 \leq \dfrac{x}{6} \leq 1$ for $x = 1, 2, 3$

(2) $\sum P(x) = P(1) + P(2) + P(3)$

$$= \tfrac{1}{6} + \tfrac{2}{6} + \tfrac{3}{6} = 1$$

Therefore, $P(x)$ is a probability distribution.

(b) Construct an experiment and sample space as follows: In a box place six balls. Label one ball with the number 1, label two balls with the number 2, and three balls with the number 3. Let the experiment be to select a ball at random. The sample space consists of the six balls. For the random variable we would let x = the number on the ball selected. It should be clear that $P(x) = x/6$ gives the probabilities associated with each value of x.

Graphic Representations

We can represent a probability distribution graphically by constructing a type of bar graph called a **probability histogram.** This is constructed by displaying the possible distinct values of the random variable along a horizontal axis. Then above each value x of the random variable we draw a vertical bar having height equal to the probability $P(x)$. We will restrict our attention to those cases where the possible values of x are whole numbers. In such cases we make the width of each bar 1.

Example 5.6
Construct a probability histogram for the distribution of Example 5.3.

Solution
The probability distribution is given by

x	$P(x)$
0	$\tfrac{1}{4}$
1	$\tfrac{1}{2}$
2	$\tfrac{1}{4}$

The probability histogram is shown in Figure 5.2.

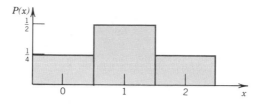

FIGURE 5.2
A Probability Histogram

Remark   In the above example, notice that the probability associated with a value x is not only the height of the bar above x, but it is also equal to the area of the bar above x, since the width of each bar is one. That is,

$$P(x) = \text{area of the bar above } x$$

This will be the case for all discrete distributions we will consider, and this seemingly insignificant observation will be of great importance in some future applications.

Exercises

5.9   An electronics firm employs 15 people whose yearly salaries (in dollars) are

25,000	25,700	25,700	26,600	27,200	27,300	27,300	27,300
27,300	27,800	27,800	28,400	28,400	29,000	29,000	

Let x be the yearly salary for any employee. Find the probability distribution for x.

5.10   Ten thousand "instant money" lottery tickets were sold. One ticket has a face value of $1000, 5 tickets have face values of $500 each, 20 tickets are worth $100 each, 500 are worth $1 each, and the rest are losers. Let x = the value of a ticket that you buy. Find the probability distribution for x.

5.11   A golfer's 18-hole scores for 25 rounds were

79	76	73	77	79	78	72	77	75	80	74	74	75
79	75	76	75	78	80	76	78	74	74	70	73	

Let x be the score. Find the probability distribution for x.

5.12   A box has five tickets numbered 1, 2, 3, 4, and 5. Two are to be randomly selected without replacement. Let x be the number of occurrences of either 4 or 5. For example, if 2 and 5 are selected, then $x = 1$. If both 4 and 5 are selected, then $x = 2$. Find the probability distribution for x.

5.13   The data below represent characteristics of the voting-age population (18 years of age and over) in the November 1988 election in the United States. Number of persons is measured in thousands. (*Source: The World Almanac and Book of Facts,* 1990, p. 339.)

	Number Voted	Number Did Not Vote
Males	47,675	36,856
Females	54,550	39,018

Find the probability distribution of x in parts (a) and (b):

(a)   A name is to be randomly selected from those who voted. Let $x = 0$ if a male is selected. Otherwise, let $x = 1$.

(b)   A name is to be randomly selected from the female population who voted. Let $x = 0$ if a female voted. Otherwise, let $x = 1$.

5.14 In a population of 3000 human beings, 1470 were classified as blood type O, 1140 as type A, 300 as type B, and the remaining as type AB. Let $x = 0, 1, 2, 3$ if a person's blood type is O, A, B, AB, respectively. Find the probability distribution for x.

In Exercises 5.15–5.19, determine whether or not each is a probability distribution. Give reasons for your answer.

5.15

x	$P(x)$
3	2
9	-1
11	-2
13	1

5.16

x	$P(x)$
1	$\frac{1}{6}$
2	$\frac{1}{6}$
3	$\frac{1}{6}$
4	$\frac{1}{6}$
5	$\frac{1}{6}$
6	$\frac{1}{6}$

5.17

x	$P(x)$
0	$\frac{2}{3}$
1	$\frac{1}{6}$
2	$\frac{1}{6}$

5.18

x	$P(x)$
-3	$\frac{1}{3}$
-1	$\frac{1}{3}$
5	$\frac{1}{3}$

5.19

x	$P(x)$
2	$\frac{1}{2}$
4	$\frac{1}{4}$
6	$\frac{1}{2}$

For the probability distributions in Exercises 5.20–5.22, construct probability histograms.

5.20 The probability distribution for the number of dots on a toss of a fair die is as follows:

x	$P(x)$
1	$\frac{1}{6}$
2	$\frac{1}{6}$
3	$\frac{1}{6}$
4	$\frac{1}{6}$
5	$\frac{1}{6}$
6	$\frac{1}{6}$

5.21 The probability distribution for the number of electronic instruments produced per hour by an assembly line is as follows:

x	$P(x)$
1	$\frac{1}{36}$
2	$\frac{3}{36}$
3	$\frac{5}{36}$
4	$\frac{7}{36}$
5	$\frac{9}{36}$
6	$\frac{11}{36}$

5.22   The probability distribution for the number of correct answers on a 10-point quiz is as follows:

x	$P(x)$
4	$\frac{4}{20}$
5	$\frac{3}{20}$
6	$\frac{2}{20}$
7	$\frac{2}{20}$
8	$\frac{2}{20}$
9	$\frac{3}{20}$
10	$\frac{4}{20}$

5.23   A business executive was sensitive to the possibility of being accused of discriminatory hiring practices due to religion. He found that of 600 employees, 204 were Catholic, 228 were Protestant, 72 were Jewish, with the rest classified as Other. Let $x = 0, 1, 2, 3$ if a person's religion is Catholic, Protestant, Jewish, or Other, respectively. Find the probability distribution and construct the probability histogram.

For Exercises 5.24–5.26, determine the probability distribution corresponding to the probability histogram.

5.24

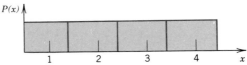

[*Hint:* Let $y = P(1) = P(2) = P(3) = P(4)$ and find y. Note that the sum of the probabilities is 1.]

5.25

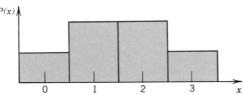

Assume that $P(0) = P(3)$, $P(1) = P(2)$, and $2P(0) = P(1)$. [*Hint:* Let $y = P(0) = P(3)$, so $2y = P(1) = P(2)$ and find y. Note that the sum of the probabilities is 1.]

5.26

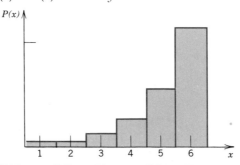

Assume that $P(1) = P(2)$, $2P(1) = P(3)$, $4P(1) = P(4)$, $8P(1) = P(5)$, $16P(1) = P(6)$. [*Hint:* Let $y = P(1)$, $2y = P(3)$, etc., and find y. Note that the sum of the probabilities is 1.]

5.4 MEAN AND VARIANCE

In Example 5.3 we discussed an experiment that consisted of a student guessing at the answers to a two-question true–false quiz. The random variable was $x =$ number of correct guesses (0, 1, or 2). The probability distribution was given by

x	$P(x)$
0	$\frac{1}{4}$
1	$\frac{1}{2}$
2	$\frac{1}{4}$

The population for this random variable would be conceived of as the string of values of x that would occur if this experiment were performed over and over by many students. The probability distribution provides a theoretical description of this population, and we ought to be able to use it to predict the mean and variance of all these values.

We will try to simulate this population by considering what the data values should theoretically look like if the experiment is performed by a large number of students, say, $N = 40{,}000$ students, and the number of correct guesses x is recorded for each student. Examining the probability distribution $P(x)$, we see that in an ideal situation we should expect 0 to occur about one-fourth of the time. Thus we expect about 10,000 zeros. Similarly, we expect about 20,000 ones and 10,000 twos.

$x =$ Number Correct	Frequency
0	10,000
1	20,000
2	10,000

Therefore, we expect the mean of all these values to be

$$\mu = \frac{\text{sum of all } x \text{ values}}{N}$$

$$= \frac{(0)(10{,}000) + (1)(20{,}000) + (2)(10{,}000)}{40{,}000}$$

$$= (0) \cdot \left(\frac{10{,}000}{40{,}000}\right) + (1) \cdot \left(\frac{20{,}000}{40{,}000}\right) + (2) \cdot \left(\frac{10{,}000}{40{,}000}\right)$$

$$= (0) \cdot (\tfrac{1}{4}) + (1) \cdot (\tfrac{1}{2}) + (2) \cdot (\tfrac{1}{4}) = 1$$

Notice that

$$(0) \cdot (\tfrac{1}{4}) = (0) \cdot P(0), \quad (1) \cdot (\tfrac{1}{2}) = (1) \cdot P(1), \quad (2) \cdot (\tfrac{1}{4}) = (2) \cdot P(2)$$

So we can write

$$\mu = 0 \cdot P(0) + 1 \cdot P(1) + 2 \cdot P(2) = 1$$

Observe that the above is the sum of the values of $x \cdot P(x)$ for each distinct value of x. We write this as

$$\mu = \sum x \cdot P(x)$$

(which works out to be 1 for the random variable under discussion).

The equation for the mean given above does not involve the population size N anywhere. In fact, no matter how large the value of N, we would be led by the above reasoning to the same formula for predicting the mean. Therefore, we will agree that the mean will be given by this formula.

We generalize the above idea in the following definition:

Definition   Given a discrete random variable x with probability distribution $P(x)$, the *mean* of the random variable x is defined as

$$\mu = \sum x \cdot P(x)$$

Note: μ is also referred to as the mean of the probability distribution $P(x)$.

We could go through a discussion very similar to the one just given to discover a suitable formula for population variance in terms of $P(x)$. However, this would be unnecessarily repetitious. Perhaps the simplest thing to do is to observe that whereas the population mean is an average of x values, the population variance is an average of values of $(x - \mu)^2$. Then the same discussion that led to a formula for the mean, μ, would lead to a similar formula for σ^2, the only difference being that instead of summing up the values of $x \cdot P(x)$, we would sum up values of $(x - \mu)^2 \cdot P(x)$. This leads to the following:

Definition   Given a discrete random variable x with probability distribution $P(x)$, the *variance* of the random variable x is defined as

$$\sigma^2 = \sum (x - \mu)^2 \cdot P(x)$$

The *standard deviation* σ is the square root of the variance. We also use the phrase variance or standard deviation of the distribution $P(x)$.

We should point out that we often do not know enough about the population to find $P(x)$. In this case, we might obtain a sample from the population and use the sample mean $\bar{x}$ to estimate μ, and s^2 to estimate σ^2.

Example 5.7

Consider the experiment of flipping a coin. Let $x =$ the number of heads showing. For this random variable calculate the mean, variance, and standard deviation.

Solution

The probability distribution is given by

x	$P(x)$
0	$\frac{1}{2}$
1	$\frac{1}{2}$

Now

$$\mu = \sum x \cdot P(x) = 0 \cdot P(0) + 1 \cdot P(1)$$
$$= (0) \cdot (\tfrac{1}{2}) + (1) \cdot (\tfrac{1}{2}) = \tfrac{1}{2}$$

By definition the variance is

$$\sigma^2 = \sum (x - \mu)^2 \cdot P(x)$$
$$= (0 - \tfrac{1}{2})^2 \cdot P(0) + (1 - \tfrac{1}{2})^2 \cdot P(1)$$
$$= (\tfrac{1}{4}) \cdot (\tfrac{1}{2}) + (\tfrac{1}{4}) \cdot (\tfrac{1}{2}) = \tfrac{1}{4}$$
$$\sigma = \sqrt{\tfrac{1}{4}} = \tfrac{1}{2}$$

There is an **alternative formula for σ^2** that is a little simpler from a computational point of view but which gives the same result. The formula is

$$\sigma^2 = \left[\sum x^2 \cdot P(x) \right] - \mu^2$$

This formula tells us to square each distinct value of x, multiply this by $P(x)$, and then add up all these products. We then subtract μ^2 from this sum. *In future work it is recommended that the alternate formula be used unless μ is a whole number.*

Example 5.8

For the random variable of Example 5.7 (number of heads showing when a coin is tossed), calculate the variance and standard deviation using the alternative formula.

Solution
We saw in Example 5.7 that $\mu = \frac{1}{2}$. Hence

$$\sigma^2 = \left[\sum x^2 \cdot P(x) \right] - \mu^2 = 0^2 \cdot P(0) + 1^2 \cdot P(1) - (\tfrac{1}{2})^2$$
$$= (0)(\tfrac{1}{2}) + (1)(\tfrac{1}{2}) - \tfrac{1}{4} = \tfrac{1}{4}$$
$$\sigma = \sqrt{\tfrac{1}{4}} = \tfrac{1}{2}$$

Sometimes it is helpful to organize our computations in the form of a table. This is especially recommended if there are more than, say, four or five values for x. We illustrate how such a table is used in the following example.

Example 5.9
Consider the following probability distribution:

x	$P(x)$
1	$\frac{4}{10}$
2	$\frac{4}{10}$
3	$\frac{1}{10}$
4	$\frac{1}{10}$

We will use Table 5.1 to find the mean, variance, and standard deviation.

TABLE 5.1

x	$P(x)$	$x \cdot P(x)$	$x^2 \cdot P(x) = x[x \cdot P(x)]$
1	$\frac{4}{10}$	$\frac{4}{10}$	$\frac{4}{10}$
2	$\frac{4}{10}$	$\frac{8}{10}$	$\frac{16}{10}$
3	$\frac{1}{10}$	$\frac{3}{10}$	$\frac{9}{10}$
4	$\frac{1}{10}$	$\frac{4}{10}$	$\frac{16}{10}$
		$\frac{19}{10}$	$\frac{45}{10}$
Totals		$\sum x \cdot P(x)$	$\sum x^2 \cdot P(x)$

$$\mu = \sum x \cdot P(x) = \tfrac{19}{10} = 1.9$$

$$\sigma^2 = \left[\sum x^2 \cdot P(x) \right] - \mu^2 = \tfrac{45}{10} - (1.9)^2 = 4.5 - 3.61 = .89$$

$$\sigma = \sqrt{.89} \doteq .94$$

Remark Note that the mean of the distribution in Example 5.9 is 1.9. This is different from the mean of the individual x values (1, 2, 3, 4), which is 2.5.

Occasionally, the mean of the distribution is the same as the mean of the individual x values. One example of when this will occur is if the distribution is symmetric. (See the distribution discussed at the beginning of this section.)

Expected Value

For the experiment of tossing a coin, suppose that you are given a dollar each time a head appears. If the coin were tossed 1000 times, ideally we would expect to see about 500 heads and 500 tails. So you would expect to win about $500 in 1000 tosses. This amounts to saying that we expect to win $\frac{1}{2}$ a dollar per toss (on the average). This is consistent with the results of Example 5.7, where we showed that when $x =$ the number of heads on a toss, the mean or average number of heads (per toss) is $\frac{1}{2}$. We often say that $\frac{1}{2}$ is the **expected value of** x. This leads to the following definition:

> **Definition** For a random variable x we define the *expected value* of the random variable, denoted by $E(x)$, to be the mean of the random variable. Therefore, for a discrete random variable,
>
> $$E(x) = \mu = \sum x \cdot P(x)$$

Example 5.10
An insurance company sells a life insurance policy with a face value of $1000 and a yearly premium of $20. If 0.1% of the policyholders can be expected to die in the course of a year, what would be the company's expected earnings per policyholder in any year?

Solution
Let $x =$ the amount of money earned by the company from an arbitrary (randomly selected) policyholder in a year. If the policyholder survives the year, then $x =$ $20. If the policyholder dies, the company must pay out $1000. This minus the $20 premium means that the company loses $980. In other words, on this policyholder the company earns -980, (i.e., $x = \$ - 980$). The probability that the policyholder dies (i.e, $x = -980$), is .001. So the probability that the policyholder lives (i.e., $x = 20$), is .999. Therefore, the probability distribution is

x	$P(x)$
20	.999
-980	.001

The expected earnings per policyholder are

$$E(x) = \sum x \cdot P(x)$$
$$= (20)(.999) + (-980)(.001)$$
$$= 19.98 - .98 = 19$$

The company can expect to earn \$19 per policyholder (on the average).

Example 5.11

Assume that the insurance company of the previous example wishes to market a \$1000 life insurance policy to a high-risk group, where the yearly mortality rate is 0.5%. What premium should the company charge in order to earn the same amount per policyholder as in the previous example, that is, \$19 per year?

Solution

Let b = yearly premium and let x = amount earned in a year from an arbitrary policyholder. If the policyholder survives the year, then the company earns $x = b$. If he dies, then the company pays out \$1000. Since the company received the premium b, this means that its net loss is $1000 - b$. In other words, the company earns $x = -(1000 - b) = b - 1000$. The probability that $x = b$, that is, the policyholder survives, is .995. The probability that $x = b - 1000$ (the policyholder does not survive) is .005. That is, the probability distribution of x is

x	$P(x)$
b	.995
$b - 1000$	.005

We want to know what value b must be so that $E(x) = 19$. Now

$$E(x) = \sum x \cdot P(x)$$
$$= (b)(.995) + (b - 1000)(.005)$$
$$= .995b + .005b - 5$$
$$= b - 5$$

But we want this to be 19,

$$b - 5 = 19$$
$$b = 24$$

Hence, if the company charges a \$24 yearly premium, it will earn \$19 per policy (on the average).

Example 5.12 (the Numbers Game*)

This game consists of selecting a three-digit number. If you guess the right number, you are paid $700 for each dollar you bet. Each day there is a new winning number. If a person bets $1 each day for 1 year, how much money can he expect to win (or lose)?

Solution

Let x = the amount won on a given day. If the correct number is guessed, then he wins $700 − $1 = $699. (Don't forget, it costs $1 to bet.) Otherwise he loses, so his winnings in this case would be $−1. There are 1000 possible three-digit numbers: 000 to 999. The probability of winning (selecting the correct three digits) is $\frac{1}{1000}$ = .001. The probability of losing is $\frac{999}{1000}$ = .999. The probability distribution is

x	$P(x)$
−1	.999
699	.001

Now we find the *expected daily winnings*. This is just $E(x)$.

$$E(x) = \sum x \cdot P(x) = (-1)(.999) + (699)(.001)$$
$$= -.999 + .699 = -\$0.30$$

Thus he can expect to lose 30 cents per day (on the average); for the year he could expect to lose $(.30)(365) = \$109.50$.

Example 5.13

An automobile dealer stocks two models of a certain make of car. Model A costs $6000 and model B costs $10,000. Of all the customers expressing interest in buying a car, 6% buy model A, 4% buy model B, and the rest (90%) fail to make a purchase.

(a) Find the expected earnings (sales) per customer.

(b) Assume that the dealer decides to sell only one model, and that the pool of prospective customers expressing interest will be the same in number regardless of which model is stocked. However, 6% will buy model A, while only 4% will buy model B. Assume, further, that it has been determined that the costs per customer would be $100 for model A and $150 for model B. (The costs are more for model B because of advertising, sales staff, auto preparation costs, etc.). Which model should be stocked?

*For a discussion of this and other casino games, see Koshy, T., *Finite Mathematics and Calculus with Applications* (Santa Monica, Calif.: Goodyear, 1979), Section 4.2.

Solution

(a) Let x = earnings (sales) for an arbitrary customer. The values of x could be $0 for no purchase, $6000 for model A, and $10,000 for model B. The probability distribution for x is

x	$P(x)$
0	.90
6,000	.06
10,000	.04

The expected earnings per customer would therefore be

$$E(x) = \sum x \cdot P(x) = (0)(.90) + (6000)(.06) + (10,000)(.04)$$
$$= 0 + 360 + 400 = \$760$$

(b) To determine which model the dealer should stock, we should compare the expected profit per customer for each model. (Keep in mind that we are assuming that the total pool of customers is the same in either case.) For model A: let v = profit on an arbitrary customer when model A is stocked. The possible values of v are -100 (no sale) or $6000 - 100 = \$5900$. Now 6% will buy model A and 94% will not. Therefore, the probability distribution for profit on model A is

v	$P(v)$
-100	.94
5900	.06

The expected value of v is

$$E(v) = \sum v \cdot P(v) = (-100)(.94) + (5900)(.06)$$
$$= -94 + 354 = \$260$$

For model B: Let w = the profit on an arbitrary customer when model B is stocked. The possible values of w are -150 or $10,000 - 150 = \$9850$; 4% will buy model B, 96% will not. Therefore, the probability distribution for profit on model B is

w	$P(w)$
-150	.96
9850	.04

The expected value of w is

$$E(w) = \sum w \cdot P(w) = (-150)(.96) + (9850)(.04)$$
$$= -144 + 394 = \$250$$

The dealer can expect profits of $260 per customer for model A and $250 per customer for model B. Therefore, model A would be the one to stock.

Exercises

For the probability distributions in Exercises 5.27–5.30, find (a) the mean, (b) the variance, and (c) the standard deviation. Also, (d) describe each of the distributions as skewed to the left, skewed to the right, or symmetric. (To organize your computations, it may help to use a table as in Example 5.9.)

5.27 x	$P(x)$	5.28 x	$P(x)$	5.29 x	$P(x)$	5.30 x	$P(x)$
0	$\frac{1}{10}$	0	$\frac{4}{10}$	0	$\frac{1}{8}$	2	$\frac{1}{4}$
1	$\frac{2}{10}$	1	$\frac{3}{10}$	1	$\frac{3}{8}$	3	$\frac{1}{4}$
2	$\frac{3}{10}$	2	$\frac{2}{10}$	2	$\frac{3}{8}$	4	$\frac{1}{4}$
3	$\frac{4}{10}$	3	$\frac{1}{10}$	3	$\frac{1}{8}$	5	$\frac{1}{4}$

5.31   The probability distribution for x, the sum of the dots when two ordinary dice are tossed, is given below. Find (a) the mean, (b) the variance, and (c) the standard deviation.

x	$P(x)$
2	$\frac{1}{36}$
3	$\frac{2}{36}$
4	$\frac{3}{36}$
5	$\frac{4}{36}$
6	$\frac{5}{36}$
7	$\frac{6}{36}$
8	$\frac{5}{36}$
9	$\frac{4}{36}$
10	$\frac{3}{36}$
11	$\frac{2}{36}$
12	$\frac{1}{36}$

5.32   A dentist has determined that the number of patients x treated in an hour is described by the probability distribution given below. Find (a) the mean, (b) the variance, and (c) the standard deviation.

x	$P(x)$
1	$\frac{2}{15}$
2	$\frac{10}{15}$
3	$\frac{2}{15}$
4	$\frac{1}{15}$

5.33   The manager of a baseball team has determined that the number of walks x issued in a game by one of the pitchers is described by the probability distribution given below. Find (a) the mean, (b) the variance, and (c) the standard deviation.

x	$P(x)$
0	$\frac{1}{20}$
1	$\frac{2}{20}$
2	$\frac{3}{20}$
3	$\frac{11}{20}$
4	$\frac{3}{20}$

5.34   An altered die has 1 dot on one face, 2 dots on three faces, and 3 dots on two faces. The die is to be tossed once. Let x be the number of dots on the upturned face. Find the mean and variance of x.

5.35   Appendix Table A describes how to select random numbers from Appendix Table B.1 (Random Numbers). Let x = a randomly selected single digit.

(a)   Assume the probability distribution of x is as follows:

$$P(x) = 1/10 \text{ for } x = 0, 1, 2, 3, 4, 5, 6, 7, 8, 9$$

(i)   Find the mean of x.   **(ii)**   Find the standard deviation of x.

(b)   Randomly select 25 digits from Appendix Table B.1. Calculate the sample mean and the sample standard deviation. Compare these sample results with the population mean and standard deviation in part (a).

5.36   A card is to be selected from an ordinary deck of 52 cards. Suppose that a casino will pay $10 if you select an ace. If you fail to select an ace, you are required to pay the casino $1.

(a)   If you play this game once, how much money does the casino expect to win?

(b)   If you play the game 26 times, how much money does the casino expect to win?

5.37   In the game of craps a player rolls two dice. If the first roll results in a sum of 7 or 11, the player wins. If the first roll results in a 2, 3, or 12, the player loses. If the sum on the first roll is 4, 5, 6, 8, 9, or 10, the player keeps rolling until he throws a 7 or the original value. If the outcome is a 7, the player loses. If it is the original value, the player wins. The probability that a player will win is .493. Suppose that a player pays $5 to a casino if he loses and is paid $4 for a win. What is the expected loss for the player if he plays (a) one game? (b) ten games?

5.38   A bus company is interested in two potential contracts, one for an express and the other for local stops. The probabilities that the bids will be accepted are 0.70 and 0.50 with costs of $500 and $750, respectively. The estimated total incomes are $6000 and $10,000, respectively. If the company were allowed only one bid, which bid should they enter?

5.39   A high school class decides to raise some money by conducting a raffle. They plan to sell 2000 tickets at $1 apiece. They will give one prize of $100, two prizes of $50, and three prizes of $25. If you plan on purchasing one ticket, what are your expected net winnings? (*Hint:* The probability of getting the $100 ticket is $\frac{1}{2000}$, of getting a $50 ticket is $\frac{2}{2000}$, and of getting a $25 ticket is $\frac{3}{2000}$.)

5.40   Megabucks, a lottery game conducted by the Massachusetts State Lottery Commission, consists of selecting 6 numbers from the 36 numbers, {1, 2, 3, 4, ..., 35, 36} (no repetitions and order doesn't matter). The commission selects by chance the 6 winning numbers. You pay $1 to play. You get a free ticket if 3 of your numbers match 3 of the winning numbers, $40 for matching 4 numbers, $400 for matching 5 numbers. Assume you get $1,000,000 if you match all 6 numbers. The sample space S consist of all possible 6-number combinations that can be selected. It can be shown that $n(S) = 1,947,792$. The number of elements in S that match 0, 1, 2, 3, 4, 5, or 6 winning numbers is given below. What are your expected winnings?

Number of Matches	Number of Possibilities
0	593,775
1	855,036
2	411,075
3	81,200
4	6,525
5	180
6	1

5.5 THE BINOMIAL PROBABILITY DISTRIBUTION

Consider the experiment of rolling a single die and recording whether the number (of dots) on the top face is odd (O) or even (E). This experiment has two possible outcomes: O or E. We can construct a new experiment by performing this basic experiment over and over a certain number of times. (In a situation like this, the basic experiment is called a *trial*.) For example, suppose that we perform this basic experiment three times. If we observed an even number, then an odd, and then an even, we would represent this outcome by EOE. The sample space for the new experiment is

$$\{OOO, EOO, OEO, OOE, OEE, EOE, EEO, EEE\}$$

The above experiment consisted of repeating a trial a certain number of times; each trial had two possible outcomes. Such an experiment is called a **binomial experiment.**

There are many practical situations that are in essence binomial experiments. For example, we may interview 10,000 voters to see how many favor candidate A. Here the trial is to interview a voter. This trial has two possible outcomes: Voter favors A or does not favor A. The trial is repeated 10,000 times. As another example, a manufacturer of transistors may select 1000 transistors, then repeatedly perform the trial of checking each transistor to see if it is defective or not.

We summarize these ideas in the following definition.

Definition   A *binomial experiment* is an experiment that has the following properties:

1. It consists of performing some basic experiment a fixed number of times, n. Each time the basic experiment is performed, we call this a *trial*.
2. Each trial is identical and has two possible outcomes. We arbitrarily call one outcome "success" (S) and the other "failure" (F). We call the probability of success $P(S) = p$ and the probability of failure $P(F) = q$. Note that $p + q = 1$. (So $q = 1 - p$.) The values of p and q do not change from one trial to another.
3. The trials are independent of one another, that is, the outcome of one trial has no bearing on the outcome of any other trial.

The random variable of interest, the *binomial random variable x*, is the number of successes in n trials. Note that x can assume the values 0, 1, 2, ... , n. The probability distribution of x is called a *binomial probability distribution*.

Example 5.14

Consider the binomial experiment of rolling a die three times. Each time we record whether the number of dots showing is odd (O) or even (E). Let $x =$ the total number of evens recorded. Find the binomial probability distribution $P(x)$.

Solution

Here the trial is to roll the die and record O or E. We will call E a success. So $S = E$ and failure $F = O$. Since half the numbers on a die are even and half are odd, $p = P(S) = \frac{1}{2}$ and $q = P(F) = \frac{1}{2}$. We have seen that the sample space is

$$\{OOO, EOO, OEO, OOE, OEE, EOE, EEO, EEE\}$$

Each outcome is equally likely and therefore has the probability $\frac{1}{8}$. The binomial random variable is $x =$ the number of successes $=$ the number of evens. We must have $x =$ 0, 1, 2, or 3.

Now $x = 0$ corresponds to OOO, so $P(0) = P(OOO) = \frac{1}{8}$. The value $x = 1$ corresponds to the event $\{EOO, OEO, OOE\}$. Then

$$P(1) = P(\{EOO, OEO, OOE\}) = \frac{3}{8}$$

The value $x = 2$ corresponds to the event $\{OEE, EOE, EEO\}$. Therefore,

$$P(2) = P(\{OEE, EOE, EEO\}) = \frac{3}{8}$$

Finally, the value $x = 3$ corresponds to *EEE*, so that $P(3) = P(EEE) = \frac{1}{8}$. Summarizing, we find that the binomial probability distribution for this situation is

x	$P(x)$
0	$\frac{1}{8}$
1	$\frac{3}{8}$
2	$\frac{3}{8}$
3	$\frac{1}{8}$

The essential mathematical ingredients of a binomial experiment are the number of trials n, the probability of success on an individual trial p, and the number of successes x. It would be convenient if we had a formula for the binomial distribution, that is, a formula for $P(x)$ using the above mathematical ingredients. Such a formula does exist.

Given a binomial experiment consisting of n trials where the probability of success on an individual trial is p and the probability of failure is q (where $q = 1 - p$), it can be shown that the probability of exactly x successes in n trials is given by the formula*

$$P(x) = \frac{n!}{x!(n-x)!} \cdot p^x \cdot q^{n-x}$$

---

*For those who have read Sections 4.5 and 4.7, it is not difficult to see why this formula works. First consider the event consisting of success on the first x trials and failure on the next $n - x$ trials:

$$\underbrace{SS\cdots S}_{x}\underbrace{FF\cdots F}_{n-x} \quad \leftarrow n \text{ trials}$$

The trials are independent. Hence the Multiplication Rule for independent events tells us the probability of this is

$$\underbrace{p\cdot p\cdots p}_{x}\underbrace{\cdot q\cdot q\cdots q}_{n-x} = p^x \cdot q^{n-x}$$

Now there are a number of different arrangements that can result in x successes and $n - x$ failures, all having the same probability, $p^x q^{n-x}$. To count how many ways this can happen, just calculate how many ways we can specify which x trials out of the n trials are the successes. In Section 4.7 we saw that this number is

$$_nC_x = \frac{n!}{x!(n-x)!}$$

So there are this many ways of getting x successes and $n - x$ failures, all having probability $p^x q^{n-x}$. Hence, the probability of x successes in n trials is

$$P(x) = \frac{n!}{x!(n-x)!} \cdot p^x \cdot q^{n-x}$$

for $x = 0, 1, 2, \ldots, n$. The symbol $n!$ (read "n factorial") is defined for any positive integer to be

$$n! = 1 \cdot 2 \cdot 3 \cdots n$$

Therefore

$$3! = 1 \cdot 2 \cdot 3 = 6$$
$$5! = 1 \cdot 2 \cdot 3 \cdot 4 \cdot 5 = 120$$
$$1! = 1$$

We will agree that $0! = 1$.

The values of $P(x)$ constitute the **binomial probability distribution.** The word *binomial* is used because the values of $P(x)$ for the various values of x are precisely the terms in the binomial expansion of $(p + q)^n$. For example, when $n = 2$, the values of $P(x)$ for $x = 0, 1, 2$ are the three terms on the right-hand side of the equation

$$(p + q)^2 = p^2 + 2pq + q^2$$

In Figure 5.3 we display probability histograms for two binomial probability distributions.

Example 5.15

Verify that the above formula for $P(x)$ does give the probability distribution for the binomial variable of Example 5.14.

Solution

For this experiment $n = 3$. Also, success = record even (E); failure = record odd (O). Thus $p = P(E) = \frac{1}{2}$ and $q = P(O) = 1 - p = \frac{1}{2}$. Finally, x = number of evens recorded in three rolls of the die. Therefore,

$$P(x) = \frac{3!}{x!(3 - x)!} \cdot \left(\tfrac{1}{2}\right)^x \cdot \left(\tfrac{1}{2}\right)^{3-x}$$

Observe that

$$\left(\tfrac{1}{2}\right)^x \cdot \left(\tfrac{1}{2}\right)^{3-x} = \left(\tfrac{1}{2}\right)^{x+3-x} = \left(\tfrac{1}{2}\right)^3 = \tfrac{1}{8}$$

and so

$$P(x) = \frac{3!}{x!(3 - x)!} \cdot \tfrac{1}{8}$$

Now

$$P(0) = \frac{3!}{0!(3 - 0)!} \cdot \tfrac{1}{8} = \frac{6}{(1) \cdot (6)} \cdot \tfrac{1}{8} = \tfrac{1}{8}$$

$$P(1) = \frac{3!}{1!(3 - 1)!} \cdot \tfrac{1}{8} = \frac{6}{(1) \cdot (2)} \cdot \tfrac{1}{8} = \tfrac{3}{8}$$

$$P(2) = \frac{3!}{2!(3 - 2)!} \cdot \tfrac{1}{8} = \frac{6}{(2) \cdot (1)} \cdot \tfrac{1}{8} = \tfrac{3}{8}$$

$$P(3) = \frac{3!}{3!(3-3)!} \cdot \tfrac{1}{8} = \frac{6}{(6) \cdot (1)} \cdot \tfrac{1}{8} = \tfrac{1}{8}$$

These results agree with the results of Example 5.14.

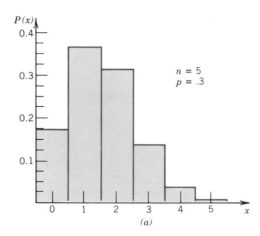

$n = 5$
$p = .3$

(a)

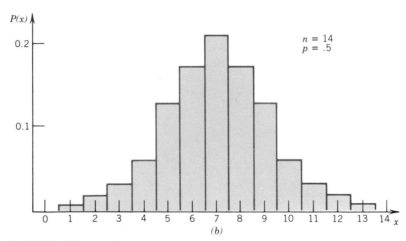

$n = 14$
$p = .5$

(b)

FIGURE 5.3
Binomial Probability Distributions

Example 5.16

Assume that when a certain hunter shoots at a pheasant, the probability of hitting it is .6. Find the probability that the hunter

(a) Will hit four out of the next five pheasants at which he shoots.

(b) Will hit at least four out of the next five.

(c) Will hit at least one of the next five.

Solution

(a)

$$\text{trial} = \text{shoots at a pheasant}$$
$$\text{success} = \text{hits the pheasant}$$
$$\text{failure} = \text{does not hit the pheasant}$$
$$P(S) = p = .6$$
$$P(F) = q = 1 - p = .4$$

The trial is repeated five times; so $n = 5$. The binomial distribution giving the probability of exactly x successes in five trials is

$$P(x) = \frac{5!}{x!(5-x)!} \cdot (.6)^x \cdot (.4)^{5-x}$$

The probability of hitting exactly four of the next five pheasants is

$$P(4) = \frac{5!}{4!(5-4)!} \cdot (.6)^4 \cdot (.4)^{5-4}$$
$$= \frac{1 \cdot 2 \cdot 3 \cdot 4 \cdot 5}{(1 \cdot 2 \cdot 3 \cdot 4) \cdot (1)} \cdot (.6)^4 \cdot (.4) = (5)(.6)^4(.4)$$
$$= .25920$$

(b) Now the probability that he hits at least four of the next five pheasants is $P(x = 4 \text{ or } 5)$. But we know that

$$P(x = 4 \text{ or } 5) = P(4) + P(5)$$
$$= .25920 + P(5)$$

and

$$P(5) = \frac{5!}{5!(5-5)!} \cdot (.6)^5 \cdot (.4)^{5-5}$$
$$= \frac{5!}{5!0!} \cdot (.6)^5 \cdot (.4)^0$$

Keeping in mind that $0! = 1$ and $(.4)^0 = 1$, we get

$$P(5) = (.6)^5 = .07776$$

Thus

$$P(x = 4 \text{ or } 5) = P(4) + P(5)$$
$$= .25920 + .07776 = .33696$$

(c) Now we find the probability that he hits at least one out of the next five pheasants. Call A the event consisting of *no* hits in the next 5 shots. Then the complement of this event, $\overline{A}$, consists of hitting at least one. Since $P(A) + P(\overline{A}) = 1$, we have

$$P(\text{hitting at least 1}) = P(\overline{A}) = 1 - P(A) = 1 - P(0)$$

Now we need only evaluate $P(0)$ and subtract from 1.

$$P(0) = \frac{5!}{0!(5 - 0)!} \cdot (.6)^0 \cdot (.4)^{5-0}$$

$$= \frac{5!}{5!} \cdot (1) \cdot (.4)^5 = (.4)^5$$

$$= .01024$$

Therefore,

$$P(\text{hitting at least one}) = 1 - P(0)$$
$$= 1 - .01024$$
$$= .98976$$

The formula for calculating binomial probabilities $P(x)$ can be rather unpleasant to use for all but very small values of n. Appendix Table B.2 gives values of $P(x)$ for various values of n and x up to $n = 25$. [In Chapter 6 we will develop a method of finding $P(x)$ for values of $n > 25$.]

To see how to use Appendix Table B.2, let us use it to find $P(x)$ when $n = 5$, $p = .60$, and $x = 4$. In Example 5.16, we have seen that this value is .259 rounded to three decimal places.

In the far left column, locate the desired value of n, in this case $n = 5$. To the right and below this, locate the row containing the desired value of x, here $x = 4$. Across the top of the table locate the column containing the desired value of p, here $p = .60$. Now where the row containing $x = 4$ and the column containing $p = .60$ intersect, we find the value of $P(4)$, namely, .259. (See Table 5.2, which shows a portion of Appendix Table B.2.)

Notice that some of the probabilities in Appendix Table B.2 are represented as $0+$. This means that the probability is so small as to be negligible, and hence may be assigned the value 0 in calculations.

TABLE 5.2

n	x	.01	.05	.10	.20	.30	.40	.50	.60	.70	.80	.90	.95	.99
.	.								.					
.	.								.					
.	.								.					
5	0								.010					
	1								.077					
	2								.230					
	3								.346					
	4								.259					
	5								.078					

Example 5.17

Thirty percent of the voters in a large voting district are veterans. If 10 voters are randomly selected, find the probability that less than half will be veterans.

Solution

This is a binomial experiment.

$$\text{trial} = \text{select voter}$$

$$\text{success} = \text{voter is a veteran}$$

$$P(S) = p = \text{proportion of voters in district who are veterans} = .30$$

$$n = 10 \text{ (repeat the trial 10 times)}$$

$$x = \text{number of veterans}$$

We want to find $P(x < 5)$. Using Appendix Table B.2, we get

$$P(x < 5) = P(0) + P(1) + P(2) + P(3) + P(4)$$

$$= .028 + .121 + .233 + .267 + .200$$

$$= .849$$

Note: Since the voting district is large, we can safely assume that the value of p remains essentially the same from one trial to the next.

Example 5.18

A company that manufactures color television sets claims that only 5% of its sets will need to be adjusted by a technician before being sold. If an appliance dealer sells 20 of these sets, how likely is it that more than 3 of them will need to be adjusted?

Solution

$$\text{trial} = \text{check one of the sets}$$

$$\text{success} = \text{it needs adjustment}$$

$$P(S) = p = .05$$

$$n = 20 \text{ (repeat the trial 20 times)}$$

$$x = \text{number out of 20 that need adjustment}$$

We want to find $P(x > 3)$. Using Appendix Table B.2, add the values of $P(x)$ for x from 4 to 20.

$$P(x > 3) = P(4) + P(5) + \cdots + P(19) + P(20)$$

$$= .013 + .002 + \text{(negligible terms)}$$

$$= .015$$

We conclude that if the manufacturer is right, it is very unlikely that more than 3 sets will need to be adjusted.

The above example provides some insight into how the binomial distribution might be used in statistical inference. For example, the manager of an appliance store chain received a large shipment of these T.V. sets, and she wished to test the manufacturer's claim that only 5% (or perhaps less) will need adjustment. She might randomly select 20 sets from the shipment and determine how many of these sets need adjustment. We can view the population as consisting of the entire shipment of T.V. sets from the manufacturer. The 20 sets then constitute a sample from the population. Based on the number of sets out of the 20 that are in need of adjustment, she will decide if there is strong evidence to suggest that more than 5% of the sets in the population will be in need of adjustment. She would want the evidence to be strong because if she concludes that more than 5% of the population need adjusting, she will return the entire shipment— a drastic step. What criterion might she use? That is, if x = the number of sets out of the 20 needing adjustment, how large should x be in order for the manager to conclude that the shipment should be returned?

Since the manufacturer claims 5% (or less) will need adjusting, it would not be alarming if we found 5% of the 20, or 1 set, needing adjusting. Perhaps 2 sets would not even be alarming. (After all, we can expect some fluctuation from one batch of 20 sets to the next.) But the larger the value of x, the more we are inclined to feel that more than 5% of the population will need adjusting. If the value of x is large enough (relative to 1), the manager would reject the manufacturer's claim (and the shipment as well). By "large enough" we mean so large that it would be highly unlikely that we would observe such a large value of x if the manufacturer's claim were true. Note that we have already seen a range of such unlikely values in Example 5.18. We said that it would be very unlikely (only a 1.5% chance) that we would get a value of x greater than 3 if in fact the manufacturer were correct. Therefore, the manager could use this as her criterion: If more than three of the sets do need adjusting, the evidence would strongly suggest that the manufacturer's claim should be rejected.

We will see in later chapters that a binomial experiment is a very important concept in a variety of statistical inference problems ranging from industrial quality control to voter preference studies. The reason for this is that *the proportion of elements in a population possessing a certain characteristic of interest may be viewed as the probability of success in a binomial experiment.* In Example 5.17 we discussed the proportion of voters in a district who were veterans. This proportion was .30. A trial consisted of selecting a voter (by chance). A success occurs if the voter selected is a veteran. The binomial experiment consisted of performing the trial 10 times. Since 30% of the voters are veterans, the probability that a voter selected on a single trial will be a veteran is .30. This is the probability of success, so that $P(S) = p = .30$. Sometimes such proportions are unknown and we can use a binomial variable to study these proportions. We will investigate these proportions, known as **population proportions,** in Chapter 7.

Mean and Variance of a Binomial Random Variable

Suppose that we consider the binomial experiment of tossing a fair coin $n = 100$ times. Let success = heads. Clearly, $p = .5$. Let $x = $ the number of heads. When the coin is tossed 100 times, how many heads would you expect to get, that is, what is your best guess? We would guess that half (or 50) would be heads. In other words, the expected value of x is 50. This can be obtained from the formula

$$E(x) = n \cdot p = 100 \cdot (\tfrac{1}{2}) = 50$$

This formula, in fact, works in general. That is, it can be proved rigorously using the formula for μ given earlier in this chapter, that **for a binomial experiment** consisting of n trials with the probability of success p, the mean or expected value of x is

$$\mu = E(x) = n \cdot p$$

It can also be proved using the formula for variance given earlier in the chapter that for a binomial variable

$$\sigma^2 = n \cdot p \cdot q$$

Thus, for the number of heads when a coin is tossed 100 times

$$\sigma^2 = n \cdot p \cdot q = (100)\left(\frac{1}{2}\right)\left(\frac{1}{2}\right) = 25$$

$$\sigma = \sqrt{25} = 5$$

Exercises

5.41   Which of the following are binomial experiments? For those that are not, indicate which part of the definition of a binomial experiment does not apply.

(a)   Tossing a fair coin 1000 times and counting the number of times a head appears.

(b)   Tossing a fair coin and counting the number of tosses before a head appears.

(c)   Checking 5 students for drug use from a class of 30 students in which 10% use drugs.

(d)   A state agency randomly selects 20 liquor stores, with replacement, and counts the number of stores involved in price fixing.

(e)   A shipment of 40 appliances contains 2 defectives. A dealer tests 15 of the appliances and counts the number that fail to meet specifications.

5.42 Consider a binomial experiment with $n = 4$. Construct probability histograms when

  (a) $p = .3$ and $p = .7$. Comment on the relationship between the two.

  (b) $p = .4$ and $p = .6$. Comment on the relationship between the two.

5.43 Consider a binomial experiment with $p = .5$. Construct probability histograms when $n = 3$ and $n = 5$. Comment on the shapes of the distributions.

5.44 Consider a binomial experiment with $n = 3$. Construct probability histograms when $p = .2$ and $p = .4$. You will see that each distribution is skewed to the right. Comment on the relative amount of skewness between them.

5.45 Consider a binomial experiment with $n = 4$, $p = .6$, and x the number of successes. Use the formula for $P(x)$ to find the probability that

  (a) $x = 0$. **(b)** $x = 1$. **(c)** $x = 2$.

5.46 Consider a binomial experiment with $n = 5$, $p = .7$, and x the number of successes. Use the formula for $P(x)$ to find the probability that

  (a) $x = 3$. **(b)** $x = 4$. **(c)** $x = 5$.

5.47 Consider a binomial experiment with $n = 11$, $p = .4$, and x the number of successes. Use Appendix Table B.2 to find the probability that

  (a) x is less than 2. **(b)** x is greater than 5 and less than 8.

  (c) x is greater than or equal to 5 and less than or equal to 8.

  (d) x equals 6. **(e)** x is greater than 0.

5.48 Consider a binomial experiment with $n = 9$, $p = .7$, and x the number of successes. Use Appendix Table B.2 to find the probability that

  (a) x is less than 3.

  (b) x equals 3.

  (c) x is greater than 3 and less than 5.

  (d) x is greater than or equal to 3 and less than or equal to 5.

  (e) x is less than 9.

5.49 A retailer decides that he will reject a large shipment of light bulbs if there is more than 1 defective bulb in a sample of size 10. If the defective rate is .10, what is the probability that the retailer will reject the shipment?

5.50 A screening examination is required of all applicants applying for a technical writing position. The examination consists of 16 questions. Each question has 5 choices, consisting of the correct answer and 4 incorrect answers. A curious applicant wonders about some probabilities if she were to randomly guess at each question.

  (a) What is the probability of getting 3 correct?

  (b) What is the probability of getting at least 2 correct?

  (c) If 50 applicants took the exam and each guessed randomly at all questions, what would you guess the mean number of correct answers to be?

5.51 Sixty percent of the voters in a large town are opposed to a proposed development. If 20 voters are selected at random, find the probability that

  (a) 10 are opposed to the proposed development.

  (b) more than 13 are opposed to the proposed development.

  (c) less than 10 are opposed to the proposed development.

5.52 A person has a 5% chance of winning a free ticket in a state lottery. If she plays the game 25 times, what is the probability she will win a free ticket at least once?

5.53 Forty percent of the student body at a large university are in favor of a ban on drinking in the dormitories. Suppose 15 students are to be randomly selected. Find the probability that

(a) 7 favor the ban.

(b) less than 4 favor the ban.

(c) more than 2 favor the ban.

5.54 Ten percent of the cans of handballs purchased at a sporting goods store are unacceptable. If 12 cans are purchased, find the probability that

(a) all of the cans are acceptable.

(b) more than 2 cans are unacceptable.

Suppose 200 cans were bought. What is the expected number of unacceptable cans?

In Exercises 5.55–5.58, find (a) the mean, (b) the variance, and (c) the standard deviation. Assume a binomial probability distribution in each case.

5.55 $n = 20, p = \frac{1}{5}$.   5.56 $n = 20, p = \frac{4}{5}$.

5.57 $n = 12, p = \frac{1}{3}$.   5.58 $n = 10, q = \frac{1}{10}$.

5.59 There are 4,230,000 eligible voters in a state, of which 147,000 are black. A sample of 100 voters is to be selected. Let x be the number of blacks. (Assume a binomial experiment.) Find (a) the mean, (b) the variance, and (c) the standard deviation of x.

5.60 Seven percent of printed circuit boards made by the SKC company are defective. A company official wishes to see if the percentage of defective boards in a current batch has decreased. A sample of size 50 is to be selected. Let x be the number of defectives. (Assume a binomial experiment.) If there is no change from the 7% figure, find (a) the mean, (b) the variance, and (c) the standard deviation of x.

5.61 Between 1972 and 1974, 15 out of 405 teachers (3.7%) hired in Hazelwood, St. Louis County, Mo., were black. In St. Louis County *plus* the nearby city of St. Louis, 15.4% of teachers were black. The Equal Employment Opportunity Commission (EEOC) sued Hazelwood for discrimination against blacks and won the case in the court of appeals. Assuming that the 405 teachers hired constitute a random sample from a population of teachers, 15.4% of whom are black,

(a) What is the mean (or expected) number of black teachers?

(b) What is the standard deviation?

(c) In the case of *Castenada* v. *Partida*, 430 U.S. 482 (1977), the "Standard Deviation Rule" was advanced. This states that if the observed value (15 teachers in the Hazelwood case) is more than 2 or 3 standard deviations below the expected value, discrimination may be present. How many standard deviations below the expected value is the observed value of 15?

(d) On appeal to the Supreme Court, the decision of the court of appeals was vacated. The Supreme Court noted that the relevant job market for teachers might well be St. Louis County alone (which does not include the city of St. Louis), where the percentage of black teachers is 5.7%. Redo parts (a), (b), and (c) using this percentage.

(e) The above analysis assumes that the selection of the 405 teachers is a random process (with respect to race). Can the selection of teachers be considered a random process?

For a discussion see DeGroot, M., S. Fienberg, and J. Kadane, *Statistics and the Law* (New York: Wiley, 1986), pp. 1–48.

5.62 Suppose that x is binomial with a probability distribution as follows:

x	$P(x)$
0	$(\frac{1}{3})^5$
1	$5(\frac{1}{3})^4(\frac{2}{3})$
2	$10(\frac{1}{3})^3(\frac{2}{3})^2$
3	$10(\frac{1}{3})^2(\frac{2}{3})^3$
4	$5(\frac{1}{3})(\frac{2}{3})^4$
5	$(\frac{2}{3})^5$

Find (a) the mean, (b) the variance, and (c) the standard deviation of x.

5.63 Suppose that x is binomial with a probability distribution as follows:

x	$P(x)$
0	$(\frac{1}{6})^4$
1	$4(\frac{1}{6})^3(\frac{5}{6})$
2	$6(\frac{1}{6})^2(\frac{5}{6})^2$
3	$4(\frac{1}{6})(\frac{5}{6})^3$
4	$(\frac{5}{6})^4$

Find (a) the mean, (b) the variance, and (c) the standard deviation of x.

In Exercises 5.64 and 5.65, assume a binomial distribution. Find n and p.

5.64 $\mu = 20, \sigma^2 = 4.$ **5.65** $\mu = 100, \sigma^2 = 75.$

5.6

USING MINITAB (OPTIONAL)

PDF Command

The PDF (probability distribution function) command can be used to obtain probabilities for a variety of probability distributions. The particular distribution is specified with a subcommand. In Example 5.16 we found the binomial probability $P(4)$ when there are 5 trials and the probability of success is .6. To compute the probability using Minitab, type

PDF 4;

BINOMIAL 5   .6.

Minitab would give the value .2592. If we wish to store this probability for later use, instead of printing it, type

PDF 4, STORE IN K1;

BINOMIAL 5 .6.

As usual we can add optional additional text for clarity, such as

PDF FOR X = 4;

BINOMIAL WITH N = 5, P = .6.

We can also find several probabilities at a time. The following program finds probabilities for a column of values:

```
MTB > SET C1
DATA> 4  5
DATA> END
MTB > PDF FOR C1;
SUBC>   BINOMIAL N = 5,   P = .6.
        K          P(X = K)
      4.00           0.2592
      5.00           0.0778
```

Assuming C1 contains the values 4, 5, the following variation of the PDF command stores the probabilities in column 2 instead of printing them:

PDF FOR C1, STORE IN C2;

BINOMIAL N = 5, P = .6.

Column 2 will contain the probabilities:

C2
0.25920	←P(4)
0.07776	←P(5)

We have seen these probabilities in Example 5.16 (where a hunter shoots at 5 pheasants). We were interested in the probability of at least 4 hits:

$$P(x = 4 \text{ or } 5) = P(4) + P(5)$$

To obtain this quickly using Minitab, we simply add one command to the above program:

PDF FOR C1, STORE IN C2;

BINOMIAL N = 5, P = .6.

SUM C2

Minitab would print the value 0.33696. To find the probability of at least one hit, we computed $1 - P(0)$ in Example 5.16. To compute this using Minitab, type

$$\text{PDF FOR 0, STORE IN K1;}$$
$$\text{BINOMIAL N} = 5, \text{P} = .6.$$
$$\text{LET K2} = 1 - \text{K1}$$
$$\text{PRINT K2}$$

If no value or column is specified in the PDF command, then all probabilities of a discrete distribution that are not too small are printed. For example, the following program gives all probabilities of a binomial distribution with $n = 5$, $p = .6$:

```
MTB > PDF;
SUBC>   BINOMIAL N = 5, P = .6.
BINOMIAL WITH N = 5 P = 0.600000
      K          P(X = K)
      0          0.0102
      1          0.0768
      2          0.2304
      3          0.3456
      4          0.2592
      5          0.0778
```

Other distributions can be specified by the subcommand. For example, consider the discrete distribution

x	$P(x)$
1	4/10
2	4/10
3	1/10
4	1/10

We could put the x values in C1 and the probabilities in C2 as follows:

```
READ INTO C1 C2
1  .4
2  .4
3  .1
4  .1
```

The program

$$\text{PDF 2;}$$
$$\text{DISCRETE, VALUES IN C1, PROBS IN C2.}$$

would return the value .4.

CDF Command

Whereas PDF gives probabilities $P(x = k)$, the CDF (cumulative distribution function) gives cumulative probabilities $P(x \leq k)$. For example,

CDF 4;

BINOMIAL N = 5, P = .6.

would return the value .9222 (along with the value 4). Therefore $P(x \leq 4) =$ .9222. As with the PDF command, we can store probabilities. For example, for the above program, if the command was

CDF 4, STORE IN K1;

the value .9222 would be stored as K1, and not printed.

In our discussion of PDF we showed how to compute $P(x = 4$ or $5)$ for a binomial distribution with $n = 5$, $p = .6$. Notice that the complement of $x = 4$ or 5 is $x \leq 3$. Hence $P(x = 4$ or $5) = 1 - P(x \leq 3)$. This suggests how to to compute the desired probability with CDF:

CDF 3, STORE IN K1;

BINOMIAL N = 5, P = .6.

LET K2 = 1 - K1

PRINT K2

As with PDF, CDF can deal with columns of values.

CDF FOR C1;

BINOMIAL N = 5, P = .6.

gives cumulative probabilities for values previously set in C1. If we don't specify a value or a column, CDF gives the cumulative probabilities for all values of x, as shown in the following printout:

```
MTB > CDF;
SUBC>   BINOMIAL N = 5, P = .6.
    BINOMIAL WITH N = 5 P = 0.600000
        K        P(X LESS OR = K)
        0              0.0102
        1              0.0870
        2              0.3174
        3              0.6630
        4              0.9222
        5              1.0000
```

INVCDF Command

We have seen that CDF gives probabilities of the form $P(x \le k)$. For example, we saw that for a binomial distribution with $n = 5$, $p = .6$, CDF 4 gave .9222:

$$P(x \le 4) = .9222$$

The INVCDF (inverse CDF) command would give the value 4. That is, the program

INVCDF .9222;
BINOMIAL N = 5, P = .6.

would print the value 4 (along with other information). To sum up: CDF answers the question

$$P(x \le k) = \underline{?}$$

for a given value of k. INVCDF answers the question

$$P(x \le \underline{?}) = p$$

where p is a given probability and $0 < p < 1$.

Thinking of distributions as abstract descriptions of populations, the INVCDF command gives percentiles. For example, if for some distribution

$$P(x \le 46) = .90$$

then the proportion of values in the population ≤ 46 is 90%. So 46 is the 90th percentile.

For discrete distributions the INVCDF may not exist for all values of p where $0 < p < 1$. For example, look at the previous printout of the CDF for a binomial distribution with $n = 5$, $p = .6$. Notice that there is no k such that

$$P(x \le k) = .8$$

So the INVCDF does not exist when $p = .8$. Observe that $P(x \le 3) = .663$ and $P(x \le 4) = .9222$, and .8 is between .663 and .9222. In a situation like this Minitab prints both 3 and 4 along with their cumulative probabilities. The following program illustrates this.

```
MTB > INVCDF .8;
SUBC>  BINOMIAL N = 5, P = .6.
     K  P(X LESS OR = K)       K  P(X LESS OR = K)
     3         0.6630          4          0.9222
```

There are a number of distributions that can be specified by the subcommand (17 in all). We will work with the normal distribution in the next chapter.

Reminder PDF, CDF, and INVCDF can all store results instead of printing them. Just remember to store constants as constants and columns as columns. For example,

INVCDF .43, STORE IN K1;

and

INVCDF C1, STORE IN C2;

Practice Quiz (Answers can be found after the Chapter Review exercises.)

For the problems in Questions 1 through 4, give the approximate values that will be returned, without using Minitab.

1. PFD 4;
   BINOMIAL N = 14, P = .3.

2. SET C1
   3, 5
   END
   PDF C1;
   BINOMIAL N = 8, P = .7.

3. CDF 2, STORE IN K1;
   BINOMIAL N = 5, P = .1.
   LET K2 = 1 − K1
   PRINT K2

4. READ INTO C1 C2
   0 .1
   1 .2
   2 .3
   3 .4
   END
   INVCDF .6;
   DISCRETE, VALUES IN C1, PROBS IN C2.

5. Write a program to find μ, σ^2, and σ for the distribution in Question 4. (*Note:* The Minitab command for finding square roots is SQRT; it works just like the MEAN command.)

Exercises

Suggested exercises for use with Minitab are 5.49, 5.50 (a, b), 5.51, 5.53, 5.54, 5.91, and 5.93.

SUMMARY

In this chapter we introduced numbers into the setting of a sample space by means of a **random variable,** which is a rule by which we can assign a number to each outcome. Presumably the number will describe some numerical property (of the outcome) that will be of interest.

A random variable is **continuous** if it can assume any value on some interval or continuous scale; it is **discrete** if the values it can assume constitute a sequence of isolated or separated points on the number axis.

We also studied the notion of the **probability distribution** for a discrete random variable x. We said $P(x)$ is the probability of observing the value x. We then studied the **mean** μ and **variance** σ^2 for a discrete random variable.

$$\mu = \sum x \cdot P(x)$$

The mean μ is also called the **expected value** of x, $E(x)$.

$$\sigma^2 = \sum (x - \mu)^2 \cdot P(x) = \left[\sum x^2 \cdot P(x) \right] - \mu^2$$

We call σ the standard deviation.

We then studied a particular discrete random variable—the binomial random variable. This counts the number of successes in a **binomial experiment** consisting of n repeated independent trials or repetitions of some basic experiment, which has two possible outcomes called success (S) and failure (F). If the probability of success on a given trial is p and the probability of failure is q, we saw that the probability of exactly x successes in n trials is given by the binomial distribution

$$P(x) = \frac{n!}{x!(n - x)!} \cdot p^x \cdot q^{n-x}$$

Finally, we said that for the binomial random variable x

$$\mu = n \cdot p$$
$$\sigma^2 = n \cdot p \cdot q$$

Review Exercises

5.66   A fair die is to be tossed. Let $x =$ the number of tosses up to and including the first time a 6 appears.

(a)   What values can x assume?

(b)   Is x a discrete or continuous random variable?

(c)   Is this a binomial experiment? Why?

5.67 A medical doctor is interested in studying the effect of a low cholesterol diet on people with high blood cholesterol levels (measured in milligrams per 100 milliliters). To study this, she obtains 50 volunteers with high blood cholesterol levels to try the diet.

(a) Describe a random variable of possible interest to the medical doctor. Is it discrete or continuous?

(b) Describe the population and sample of interest.

In Exercises 5.68–5.72 determine whether or not a probability distribution is given. Give a reason for your answer.

5.68

x	$P(x)$
0	$\frac{1}{4}$
4	$\frac{3}{4}$

5.69

x	$P(x)$
2	$\frac{2}{3}$
-4	$\frac{1}{6}$
6	$\frac{1}{6}$

5.70

x	$P(x)$
0	$\frac{1}{2}$
1	$\frac{1}{2}$
2	$\frac{1}{2}$

5.71

x	$P(x)$
0	$\frac{1}{5}$
1	$\frac{2}{5}$
2	$\frac{3}{5}$
3	$\frac{1}{5}$
4	$-\frac{2}{5}$

5.72

x	$P(x)$
3	1

5.73 Determine (a) the mean, (b) the variance, and (c) the standard deviation of the following probability distribution:

x	$P(x)$
2	$\frac{4}{6}$
3	$\frac{1}{6}$
13	$\frac{1}{6}$

5.74 Determine (a) $P(0)$, (b) the mean, (c) the variance, and (d) the standard deviation of the following probability distribution:

x	$P(x)$
0	?
2	$\frac{1}{2}$
5	$\frac{2}{5}$

5.75 Determine (a) $P(3)$, (b) the mean, (c) the variance, and (d) the standard deviation of the following probability distribution:

x	$P(x)$
1	$\frac{2}{12}$
2	$\frac{5}{12}$
3	?
6	$\frac{2}{12}$

5.76   Find the value of c so that a probability distribution is determined.

$$P(x) = \begin{cases} c, \text{ for } x = 1 \\ 2c, \text{ for } x = 2 \\ 3c, \text{ for } x = 3 \\ 4c, \text{ for } x = 4 \end{cases}$$

5.77   Find the value of c so that a probability distribution is determined.

$$P(x) = \begin{cases} c, \text{ for } x = 0 \\ 2c, \text{ for } x = 1 \end{cases}$$

5.78   Find the value of c so that a probability distribution is determined.

$$P(x) = \frac{cx}{6}, \text{ for } x = 1, 2$$

5.79   In the past year, 10 percent of RGH Airlines flights violated at least one safety code. A government inspector decided to monitor one flight per day for the next 3 days. Let x be the number of flights that violate at least one safety code. Find each of the following:

   (a)   The probability distribution.   **(b)**   The mean of x.

   (c)   The variance of x.   **(d)**   The standard deviation of x.

5.80   Three cards are numbered 1, 2, 3, and the experiment consists of turning the cards over one by one. A match occurs if the number of the card matches the number of the turn the card appears. So if the cards turn up 2, 1, 3, in that order, there is one match, namely, the 3. Let x be the number of matches.

   (a)   Find the probability distribution. (*Hint:* List the 6 possible outcomes.)

   (b)   If 60 people were to independently perform the experiment, what would you expect the approximate mean to be?

5.81   The probability distribution for the number of incoming phone calls x over a 5-minute period at a switchboard for a small business is

x	$P(x)$
0	.3679
1	.3679
2	.1839
3	.0613
4	.0153
5	.0031
6	.0005
7	.0001

Find each of the following:

   (a)   The mean of x.   **(b)**   The variance of x.

   (c)   The standard deviation of x.   **(d)**   Construct a probability histogram.

   (e)   Over a 5-minute period, what is the probability of at least one incoming phone call?

5.82   A *binomial* probability distribution for the number of successes x is

x	$P(x)$
0	.2401
1	.4116
2	.2646
3	.0756
4	.0081

Find each of the following:

(a)   The mean of x.   **(b)**   The variance of x.

(c)   The standard deviation of x.   **(d)**   Construct a probability histogram.

[*Hint:* Find q from the fact that $q^4 = P(0)$. Now use the binomial shortcut formulas for μ and σ^2.]

5.83   A *binomial* probability distribution for the number of successes x is

x	$P(x)$
0	.36
1	.48
2	.16

Find each of the following:

(a)   The mean of x.   **(b)**   The variance of x.

(c)   The standard deviation of x.   **(d)**   Construct a probability histogram.

[*Hint:* Use the binomial shortcut formulas and the fact that $q^2 = P(0)$.]

5.84   Match the three binomial probability histograms below with the correct p: $p = .3$, $p = .5$, $p = .8$.

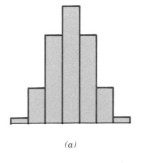 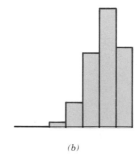 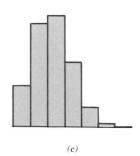

(a)   (b)   (c)

5.85   A roulette wheel has 37 equally spaced slots that are numbered 0, 1, 2, ... , 36. In advance of the wheel being spun, a player picks one of the numbers. If the wheel is spun and stops on the number, the player wins $35. If the wheel stops on one of the other numbers, the player pays $1. What is the player's expected winnings (per game)?

5.86 In the game of craps a player rolls two dice. If the first roll results in a sum of 7 or 11, the player wins. If the first roll results in a 2, 3, or 12, the player loses. If the sum on the first roll is 4, 5, 6, 8, 9, or 10, the player keeps rolling until he throws a 7 or the original value. If the outcome is a 7, the player loses. If it is the original value, the player wins. The probability that a player will win is .493. If the player nets $4 for a win, what should he pay to play the game in order that the game favor neither the player nor the casino?

5.87 A friend proposes the following game. He is to select one card from an ordinary deck of cards. You are to pay him $10 if he selects an ace and $5 if he selects a 2 or 3. What should he pay you if he selects neither an ace, 2, nor 3 in order that the game not favor either person?

5.88 The probability that a new flu vaccine prevents the flu is 0.80. Twelve people are vaccinated with the new vaccine at a local clinic. What is the probability that

(a) At least 8 will not get the flu?

(b) Two of the people will get the flu?

5.89 Assume that the probability of a boy being born is 0.50. If a couple plan on having six children, find the probability that

(a) Exactly half are boys.   (b) All are boys.

(c) All are boys or all are girls.   (d) At least one is a boy.

(Assume that the sex of any child is independent of the sex of the other children.)

5.90 Thirty percent of all voters in a large city are Independents. If 15 voters are to be selected by chance, find the probability that

(a) Exactly 4 are independent.

(b) No more than 3 are independent.

(c) At least 10 are independent.

5.91 Seventy percent of the complaints received by a consumer protection agency are investigated by the agency. For the next 7 complaints received by the agency, find the probability that

(a) All will be investigated.

(b) One will be investigated.

(c) No more than 2 will be investigated.

5.92 In order for an electronic system to function, each of 12 components must work. Assume the components function independently of each other, and each has a probability .99 of not breaking down for one year, when new. Suppose you buy a new system. Find the probability that the system does not break down in the first year.

5.93 Eighty-five percent of the dishwashers manufactured by a large company do not need repairs for the first 2 years. If five of the dishwashers are to be selected by chance, find the probability that

(a) All five will not need repairs for 2 years.

(b) At least three will need repairs within 2 years.

In Exercises 5.94–5.97, find (a) the mean, (b) the variance, and (c) the standard deviation of the number of successes x.

5.94 Use Exercise 5.88 with x, the number of people who don't get the flu.

5.95 Use Exercise 5.90 with x, the number of voters who are Independent.

5.96 Use Exercise 5.91 with x, the number of complaints that will be investigated.

5.97 Use Exercise 5.93 with x, the number of dishwashers that will not need repairs in the next 2 years.

Answers to Practice Quiz

1. .229

2. .047
   .254

3. .009

4. 2

5. LET K1 = SUM(C1*C2)
   LET K2 = SUM(C1**2*C2) − K1**2
   LET K3 = SQRT(K2)
   PRINT K1 − K3
   (*Note:* K1 = μ, K2 = σ^2, K3 = σ.)

Notes

DeGroot, M. H., S. E. Fienberg, and J. B. Kadane, ed., *Statistics and the Law* (New York: John Wiley, 1986).

The World Almanac and Book of Facts (New York: Newspaper Enterprise Association, 1989).

The World Almanac and Book of Facts (New York: Newspaper Enterprise Association, 1990).

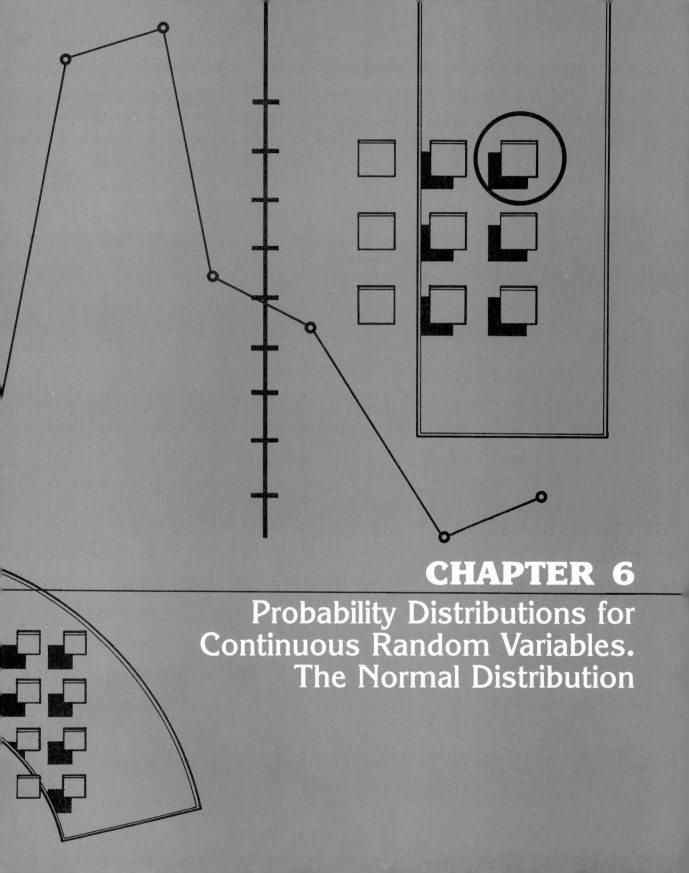

CHAPTER 6

Probability Distributions for
Continuous Random Variables.
The Normal Distribution

6.1 INTRODUCTION

As we have said a number of times so far, inferential statistics is concerned with making judgments about populations, and we carry on this study by examining samples from the populations. In Chapter 5 we indicated that we would find it convenient to introduce a concept that summarizes the essential features of a particular population of data values. This was called a "probability distribution."

In Chapter 5 we dealt with probability distributions for discrete random variables. In this chapter we deal with *probability distributions for continuous random variables* (also called continuous probability distributions). We will then study a very important class of continuous random variables—**normal random variables**—along with their probability distributions. We will show how a probability associated with a binomial random variable may be quickly approximated by a probability associated with an appropriate normal random variable. Finally, we will study an important theorem that is useful in statistical inference. It is called the **Central Limit Theorem.**

CONTINUOUS PROBABILITY DISTRIBUTIONS

Recall that a continuous random variable is one that can theoretically take on any value on some line interval (see Section 5.2). Its probability distribution should somehow enable us to find probabilities associated with the random variable. The way this information is conveyed, however, is somewhat different for continuous random variables than for discrete random variables.

We saw in Chapter 5 that if b is some value of a discrete random variable, $P(b)$ is the probability that the value b would be observed. If we have a continuous random variable whose values are represented by the symbol x, we often use a symbol such as $f(x)$, read "f of x," to represent the probability distribution. We can think of $f(x)$ as some mathematical expression involving x so that for each value of x, the expression determines some number $f(x)$. But how does this number relate to probability? Unfortunately, $f(x)$ does *not* give us the probability that the value x will be observed.

To understand how a probability distribution for a continuous random variable enables us to find probabilities, it is important to understand a certain relationship between probability and area. Figure 6.1 is a histogram for a grouped frequency distribution for 25 values of a certain continuous random variable. The frequencies are displayed above each bar.

Suppose that we were interested in the probability that a randomly observed value of x will fall between 13.5 and 21.5. We denote this probability by $P(13.5 < x < 21.5)$. In our sample there are 14 data values between 13.5 and

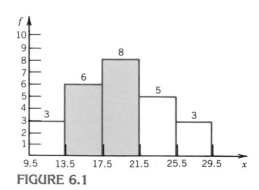

FIGURE 6.1

21.5 (out of a total of 25 data values). Therefore, it would be natural to estimate this probability by $\frac{14}{25} = .56$

Observe that the width of each bar (the class width) is four units. The area of any bar is the frequency times the class width. The area of the bars between 13.5 and 21.5 is

$$(6)(4) + (8)(4) = 56$$

The total area of all the bars is

$$(3)(4) + (6)(4) + (8)(4) + (5)(4) + (3)(4) = 100$$

Notice that

$$P(13.5 < x < 21.5) \doteq .56 = \frac{56}{100} = \frac{\text{area of bars between 13.5 and 21.5}}{\text{total area of all the bars}}$$

This is only a very rough estimate of the desired probability, since we are dealing with a sample of only 25 data values.

Now suppose that we obtained more and more data values and made the class width smaller and smaller. The histogram could be expected to smooth out and approach some smooth curve, as in Figure 6.2.

Based on the foregoing discussion, we see that a probability $P(a < x < b)$ can be approximated by the ratio of the total area of the bars between a and b divided by the total area of all the bars. (See Figure 6.2.) Examining the curve in Figure 6.2, it should be clear that this probability can also be obtained by calculating the area between a and b under the curve divided by the total area under the curve. Assuming that there are mathematical techniques for finding such a curve and areas such as the ones discussed (and there are such techniques), this curve can be quite useful in obtaining probabilities associated with the random variable x. Actually, to avoid the necessity of dividing by the total area under the curve when finding probabilities, statisticians prefer to work with a curve having the same shape but which is adjusted so that the total area under the curve is 1. Then the probability $P(a < x < b)$ will simply be the area between a and b under the curve. See the curve in the xy plane in Figure 6.3.

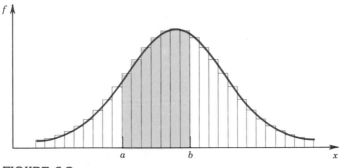

FIGURE 6.2

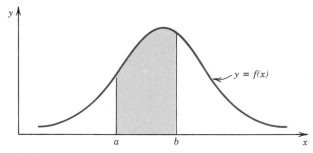

FIGURE 6.3
$P(a < x < b)$ = **Area of Shaded Region**

Suppose that the curve in Figure 6.3 has the equation $y = f(x)$, where $f(x)$ is some mathematical expression involving x. This means that this curve is the graph of the equation; that is, it is the collection of all points (x, y) that satisfy the equation. Then the expression $f(x)$ is called the **probability distribution** (or *probability density function*) for the continuous random variable.

The particular curve that we get and the form that the expression $f(x)$ takes will depend on the random variable under discussion. The technique for finding the probability distribution $f(x)$ for a continuous random variable is beyond the scope of this book. Suffice it to say that statisticians have found a number of such commonly encountered probability distributions and have developed tables containing probabilities associated with some of these distributions. Part of our task in statistics is learning which table applies in a given situation.

The following example is one that can be worked out "from scratch" without the use of tables.

Example 6.1
Let x represent the amount of ice cream (in hundreds of gallons) sold by the Kenmore drive-in restaurant on any selected day. Due to limitations in storage, the maximum amount of ice cream that can be kept on hand on any given day is 200 gallons. Thus $0 \le x \le 2$. Further, it is known that the probability distribution for x is given by the formula $f(x) = x/2$. Sketch the graph of this probability distribution and find the probability that on a selected day less than 100 gallons of ice cream will be sold; that is, find the probability that x falls between 0 and 1.

Solution
First we will sketch the graph of the equation $y = x/2$. This graph will be a straight line, so that we need only find two points to determine its location. Note that when $x = 0$, $y = 0$, so that $(0, 0)$ is on the line. Also, when $x = 2$, $y = 1$, so that $(2, 1)$ is on the line. The line is shown in Figure 6.4. The desired probability $P(0 < x < 1)$ is the area under the graph between 0 and 1; it is shown as the shaded region in Figure 6.4. Since this region is a triangle, its area is one-

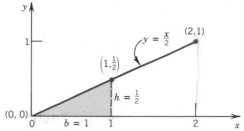

FIGURE 6.4
Probability Distribution for x, the
Amount of Ice Cream Sold in Hundreds
of Gallons (Example 6.1)

half the base times the height, that is $\frac{1}{2}b \cdot h$. It should be clear from Figure 6.4 that $b = 1$ and $h = \frac{1}{2}$. Hence

$$P(0 < x < 1) = \frac{1}{2} \cdot b \cdot h = (\tfrac{1}{2})(1)(\tfrac{1}{2}) = \tfrac{1}{4}$$

Therefore, the probability of selling less than 100 gallons of ice cream is .25.

Now we summarize some of the important properties of a continuous probability distribution.

1. For a continuous probability distribution, $f(x) \geq 0$ for all values x of the random variable.

2. The total area under the graph of the probability distribution [i.e., under the graph of the equation $y = f(x)$] is 1.

3. The probability that an observed value of x will fall between a and b, $P(a < x < b)$, is the area between a and b under the graph. We can generalize this result. Sometimes we work with regions of x values that are not just single intervals (e.g., we may be interested in a region that consists of two separate intervals). For the regions that will interest us, the probability that x will fall in the region will be equal to the area directly above the region and under the graph.

Note that for a continuous random variable x,

$$P(a \leq x \leq b) = P(a < x < b)$$

This is because adding a and b to the interval does not increase the area directly above the interval.

Exercises

6.1   Find the following probabilities using the graph below of a continuous probability distribution.

    (a) $P(x < 1/2)$ **(b)** $P(1/4 < x < 7/4)$

    (c) $P(x \geq 3/5)$ **(d)** $P(x < 1/5 \text{ or } x > 2/5)$

    (e) $P(x < 2)$ **(f)** $P(x = 3/4)$

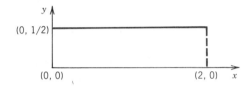

6.2   Assume that the time of birth of a New Year's baby at a city hospital will occur at some random time between midnight and 2:00 A.M. Let x be the number of minutes after midnight that the baby will be born. So $0 \leq x \leq 120$. The graph of the probability distribution follows. Note that x is said to be uniformly distributed.

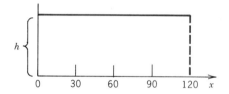

    (a)   What is the area of the rectangle? What is the value of h?

    (b)   What is the probability that the New Year's baby will be born

        (i)   Before 12:30 A.M.?

        (ii)   After 1:15 A.M.?

        (iii)   Between 12:45 A.M. and 1:00 A.M.?

        (iv)   Before 12:15 A.M. or after 1:30 A.M.?

    (c)   Express parts (i), (ii), and (iii) of part (b) in one of the forms $P(a < x < b)$, $P(x < a)$, or $P(x > a)$.

6.3   Scores on a standardized test have a mean of 500 and a standard deviation of 100. Let x represent a student's test score. In the following probability distribution, assume a symmetric distribution.

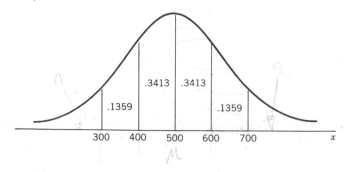

What is the probability that a randomly selected student will score

(a) Higher than 700 or less than 300?

(b) Higher than the mean?

(c) Between 600 and 700?

(d) Lower than 600?

(e) Between 300 and 500?

(f) Within one standard deviation of the mean?

(g) Within two standard deviations of the mean?

6.4 A psychologist studied the length of time x, in seconds, required for completion of a project by first graders in a large city. The probability distribution is as follows:

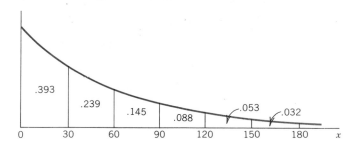

What percentage of students completed the project

(a) In less than 60 seconds?

(b) Between 60 and 120 seconds?

(c) In more than 180 seconds?

(d) In more than 90 seconds?

(e) Now suppose that the psychologist is going to give the problem to a new group of first graders. She is going to allow them 2 minutes. About what percentage of these children are not expected to complete the project?

6.5 Below is the graph of a continuous probability distribution.

    (a) Find the value of w.   **(b)** Find $P(x < 1)$.

    (c) Find $P(x > 3)$.   **(d)** Find $P(x > 3/2)$.

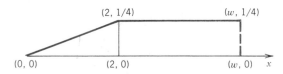

6.3   THE NORMAL DISTRIBUTION

Perhaps the most important class of continuous random variables in statistics is the class of **normal random variables.** The probability distribution of a normal random variable is called a **normal distribution.** Many naturally occurring random variables are normal or very nearly so. For example, IQ's, heights of humans, Scholastic Aptitude Test scores, all have approximately normal distributions. We often say they are *normally distributed* or have *normal populations.* The fact that many variables occurring in nature are normally distributed is no accident, and we will attempt to show why this is so later in the chapter. For now, we will discuss some properties of normal distributions and investigate how one finds probabilities associated with normal random variables.

Properties of a Normal Distribution

A normally distributed random variable x with mean μ and standard deviation σ has the following properties:

1.   Its probability distribution is given by the formula

$$f(x) = \frac{e^{-(x-\mu)^2/2\sigma^2}}{\sigma\sqrt{2\pi}}$$

where x can take on any numerical value. Also $\pi \doteq 3.1416$ and $e \doteq 2.7183$. This formula may appear threatening, but there is no need for concern, because we will not have to deal with it directly. This is because normal probabilities can be obtained by the use of a table, as we shall see in the next section.

2.   The graph of a normal probability distribution is a bell-shaped curve that has its peak above the value $x = \mu$. Also, it is symmetric with respect to this value; that is, the portion of the curve to the right of μ is a mirror image of the portion to the left of μ. The shape of a normal distribution is given in Figure 6.5

3.   The mean, median, and mode for a normal population all have the same value.

   Some additional normal curves are displayed in Figure 6.6. Note how the systolic blood pressure curves indicate that the younger males not only tend to have lower blood pressure, but the variability within the younger group is less than that of the older group.

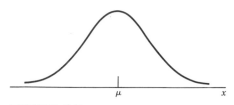

FIGURE 6.5
A Normal Probability Distribution

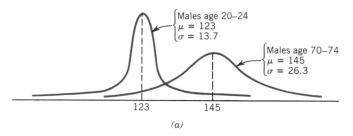

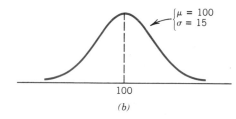

FIGURE 6.6
Some Normal Distributions. (*a*) **Systolic Blood Pressure**
(*Source for Parameters:* **Lasser, R. P., and A. M. Master,**
Geriatrics **14:347, 1959).** (*b*) **IQ Scores**

6.4 THE STANDARD NORMAL DISTRIBUTION

There is a normal random variable that stands out in importance above all other normal variables. It is important because we can use it to find probabilities associated with any normal variable. This random variable is called the **standard normal variable,** which we now define.

> **Definition** The *standard normal random variable* is that normal variable with mean 0 and standard deviation 1. Its values are usually represented by the symbol z.

Probabilities associated with the standard normal variable can be found by the use of Appendix Table B.3. Suppose we want to find the probability that a value of the standard normal variable will fall between 0 and some specific positive value of z. This is the area under the standard normal curve, portrayed in Figure 6.7. Appendix Table B.3 gives the value of such areas for various values of z. The following examples show how to find any probabilities associated with a standard normal variable.

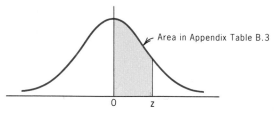

FIGURE 6.7
The Standard Normal Distribution

Example 6.2

The Barton Company manufactures automobile charge indicator gauges. Under test conditions these gauges are supposed to read 0 volts. However, some gauges read slightly more than 0 volts and some slightly less. Let z = voltage reading of a gauge under test conditions (So there is a value of z associated with each gauge). Past experience has shown that z is approximately normally distributed with a mean of 0 volts and a standard deviation of 1 volt. If one of these gauges is selected, find the probability that its voltage reading will be anywhere between 0 and 1.43 volts.

Solution

Notice that z has the standard normal distribution. We want to find $P(0 < z < 1.43)$. Using Appendix Table B.3, locate the number in the left-hand column that contains the same units and tenth's digits as 1.43. This is 1.4. Then in the row across the top locate the number that has the same hundredth's digit. This is .03. Simply put, this corresponds to decomposing 1.43 as follows:

$$1.43 = 1.4 + .03$$

Now look at the intersection of the row containing 1.4 and the column containing .03. This number is .4236. (See Figure 6.8.) This means that

$$P(0 < z < 1.43) = .4236$$

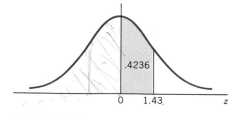
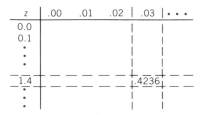

FIGURE 6.8
$P(0 < z < 1.43) = .4236$

Example 6.3
For the random variable of Example 6.2 find

(a) $P(z < 1.43)$.   **(b)** $P(-1.43 < z < 0)$.

Solution

(a) The probability $P(z < 1.43)$ is the area to the left of 1.43. This is the same as the area to the left of 0 plus the area from 0 to 1.43. See Figure 6.9a. In the previous example, we saw that the area from 0 to 1.43 was .4236. The total area under the curve is 1. The area to the left of 0 equals the area to the right of 0, and hence each has area .5. Therefore

$$P(z < 1.43) = .5 + .4236 = .9236$$

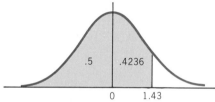

FIGURE 6.9a
$P(z < 1.43) = .5 + .4236 = .9236$

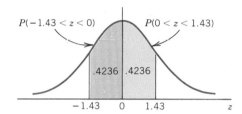

FIGURE 6.9b

(b) Note that -1.43, being negative, is not one of the values of z in the table. However, we can use the symmetry of the normal distribution to obtain the desired probability. The probability $P(-1.43 < z < 0)$ is equal to the area under the standard normal distribution between -1.43 and 0. But by symmetry, this area is the same as the area under the curve between 0 and 1.43. (See Figure 6.9b.) This latter area is equal to the probability $P(0 < z < 1.43)$, which by Example 6.2 is .4236. Thus

$$P(-1.43 < z < 0) = P(0 < z < 1.43) = .4236$$

Example 6.4
For the random variable of Example 6.2 find the probability $P(-.57 < z < 1.12)$.

Solution
As long as the reader keeps in mind the identification of probabilities with areas, the following steps should be clear on the basis of geometric intuition. (See Figure 6.10.) Now the area under the curve from $-.57$ to 1.12 is the area from $-.57$ to 0 plus the area from 0 to 1.12:

$$P(-.57 < z < 1.12) = P(-.57 < z < 0) + P(0 < z < 1.12)$$

By symmetry the area from $-.57$ to 0 is the same as the area from 0 to .57.

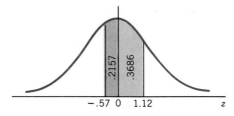
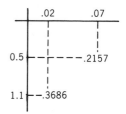

FIGURE 6.10
$P(-.57 < z < 1.12) = .2157 + .3686 = .5843$

Hence,

$$P(-.57 < z < 0) = P(0 < z < .57)$$

Now we may look up these probabilities in Appendix Table B.3, where we find $P(0 < z < .57) = .2157$ and $P(0 < z < 1.12) = .3686$. Hence

$$P(-.57 < z < 1.12) = .2157 + .3686 = .5843$$

Example 6.5
For the random variable of Example 6.2 find $P(1.12 < z < 1.41)$.

Solution
We must express this in terms of probabilities of the form $P(0 < z < a)$ so that we can use Appendix Table B.3. From Figure 6.11 we see that the area from 1.12 to 1.41 is equal to the area from 0 to 1.41 minus the area from 0 to 1.12. Therefore, using this fact and Appendix Table B.3, we get

$$P(1.12 < z < 1.41) = P(0 < z < 1.41) - P(0 < z < 1.12)$$
$$= .4207 - .3686$$
$$= .0521$$

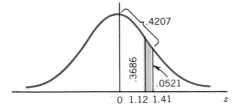
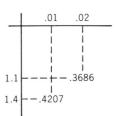

FIGURE 6.11
$P(1.12 < z < 1.41) = .4207 - .3686 = .0521$

Example 6.6
For the random variable of Example 6.2 find
(a) $P(z > 1.28)$. **(b)** $P(z > 1.28 \text{ or } z < -1.28)$.

Solution

(a) Again we would like to express this probability in terms of a probability of the form $P(0 < z < a)$ so that we can use Appendix Table B.3. The area to the right of 0 is .5. Therefore, the area to the right of 1.28 (which is what we want) is equal to .5 minus the area from 0 to 1.28. (See Figure 6.12.) Thus

$$P(z > 1.28) = .5 - P(0 < z < 1.28)$$
$$= .5 - .3997 = .1003$$

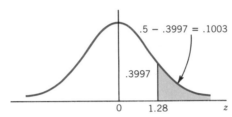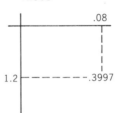

FIGURE 6.12
$P(z > 1.28) = .5 - .3997 = .1003$

(b) $P(z > 1.28 \text{ or } z < -1.28)$ is the probability that a value of z will be either greater than 1.28 or less than -1.28. This probability will be the area under the curve directly above the region (on the z axis) consisting of z values to the left of -1.28 together with z values to the right of 1.28. (See Figure 6.13.)

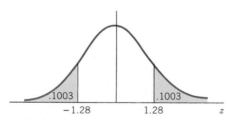

FIGURE 6.13
$P(z > 1.28 \text{ or } z < -1.28) = .2006$

To find this area, we just add up the areas of the two tails (shaded). We have already found the area of the right tail in part (a) of this example to be .1003. By symmetry, the left tail has the same area, so that

$$P(z > 1.28 \text{ or } z < -1.28) = .1003 + .1003 = .2006$$

Example 6.7

Find the value on the z axis, such that the area under the curve to the right of it is .025. In other words,

$$P(z > \underline{?}) = .025$$

Solution

If the area to the right of the desired z value is .025, then the area from 0 to this value is: $.5 - .025 = .475$. (See Figure 6.14.) Therefore,

$$P(0 < z < \underline{?}) = .475$$

Now in Appendix Table B.3 we locate the area .475 (not .025). We then read the desired z value to be $1.9 + .06 = 1.96$. (See Figure 6.14.)

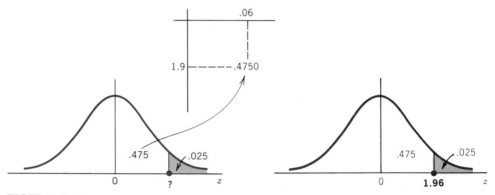

FIGURE 6.14
$P(z > 1.96) = .025$

Notation

In future work it will be helpful to be able to relate a value of z to an area (or probability) under the standard normal curve, by means of notation. The area we use is the area under the standard normal curve to the right of the z value. For example, the value of z such that the area under the curve to the right of it is .025, is denoted by z(.025), read "z of .025." [*Warning: z(.025) does not represent multiplication.*]

$$z(.025) = \text{that value of } z \text{ such that the area}$$
$$\text{under the standard normal curve}$$
$$\text{to the right of it is .025}$$

From Figure 6.14 we can see that this value is 1.96:

$$z(.025) = 1.96$$

Example 6.8

Find the following values of z:

(a) $z(.10)$ **(b)** $z(.05)$ **(c)** $z(.975)$ **(d)** $z(.80)$

Solution

(a) $z(.10)$ is displayed in Figure 6.15.

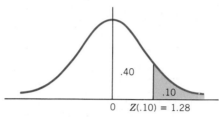 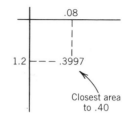

FIGURE 6.15

We will use Appendix Table B.3 to find the value. But remember that Appendix Table B.3 relates a specific value of z to the area under the curve from 0 to that specific value. The area from 0 to $z(.10)$ is .40, so that we must look up the value of z corresponding to the area .40 in Appendix Table B.3. We see that the area closest to .40 is .3997. The z value corresponding to .3997 is 1.28,

$$P(0 < z < 1.28) = .3997$$

Therefore $z(.10) = 1.28$.

(b) $z(.05)$ is displayed in Figure 6.16. The area between 0 and $z(.05)$ is .45. Therefore we look up the z value corresponding to .45. In Appendix Table B.3 we see that the area .45 is halfway between .4495 and .4505. When this occurs, it is customary to use the larger value of z so that $z(.05) = 1.65$.

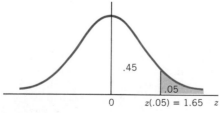 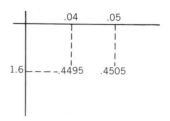

FIGURE 6.16

(c) Keeping in mind the symmetry of the standard normal distribution, we see from Figure 6.17 that $z(.975) = -z(.025)$. We have already seen that $z(.025) = 1.96$; therefore

$$z(.975) = -1.96$$

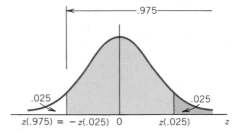

FIGURE 6.17

(d) $z(.80)$ is displayed in Figure 6.18.

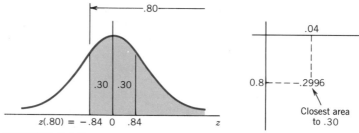

FIGURE 6.18

Because of the symmetry of the standard normal distribution, we should look up the z value corresponding to .30 in Appendix Table B.3 and then take the negative of this value. The closest area to .30 in Appendix Table B.3 is .2996. The value of z corresponding to this is .84. Thus

$$z(.80) = -.84$$

Remark Notice that all the examples of this section were solved by drawing the normal curve and the area corresponding to the desired probability. When working a normal probability problem, **always draw a picture**—one picture's worth a thousand words.

Exercises

6.6 Suppose z represents the standard normal variable. If a value is selected at random from the z distribution, find the probability that z is

(a) Less than 0.

(b) Between 0 and $-.67$.

(c) Between -2.3 and -1.45.

(d) Between $-.73$ and 2.31.

(e) Less than 1.96.

(f) Within one standard deviation of the mean.

(g) Within three standard deviations of the mean.

6.7 Suppose z is the standard normal variable. If a value is selected at random from the z distribution, find the probability that z is

(a) Between 0 and .67.

(b) Between 1.65 and 2.1.

(c) Between -2.1 and 1.7.

(d) Larger than -1.86.

(e) Larger than 2 or less than -2.

(f) Within two standard deviations of the mean.

6.8 Assume the standard normal distribution. Fill in the blanks.

(a) $P(z < \underline{\ \ }) = .9772.$ **(b)** $P(z < \underline{\ \ }) = .0668.$

(c) $P(z > \underline{\ \ }) = .5.$ **(d)** $P(z > \underline{\ \ }) = .9599.$

6.9 Assume the standard normal distribution. Fill in the blanks.

(a) $P(z < \underline{\ \ }) = .9573.$ **(b)** $P(z < \underline{\ \ }) = .1075.$

(c) $P(z > \underline{\ \ }) = .0793.$ **(d)** $P(z > \underline{\ \ }) = .9929.$

(e) $z(.02) = \underline{\ \ }$ **(f)** $z(.75) = \underline{\ \ }$

6.10 Evaluate each of the following:

(a) $z(.01).$ **(b)** $z(.95).$ **(c)** $z(.90).$

(d) $z(.005).$ **(e)** $z(.20).$

6.11 Below is the standard normal curve. Find the areas A, B, C, and D.

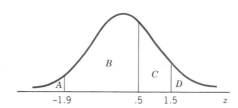

6.12 Below is the standard normal curve and some areas under the curve. Find the z values for a, b, and c.

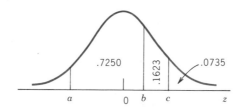

6.5 MORE ON NORMAL PROBABILITY

We will now show how probabilities associated with any normal variable may be found by the use of a standard normal variable. To see how this is done, consider a normal variable x with mean μ and standard deviation σ. Recall the formula for the standard score (or z score) given in Section 3.5:

$$z = \frac{x - \mu}{\sigma}$$

Each value of x determines a value of z, so that z is itself a random variable. In fact, all z does is measure whatever x is measuring in different units, namely, the number of standard deviations that x is above or below the mean μ. It follows that z should have the same type of probability distribution as x. Also, if we call the mean of z, μ_z, it seems reasonable that μ_z can be obtained by replacing x by its mean in the formula for z.

$$\mu_z = \frac{\mu - \mu}{\sigma} = 0$$

It can also be shown that if we let σ_z be the standard deviation of z, then $\sigma_z = 1$.

Putting all these ideas together, we can say that if x is normal with mean μ and standard deviation σ, then

$$z = \frac{x - \mu}{\sigma}$$

is a normal random variable with mean 0 and standard deviation 1; that is, it is a standard normal variable. To see how we can make use of this fact to find probabilities associated with any normal variable, consider the following example.

Example 6.9
The scores of males on the 1974 mathematical scholastic aptitude test (MSAT) were approximately normally distributed with mean $\mu = 500$ and standard deviation $\sigma = 100$ points (approximately). Find the proportion of males who received scores

(a) Between 500 and 600. **(b)** Between 400 and 600.

Solution
(a) If we keep in mind that proportions and probabilities are really the same thing, it should be clear that we will have the proportion of scores between 500 and 600 if we simply find the probability that a randomly selected score falls between 500 and 600, that is, $P(500 < x < 600)$. Consider the z score of x:

$$z = \frac{x - \mu}{\sigma} = \frac{x - 500}{100}$$

When $x = 500$,

$$z = \frac{500 - 500}{100} = 0$$

When $x = 600$,

$$z = \frac{600 - 500}{100} = 1$$

Therefore, to say that a value of x is between 500 and 600 is the same as saying its z score is between 0 and 1. Hence the probability that an x value is between 500 and 600 is the same as the probability its z score is between 0 and 1.

$$P(500 < x < 600) = P(0 < z < 1)$$

This is depicted in Figure 6.19a.

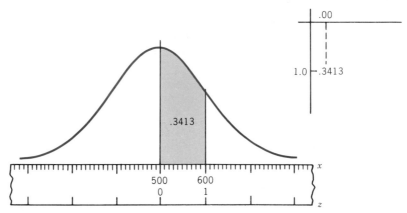

FIGURE 6.19a

Keep in mind that changing from x to z is just changing to different units of measurement, from points to standard deviations.

$$(x = 600 \text{ points}) \rightarrow (z = 1 \text{ standard deviation above the mean})$$

We may now look up the probability $P(0 < z < 1)$ in Appendix Table B.3. We find

$$P(500 < x < 600) = P(0 < z < 1) = .3413$$

So about 34% of scores were between 500 and 600.

(b)   While 600 is 100 points above the mean, 400 is 100 points below the mean. Examining Figure 6.19b, we see that due to symmetry of the normal distribution, the area from 400 to 500 is the same as the area from 500 to 600. We computed this area in part (a) to be .3413. Hence,

$$P(400 < x < 600) = 2(.3413) = .6826$$

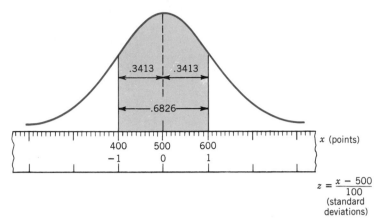

FIGURE 6.19*b*
$P(400 < x < 600) = .6826$

Therefore about 68% of the MSAT scores fall between 400 and 600, which is within one standard deviation (100 points) of the mean.

Example 6.9 shows how we can find probabilities associated with a normal variable x. We use the standard score for x:

$$z = \frac{x - \mu}{\sigma}$$

which enables us to express the values of x in terms of standard deviations, z. Using this, we can take a probability statement involving x and change it to an equivalent statement in terms of z. Since z is a standard normal variable, we can then use Appendix Table B.3 to find the desired probability.

Example 6.10

The heights of adult females are normal with mean approximately 64 inches and standard deviation 2.5 inches. Find the proportion of females who are

(a)   Between 65 and 68 inches tall.   **(b)**   Shorter than 61 inches.

Solution

(a)   Let x represent the height of a randomly selected female. The proportion we are looking for is just the probability that x will be between 65 and 68, that is, $P(65 < x < 68)$. Consider the standard score

$$z = \frac{x - 64}{2.5}$$

This is standard normal. The z scores corresponding to $x = 65$ and $x = 68$ are

$$z = \frac{65 - 64}{2.5} = \frac{1}{2.5} = .40 \qquad z = \frac{68 - 64}{2.5} = 1.60$$

To say that x is between 65 and 68 is equivalent to saying its z score is between .40 and 1.60. Therefore

$$P(65 < x < 68) = P(.40 < z < 1.60)$$

This is depicted in Figure 6.20a.

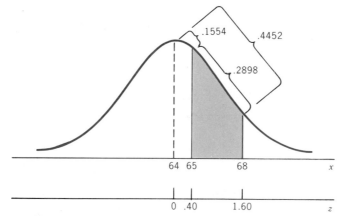

FIGURE 6.20a
$P(65 < x < 68) = .4452 - .1554 = .2898$

From Appendix Table B.3 we find

$$P(.40 < z < 1.60) = P(0 < z < 1.60) - P(0 < z < .40)$$
$$= .4452 - .1554 = .2898$$

This means that approximately 29% of females are between 65 and 68 inches tall.

(b) We now find $P(x < 61)$. When $x = 61$

$$z = \frac{61 - 64}{2.5} = \frac{-3}{2.5} = -1.20$$

So

$$P(x < 61) = P(z < -1.20)$$

See Figure 6.20b. The area to the left of -1.20 can be found by subtracting the area under the curve between -1.20 and 0 from .5. The area from

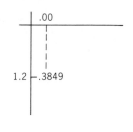

FIGURE 6.20*b*
$P(x < 61) = .5 - .3849 = .1151$

-1.20 to 0 is the same as the area from 0 to 1.20, which from Appendix Table B.3 is .3849. (See Figure 6.20*b*.) Thus

$$P(x < 61) = P(z < -1.20)$$
$$= .5 - .3849 = .1151$$

Therefore about 11.5% of females are shorter than 61 inches.

Example 6.11
Recall that the 1974 MSAT scores for males were approximately normal with $\mu = 500$, $\sigma = 100$. For this population find

(a)   The 90th percentile, P_{90}.   **(b)**   The 30th percentile, P_{30}.
(c)   The percentile rank of the score 567.

Solution

(a)   We will first find P_{90} in terms of z and then, using the relation

$$z = \frac{x - 500}{100}$$

we will solve for x, thus getting the 90th percentile in terms of x. Now P_{90} separates the top 10% of scores from the bottom 90%. Look at Figure 6.21*a*. First find the number that satisfies

$$P(0 < z < \underline{?}) = .40$$

We look up the closest probability to .40 in Appendix Table B.3; this is .3997. The corresponding z value is 1.28. This is the 90th percentile expressed in terms of z. We want this in terms of x. So we substitute and solve for x

$$1.28 = \frac{x - 500}{100}$$

$$128 = x - 500$$

$$500 + 128 = x$$

$$x = 628$$

So $P_{90} = 628$.

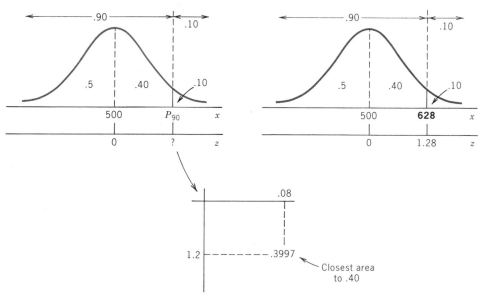

FIGURE 6.21a
The 90th Percentile = 628

(b) The situation for P_{30} is shown in Figure 6.21*b*. First look up the number in Appendix Table B.3 that satisfies

$$P(0 < z < \underline{?}) = .20$$

The z score we want will be the negative of the one found. The closest probability to .20 is .1985, which corresponds to a z score of .52. So the value we want is $z = -.52$. Substituting and solving for x,

$$-.52 = \frac{x - 500}{100}$$

$$-52 = x - 500$$

$$500 - 52 = x$$

$$x = 448$$

So $P_{30} = 448$.

(c) To find the percentile rank of 567, we must find the proportion of scores less than 567.

$$P(x < 567) = \underline{?}$$

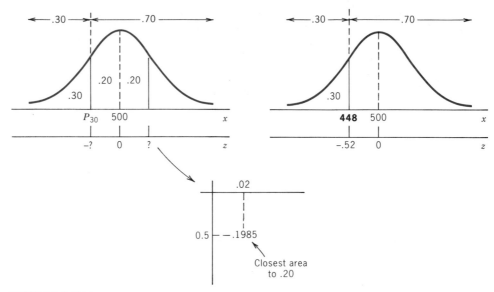

FIGURE 6.21*b*

When $x = 567$

$$z = \frac{567 - 500}{100} = .67$$

Figure 6.21*c* shows that

$$P(x < 567) = P(z < .67) = .5 + .2486 = .7486$$

This means that the percentile rank of 567 is 74.86, or approximately 75. Thus 567 is approximately the 75th percentile or third quartile.

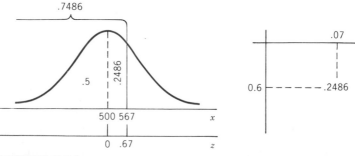

FIGURE 6.21*c*

Example 6.12

A large brass company had a labor contract that required any employee layoffs to be based on the employee's years of service with the company (seniority). A cutoff value is determined and any employee whose seniority is less than the value is laid off.

(a) At one plant seniority is normal with mean $\mu = 15$ years and $\sigma = 5$ years. If 18% are to be laid off, what is the cutoff value?

(b) At another plant seniority is normal with $\sigma = 3$ years. If 35% are laid off and the cutoff value is 9 years, what is the mean seniority μ?

Solution

(a) The standard score for x years of seniority is

$$z = \frac{x - 15}{5}$$

The situation is depicted graphically in Figure 6.22a. First look up the value in Appendix Table B.3 that satisfies

$$P(0 < z < \underline{?}) = .32$$

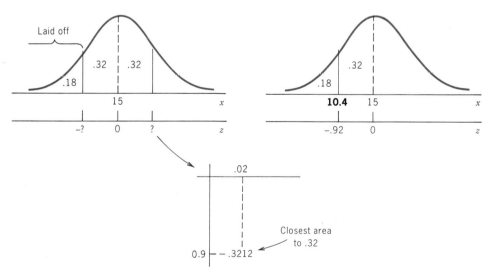

FIGURE 6.22a

Then take the negative of this value, getting $z = -.92$. Substitute and solve for x:

$$-.92 = \frac{x - 15}{5}$$

$$-4.6 = x - 15$$

$$15 - 4.6 = x$$

$$x = 10.4$$

So those with less than 10.4 years of seniority are laid off.

(b)   Here the standard score for x years of seniority is

$$z = \frac{x - \mu}{3}$$

We will find μ. We know that those with less than 9 years seniority are laid off, and this constitutes 35% of employees. (See Figure 6.22b.) As indicated in Figure 6.22b, we use Appendix Table B.3 to find the value that satisfies

$$P(0 < z < \underline{?}) = .15$$

Then take the negative of this, getting $z = -.39$. Now we know that this corresponds to a seniority of $x = 9$ years. If we substitute $z = -.39$ and $x = 9$ in the equation $z = (x - \mu)/3$, we can solve for μ.

$$-.39 = \frac{9 - \mu}{3}$$

$$-1.17 = 9 - \mu$$

$$\mu - 1.17 = 9$$

$$\mu = 9 + 1.17 = 10.17 \text{ years}$$

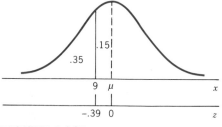

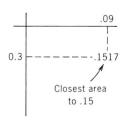

FIGURE 6.22b

In Example 6.9 we saw that the 1974 MSAT scores for men were normally distributed with $\mu = 500$ and $\sigma = 100$. If x represents an arbitrary score, we showed

$$P(400 < x < 600) = .6826$$

Now

$$400 = 500 - 100 = \mu - \sigma$$
$$600 = 500 + 100 = \mu + \sigma$$

Thus

$$P(\mu - \sigma < x < \mu + \sigma) = .6826$$

This means that about 68% of the MSAT scores will fall within one standard deviation of the mean. If we were to calculate the probability of a score being within two standard deviations of the mean, that is, the proportion of scores between $\mu - 2\sigma$ and $\mu + 2\sigma$, we would find it to be .9544. If we calculated the proportion of scores within three standard deviations of the mean, that is, between $\mu - 3\sigma$ and $\mu + 3\sigma$, we would find it to be .9974.

It turns out that these results hold for any normal random variable. For example, no matter what values we have for μ and σ, one standard deviation above the mean corresponds to $z = 1$ and one standard deviation below the mean corresponds to $z = -1$. So the proportion of data within one standard deviation of the mean is

$$P(-1 < z < 1) = .6826$$

as we saw in Example 6.9. Therefore if x is any normal variable with mean μ and standard deviation σ, then

$$P(\mu - \sigma < x < \mu + \sigma) = .6826$$
$$P(\mu - 2\sigma < x < \mu + 2\sigma) = .9544$$
$$P(\mu - 3\sigma < x < \mu + 3\sigma) = .9974$$

So for a normal population we expect:

Between	Percentage of Data
$\mu - \sigma$ and $\mu + \sigma$	68.26
$\mu - 2\sigma$ and $\mu + 2\sigma$	95.44
$\mu - 3\sigma$ and $\mu + 3\sigma$	99.74

This is the basis for the Empirical Rule discussed in Chapter 3. This rule says that if we have a large sample from a normal population and we use $\bar{x}$ and s as estimates for μ and σ, then the same proportions should hold for the sample, at least approximately.

Exercises

6.13 Let x be a normal random variable with mean 46 and standard deviation 4. What percentage of values are

(a)   Larger than 46?   (b)   Larger than 50?

(c)   Larger than 40?   (d)   Less than 38?

(e)   Less than 49?   (f)   Between 45 and 49?

(g)   Between 50 and 54?   (h)   Larger than 56 or less than 46?

(i)   Within 1.5 standard deviations of the mean?

(j)   Outside of 2.3 standard deviations of the mean?

6.14 Let x be a normal random variable with mean 80 and standard deviation 12. If a value is randomly selected from the x distribution, find each of the following probabilities:

(a) $P(80 < x < 92)$.   (b) $P(71 < x < 80)$.   (c) $P(x > 83)$.

(d) $P(x > 56)$.   (e) $P(x < 92)$.   (f) $P(x < 62)$.

(g) $P(89 < x < 95)$.   (h) $P(53 < x < 59)$.

(i) $P(65 < x < 98)$.   (j) $P(x = 80)$.

6.15 Consider a normal population with mean 200 and standard deviation 25. Find the following:

(a)   The third quartile.

(b)   The 35th percentile.

(c)   The percentile rank of the data value 211.

(d)   The percentile rank of the data value 150.

6.16 Below is a normal curve for a normal random variable x with mean 70 and standard deviation 10. Find the areas A, B, C, and D.

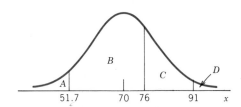

6.17 Below is a normal curve for a normal random variable x with mean 85 and standard deviation 8. Find the x values a and b.

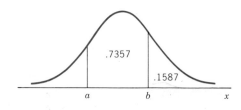

6.18   The lengths x of the nails in a large shipment of nails received by a carpenter are approximately normally distributed with mean 2 inches and standard deviation .1 inch.

   (a)   If a nail is randomly selected, find $P(1.8 < x < 2.07)$.

   (b)   What proportion of nails lie within 1 standard deviation of the mean?

   (c)   The carpenter cannot use a nail shorter than 1.75 inches or longer than 2.25 inches. What percentage of the shipment of nails will the carpenter be able to use?

6.19   The length of time it takes for a ferry to reach a summer resort from the mainland is approximately normally distributed with mean 2 hours and standard deviation 12 minutes. Over many past trips, what proportion of times has the ferry reached the island in

   (a)   Less than 1 hour and 45 minutes?

   (b)   More than 2 hours and 5 minutes?

   (c)   Between 1 hour, 50 minutes and 2 hours, 20 minutes?

6.20   Assume that the scores of males on the 1974 mathematical scholastic aptitude test (MSAT) were normally distributed with mean 500 and standard deviation 100.

   (a)   What score indicates a percentile rank of 95?

   (b)   The middle 40% of the distribution is bounded by what two scores?

   (c)   If 1000 new students are to take the exam, how many are expected to score more than 650?

6.21   The length of life x (in months) of a hair dryer is approximately normally distributed with mean 96 and standard deviation 18.

   (a)   The manufacturer decides to guarantee the product for 5 years. What percentage of the product will fail to satisfy the guarantee?

   (b)   The manufacturer decides to replace only 1% of all hair dryers. What should the length (in months) of the guarantee be?

6.22   Let x be the number of minutes after 11 o'clock a bus leaves the bus station. Assume that the distribution of times is approximately normal with mean 15 and standard deviation 4.

   (a)   If a person gets to the bus station at 11:10, what is the probability the person has missed the bus?

   (b)   If a person is willing to risk a 20% chance of not making the bus, what is the maximum number of minutes after 11 o'clock that the person can reach the station?

   (c)   What time should the person reach the station in order to have a 50–50 chance of catching the bus?

6.23   Assume the number of hours a product will function is approximately normally distributed.

   (a)   If the standard deviation were 70 and 10% of the product will break down in less than 700 hours, what would the mean be?

   (b)   If the mean time were 800 hours and 20% will function for more than 850 hours, what would the standard deviation be?

6.24 The sample mean and sample standard deviation of 35 data values are 250.77 and 6.62, respectively. The data and frequency histogram are given below.

(a) Find the percentage of data values within one, two, and three standard deviations of the mean. Compare with the Empirical Rule.

(b) Are the results of part (a) and the histogram consistent with sampling from a normal population?

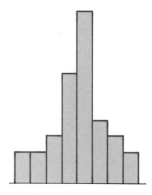

236	238	238	241	244	246	246
247	247	247	248	248	248	249
251	251	251	251	251	252	252
252	253	253	254	254	255	256
256	256	259	260	261	263	263

6.6 NORMAL APPROXIMATION TO THE BINOMIAL DISTRIBUTION

Recall that in Section 5.5 we discussed the binomial probability distribution

$$P(x) = \frac{n!}{x!(n-x)!} \cdot p^x \cdot q^{n-x} \qquad x = 0, 1, 2, \dots, n$$

Values of $P(x)$ have been worked out in Appendix Table B.2. But notice that this table only goes up to $n = 25$. For large values of n the formula for $P(x)$ can be quite unpleasant to work with. Even hand calculators are not much help. They do not have the capacity to handle factorials of large numbers. However, there is a way to approximate values of $P(x)$ for large values of n using normal probabilities. We will now show how this can be done by means of a concrete example.

A sociologist conducted interviews with 16 registered voters. The voters were randomly selected from a list of voters. Assume that half the voters on the list are Democrats. The process of selecting the 16 voters can be viewed as a binomial experiment. A trial is randomly selecting a voter. Let a success consist of selecting a Democrat. The probability of a success is the proportion of voters who are Democrats, in this case $\frac{1}{2}$. So

$$P(S) = p = \tfrac{1}{2}$$

The trial is repeated 16 times and x is the number of Democrats selected. (Assume the list is large enough so that the probability of success does not change ap-

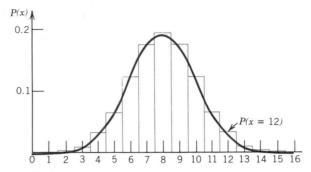

FIGURE 6.23a
Normal Approximation to a Binomial Distribution
with $n = 16$, $p = q = \frac{1}{2}$

preciably from one trial to the next, and so $p = \frac{1}{2}$ for each trial. This means that the trials are independent.) Therefore, we have a binomial experiment with $n = 16$ and $p = q = \frac{1}{2}$.

In Figure 6.23a we have sketched the probability distribution for x.
The area of each bar represents the probability of observing the value at the base of the bar. The bars taken together appear to resemble a bell shaped curve; we have superimposed a normal curve to emphasize this. But which normal curve is this? This is the normal curve of the distribution that has the same mean and standard deviation as the binomial variable. Now recall that for the binomial random variable

$$\mu = np = (16)(\tfrac{1}{2}) = 8$$

$$\sigma = \sqrt{npq} = \sqrt{(16)(\tfrac{1}{2})(\tfrac{1}{2})} = \sqrt{4} = 2$$

Now how can the normal curve be used to find binomial probabilities? Let us examine the probability of 12 successes, that is, $P(12)$. This is the area of the bar whose base stretches from 11.5 to 12.5. If you wish to look up the answer in Appendix Table B.2, you will find that it is .028. But let's see if we can get the answer by the use of the normal curve. It would seem natural to estimate the area of the bar by the area under the normal curve from 11.5 to 12.5. (See Figure 6.23b.)

Therefore, let us regard the binomial variable x as approximately normal, and compute the normal probability $P(11.5 < x < 12.5)$ and see just how close we come to the answer of .028. Now the standard score for x is

$$z = \frac{x - \mu}{\sigma} = \frac{x - 8}{2}$$

When $x = 11.5$,

$$z = \frac{11.5 - 8}{2} = 1.75$$

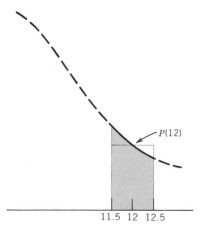

FIGURE 6.23*b*
Estimating *P*(12) by Area under a
Normal Curve (Shaded)

When $x = 12.5$,

$$z = \frac{12.5 - 8}{2} = 2.25$$

Thus $11.5 < x < 12.5$ is equivalent to $1.75 < z < 2.25$. Hence

$$P(11.5 < x < 12.5) = P(1.75 < z < 2.25)$$

(See Figure 6.23*c*.) Using Appendix Table B.3, we see that

$$P(1.75 < z < 2.25) = P(0 < z < 2.25) - P(0 < z < 1.75)$$
$$= .4878 - .4599 = .0279$$

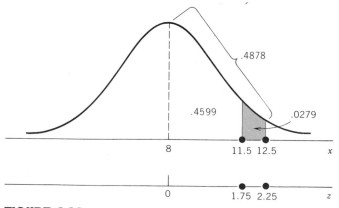

FIGURE 6.23*c*
$P(11.5 < x < 12.5) = .4878 - .4599 = .0279$

Therefore,

$$P(11.5 < x < 12.5) = .0279$$

Note that, rounded to three places, this gives the same result as Appendix Table B.2.

This shows how a binomial distribution may be approximated by an appropriate normal distribution. One thing that did not come out in the above discussion is the fact that it can be shown that the approximation is valid only when n is sufficiently large. It is known that n will be sufficiently large if $np \geq 5$ and $nq \geq 5$. For example, if $p = \frac{1}{2}$, then we would want n to be such that $np = n(\frac{1}{2}) \geq 5$ and $nq = n(\frac{1}{2}) \geq 5$. Therefore, we would want to have $n \geq 10$.

We summarize and generalize the above results as follows:

---

1. A binomial random variable x may be thought of as having an approximately normal distribution with mean np and standard deviation $\sqrt{npq}$ if n is sufficiently large. Note that n will be sufficiently large if both np and nq are at least 5.

2. When approximating a binomial probability with a normal probability, we must use the so-called *continuity correction*. For example, to find the probability of 12 successes, calculate the normal probability $P(11.5 < x < 12.5)$. To find the probability of 9 or 10 or 11 successes, calculate the normal probability $P(8.5 < x < 11.5)$.

Warning: The continuity correction discussed above should only be used when approximating a binomial probability with a normal probability. Don't use the continuity correction with other normal probability problems.

Example 6.13
Find the probability that between 30 and 35 of the next 50 births at a particular hospital will be boys. (By this we mean more than 30 but less than 35.)

Solution
Notice that the 50 births can be thought of as a binomial experiment.

$$\text{trial} = \text{a birth}$$
$$\text{success} = \text{the child is a boy}$$
$$p = P(\text{success}) = .5$$
$$q = 1 - p = .5$$
$$n = 50$$
$$x = \text{number of successes (boys)}$$

We want to find the probability that x is between 30 and 35 (i.e., 31 or 32 or 33 or 34). Now

$$np = (50)(.5) = 25 \quad \text{and} \quad nq = (50)(.5) = 25$$

Since np and nq are both at least 5, we may view x as approximately normal with

$$\mu = np = (50)(.5) = 25$$
$$\sigma = \sqrt{npq} = \sqrt{(50)(.5)(.5)} = \sqrt{12.5} \doteq 3.54$$

The desired probability is the sum of the areas of the rectangles above 31, 32, 33, and 34. This is approximated by the area under the normal curve between 30.5 and 34.5. So we must find the probability $P(30.5 < x < 34.5)$. (See Figure 6.24.) Let

$$z = \frac{x - \mu}{\sigma} = \frac{x - np}{\sqrt{npq}} = \frac{x - 25}{3.54}$$

This is approximately standard normal. When $x = 30.5$,

$$z = \frac{30.5 - 25}{3.54} \doteq 1.55$$

When $x = 34.5$,

$$z = \frac{34.5 - 25}{3.54} \doteq 2.68$$

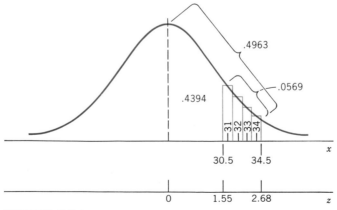

FIGURE 6.24
$P(30.5 < x < 34.5) = .4963 - .4394 = .0569$

Therefore,

$$P(30.5 < x < 34.5) = P(1.55 < z < 2.68)$$
$$= P(0 < z < 2.68) - P(0 < z < 1.55)$$
$$= .4963 - .4394 = .0569$$

This is the desired probability.

Example 6.14

A roulette wheel has 38 slots numbered 0, 00, and 1 through 36. Slots 0 and 00 are green, and of the remaining slots half are red and half black. If you place a bet of $1 on black and the ball drops in a black slot, you receive $1 (plus your original bet). If this game is played 100 times, find the probability of at least breaking even, that is, of winning at least 50 times.

Solution

This is a binomial experiment.

$$\text{trial} = \text{play the game}$$
$$\text{success} = \text{the ball drops on black number}$$
$$p = P(\text{success}) = \tfrac{18}{38} \doteq .47$$
$$q = 1 - p = .53$$
$$n = 100$$
$$x = \text{number of successes}$$

We must find the probability that x is 50 or more. Now $np = 47$ and $nq = 53$ so that x may be viewed as approximately normal with

$$\mu = np = (100)(.47) = 47$$
$$\sigma = \sqrt{npq} = \sqrt{(100)(.47)(.53)} = \sqrt{24.91} \doteq 4.99$$

When x is viewed as normal, the desired probability is obtained by calculating $P(x > 49.5)$. (We used the continuity correction.) The standard score for x is

$$z = \frac{x - \mu}{\sigma} = \frac{x - 47}{4.99}$$

This will be approximately standard normal. When $x = 49.5$,

$$z = \frac{49.5 - 47}{4.99} = .50$$

Hence

$$P(x > 49.5) = P(z > .5)$$
$$= .5 - P(0 < z < .5)$$
$$= .5 - .1915 = .3085$$

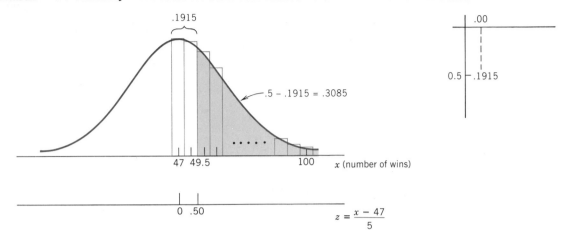

FIGURE 6.25
$P(x > 49.5) = .5 - .1915 = .3085$

Thus there is about a 31% chance of at least breaking even. (See Figure 6.25.) *Note:* The reader may ask why we did not calculate $P(49.5 < x < 100.5)$. The reason we did not do this is that the area under the curve to the right of 100.5 is practically 0. Therefore, including this area (probability) by calculating $P(x > 49.5)$ does not lead to a serious error. And it is easier to calculate $P(x > 49.5)$.

Example 6.15
A state health official wished to study the death rate from cancer in a Massachusetts town situated near a toxic waste dump. The death certificates of 200 randomly selected individuals in this town were examined, and it was found that 58 had died of cancer. It is known that 23% of all deaths in the state are due to cancer. If the death rate from cancer in this town were the same as for the state, find the probability that more than 57 out of 200 people in the town would die of cancer. What conclusions would you draw?

Solution
Checking each death certificate is a trial. Success is a death from cancer. If the death rate from cancer in the town is the same as for the state, then $p = P(S) = .23$ and $q = .77$. The trial is repeated $n = 200$ times, so we have a binomial experiment. The reader may verify that both np and nq are at least 5. Thus the number of successes x may be viewed as approximately normal with

$$\mu = np = (200)(.23) = 46$$

$$\sigma = \sqrt{npq} = \sqrt{(200)(.23)(.77)} = \sqrt{35.42} \doteq 5.95$$

Therefore

$$z = \frac{x - \mu}{\sigma} = \frac{x - 46}{5.95}$$

is approximately standard normal.

Let us find the probability of observing more than 57 deaths due to cancer out of 200 deaths. Using the continuity correction, we want $P(x > 57.5)$. When $x = 57.5$,

$$z = \frac{57.5 - 46}{5.95} \doteq 1.93$$

Then using Appendix Table B.3,

$$P(x > 57.5) = P(z > 1.93) = .5 - .4732 = .0268$$

Therefore, it would be very unlikely (less than a 3% chance) that we would observe more than 57 out of 200 deaths from cancer if the death rate from cancer in this town were really only 23%. Since the observed number of deaths was 58, it would seem that the death rate from cancer in this town is higher than the rate for the entire state.

Exercises

In Exercises 6.25–6.30 assume a binomial experiment. Using (a) Appendix Table B.2 and (b) the normal approximation to the binomial, do the following.

6.25   Find the probability of at least 10 successes, where $n = 13$ and $p = .5$.

6.26   Find the probability of at most 7 successes, where $n = 16$ and $p = .4$.

6.27   Find the probability of exactly 14 successes, where $n = 20$ and $p = .7$.

6.28   Find the probability of less than 10 successes, where $n = 14$ and $p = .4$.

6.29   Find the probability of between 5 and 7 successes inclusive, where $n = 15$ and $p = .6$.

6.30   Find the probability of exactly 10 successes, where $n = 12$, $p = .5$.

6.31   A circuit board retailer receives a shipment of 800 boards. The manufacturer claims that only 1% of the boards are defective. If the claim is true, find the probability that the shipment contains 15 or more defective circuit boards.

6.32   A civil service multiple choice examination (four choices per problem) consists of 100 questions. Suppose an applicant randomly guesses the answer to every question. Find the probability that the applicant will answer at least 35 questions correctly.

6.33   A city planning board denied a developer's request to increase the size of an existing mall. The board members claimed that they had the support of 60% of the voters.

   (a)   Suppose that 400 voters are selected and the planning board's claim is true. What is the probability that less than 211 voters will support the board's decision?

(b)   Suppose that a sample of 400 voters shows that 210 voters supported the board's decision. Do you feel that the board's claim is valid? Why? (*Hint:* If the board is correct, would it be unlikely that fewer than 211 voters would support the board's decision?)

6.34   Suppose 65% of 45 year olds with a diagnosed illness die between the ages of 45 to 50. Assume 500 45 year olds have the disease.

(a)   What is the probability that between 305 and 345 inclusive will die within the next 5 years?

(b)   The 500 were given a new treatment and 295 died instead of the expected 325. Do you think 65% is still a plausible percentage for a population receiving the treatment? (*Hint:* If 65% were still correct, how likely is it that less than 296 would die?)

6.35   A hospital administrator claimed that an unusually large number of girls were born at the hospital in the preceding year. A check of the records showed that 120 girls and 75 boys had been born at the hospital the previous year. Do you think the administrator's claim is reasonable? Why?

6.36   Between 1972 and 1974, 15 out of 405 teachers (3.7%) hired in Hazelwood, St. Louis County, Mo., were black. The Equal Employment Opportunity Commission (EEOC) sued Hazelwood for discrimination under Title VII of the Civil Rights Act of 1964.

(a)   The percentage of black school teachers in St. Louis County, plus the city of St. Louis, was 15.4%. Assuming a random sample of size 405 from a population of qualified teachers containing 15.4% blacks, find the approximate probability of observing fewer than 16 black teachers. Does your answer support the decision of a court of appeals that found discrimination against blacks?

(b)   The Supreme Court vacated the decision of the court of appeals. It noted that the relevant job market might be St. Louis County alone (which does not include the city of St. Louis), where the percentage of black teachers was only 5.7%. Redo part (a) using this percentage.

For a discussion see DeGroot, M., S. Fienberg, and J. Kadane, *Statistics and the Law* (New York: Wiley, 1986), pp. 1–48.

6.7 THE CENTRAL LIMIT THEOREM

Suppose that a company is considering publishing a book that prepares students for the Mathematical Scholastic Aptitude Test (MSAT). (We will let the variable x represent individual scores on the test.) To investigate the effectiveness of the book, 50 students are selected by chance and are asked to use the book to study for the examination. Their scores are then obtained and the sample mean $\bar{x}$ is computed. Now the average for all students on the MSAT is 476 (as of 1991). If the 50 students had an average score of, say, 750 on the test, we would conclude that the book was quite effective. Of course, we might have gotten 50 very bright students by chance but this seems unlikely. In other words, we concluded that the book is effective because it seems highly unlikely that the average score of the 50 students would be so high if the book were not effective.

In the above discussion we had to know something about what values of $\bar{x}$ were unlikely, and we did so on the basis of some general knowledge we had about SAT scores. However, it is not always so easy. We would like to have a better way of determining which values of $\bar{x}$ are likely and which are not.

The value of $\bar{x}$ could change from one sample of 50 students to the next. The fact is that $\bar{x}$ is a random variable. *What we are really interested in is the probability distribution of $\bar{x}$.* There are many situations in statistical inference similar to the one discussed above, where we need to know about the probability distribution of a sample mean $\bar{x}$ to make inferences about a random variable x.

It turns out that for any random variable x a result known as the **Central Limit Theorem** enables us to find probabilities associated with $\bar{x}$, because it tells us that under suitable conditions $\bar{x}$ will be approximately normal (even when x is not normal). Before stating this result more precisely, we investigate a concrete example.

Suppose we have a box containing three tags labeled 1, 3, and 5. Consider the experiment of selecting a tag at random, that is, in such a way that each tag has equal likelihood of being selected. Let x = the number on the tag selected. Since each number has an equal likelihood of being selected, $P(x) = \frac{1}{3}$ for $x =$ 1, 3, or 5. Now consider the mean of each sample of size $n = 2$ that can be drawn with replacement. To obtain such a sample, we randomly select a tag, record its value, return it to the box, and then repeat the process. Let us find the probability distribution for $\bar{x}$. [To distinguish the distribution of $\bar{x}$ from that of x, we will use the symbol $Q(\bar{x})$ when dealing with the probability distribution of $\bar{x}$]. There are nine possible samples.

Possible Samples	$\bar{x}$
1, 1	1
1, 3	2
1, 5	3
3, 1	2
3, 3	3
3, 5	4
5, 1	3
5, 3	4
5, 5	5

Each sample has an equal likelihood of occurring and so has the probability $\frac{1}{9}$. Using this fact, we may compute the probabilities associated with various

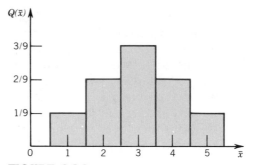

FIGURE 6.26
Probability Histogram for $\bar{x}$

values of $\bar{x}$. For example, the value $\bar{x} = 4$ can occur in two ways; therefore, $Q(4) = \frac{2}{9}$. Now we give the probability distribution of $\bar{x}$.

$\bar{x}$	$Q(\bar{x})$
1	$\frac{1}{9}$
2	$\frac{2}{9}$
3	$\frac{3}{9}$
4	$\frac{2}{9}$
5	$\frac{1}{9}$

The probability histogram of $Q(\bar{x})$ is shown in Figure 6.26.

Now we find the mean μ and variance σ^2 for x and compare them with the mean $\mu_{\bar{x}}$ and variance $\sigma_{\bar{x}}^2$ for $\bar{x}$.

$$\mu = \sum x \cdot P(x) = (1)(\tfrac{1}{3}) + (3)(\tfrac{1}{3}) + (5)(\tfrac{1}{3}) = 3$$

$$\sigma^2 = \left[\sum x^2 \cdot P(x) \right] - \mu^2 = (1)^2(\tfrac{1}{3}) + (3)^2(\tfrac{1}{3}) + (5)^2(\tfrac{1}{3}) - (3)^2$$

$$= \tfrac{1}{3} + \tfrac{9}{3} + \tfrac{25}{3} - 9$$

$$= \tfrac{35}{3} - 9 = \tfrac{8}{3}$$

$$\mu_{\bar{x}} = \sum \bar{x} \cdot Q(\bar{x})$$

$$= (1)(\tfrac{1}{9}) + (2)(\tfrac{2}{9}) + (3)(\tfrac{3}{9}) + (4)(\tfrac{2}{9}) + (5)(\tfrac{1}{9})$$

$$= \frac{1 + 4 + 9 + 8 + 5}{9} = \tfrac{27}{9} = 3$$

$$\sigma_{\bar{x}}^2 = \left[\sum \bar{x}^2 \cdot Q(\bar{x}) \right] - \mu_{\bar{x}}^2$$

$$= (1)^2(\tfrac{1}{9}) + (2)^2(\tfrac{2}{9}) + (3)^2(\tfrac{3}{9}) + (4)^2(\tfrac{2}{9}) + (5)^2(\tfrac{1}{9}) - (3)^2$$

$$= \tfrac{1}{9} + \tfrac{8}{9} + \tfrac{27}{9} + \tfrac{32}{9} + \tfrac{25}{9} - 9 = \tfrac{93}{9} - 9 = \tfrac{4}{3}$$

Note that

- $\mu_{\bar{x}} = \mu$.

- $\sigma_{\bar{x}}^2 = \sigma^2/2$. Observe that since $n = 2$ was the sample size, we can say that $\sigma_{\bar{x}}^2 = \sigma^2/n$. (So $\sigma_{\bar{x}} = \sigma/\sqrt{n}$.)

- We should also note that the probability histogram in Figure 6.26 is somewhat bell shaped. If the sample size were to increase, the values of $\bar{x}$ would start to fill in a portion of the axis like continuous data. For example, for samples of size $n = 3$ drawn from the three tags, the possible values of $\bar{x}$ are 1, $1\frac{2}{3}$, $2\frac{1}{3}$, 3, $3\frac{2}{3}$, $4\frac{1}{3}$, and 5. A histogram would more and more resemble a normal curve as the sample size increased. This suggests that when n is large enough, $\bar{x}$ will be approximately normal. Figure 6.27 depicts a computer-generated frequency histogram describing 100 sample means for samples of size 30. (The values on the horizontal axis are the midpoints of the classes.)

The above ideas generalize to other random variables. Before stating this generalization, we give a formal definition of a random sample.

> **Definition**   A *random sample* of size n is a sample of size n obtained from a population in such a manner that all samples of the same size from the population have an equal likelihood of being selected. Any method of obtaining random samples is called *random sampling*.

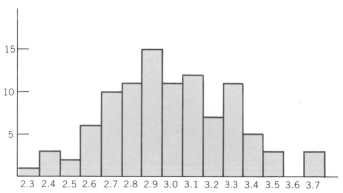

FIGURE 6.27
Histogram of 100 Sample Means for Samples of Size $n = 30$

Central Limit Theorem   Let x represent data values from a population with mean μ and standard deviation σ, and let $\bar{x}$ represent the sample mean defined for random samples of size n. Then $\bar{x}$ is a random variable with the following properties:

1.   The mean of $\bar{x}$ is $\mu_{\bar{x}} = \mu$.
2.   The standard deviation of $\bar{x}$ is $\sigma_{\bar{x}} = \sigma/\sqrt{n}$.
3. $\bar{x}$ will be approximately normal when n is sufficiently large ($n \geq 30$ will usually be sufficient). The larger the value of n, the closer the distribution of $\bar{x}$ will be to normality.

Figure 6.28 illustrates the Central Limit Theorem for a particular population. Two points should be mentioned in connection with the Central Limit Theorem.

1.   The quantity $\sigma_{\bar{x}}$ is sometimes called the *standard error of the mean.*

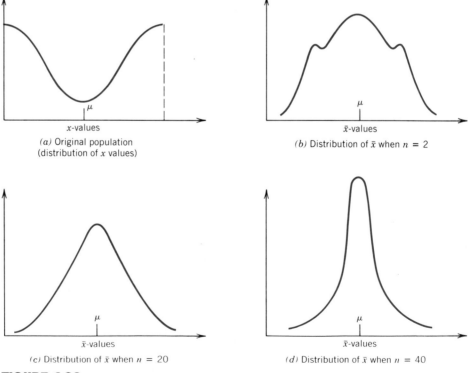

x-values
(a) Original population
(distribution of x values)

$\bar{x}$-values
(b) Distribution of $\bar{x}$ when $n = 2$

$\bar{x}$-values
(c) Distribution of $\bar{x}$ when $n = 20$

$\bar{x}$-values
(d) Distribution of $\bar{x}$ when $n = 40$

FIGURE 6.28
As the Sample Size n Becomes Larger, Distribution of $\bar{x}$ Approaches Normality and $\sigma_{\bar{x}}$ Becomes Smaller

2. The Central Limit Theorem states that $\bar{x}$ will be approximately normal for "n sufficiently large." We said that $n \geq 30$ will usually suffice. We use 30 as a dividing point between large and small samples. So if $n \geq 30$, the sample is said to be *large*. If $n < 30$, the sample is *small*. Note that we don't require x to be normal for $\bar{x}$ to be (approximately) normal. We don't even require x to be continuous. However, if x happens to be normal (or approximately normal), then $\bar{x}$ will be normal (or approximately normal) for all values of n.

It is of interest to note that the Central Limit Theorem can be offered as a possible explanation of the fact that so many variables encountered in nature are normal. Take the variable: height of an adult male American. Height is influenced by many factors: the height of one's ancestors as well as other genetic and environmental factors. Perhaps we can think of height as a sort of mean of all these factors. If height is a mean, this would explain why it is approximately normal.

Consider the following example: Let x be the number of dots showing on the toss of a die. (The probability histogram for x is shown in Figure 6.29.) Now look at the data in Table 6.1. This table contains 25 random samples of size $n = 10$ obtained by tossing the die. Each sample is obtained by tossing the die 10 times and recording the number (x) on the face showing each time. A frequency histogram for the sample means is given in Figure 6.30. The shape suggests normality, even though the sample size ($n = 10$) is less than 30.

The mean of the 25 values of $\bar{x}$ in Table 6.1 is 3.52 and the standard deviation is .53. The 25 samples in Table 6.1 do not constitute all possible random samples. (There are over 60 million possibilities.) But we can use 3.52 and .53 as estimates for the mean and standard deviation of *all* the possible values of $\bar{x}$, denoted by $\mu_{\bar{x}}$ and $\sigma_{\bar{x}}$, respectively. Let's see how these estimates check with the values predicted by the Central Limit Theorem.

For the random variable $x = $ the number of dots showing on the die, the probability distribution is

$$P(x) = \tfrac{1}{6} \qquad \text{for } x = 1, 2, 3, 4, 5, 6$$

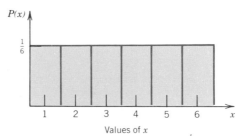

FIGURE 6.29
Probability Histogram for $x = $ Number of Dots Observed when a Die Is Tossed

TABLE 6.1
Sample Obtained by Tossing a Die

Sample Number	Samples of Size $n = 10$										Sample Mean $\bar{x}$
1	1	3	2	5	3	3	6	5	2	1	3.1
2	3	5	6	3	4	6	3	5	4	6	4.5
3	5	1	1	3	2	2	4	4	5	3	3.0
4	4	2	4	5	6	4	4	3	2	6	4.0
5	2	3	2	2	5	6	1	5	5	3	3.4
6	3	4	6	4	3	5	3	3	6	6	4.3
7	1	2	5	2	3	6	1	5	3	3	3.1
8	3	1	1	3	1	1	6	5	4	3	2.8
9	4	4	3	6	2	6	5	1	2	1	3.4
10	3	6	1	4	6	4	3	2	3	6	3.8
11	6	5	3	5	1	1	6	1	2	1	3.1
12	1	6	3	3	6	2	6	5	4	1	3.7
13	3	4	2	6	3	6	6	3	1	6	4.0
14	5	1	3	3	5	1	1	1	2	1	2.3
15	3	1	6	1	5	4	5	5	1	2	3.3
16	4	6	3	5	3	3	4	2	6	1	3.7
17	4	6	4	3	2	3	4	3	3	1	3.3
18	5	2	6	6	2	2	5	1	6	2	3.7
19	2	6	6	6	5	4	1	3	5	3	4.1
20	2	4	1	6	2	3	5	6	3	3	3.5
21	3	2	1	3	5	5	3	2	5	4	3.3
22	4	5	5	3	5	5	4	2	4	6	4.3
23	2	3	4	1	2	6	2	4	3	6	3.3
24	4	5	5	3	3	2	6	4	3	5	4.0
25	3	2	2	6	3	2	3	4	1	3	2.9

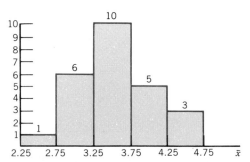

FIGURE 6.30
Frequency Histogram for the Sample Means of Table 6.1

Now

$$\mu = \sum x \cdot P(x)$$

$$= (1)(\tfrac{1}{6}) + (2)(\tfrac{1}{6}) + (3)(\tfrac{1}{6}) + (4)(\tfrac{1}{6}) + (5)(\tfrac{1}{6}) + (6)(\tfrac{1}{6}) = 3.5$$

$$\sigma^2 = \left[\sum x^2 \cdot P(x)\right] - \mu^2 = (1)(\tfrac{1}{6}) + (4)(\tfrac{1}{6})$$

$$+ (9)(\tfrac{1}{6}) + (16)(\tfrac{1}{6}) + (25)(\tfrac{1}{6}) + (36)(\tfrac{1}{6}) - (3.5)^2 \doteq 2.92$$

so that

$$\sigma = \sqrt{2.92} \doteq 1.71$$

Therefore, according to the Central Limit Theorem

$$\mu_{\bar{x}} = \mu = 3.5$$

$$\sigma_{\bar{x}} = \frac{\sigma}{\sqrt{n}} = \frac{1.71}{\sqrt{10}} \doteq .54$$

These true values and the estimates agree quite closely.

Probabilities associated with values of a sample mean may be found by viewing $\bar{x}$ as normal (when the sample size n is at least 30) and using the mean and standard deviation given in the Central Limit Theorem. Before giving some examples, there is an important point to be made. The reader may have noted a similarity between this section and the last section, where we said that a binomial variable may be viewed as (approximately) normal under certain conditions. There is an important difference, however. You will note that *when finding the probabilities associated with $\bar{x}$ we will not make any continuity correction.* The only time we will make a continuity correction in this book is when we approximate binomial probabilities by using a normal distribution.

Example 6.16
Let x represent the scores on a mathematics achievement test given to eighth grade students. Suppose that for the population of all scores on the test, the mean $\mu = 50$ and standard deviation $\sigma = 15$. If a random sample of 100 values of x is to be obtained, find the probability that the sample mean $\bar{x}$ will be between 51 and 53; that is, find $P(51 < \bar{x} < 53)$.

Solution
According to the Central Limit Theorem we may view $\bar{x}$ as approximately normal with

$$\mu_{\bar{x}} = \mu = 50 \quad \text{and} \quad \sigma_{\bar{x}} = \frac{\sigma}{\sqrt{n}} = \frac{15}{\sqrt{100}} = 1.5$$

Therefore,

$$z = \frac{\bar{x} - \mu_{\bar{x}}}{\sigma_{\bar{x}}} = \frac{\bar{x} - \mu}{\sigma/\sqrt{n}} = \frac{\bar{x} - 50}{1.5}$$

is approximately standard normal. When $\bar{x} = 51$,

$$z = \frac{51 - 50}{1.5} \doteq .67$$

When $\bar{x} = 53$,

$$z = \frac{53 - 50}{1.5} = 2$$

Therefore, using Appendix Table B.3,

$$P(51 < \bar{x} < 53) = P(.67 < z < 2)$$
$$= P(0 < z < 2) - P(0 < z < .67)$$
$$= .4772 - .2486 = .2286$$

See Figure 6.31.

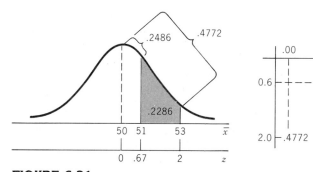

FIGURE 6.31
$P(51 < \bar{x} < 53) = .4772 - .2486 = .2286$

Example 6.17
The mean income of all wage earners in a large city is $30,000 with a standard deviation of $3000. If 40 wage earners are selected at random, find the probability that the mean of their salaries will be less than $30,750.

Solution
Let x represent salaries. Now $n = 40$, $\mu = 30,000$, and $\sigma = 3000$. By the Central Limit Theorem, $\bar{x}$ is approximately normal and

$$\mu_{\bar{x}} = \mu = 30,000$$

$$\sigma_{\bar{x}} = \frac{\sigma}{\sqrt{n}} = \frac{3000}{\sqrt{40}} \doteq 474.34$$

Therefore,

$$z = \frac{\bar{x} - \mu_{\bar{x}}}{\sigma_{\bar{x}}} = \frac{\bar{x} - 30{,}000}{474.34}$$

is approximately standard normal. When $\bar{x} = 30{,}750$,

$$z = \frac{30{,}750 - 30{,}000}{474.34} \doteq 1.58$$

Hence

$$P(\bar{x} < 30{,}750) = P(z < 1.58)$$
$$= .5 + .4429 = .9429$$

(See Figure 6.32.)

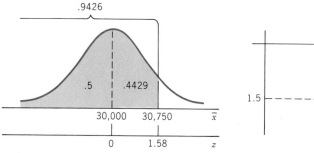

FIGURE 6.32
$P(\bar{x} < \$30{,}750) = .5 + .4429 = .9429$

Example 6.18

A company has been producing a 60-watt light bulb that has shown a mean life of 750 hours with a standard deviation of 30 hours. An engineer developed a new process for producing the bulb. It was felt that bulbs produced by the new process would show the same standard deviation but possibly a longer mean lifetime. Thirty-six bulbs produced by the new process were tested and showed a (sample) mean lifetime of 765 hours.

(a) If the (population) mean lifetime μ of bulbs produced by the new process were still 750 hours, find the probability of getting a value of the sample mean as large or larger than 765.

(b) What conclusions might be drawn from the answer to part (a)?

Solution

(a)   We use x to represent the lifetime of an arbitrary bulb produced by the new process. A sample of size $n = 36$ gave $\bar{x} = 765$. We must find the probability of getting a value of $\bar{x} \geq 765$. We are assuming for the moment that $\mu = 750$. We are also given that $\sigma = 30$. By the Central Limit Theorem, $\bar{x}$ is approximately normal with

$$\mu_{\bar{x}} = \mu = 750 \qquad \text{and} \qquad \sigma_{\bar{x}} = \frac{\sigma}{\sqrt{n}} = \frac{30}{\sqrt{36}} = \frac{30}{6} = 5$$

so that

$$z = \frac{\bar{x} - \mu_{\bar{x}}}{\sigma_{\bar{x}}} = \frac{\bar{x} - 750}{5}$$

is approximately standard normal. When $\bar{x} = 765$,

$$z = \frac{765 - 750}{5} = 3$$

From Appendix Table B.3,

$$P(\bar{x} \geq 765) = P(z \geq 3) = .5 - P(0 < z < 3) = .5 - .4987 = .0013$$

See Figure 6.33.

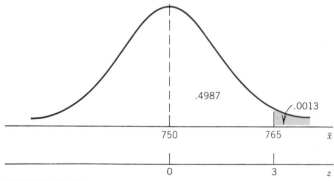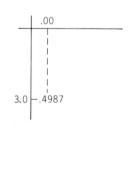

FIGURE 6.33
$P(\bar{x} \geq 765) = .5 - .4987 = .0013$

(b)   Notice that the probability obtained in part (a) is very small. It tells us that if the mean lifetime μ were only 750 hours, it would be very unlikely that we would observe a sample mean lifetime as large or larger than 765 hours (only about a 0.1% chance). Therefore, since we did observe such a large value for $\bar{x}$, we would be inclined to conclude that μ is not really 750 hours, but in fact more. In other words, the new process appears to increase the mean lifetime. So we have decided that a figure of $\bar{x} = 765$ hours should not be regarded as a routine chance fluctuation above an average of 750 hours, but instead should be interpreted as reflecting a real difference between the old and new processes.

Remark In the above example we used a probability value to make a decision about a population. We concluded that the mean lifetime μ for the population of bulbs produced by the new process would be more than 750 hours. (A similar technique was used in Example 6.14.) The probability .0013 used to make our decision is called a P value. We will have more to say about P values in Chapter 7, including a formal definition.

Exercises

6.37 This exercise demonstrates the concept of $\bar{x}$ as a random variable. Suppose that a large urn contains a collection of balls, $\frac{1}{3}$ of them numbered "0," $\frac{1}{3}$ of them numbered "2," and $\frac{1}{3}$ of them numbered "4." An experiment consists of selecting a ball, noting its number, putting the ball back into the urn, selecting another ball, and noting its number. There are nine distinct ordered pairs of numbers that can be obtained, namely, (0, 0), (0, 2), (0, 4), (2, 0), (2, 2), (2, 4), (4, 0), (4, 2), and (4, 4). For example, a (2, 4) means that a 2 was selected first, followed by a 4. Note that $\bar{x} = 3$ for the pair (2, 4).

(a) Obtain the probability distribution for $\bar{x}$.

(b) Find $\mu_{\bar{x}}$ and $\sigma_{\bar{x}}^2$.

Let $x = $ the number of the ball obtained on any selection. Then the probability distribution for x is given by

x	$P(x)$
0	$\frac{1}{3}$
2	$\frac{1}{3}$
4	$\frac{1}{3}$

(c) Find (i) μ, and (ii) σ^2 with formulas from Chapter 5. Compare with part (b) and note that $\mu = \mu_{\bar{x}}$ and $\sigma_{\bar{x}}^2 = \sigma^2/n$, where $n = 2$, the sample size.

6.38 A die has one side marked 1, two sides marked 2, and three sides marked 3. Let x be the number showing when the die is rolled once. Then the probability distribution for x is

x	$P(x)$
1	$\frac{1}{6}$
2	$\frac{2}{6}$
3	$\frac{3}{6}$

(a) Verify that $\mu = \frac{7}{3}$ and $\sigma^2 = \frac{5}{9}$.

Now consider the sample mean $\bar{x}$ for random samples of size 49.

(b) What is the largest value that $\bar{x}$ can assume? The smallest?

(c) According to the Central Limit Theorem, what are the values of $\mu_{\bar{x}}$ and $\sigma_{\bar{x}}^2$?

(d) What can be said about the distribution of sample means? Why?

6.39 A population has mean 325 and variance 144. Suppose the distribution of sample means is generated by random samples of size 36.

(a) Find $\mu_{\bar{x}}$, and $\sigma_{\bar{x}}$.   **(b)** Find $P(\bar{x} < 323)$.

(c) Find $P(320 < \bar{x} < 322)$.   **(d)** Find $P(\bar{x} > 328)$.

(e) Find $P(321 < \bar{x} < 327)$.

6.40 A normal population has mean 200 and standard deviation 100. Suppose that the distribution of sample means is generated by samples of size $n = 100$.

(a) Find $\mu_{\bar{x}}$.   **(b)** Find $\sigma_{\bar{x}}$.

(c) Find $P(195 < \bar{x} < 205)$.   **(d)** Find $P(\bar{x} > 210)$.

(e) If one x value is selected, find $P(195 < x < 205)$. Compare with part (c).

(f) If one x value is selected, find $P(x > 210)$. Compare with part (d).

6.41 Refer to Exercise 6.40. Suppose that the distribution of sample means is generated by samples of size $n = 400$.

(a) Find $\mu_{\bar{x}}$.

(b) Find $\sigma_{\bar{x}}$.

(c) Without calculating, do you believe that $P(195 < \bar{x} < 205)$ is smaller, equal to, or larger than the answer obtained in Exercise 6.40, part (c)? Now calculate $P(195 < \bar{x} < 205)$.

(d) Without calculating, do you believe that $P(\bar{x} > 210)$ is smaller, equal to, or larger than the answer obtained in Exercise 6.40, part (d)? Now calculate $P(\bar{x} > 210)$.

6.42 At a city high school, past records indicate that students have a mean of 510 and a standard deviation of 90 for MSAT scores. One hundred students in the high school are to take the test. What is the probability that their mean score will be

(a) More than 530?   **(b)** Less than 500?   **(c)** Between 495 and 515?

6.43 At a large factory, the mean wage is $42,500 and the standard deviation is $2000. What is the probability that the mean wage of 75 randomly selected workers will exceed $43,000?

6.44 A light bulb manufacturer claims that the mean life of his product is 1200 hours. Assume that the standard deviation is 120 hours. A consumer agency decides to randomly select 35 bulbs and will reject the claim if $\bar{x} < 1160$ hours. If the manufacturer's claim is true, what is the probability that the claim will be rejected?

6.45 A grocery store produce manager is told by a wholesaler that the apples in a large shipment have a mean weight of 6 ounces and a standard deviation of 1 ounce. The manager is going to randomly select 100 apples.

(a) Assuming the wholesaler's claim is true, find the probability that the mean of the sample is more than 5.9 ounces.

(b)   The manager decides to return the shipment if the mean weight of the sample is less than 5.75 ounces. Assuming the wholesaler's claim is true, find the probability that the shipment of apples will be returned.

(c)   Suppose the retailer is willing to risk a 1% chance of returning the shipment if the wholesaler's claim is true. Let W be the mean weight of the sample below which the shipment will be returned. Find the value of W.

6.46   A biology teacher had noted that students from her past classes had a mean of 74 and standard deviation 14 on a standardized examination. The teacher decided to use a new book. She felt this would increase the mean, but that the standard deviation would remain at 14. Using the new book, a class of 50 students had a mean of 76 on the standardized examination.

(a)   Find the probability that a class of 50 students using the new book will have a (sample) mean as large or larger than 76 if the new book were equivalent to the old book (i.e., if $\mu = 74$ and $\sigma = 14$ for all students using the new book.)

(b)   Do you feel that there is evidence to support the teacher's claim that the new book is superior? [*Hint:* Use part (a).]

6.8 USING MINITAB (OPTIONAL)

For discrete distributions PDF gives values of the probability distribution $P(x)$, which are probabilities. For continuous distributions, PDF also gives values of the probability distribution, represented by the symbol $f(x)$. We have seen, however, that values of $f(x)$ are not probabilities. Probabilities are areas under ·the graph. Hence, the PDF command is not quite as useful for continuous distributions.

However, the CDF command is quite useful. Figure 6.34 shows the area computed from a normal distribution by the CDF command.

In Example 6.10 we saw that heights of females are approximately normal

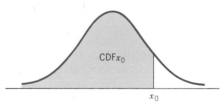

FIGURE 6.34
CDFx_0 Is the Area under the Curve to Left of x_0

with $\mu = 64$ inches, $\sigma = 2.5$ inches. To find the proportion of females shorter than 68 inches, we could type

CDF 68;

NORMAL MU = 64, SIGMA = 2.5.

Minitab would give the value .9452. This can be understood by looking at Figure 6.20*a* and Example 6.10. There we saw that the area from 64 to 68 was .4452. But the area to the left of 64 is .5. So the area to the left of 68 is .5 + .4452 = .9452.

In Example 6.10 we found the proportion of females between 65 and 68 inches. Figure 6.35 suggests how to do this using Minitab.

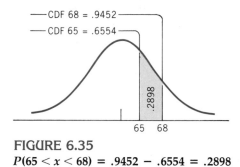

FIGURE 6.35
$P(65 < x < 68) = .9452 - .6554 = .2898$

The following Minitab program computes the desired probability. Note the abbreviated form of the subcommand—we only need the word NORMAL plus two numbers: the mean and standard deviation.

```
MTB > SET C1
DATA> 65  68
DATA> END
MTB > CDF C1 STORE IN C2;
SUBC>  NORMAL 64  2.5.
MTB > LET K1 = C2(2) - C2(1)
MTB > PRINT K1
K1       0.289779
```

C2
0.655422
0.945201

We have seen in Figure 6.35 that the proportion of females less than 68 inches is .9452. So CDF 68 = .9452. Therefore according to the definition of the INVCDF command described in the previous chapter

INVCDF .9452 = 68

Another way to look at this is that 68 is the 94.52 percentile. Thus the INVCDF command can be used to find percentiles.

In Example 6.11, we considered percentiles for the scores of males on the MSAT. The following program gives the 90th percentile, 628:

```
MTB > INVCDF .90;
SUBC>   NORMAL 500  100.
       0.9000  628.1552
```

INVCDF can also be used to find values of the standard normal variable such as $z(.025)$. We have seen that $z(.025) = 1.96$. The area to the right of $z(.025)$ under the standard normal curve is .025. (See Figure 6.14.) Hence the area to the left of $z(.025)$ is $1 - .025 = .975$. This means that

$$z(.025) = \text{INVCDF } .975$$

The following program returns the value of $z(.025)$:

INVCDF .975;

NORMAL 0,1.

Note: If no subcommand is used with PDF, CDF, or INVCDF, Minitab assumes a standard normal distribution. Therefore we didn't need the subcommand above.

Simulation

The RANDOM command can be used to generate samples from various distributions (populations). The distribution to be used is specified by a subcommand. The following program generates a sample of 100 values of a binomial variable for which $n = 16$, $p = .5$. Recall that $\mu = 8$, $\sigma = 2$ for such a binomial variable. In Section 6.6, we examined this binomial distribution and noted that it is approximated by a normal distribution. The following printout reflects this:

```
MTB > RANDOM 100 VALUES INTO C1;
SUBC>   BINOMIAL N = 16, P = .5.
MTB > MEAN C1
    MEAN    =      7.9100
MTB > STDEV C1
    ST.DEV. =      1.9648
MTB > HISTOGRAM FOR C1;
SUBC>   INCREMENT = 1;
SUBC>   START AT .5.

Histogram of C1  N = 100

Midpoint  Count
    0.50      0
    1.50      0
    2.50      0
    3.50      1  *
```

```
4.50      3   ***
5.50      6   ******
6.50     12   ************
7.50     21   *********************
8.50     23   ***********************
9.50     11   ***********
10.50     14   **************
11.50      6   ******
12.50      1   *
13.50      2   **
```

Note the bell-shaped histogram.

The RANDOM command can place data in several columns at once. Consider the following printout:

```
MTB > RANDOM 10 VALUES INTO C1 — C5;
SUBC>  NORMAL MU = 60,  SIGMA = 4.
MTB > PRINT C1 — C5
ROW      C1         C2         C3         C4         C5
 1    56.4312    60.8153    49.8243    57.9456    59.7937
 2    53.1379    68.4529    55.1153    63.0129    67.7372
 3    64.8729    61.1107    61.2268    63.6500    62.0051
 4    61.7565    60.5206    62.4910    67.2951    65.1568
 5    58.6765    60.5093    60.4830    60.2945    55.3945
 6    59.3649    54.5494    57.1270    62.5233    59.1065
 7    66.5755    61.5291    63.1417    61.2651    64.9362
 8    55.6163    66.3027    62.5761    62.4052    57.1530
 9    61.8896    56.2025    51.5280    56.2052    58.5584
10    54.5946    53.6515    58.8242    61.5699    62.0890
```

There are two ways to look at this. We can view this as 5 columns, each containing a sample of size 10; or 10 rows, each containing a sample of size 5. In some ways the latter interpretation is more useful, because Minitab has some very nice commands that operate on rows. The row mean command RMEAN is one of these. We could have added the following command after the subcommand:

RMEAN FOR C1–C5 PUT IN C6

This command computes the mean for each row and puts it in C6. Column 6 then contains 10 values of $\bar{x}$ for samples of size 5.

In Section 6.7 we simulated 25 samples of 10 rolls of a die and then looked at the 25 sample means. Let's use Minitab to generate 50 samples of size 40. We could then obtain 50 values of $\bar{x}$ using the RMEAN command. According to the Central Limit Theorem, $\bar{x}$ should be approximately normal. First we store the values for x (the number on the roll of a die) in C1 and the probabilities $P(x)$ in C2. Note that $P(x) = 1/6 \doteq .16667$. Then we simulate 50 samples of size 40.

```
MTB > READ INTO C1 C2
DATA> 1   .16667
DATA> 2   .16667
DATA> 3   .16667
DATA> 4   .16667
DATA> 5   .16667
DATA> 6   .16667
DATA> END
      6 ROWS READ
MTB > RANDOM 50 VALUES INTO C3 - C42;
SUBC>  DISCRETE, VALUES IN C1, PROBS IN C2.
MTB > RMEAN C3 - C42, PUT IN C43
MTB > MEAN FOR C43
  MEAN   =     3.4565
MTB > STDEV FOR C43
   ST.DEV. =   0.28695
MTB > HISTOGRAM FOR C43

Histogram of C43   N = 50

Midpoint  Count
    2.8      2  **
    3.0      3  ***
    3.2     10  **********
    3.4     10  **********
    3.6     16  ****************
    3.8      7  *******
    4.0      1  *
    4.2      1  *
```

In Section 6.7 we saw that for the number observed on the roll of a die,

$$\mu_{\bar{x}} = \mu = 3.5 \qquad \sigma_{\bar{x}} = \frac{\sigma}{\sqrt{n}} = \frac{1.71}{\sqrt{40}} \doteq .27$$

We see that the mean for C43 is 3.4565 and the standard deviation is .28695. This is consistent with the Central Limit Theorem. So is the histogram, which looks fairly normal.

Practice Quiz (Answers can be found after the Chapter Review exercises.)

Without using Minitab, give the approximate values that will be returned by the programs in Questions 1 through 4. Also indicate what the values represent.

1. CDF 70, STORE IN K1;
      NORMAL 50 10.
    LET K2 = 1 − K1
    PRINT K2

2. SET C1
    60 70
    END
    CDF C1, STORE IN C2;
        NORMAL 50 10.
    LET K1 = C2(2) − C2(1)
    PRINT K1
3. INVCDF .75;
        NORMAL 50 10.
4. INVCDF .90
5. Write a single program to compute $z(.01)$, $z(.025)$, $z(.05)$, and $z(.10)$.

Exercises

Suggested exercises for use with Minitab: 6.7 (a, b, c, d), 6.9, 6.10 (a, c, e), 6.13 (c, e, f, h), 6.15, 6.21, 6.42, 6.43, and 6.44.

SUMMARY

In this chapter we studied **continuous probability distributions** (i.e., probability distributions for continuous random variables). The most important class of continuous probability distributions is the class of **normal distributions.** The **standard normal random variable** z is the normal variable that has mean 0 and standard deviation 1. Appendix Table B.3 can be used to find probabilities associated with z. If x is a normal variable with mean μ and standard deviation σ, we can find probabilities associated with x by using the standard score of x, $z = (x - \mu)/\sigma$, which has the standard normal distribution.

We showed how a binomial distribution may be approximated by a normal distribution with mean $\mu = np$ and standard deviation $\sigma = \sqrt{npq}$ when both np and nq are at least 5.

Finally, we studied the **Central Limit Theorem.** This important theorem tells us that if x is a random variable with mean μ and standard deviation σ, then the sample mean $\bar{x}$ for random samples of size n has the following properties:

1. $\mu_{\bar{x}} = \mu$.
2. $\sigma_{\bar{x}} = \sigma/\sqrt{n}$.
3. When $n \geq 30$, $\bar{x}$ will be approximately normal.

We will see in later chapters that the Central Limit Theorem is extremely important in the study of statistical inference.

Review Exercises

6.47   Let x represent the time in hours required to repair a machine. The probability distribution follows:

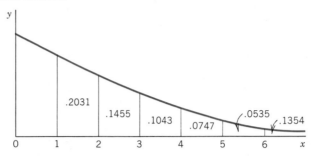

What is the probability that the machine will be repaired in

(a)   Less than 1 hour?   **(b)**   More than 6 hours?

(c)   Between 1 and 3 hours?   **(d)**   More than 1 hour?

(e)   Either less than 2 hours or more than 3 hours?

6.48   Let x be a continuous random variable with the following probability distribution:

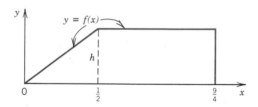

Assume that $P(0 < x < \frac{1}{2}) = \frac{1}{8}$.

(a)   What is the value of h?

(b)   Find the expression $f(x)$. [*Hint:* Find $f(x)$ when $0 \le x < \frac{1}{2}$, then $f(x)$ when $\frac{1}{2} \le x \le \frac{9}{4}$.]

(c)   Find $P(\frac{1}{2} < x < \frac{9}{4})$.   **(d)**   Find $P(1 < x < \frac{9}{4})$.

(e)   Find $P(0 < x < \frac{7}{4})$.   **(f)**   Find $P(0 \le x < \frac{1}{4})$.

6.49   Let x be a continuous random variable with the following probability distribution:

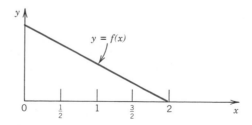

Suppose that $f(0) = 1$, $f(\frac{1}{2}) = \frac{3}{4}$, $f(1) = \frac{1}{2}$, $f(\frac{3}{2}) = \frac{1}{4}$, $f(2) = 0$. An x value is randomly selected. Find

(a) $P(0 < x < 1)$.   **(b)** $P(\frac{1}{2} < x < 1)$.

   (c) $P(x > \frac{3}{2})$.   **(d)** $P(x < \frac{1}{2})$.

   [*Hint:* Use the idea of area under the curve representing probability. Then use the formula for area of a triangle and the area of a rectangle.]

6.50   Let x be the number of years a person teaches in a public school system. The probability distribution follows:

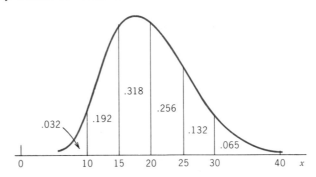

   Find the percentage of teachers who teach in the system

   (a)   At most 15 years.

   (b)   Between 10 and 25 years.

   (c)   More than 30 years.

6.51   Let x represent the length of life (in years) for an electrical component. The probability distribution for x follows:

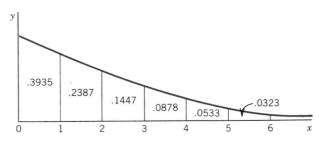

   (a)   What is the total area under the curve?

   What is the probability that a randomly selected component will last

   (b)   Longer than 6 years?   **(c)**   Between 2 and 4 years?

   (d)   Longer than 1 year?   **(e)**   Less than 2 years?

6.52   A computer program simulates random sampling from the standard normal distribution. Suppose that 1000 random values are to be obtained. Approximately how many of these values would you expect to be

   (a)   Larger than 2?   **(b)**   Smaller than 1.5?

   (c)   Between -1.2 and -1.1? **(d)**   Larger than 2.5 or less than -2.5?

6.53   Evaluate each of the following:

   (a) $z(.025)$.   **(b)** $z(.25)$.   **(c)** $z(.50)$.   **(d)** $z(.85)$.

6.54 Let x be a normal random variable with mean 44 and standard deviation 6. If a value is randomly selected from the x distribution, find each of the following probabilities:

(a) $P(38 < x < 47)$. **(b)** $P(48 < x < 54)$.

(c) $P(x > 59)$. **(d)** $P(x < 52)$.

6.55 Let x be a normal random variable with mean 180 and standard deviation 20. Find the following:

(a) The 10th percentile.

(b) The 95th percentile.

(c) The percentile rank of $x = 190$.

(d) The percentile rank of $x = 140$.

6.56 The I.Q.'s of kindergarten children in a school district are approximately normally distributed with mean 105 and standard deviation 16. Find the probabilities corresponding to A, B, C, and D under the probability curve below. Note that x represents an I.Q. score.

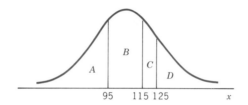

6.57 A plumber has found that the length of time x, in minutes, for installing a bathtub is approximately normally distributed with mean 160 and standard deviation 25.

(a) What percentage of bathtubs are installed within 3 hours?

(b) Find the length of time T so that only 5% of the bathtubs take longer than T to install.

6.58 A radio manufacturer believes that the length of life x (in years) of the WXX2 model radio is approximately normal with mean 12 and standard deviation 2.5.

(a) A customer buys a new radio. Find the probability that the radio will not last 15 years.

(b) What percentage of radios will last longer than 14 years?

(c) Suppose the manufacturer guarantees the radio for 7 years. What percentage of radios will have to be replaced?

(d) Suppose the manufacturer is willing to replace 5% of the radios. What should be the length of the guarantee?

6.59 The weights of chicken lobsters sold by a wholesale seafood company are approximately normally distributed with a mean of 1.5 pounds and a standard deviation of .2 pounds. Find the proportion of lobsters

(a) Weighing more than 2 pounds.

(b) Weighing between 1.5 and 2 pounds.

(c) Weighing less than 1.75 pounds.

6.60   You need a 15-ampere fuse for the electrical system in your house and have decided to buy either brand A or brand B. The length of life for brand A is approximately normal with mean 1000 days and standard deviation 30 days. The length of life for brand B is approximately normal with mean 990 days and standard deviation 10 days. Assume that you would be completely satisfied if the fuse you buy lasts longer than 980 days. Which fuse should you buy? Why?

In Exercises 6.61–6.66 assume a binomial experiment. Using (a) Appendix Table B.2 and (b) the normal approximation to the binomial, find the following:

6.61   The probability of no more than 3 successes, where $n = 18$ and $p = .3$.

6.62   The probability of at least 8 successes, where $n = 16$ and $p = .6$.

6.63   The probability of exactly 7 successes, where $n = 15$ and $p = .5$.

6.64   The probability of between 8 and 12 successes inclusive, where $n = 20$ and $p = .7$.

6.65   The probability of less than 6 successes, where $n = 12$ and $p = .5$.

6.66   The probability of between 8 and 11 successes inclusive, where $n = 16$ and $p = .5$.

6.67   Assume a binomial experiment with $p = .5$ and $n = 400$. Let x be the number of successes. Approximate each of the following probabilities:

   (a) $P(x \geq 210)$.   (b) $P(180 \leq x \leq 220)$.
   (c) $P(170 \leq x \leq 230)$.   (d) $P(x < 185)$.
   (e) $P(x = 200)$.   (f) $P(x = 210)$.

6.68   Using past data, an airline believes that 8% of the people who make reservations for a certain flight will not appear. The seating capacity for the flight is 300; the airline sells 315 tickets. What is the probability that everyone who shows up has a seat on the flight?

6.69   Assume that 21.6 % of Californians under the age of 65 have no health insurance (*Source: Newsweek,* July 31, 1989). Find the probability that more than 40 of 200 randomly selected Californians will have no health insurance.

6.70   The heights (x) of players in a division of high school football teams are approximately normal with mean 71 inches and standard deviation 2.5 inches. Consider the distribution of sample means with sample size $n = 100$.

   (a)   Find $\mu_{\bar{x}}$.
   (b)   Find $\sigma_{\bar{x}}$.
   (c)   What percentage of sample means are larger than 70.5?
   (d)   What percentage of heights are more than 70.5?

6.71   The treatment time x of patients with an eye disease is approximately normal with mean 70 minutes and standard deviation 9 minutes. In parts (a) and (b), find the proportion of treatment times

   (a)   Less than 79 minutes.   (b)   Between 58 and 82 minutes.

   For parts (c) and (d) assume a sample of 36 treatment times are selected. Find

   (c) $P(67 < \bar{x} < 73)$.   (d) $P(\bar{x} > 73)$.

6.72 A dairy claimed that the mean amount in their milk containers was 128 ounces. Let x be the number of ounces of milk per container, and assume that x is normally distributed with standard deviation 1 ounce. If the claim is true, what percentage of containers will have

(a) Less than 126 ounces?

(b) More than 129 ounces?

(c) Between 127.5 and 130.5 ounces?

A random sample of 25 containers gave a sample mean of 127.4 ounces.

(d) Find $P(\bar{x} < 127.4)$.

(e) Using part (d) do you feel that there is evidence that the true mean is less than 128 ounces? Why?

6.73 A digit from 0 to 9 is to be randomly selected. Let x be the value of the digit selected. The probability distribution for x is given by $P(x) = 1/10$ for $x = 0, 1, 2, 3, 4, 5, 6, 7, 8,$ and 9.

(a) Find μ and σ^2. (You may already have computed these in Exercise 5.35.)

Suppose n digits are to be randomly selected. Let $\bar{x}$ be the mean of these x values.

(b) Find $\mu_{\bar{x}}$ and $\sigma_{\bar{x}}^2$ for each of the following values of n:

(i) $n = 36.$ (ii) $n = 49.$ (iii) $n = 100.$

(c) For each of the following values of n, find $P(4 < \bar{x} < 5)$.

(i) $n = 36.$ (ii) $n = 49.$ (iii) $n = 100.$

Compare and explain your results.

(d) For each of the following values of n, find $P(\bar{x} > 4.75)$.

(i) $n = 36.$ (ii) $n = 49.$ (iii) $n = 100.$

Compare and explain your results.

Answers to Practice Quiz

1. .0228. This is the normal probability $P(x > 70)$.

2. .1359. This is the normal probability $P(60 < x < 70)$.

3. 56.7. $P(x \leq 56.7) = .75$. So 56.7 is the 75th percentile.

4. 1.28. There is no subcommand, so Minitab assumes a standard normal distribution. Note that $P(z \leq 1.28) = .90$, and so $P(z \geq 1.28) = 1 - .90 = .10$. Hence, $z(.10) = 1.28$.

5. SET C1
   .99, .975, .95, .90
   END
   INVCDF C1;
     NORMAL 0   1.

   Note: $z(\alpha) = $ INVCDF of $(1 - \alpha)$.

Notes

DeGroot, M. H., S. E. Fienberg, and J. B. Kadane, ed., *Statistics and the Law* (New York: Wiley, 1986).

Lasser, R. P., and A. M. Master, *Geriatrics* 14: 345–360 (1959).

Newsweek (New York: Newsweek, Inc., July 31, 1989).

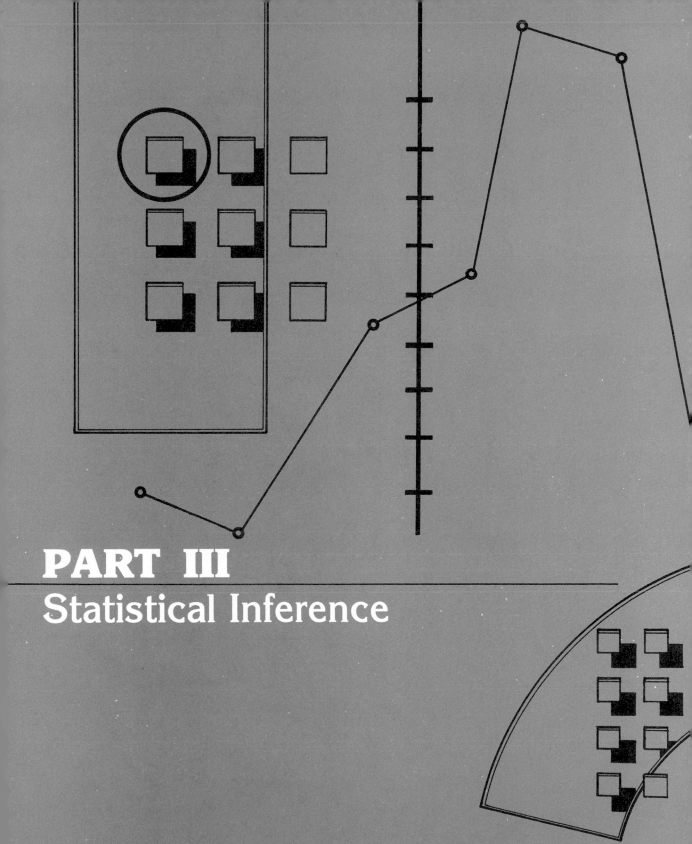

PART III
Statistical Inference

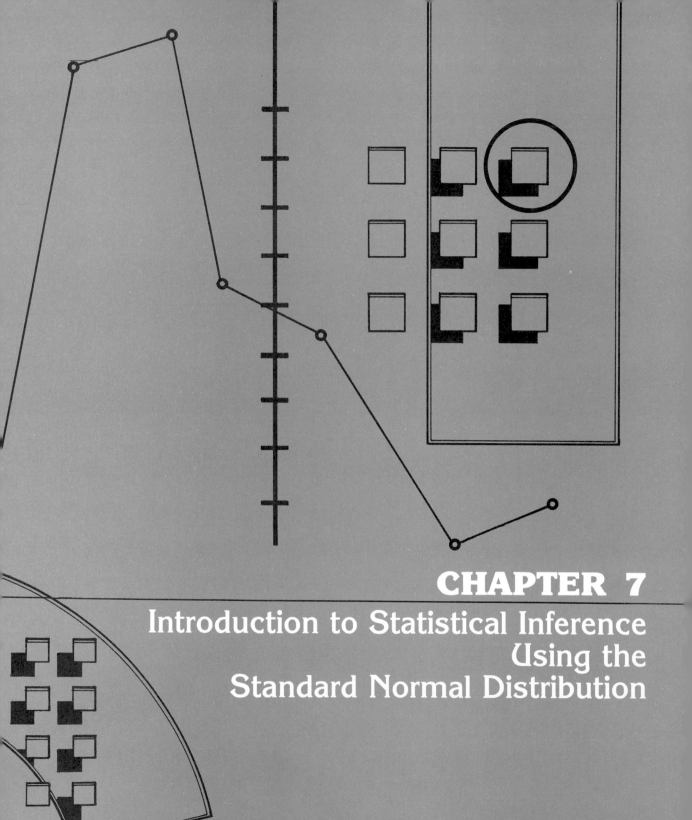

CHAPTER 7
Introduction to Statistical Inference
Using the
Standard Normal Distribution

7.1 INTRODUCTION

We have seen that statistical inference is the process of making judgments about a population based on properties of a sample from the population. There are two types of statistical inference. **Estimation** involves approximating the value of an unknown parameter. (Recall that a parameter is a number describing some numerical property of a population.) For example, we might be interested in obtaining an estimate of the mean value of all homes in Cleveland.

The other type of inference is called **decision making.** This involves choosing between two opposing statements concerning a population. These statements are called **hypotheses.** For example, we may wish to decide if the mean value of all homes in Cleveland is more than $80,000 or if it is not more than $80,000.

In this chapter we will introduce the two types of inference. The Central Limit Theorem (see Section 6.7) plays an important part in this study. Also, the normal approximation to the binomial distribution will be useful (see Section 6.6). The reader may wish to review these two sections before proceeding further.

We saw in Chapter 6 that in order for the Central Limit Theorem to hold or to be able to approximate a binomial distribution by a normal distribution, the sample size n should be "sufficiently large." For this reason we call the methods of this chapter *large-sample methods*. In the next chapter we will discuss inference based on small samples.

ESTIMATING A POPULATION MEAN

Let us investigate how a population mean may be estimated from a sample of data values representing blood pressures.

Blood pressure is measured in millimeters. A reading of 120 millimeters corresponds to a blood pressure that will support a column of mercury 120 millimeters high. In measuring blood pressure two readings are important. The "systolic pressure" is the blood pressure when the heart muscle is contracting. The "diastolic pressure" is the blood pressure when the heart muscle is relaxed (between beats). Both these values are important. For young adults the mean for systolic pressure is about 120 millimeters and for diastolic it is about 74 millimeters. This combination is expressed as 120/74. Blood pressure tends to increase with age. For males age 35–59 years, mean blood pressure readings are about 133/84. Elevated blood pressure, if untreated, can lead to serious medical problems such as heart disease or stroke.

It is felt by some that regular exercise tends to lower one's blood pressure. Researchers Haskell, Stern, Lewis, and Perry at the Stanford University School of Medicine have studied blood pressures and various other characteristics of male and female runners (Haskell et al., 1977, p. 148). A sample of 41 male runners age 35–59 years showed a sample mean systolic blood pressure of $\bar{x} =$ 123. These 41 runners can be thought of as a sample from the population of all male runners age 35–59 years.* Suppose we wish to find the mean systolic blood pressure μ for this population. Since it is not feasible to obtain all the data values in this population, it would seem natural to use the sample mean 123 for the 41 runners as an estimate for the population mean μ. This is called a **point estimate.**

---

*All runners in the sample lived in the Palo Alto, California, area so we should probably restrict the population to male runners age 35–59 in this geographic area.

> **Definition**   The sample mean $\bar{x}$ is called a *point estimate* for the population
> mean μ.

Sometimes we hear statements like: "The average price of all new auto-mobiles is between $13,000 and $15,000," or "The average age of all Americans is between 25 and 30." These are examples of **interval estimates.** Actually, in statistics it is customary to give not only an interval estimate for a parameter, but also the probability that the method used to find this interval will lead to an interval that contains the parameter. This probability is called the **level of confidence** and the resulting interval is called a **confidence interval.**

We will explore this idea with results of the Stanford study already discussed. Suppose we want to find an interval that contains the mean systolic pressure μ of all male runners age 35–59 with a level of confidence .95. This is a 95% confidence interval. Now by the Central Limit Theorem (see Section 6.7), the values of the sample mean $\bar{x}$ for random samples of size n have a distribution that is approximately normal with

$$\mu_{\bar{x}} = \mu \qquad \text{and} \qquad \sigma_{\bar{x}} = \sigma/\sqrt{n}$$

when n is sufficiently large ($n \geq 30$ will usually suffice). Hence the standard score

$$z = \frac{\bar{x} - \mu_{\bar{x}}}{\sigma_{\bar{x}}} = \frac{\bar{x} - \mu}{\sigma/\sqrt{n}}$$

is approximately standard normal. We are dealing with the case where $n = 41$. We will assume that, although μ is unknown, the (population) standard deviation σ for the runners is no different from that of other males aged 35–59. This value is about 17 millimeters. (In Chapter 8 we will see why this is a reasonable assumption in this case.) From this it follows that $\sigma_{\bar{x}} = \sigma/\sqrt{n} = 17/\sqrt{41} \doteq 2.65$. Therefore,

$$z = \frac{\bar{x} - \mu}{2.65}$$

is approximately standard normal. Recall that z measures the number of standard deviations by which a value of $\bar{x}$ is above or below the mean. (A positive value indicates that $\bar{x}$ is above the mean; a negative value indicates that $\bar{x}$ is below the mean.)

Perhaps if we found a type of confidence interval for z, we could use the relation between z and μ to find a confidence interval for μ. Now from Figure 7.1 we see that

$$P(-1.96 < z < 1.96) = .95$$

[To understand this, it may help to recall that the area under the standard normal curve to the right of 1.96 is .025. That is, $z(.025) = 1.96$. See Example 6.7.]

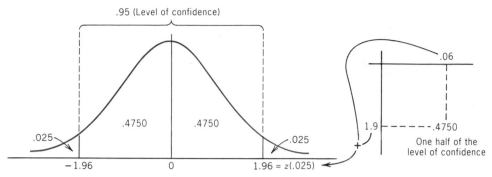

FIGURE 7.1

Therefore, 95% of all values of z will be between -1.96 and 1.96. This means that 95% of all values of $\bar{x}$ will be within 1.96 standard deviations of the mean μ. One standard deviation is $\sigma_{\bar{x}} = 2.65$ millimeters. Therefore, 1.96 standard deviations is $(1.96)(2.65) \doteq 5.19$ millimeters. So for any randomly obtained value of $\bar{x}$, the probability is .95 that $\bar{x}$ will be within 5.19 millimeters of μ. The probability is .95 that the distance from μ to $\bar{x}$ is less than 5.19, that is, μ is between $\bar{x} - 5.19$ and $\bar{x} + 5.19$.

$$\bar{x} - 5.19 < \mu < \bar{x} + 5.19$$

For any randomly obtained value of $\bar{x}$ we are 95% sure that the above inequality is correct. In particular, we can substitute the value of $\bar{x}$, namely, 123 millimeters, obtained from the Stanford study, into this inequality, getting

$$123 - 5.19 < \mu < 123 + 5.19$$
$$117.81 < \mu < 128.19$$

This is our desired 95% confidence interval. This confidence interval is shown in Figure 7.2.

Recall that we said that the mean systolic blood pressure of *all* males age 35–59 is 133. Notice that our confidence interval for the mean pressure for runners (117.81 to 128.19) does not contain 133; it is below 133. Thus we can be 95% sure that the mean systolic blood pressure of male runners age 35–59 is less than the mean for males in general (of comparable age).

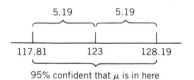

FIGURE 7.2

Formula for Confidence Interval for μ

Suppose we wish to find a confidence interval for the mean μ for a random variable x. It is customary to call the level of confidence $1 - \alpha$. (For example, if our level of confidence is .95, then $1 - \alpha = .95$. So $\alpha = .05$.) Let's examine the 95% confidence interval we developed in the preceding discussion and see if we can use this to find out what a $1 - \alpha$ confidence interval should look like, in general. We will proceed by analogy. The 95% confidence interval was

$$123 - 5.19 < \mu < 123 + 5.19$$

For this confidence interval, recall that

level of confidence $= .95 = 1 - \alpha$ (so $\alpha = .05$)

$123 = \bar{x}$

$5.19 = (1.96)(2.65)$ where $1.96 = z(.025) = z\left(\dfrac{\alpha}{2}\right)$ and $2.65 = \sigma_{\bar{x}} = \dfrac{\sigma}{\sqrt{n}}$

Therefore,

$$5.19 = z\left(\frac{\alpha}{2}\right) \cdot \frac{\sigma}{\sqrt{n}}$$

Now we can rewrite the above confidence interval symbolically as

$$\bar{x} - z\left(\frac{\alpha}{2}\right) \cdot \frac{\sigma}{\sqrt{n}} < \mu < \bar{x} + z\left(\frac{\alpha}{2}\right) \cdot \frac{\sigma}{\sqrt{n}}$$

This, in fact, is the general form for a $1 - \alpha$ confidence interval for μ.

A $1 - \alpha$ confidence interval for a population mean μ when the sample size n is large ($n \geq 30$) is

$$\bar{x} - z\left(\frac{\alpha}{2}\right) \cdot \frac{\sigma}{\sqrt{n}} < \mu < \bar{x} + z\left(\frac{\alpha}{2}\right) \cdot \frac{\sigma}{\sqrt{n}}$$

Sometimes the interval is described by its endpoints: $\bar{x} \pm z(\alpha/2) \cdot \sigma/\sqrt{n}$

$1 - \alpha$ confidence interval

$$\bar{x} - z\left(\frac{\alpha}{2}\right) \cdot \frac{\sigma}{\sqrt{n}} \qquad \bar{x} \qquad \bar{x} + z\left(\frac{\alpha}{2}\right) \cdot \frac{\sigma}{\sqrt{n}}$$

The level of confidence $1 - \alpha$ is the probability that an interval constructed in this manner will contain μ.

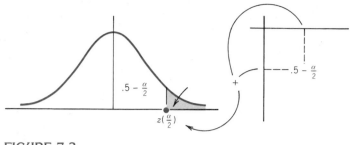

FIGURE 7.3

Finding $z\left(\dfrac{\alpha}{2}\right)$

Remark   The reader is reminded that the symbol $z(\alpha/2)$ represents a value of z (called "z of $\alpha/2$") and does *not* represent a multiplication. If you are unsure about the meaning of this symbol, review Example 6.8. Some common values of $z(\alpha/2)$ are as follows:

Level of Confidence $1 - \alpha$	$z\left(\dfrac{\alpha}{2}\right)$
90%	1.65
95%	1.96
98%	2.33
99%	2.58

Since the Central Limit Theorem was the basis of our confidence interval, we should be sure that the sample size n is sufficiently large before we use it. That is, we should have $n \geq 30$. If σ is *unknown, we can use the sample standard deviation s as an estimate for σ as long as the sample size is large ($n \geq 30$). If the random variable x is normal or approximately so, then the above confidence interval may be used, even when the sample size is small ($n < 30$). But in this case σ should be known because it would be inappropriate to estimate it by s when n is small.

When the above conditions are not met, it may still be possible to obtain a confidence interval for μ, as we shall see in Chapter 8.

Example 7.1

Thirty-six automobiles of the same model are driven the same distance. The gas mileage for each is recorded. The results give $\bar{x} = 18$ miles per gallon and $s = 3$ miles per gallon. Give a 90% confidence interval for the mean mileage μ for all autos of this model.

Solution

$1 - \alpha = .90,\ \alpha = .10$

$z\left(\dfrac{\alpha}{2}\right) = z(.05) = 1.65$

$\bar{x} = 18$

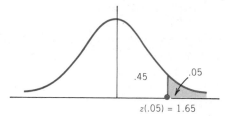

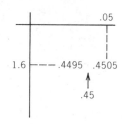

FIGURE 7.4

σ is unknown, but since n is large, we can use $s = 3$ as an estimate. Therefore,

$$\frac{\sigma}{\sqrt{n}} \doteq \frac{s}{\sqrt{n}} = \frac{3}{\sqrt{36}} = \frac{3}{6} = .5$$

Thus the confidence interval

$$\bar{x} - z\left(\frac{\alpha}{2}\right) \cdot \frac{\sigma}{\sqrt{n}} < \mu < \bar{x} + z\left(\frac{\alpha}{2}\right) \cdot \frac{\sigma}{\sqrt{n}}$$

becomes

$$18 - (1.65)(.5) < \mu < 18 + (1.65)(.5)$$

$$18 - .83 < \mu < 18 + .83$$

$$17.17 < \mu < 18.83$$

Therefore we are 90% confident that the mean gas mileage μ is between 17.17 and 18.83 miles per gallon.

The following question often arises at this point: why settle for a 90% or 95% confidence interval, when we can find a 99% or even a 99.99% confidence interval? The answer is that the higher our level of confidence, the wider the confidence interval. Obviously, the wider the confidence interval, the less useful it will be as an estimate for μ. For example, if we had asked for a 99% confidence interval in Example 7.1, then $1 - \alpha = .99$, $\alpha = .01$, $z(\alpha/2) = z(.005) = 2.58$. The 99% confidence interval is then

$$16.71 < \mu < 19.29$$

Hence there is a trade-off between level of confidence and accuracy: the higher the confidence, the less accuracy; the more accuracy, the lower the confidence. A level of confidence often used is 95%; apparently this is regarded in many situations as a happy medium.

Maximum Error of Estimate

If we obtain a sample of data values of a random variable x, we have said that a natural point estimate for the mean μ is the sample mean $\bar{x}$. We can get an idea of how good an estimate this is by examining a confidence interval for μ.

We have said that a $1 - \alpha$ confidence interval for μ is described by the endpoints

$$\bar{x} \pm z\left(\frac{\alpha}{2}\right) \cdot \frac{\sigma}{\sqrt{n}}$$

In Example 7.1 this gave us the 90% confidence interval

$$18 \pm .83$$

We display this graphically in Figure 7.5(a).

If we were to estimate the mean gas mileage μ by 18 miles per gallon (the sample mean), we can be 90% confident that the amount by which our estimate is in error will be less than .83 miles per gallon. We say that .83 is the maximum error of the estimate with 90% level of confidence. This idea can be generalized, as suggested by Figure 7.5(b).

Definition   When estimating μ by $\bar{x}$ (from a large sample), *the maximum error of the estimate,* with level of confidence $1 - \alpha$, is

$$E = z\left(\frac{\alpha}{2}\right) \cdot \frac{\sigma}{\sqrt{n}}$$

Here again, when σ is unknown, we can estimate it by s, as long as $n \geq 30$. Note that the endpoints of the confidence interval can be written

$$\bar{x} \pm E$$

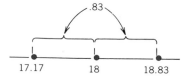

(a)
90% Confidence Interval for
Mean Gas Mileage in Example 8.1.

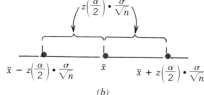

(b)
$1 - \alpha$ Confidence Interval
for μ.

FIGURE 7.5

Confidence Intervals. (*a*) 90% Confidence Interval for Mean Gas Mileage in Example 7.1. (*b*) 1 − α Confidence Interval for μ

Example 7.2

Let x = the amount of suspended impurities in the air in micrograms per cubic meter as measured by an Environmental Protection Agency sampling station. Forty-five readings of x were randomly obtained over a 1-year period, giving $\bar{x}$ = 52.1 and s = 5.3. If $\bar{x}$ is used to estimate μ, give the maximum error of the estimate with 95% level of confidence.

Solution

$$1 - \alpha = .95, \qquad \alpha = .05$$

We have seen that $z(\alpha/2) = z(.025) = 1.96$. It is given that

$$\frac{\sigma}{\sqrt{n}} \doteq \frac{s}{\sqrt{n}} = \frac{5.3}{\sqrt{45}} \doteq .79$$

Note that we have estimated σ by s. Now

$$E = z\left(\frac{\alpha}{2}\right) \cdot \frac{\sigma}{\sqrt{n}} \doteq (1.96)(.79) \doteq 1.55$$

measured in micrograms per cubic meter.

Determining the Sample Size

A professor wanted to estimate the mean score μ of all students at her university on a national mathematics placement test that was being considered for use at the school. The national mean was 70 with a standard deviation of 10 points. She felt that the standard deviation for her institution would be the same (so σ = 10), but that the mean would be higher. She intended to give the exam to a sample of students at her institution, and use the result ($\bar{x}$) to estimate the mean score μ for all students at the school. She wanted to decide how many students should be tested, so that the maximum error of the estimate (E) would be 2 points with 95% confidence. Now

$$E = z\left(\frac{\alpha}{2}\right) \cdot \frac{\sigma}{\sqrt{n}}$$

where

$$E = 2$$
$$\sigma = 10$$
$$1 - \alpha = .95 \ (\alpha = .05)$$
$$z\left(\frac{\alpha}{2}\right) = z(.025) = 1.96$$
$$n = \underline{?}$$

We can substitute the known values in the equation for E and solve for n.

$$2 = (1.96) \cdot \frac{10}{\sqrt{n}}$$

$$2 = \frac{19.6}{\sqrt{n}}$$

Multiplying both sides by $\sqrt{n}$ and dividing both sides by 2, we get

$$\sqrt{n} = \frac{19.6}{2} = 9.8$$

Now we square both sides

$$n = (9.8)^2 = 96.04$$

Thus the professor should test at least 97 students. Note that we round *up* to be on the safe side.

Notice that in the above discussion, if we trace each step backward, we find that

$$n = (9.8)^2 = \left(\frac{19.6}{2}\right)^2 = \left[\frac{(1.96)(10)}{2}\right]^2 = \left[\frac{z(\alpha/2) \cdot \sigma}{E}\right]^2$$

The above observation enables us to generalize our result as follows:

> When estimating μ by $\bar{x}$, if we want our level of confidence to be $1 - \alpha$ that our error is less than an amount E, the sample size should be at least
>
> $$n = \left[\frac{z(\alpha/2) \cdot \sigma}{E}\right]^2$$

Notice that σ must be known to employ this formula, which unfortunately is usually not the case. (Obviously, we can't estimate σ by s in this situation because we don't have the sample on hand yet.) However, based on past experience, the researcher might have an idea of the approximate value of σ. If so, this value would be used in the above formula.

Example 7.3

A machine is designed to produce rubber gaskets with a mean thickness of .125 inch; the standard deviation is $\sigma = .01$ inch. It is thought that the mean has changed (but σ remains the same). How many gaskets should be measured in order to be able to estimate the new mean μ by the sample mean $\bar{x}$ with maximum error of the estimate .001 inch and 90% level of confidence?

Solution

$$1 - \alpha = .90 \qquad (\alpha = .10)$$

In Example 7.1 we saw that $z(\alpha/2) = z(.05) = 1.65$. Now

$$\sigma = .01$$

$$E = .001$$

$$n = \left[\frac{z(\alpha/2) \cdot \sigma}{E}\right]^2 = \left[\frac{(1.65)(.01)}{.001}\right]^2 = 272.25$$

Thus the sample size should be at least 273. Note again that we round up to the next integer to be on the safe side.

Exercises

7.1   In each part below, **(i)** find the maximum error of the estimate, and **(ii)** construct a confidence interval for the population mean μ. Note that n refers to the sample size.

	$\overline{x}$	s	n	Percentage Confidence Interval
(a)	125	16	64	95
(b)	206	25	100	99
(c)	154	3	81	90
(d)	309	50	225	95
(e)	40	7	49	99
(f)	78	6	144	90

7.2   Fifty samples were randomly selected from a population with mean $\mu = 200$ and variance $\sigma^2 = 100$. For each sample, 80, 90, 95, and 99% confidence intervals for μ were constructed.

(a)   For each level of confidence (80, 90, 95, and 99%), about how many of the 50 confidence intervals should contain the number $\mu = 200$?

(b)   The table below gives the 11 samples in which the 80% confidence interval did not contain the value $\mu = 200$. (*Note:* N − interval did not contain $\mu = 200$; Y = interval did contain $\mu = 200$.) Fill in the 99% column with either a Y, N, or ? (can't be determined from the information given).

Sample #	80%	90%	95%	99%
11	N	N	Y	
14	N	N	N	
17	N	N	Y	
22	N	N	Y	
24	N	N	Y	
30	N	N	N	
35	N	N	Y	
40	N	N	Y	
46	N	Y	Y	
48	N	Y	Y	
50	N	Y	Y	

7.3 **(a)** Sixty pieces of a plastic are randomly selected, and the breaking strength of each piece is recorded in pounds per square inch. Suppose that $\bar{x} = 26$ and $s = 1.5$ pounds per square inch. Find a 99% confidence interval for the mean breaking strength μ.

(b) If you were to obtain 200 99% confidence intervals for μ, about how many can be expected to contain μ?

7.4 A medical doctor wanted to estimate the mean length of time μ that a patient had to wait to see him after arriving at the office. A random sample of 50 patients showed a mean waiting time of 23.4 minutes and a standard deviation of 7.1 minutes. Find a 95% confidence interval for μ.

7.5 A study was done to estimate the mean annual growth μ in a population of *Conus pennaceus* trees in Hawaii. For those with an initial size of 2.41–2.60 centimeters, a sample of size 33 yielded a mean annual growth of .72 centimeters with a standard deviation of .31 centimeter. Find a 90% confidence interval for the population mean μ of annual growth (of those with an initial size of 2.41–2.60 centimeters) (*Source:* Perron, 1983, p. 55).

7.6 A city assessor wished to estimate the mean income per household. The previous mean income was $25,300. A random sample of 40 households in the city showed a mean income of $29,400 with a standard deviation of $6,325.

(a) Find a 90% confidence interval for the population mean income per household μ.

(b) Based on your answer in part (a), would the assessor conclude the mean income had increased over the previous estimate of $25,300?

7.7 Noise level tests were done on 40 new light rail vehicles (LRV's—the new name for trolley cars). The results of the test gave a sample mean of 65 decibels with a sample standard deviation of 6 decibels.

(a) Find a 90% confidence interval for the mean decibel level μ for this type of transit vehicle.

(b) What is the maximum error of the estimate at the 90% level?

(c) Based on your answer in part (a), would you conclude that the new LRV's are quieter on the average than older-type trolley cars that had a mean decibel level of 80?

7.8 An electrical company tested a new type of transformer oil to be used in its transformers. Thirty-five readings of dielectric strength were obtained. Dielectric strength is potential (in Kilovolts per centimeter of thickness) necessary to cause a disruptive discharge of electricity through an insulator. The results of the test gave $\bar{x} = 77$ kv, $s = 8$ kv.

(a) Find a 95% confidence interval for the mean dielectric strength of the oil.

(b) The old transformer oil had a mean dielectric strength of 75 kv. Would you conclude that the new oil has a higher mean dielectric strength on the basis of your answer in part (a)?

7.9 A union official wanted to estimate the mean hourly wage μ of its members. A random sample of 100 members gave $\bar{x} = \$18.30$ with $s = \$3.25$ per hour.

(a) Find an 80% confidence interval for μ.

(b) Find a 95% confidence interval for μ.

(c) If you were to construct a 90% confidence interval for μ (do not construct it), would the length of the interval be longer or shorter than the length of the 80% confidence interval? Longer or shorter than the length of the 95% confidence interval?

7.10 A transit official wanted to estimate the mean time μ for a bus trip between two cities. A random sample of 50 such trips gave $\bar{x} = 150$ minutes and $s = 15$ minutes.

(a) Find a 90% confidence interval for μ.

(b) Find a 99% confidence interval for μ.

(c) If you were to construct a 95% confidence interval for μ (do not construct it), would the length of the interval be longer or shorter than the length of the 99% confidence interval? Longer or shorter than the length of the 90% confidence interval?

7.11 Match the confidence interval with the appropriate level of confidence. Assume the same sample size and sample standard deviation in each case.

Confidence Interval	Level of Confidence
10 ± 3	90%
10 ± 4.5	95%
10 ± 5	80%
10 ± 8	99%

7.12 A newspaper article stated that a random sample of 144 summer utility bills gave $\bar{x} = 150$ and $s = 36$ dollars. The article gave an interval estimate of the population mean μ as 150 ± 4.95 but did not mention the level of confidence. What is the level of confidence?

7.13 For males living on the island of Crete, a 7-day study of 33 men's diets gave a sample mean of 41.8% and a sample standard deviation of 5.7% calories from fats (*Source:* Keys, 1970, p. I–166).

(a) Find a 95% confidence interval for the mean μ of the population from which the men were selected.

(b) What is the maximum error of estimate for μ?

7.14 For males living in an area of Greece called Montegiorgio, a 7-day study of 34 men's diets gave a sample mean of 23.9% and a sample standard deviation of 4.6% calories from fats (*Source:* Keys, 1970, p. I–166).

(a) Find a 95% confidence interval for the mean μ of the population from which the men were selected.

(b) What is the maximum error of estimate for μ?

7.15 An educator wishes to estimate the mean number of hours μ that 10-year-old children in a city watch television per day. How large a sample is needed if the educator wants to estimate μ to within .5 hours with 90% confidence? Use $\sigma = 1.75$.

7.16 How many households in a large town should be randomly sampled to estimate the mean number of dollars spent per household (per week) on food supplies to within $3 with 80% confidence? Assume a standard deviation of $15.

7.17 Consider a population with unknown mean μ and population standard deviation $\sigma = 20$.

(a) How large a sample size is needed to estimate μ to within four units with 90% confidence?

(b) Suppose that you were to estimate μ to within four units with 95% confidence. Without calculating, would the sample size required be larger or smaller than that found in part (a)?

(c) Suppose that you were to estimate μ to within two units with 90% confidence. Without calculating, would the sample size required be larger or smaller than that found in part (a)?

7.18 Consider a population with unknown mean μ and population standard deviation $\sigma = 15$.

(a) How large a sample size is needed to estimate μ to within 5 units with 95% confidence?

(b) Suppose you were to estimate μ to within five units with 90% confidence. Without calculating, would the sample size required be larger or smaller than that found in part (a)?

(c) Suppose you were to estimate μ to within six units with 95% confidence. Without calculating, would the sample size required be larger or smaller than that found in part (a)?

7.19 A journal article stated that a random sample from a population gave a sample standard deviation of $s = 30$ and a 95% confidence interval for the population mean μ as lying between 18.58 and 26.42.

(a) What is the point estimate for μ?

(b) What is the maximum error of estimate?

(c) What is the sample size?

7.20 Suppose that a random sample of size 100 from a population gave $\Sigma x = 3500$ and $\Sigma x^2 = 370,000$.

(a) What is a point estimate for the population mean μ?

(b) Find a 95% confidence interval for μ.

(c) What is the maximum error of estimate for μ?

7.21 The main reason for deaths in the first month of life is low birthweight. An administrator at Claybak Memorial Hospital (CMH) obtained the following 37 weights (in ounces):

100.2	102.4	82.6	79.4	132.4	107.9	120.1	119.4
63.9	137.3	135.1	143.4	128.9	78.6	144.7	131.4
117.4	114.8	108.3	109.8	122.0	65.4	81.6	
101.1	73.6	120.8	105.8	137.6	96.8	134.0	
95.2	127.8	88.9	67.9	114.0	79.4	84.7	

(a) Find a 95% confidence interval for the mean birthweight at CMH. (*Note:* $\Sigma x = 3954.6$ and $\Sigma x^2 = 442,887.6$.)

(b) The administrator felt that babies weighing less than 88 ounces were abnormally small. Can the administrator feel confident (at the 95% level of confidence) that babies born at CMH are not abnormally small on average?

7.22 In Exercise 7.21 the administrator believed that the mean birthweight of babies nationwide is 120 ounces. At the 95% level, what does the administrator conclude about mean birthweights at CMH compared with mean birthweights nationwide?

7.23 A restaurant owner believed that customer spending was below normal at tables manned by one of the waiters. The owner sampled 36 receipts from the waiter's tables and got the following amounts (rounded to the nearest dollar):

47	46	56	70	52	58	48	57	49	61	52	40
60	22	74	59	60	30	61	44	62	41	53	57
50	52	57	59	69	51	58	56	44	36	47	51

(a) Find a 90% confidence interval for the mean amount of money spent at the waiter's tables.

(b) At the 90% confidence level, does it appear that the mean intake at the waiter's tables is smaller than the restaurant average of $55?

7.24 The 36 data values below were randomly generated. Find a 95% confidence interval for the mean μ of the population from which the data were selected.
(*Note:* $\Sigma x = 2565$ and $\Sigma x^2 = 367,763$.) The data are

19	58	130	319	163	29	18	5	38	47	150	31
72	26	260	123	74	20	8	18	45	3	67	12
1	91	28	61	17	95	153	9	76	128	143	28

7.3 ESTIMATING A POPULATION PROPORTION

In this section we will discuss how to estimate the (unknown) proportion p of those elements in a population possessing a certain characteristic of interest. The quantity p is called a **population proportion.**

For example, a pollster wishes to estimate the proportion of all voters in a state who favor a proposal to limit property taxes (call it Proposition A). This proposition is to appear on the ballot in an upcoming election. The collection of all voters in the state constitutes a population. A voter possesses the characteristic of interest if he or she favors Proposition A, and the proportion of all voters favoring A is a population proportion p. Suppose the pollster obtains a random sample of 1000 voters from the population and finds that 540 of them favor Proposition A. A point estimate for p would be $540/1000 = .54$. This is called a **sample proportion.** This result should prove reassuring to the supporters of Proposition A. However, we should keep in mind that there will usually be an error when we use a point estimate for a population parameter. In other words, we might estimate that 54% of the voters will favor Proposition A and so A ought to pass. But if we are off by more than 4 percentage points, it may not pass.

Perhaps a confidence interval for p would be more appropriate. We can view the interviewing of the 1000 voters as a binomial experiment.

$$\text{trial} = \text{randomly select a voter}$$
$$\text{success} = S = \text{voter selected favors Proposition A}$$
$$\text{failure} = F = \text{voter selected does not favor Proposition A}$$
$$P(S) = p = \text{proportion of all voters who favor Proposition A}$$
$$\text{(this is what we want to estimate)}$$
$$P(F) = q = \text{proportion of all voters who do not favor}$$
$$\text{Proposition A} = 1 - p$$
$$n = \text{number of trials, that is, the number of voters in the sample}$$
$$x = \text{number of successes, that is, the number of voters in}$$
$$\text{sample who favor Proposition A (this can vary from}$$
$$\text{one sample to the next)}$$

Using the observed value of x from the sample, we can estimate p by x/n. This is a point estimate for p denoted by $\hat{p}$:

$$\hat{p} = \frac{x}{n}$$

We have seen in Section 6.6 that when n is sufficiently large, the variable x may be regarded as approximately normal. We said that n will be large enough if both np and nq are at least 5. The problem here is that we do not know p (or $q = 1 - p$). If we wish to take advantage of the normal approximation in our study of p, how do we know if the sample size is large enough? There is an alternative rule of thumb that may be used: *n will be considered large enough if the observed number of successes x and the number of failures n − x are both at least 5.**

We also saw in Section 6.6 that the mean value of x is np and the standard deviation is $\sqrt{npq}$. Just as x can vary from one sample to the next, so can $\hat{p}$. It seems plausible, and is indeed the case, that $\hat{p}$ is also approximately normal (when n is large enough) with

$$\text{mean} = \frac{np}{n} = p$$

$$\text{standard deviation} = \frac{\sqrt{npq}}{n} = \sqrt{\frac{npq}{n^2}} = \sqrt{\frac{pq}{n}}$$

It follows that

$$z = \frac{\hat{p} - p}{\sqrt{pq/n}}$$

is approximately standard normal.

---

*The reason for this is that for a given sample we may approximate p by $\hat{p}$ and q by $\hat{q}$, where $\hat{q} = 1 - \hat{p}$. Then n will be large enough if both $n\hat{p}$ and $n\hat{q}$ are at least 5. But

$$n\hat{p} = n\left(\frac{x}{n}\right) = x \quad \text{and} \quad n\hat{q} = n(1 - \hat{p}) = n\left(1 - \frac{x}{n}\right) = n - n\left(\frac{x}{n}\right) = n - x$$

So the observed x and $n - x$ should both be at least 5.

Recall in the last section that we used a standard normal expression to develop a $1 - \alpha$ confidence interval for μ. Notice the relationship

$$\overbrace{\frac{\bar{x} - \mu}{\sigma/\sqrt{n}}} \qquad \bar{x} \pm z\left(\frac{\alpha}{2}\right) \cdot \frac{\sigma}{\sqrt{n}}$$

Proceeding by analogy, we might guess what a $1 - \alpha$ confidence interval for p would look like:

$$\overbrace{\frac{\hat{p} - p}{\sqrt{\frac{pq}{n}}}} \qquad \hat{p} \pm z\left(\frac{\alpha}{2}\right) \cdot \sqrt{\frac{pq}{n}}$$

However, we don't know the values of p and q in the above expression. Therefore we will approximate them by $\hat{p} = x/n$ and $\hat{q} = 1 - \hat{p}$. This gives the $1 - \alpha$ **confidence interval for p:**

$$\hat{p} \pm z\left(\frac{\alpha}{2}\right) \cdot \sqrt{\frac{\hat{p}\hat{q}}{n}}$$

When estimating p by $\hat{p}$, the **maximum error of the estimate** with confidence $1 - \alpha$ is

$$E = z\left(\frac{\alpha}{2}\right) \cdot \sqrt{\frac{\hat{p}\hat{q}}{n}}$$

We summarize these results as follows:

A point estimate for a population proportion p is the sample proportion $\hat{p} = x/n$. When the sample size n is sufficiently large, a $1 - \alpha$ confidence interval for p is

$$\hat{p} - z\left(\frac{\alpha}{2}\right) \cdot \sqrt{\frac{\hat{p}\hat{q}}{n}} < p < \hat{p} + z\left(\frac{\alpha}{2}\right) \cdot \sqrt{\frac{\hat{p}\hat{q}}{n}}$$

and if we estimate p by $\hat{p}$, the maximum error of the estimate with confidence $1 - \alpha$ is

$$E = z\left(\frac{\alpha}{2}\right) \cdot \sqrt{\frac{\hat{p}\hat{q}}{n}}$$

n will be sufficiently large if it is large enough so that both x and $n - x$ are at least 5. In terms of E, the endpoints of the confidence interval are:

$$\hat{p} \pm E$$

Example 7.4

Assume that the pollster (in the previous discussion) interviews 1000 voters and finds that 540 favor Proposition A.

(a) Find a point estimate for p (the proportion of all voters who favor Proposition A).

(b) Find a 95% confidence interval for p.

(c) If the point estimate is used to estimate p, find the maximum error of the estimate (with 95% confidence).

(d) Comment on how these results should be interpreted.

Solution

Note that $x = 540$ and $n - x = 1000 - 540 = 460$ are both at least 5; therefore we can use the procedure discussed above.

(a) As we stated previously, a point estimate for p is

$$\hat{p} = \frac{x}{n} = \frac{540}{1000} = .54$$

(b) To find a 95% confidence interval, we set

$$1 - \alpha = .95 \qquad\qquad \hat{p} = .54$$
$$\alpha = .05 \qquad\qquad \hat{q} = .46$$
$$z\left(\frac{\alpha}{2}\right) = z(.025) = 1.96 \qquad\qquad n = 1000$$

The 95% confidence interval is

$$\hat{p} - z\left(\frac{\alpha}{2}\right) \cdot \sqrt{\frac{\hat{p}\hat{q}}{n}} < p < \hat{p} + z\left(\frac{\alpha}{2}\right) \cdot \sqrt{\frac{\hat{p}\hat{q}}{n}}$$

or

$$.54 - (1.96) \cdot \sqrt{\frac{(.54)(.46)}{1000}} < p < .54 + (1.96) \cdot \sqrt{\frac{(.54)(.46)}{1000}}$$
$$.54 - .03 < p < .54 + .03$$
$$.51 < p < .57$$

(c) Since the confidence interval is $.54 \pm .03$,

$$E = .03$$

(d) Part (b) suggests that the pollster can be 95% sure that the percentage of all voters who favor Proposition A is between 51% and 57%, which indicates that Proposition A will probably pass. Another way of looking at this is given by part (c). This says that if we estimate p by $\hat{p} = (x/n) = .54$, we can be 95% sure that this estimate will be in error by less than .03. In other words, if we predict the percentage of voters favoring Proposition A to be 54%, we are 95% sure that we will be less than 3 percentage points off. Again this suggests that Proposition A will pass.

We can determine the sample size necessary to be able to estimate p by $\hat{p}$ in such a way that the maximum error of our estimate will be a given value E with a given level of confidence $1 - \alpha$ by solving for n in the equation

$$E = z\left(\frac{\alpha}{2}\right) \cdot \sqrt{\frac{\hat{p}\hat{q}}{n}}$$

We square both sides, getting

$$E^2 = \left[z\left(\frac{\alpha}{2}\right)\right]^2 \cdot \frac{\hat{p}\hat{q}}{n}$$

and multiply by n/E^2, getting

$$n = \left[\frac{z(\alpha/2)}{E}\right]^2 \cdot \hat{p} \cdot \hat{q}$$

But there is a problem here. Since the sample has not yet been obtained, we do not know the values of $\hat{p}$ and $\hat{q}$ (where $\hat{q} = 1 - \hat{p}$).* However, it is known that regardless of the values of $\hat{p}$ and $\hat{q}$, the value of $\hat{p} \cdot \hat{q}$ will never be more than $\frac{1}{4}$. Therefore, to be on the safe side, we should take the sample size to be at least

$$n = \left[\frac{z(\alpha/2)}{E}\right]^2 \cdot \frac{1}{4}$$

To summarize:

> When estimating a population proportion p by a sample proportion $\hat{p}$, if we wish the maximum error of our estimate to be (some given value) E with level of confidence $1 - \alpha$, we should choose the sample size to be at least
>
> $$n = \left[\frac{z(\alpha/2)}{E}\right]^2 \cdot \frac{1}{4}$$

*If we have a rough idea as to the values of p and q, we can use these values in place of $\hat{p}$ and $\hat{q}$, but often we have no such information.

Example 7.5

A survey is to be made to estimate the proportion p of Americans who favor wage and price controls. How many people should be interviewed in order for our estimate to be within 3 percentage points of p with 98% confidence?

Solution

$$\text{maximum error of the estimate, } E = .03$$
$$\text{level of confidence, } 1 - \alpha = .98$$

$$\alpha = .02$$
$$z\left(\frac{\alpha}{2}\right) = z(.01) = 2.33$$

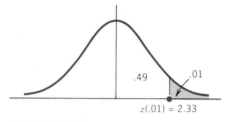

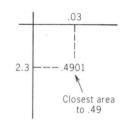

$z(.01) = 2.33$

FIGURE 7.6

To find the necessary number of interviews, we use

$$n = \left[\frac{z(\alpha/2)}{E}\right]^2 \cdot \frac{1}{4} = \left(\frac{2.33}{.03}\right)^2 (.25) \doteq 1508.03$$

Therefore, at least 1509 people should be interviewed. (Note that we round up to be on the safe side.)

Example 7.6

The A. C. Nielsen Company conducts surveys to determine the proportion of households viewing various television shows. A device called an audometer is placed on a number of T.V. sets throughout the country and information on which shows are watched is fed to a computer. In order to be able to estimate the proportion of all households viewing a certain show within 2 percentage points with 90% confidence, how many households should be surveyed (if a random sample were used)?

Solution

$$\text{maximum error of the estimate, } E = .02$$
$$\text{level of confidence, } 1 - \alpha = .90$$
$$\alpha = .10$$
$$z\left(\frac{\alpha}{2}\right) = z(.05) = 1.65$$
$$n = \left[\frac{z(\alpha/2)}{E}\right]^2 \cdot \frac{1}{4} = \left(\frac{1.65}{.02}\right)^2 (.25) \doteq 1701.56$$

Therefore, at least 1702 households should be surveyed.

Exercises

7.25 Fill in the blanks and construct a confidence interval for a population proportion p in each of the following. Note that n refers to the sample size and x the number of successes.

	$\hat{p}$	x	n	Percentage Confidence Interval
(a)		200	400	95
(b)	.25		900	90
(c)		160	225	99
(d)	.82		1250	95
(e)		592	1600	90
(f)	.61		1000	80

7.26 Officials in a resort town wanted to estimate the proportion p of voters who would support a proposal restricting the number of housing starts per year. A random sample of 150 voters showed that 105 supported the proposal. Find a 99% confidence interval for p.

7.27 A psychologist gave a test measuring anxiety level, the State–Trait Anxiety Inventory (STAI), to a sample of 50 abused children, and found that 45 had above-average anxiety levels. Find a 90% confidence interval for the true proportion of abused children with above-average anxiety levels.

7.28 A dentist wanted to estimate the proportion p of patients who paid their bill within a reasonable period of time. A random sample of 90 records showed that 73 patients paid their bill to the satisfaction of the dentist. Find a 90% confidence interval for p.

7.29 In a study initiated in 1940 a research group followed the lives of 200 Harvard graduates and 400 inner-city, working-class men from Boston and Cambridge, Massachusetts (*Source:* Vaillant, 1983, p. 27). Eventually, 26 from the Harvard group and 110 from the inner-city, working-class group became alcoholics. Assuming that these are random samples from the collection of Harvard graduates and inner-city, working-class men from Boston and Cambridge,

(a) Find a 95% confidence interval for the proportion p of Harvard graduates that eventually became alcoholic.

(b) Find a 95% confidence interval for the proportion p of the inner-city, working-class that eventually became alcoholic.

7.30 In the study discussed in Exercise 7.29, 96 of the 110 working-class alcoholics reported that they had experienced "family or friends' complaints." Find a 95% confidence interval for the proportion p of working-class alcoholics that have experienced "family or friends' complaints."

7.31 The Aid to Families with Dependent Children (AFDC) program has an overall error rate (in determining eligibility) of 4%. The state of California uses sampling to monitor its counties to see if they exceed the 4% (which can result in economic sanctions). In one county, 9 cases out of 150 were found to be in error.

(a) Find a 95% confidence interval for the error rate for the county (proportion of all cases in error).

(b)   In 1982 the California legislature mandated that a 95% confidence interval be used in studying error rate. Based on your answer in part (a), would you conclude that the error rate for the county is above the 4% rate?

7.32   Candidates A and B are opponents for political office. Candidate A's pollster conducts a poll 1 week before election and finds that 165 of 300 potential voters say they will vote for candidate A. At the time of the poll, can we be

(a)   80% confident that candidate A will win?

(b)   98% confident that candidate A will win?

Use the results of parts (a) and (b), and do not calculate, to answer the following:

(c)   Can candidate A be
   (i)   70% confident that he will win?
   (ii)   99% confident that he will win?

7.33   To estimate the proportion p of passengers who had purchased tickets for more than \$400 over a year's time, an airline official obtained a random sample of 75. The number of those purchasing tickets for more than \$400 was 45.

(a)   What is a point estimate for p?

(b)   Find a 95% confidence interval for p.

(c)   What is the maximum error of estimate for p?

7.34   A city council commissioned a statistician to estimate the proportion p of voters in favor of a proposal to build a new library. The statistician obtained a random sample of 200 voters, with 112 indicating approval of the proposal.

(a)   What is a point estimate for p?

(b)   Find a 90% confidence interval for p.

(c)   What is the maximum error of estimate for p?

7.35   A union official wanted to get an idea of whether a majority of workers at a large corporation would favor a contract proposal. She surveyed 500 workers and found that 260 favored the proposal.

(a)   Find a 95% confidence interval for the proportion of all the workers who favor the contract proposal.

(b)   Find the maximum error of the estimate.

(c)   Based on the results of part (a), can we conclude that the contract will be ratified by the membership?

7.36   A study of 75 lakes in Massachusetts indicated that 82% of these lakes may at some point be affected by acid rain. Assume that there are a large number of lakes in Massachusetts and that the sample can be considered random.

(a)   Find a 95% confidence interval for the proportion p of lakes in Massachusetts that may at some point be affected by acid rain.

(b)   What is the maximum error of estimate?

7.37   A superintendent of a city school system wants to estimate the proportion p of parents who believe the school system is providing an adequate education. How large a random sample is needed to estimate p to within three percentage points with 90% confidence?

7.38 A town official wants to estimate the proportion p of voters who favor the granting of a variance in order that a builder can construct a health spa in a residential area. How large a random sample is required to estimate p to within four percentage points with 95% confidence?

7.39 To estimate the proportion p of voters favoring a nuclear freeze in your voting district, how large a random sample is needed to estimate p to within two percentage points with

(a) 90% confidence? **(b)** 99% confidence?

7.40 Do parts (a), (b), and (c) without calculating. Once you have answered parts (a), (b), and (c), then calculate and compare with Exercise 7.39.

(a) Suppose that you were to estimate the proportion p in Exercise 7.39 to within two percentage points with 95% confidence.

   (i) Would the required sample size be larger, the same as, or smaller than that for 90% confidence?

   (ii) Would the required sample size be larger, the same as, or smaller than that for 99% confidence?

(b) Suppose that you were to estimate the proportion p of items in a population to within 1 percentage point with 90% confidence. Would the required sample size be larger, the same as, or smaller than that found in Exercise 7.39 (a)?

(c) Suppose that you were to estimate the proportion p of items in a population to within four percentage points with 90% confidence. Would the required sample size be larger, the same as, or smaller than that found in Exercise 7.39 (a)?

7.4 DECISION MAKING CONCERNING A POPULATION MEAN

Suppose that you are told that there is a male human being in the next room and you are asked to decide whether he is an adult or a child based on one piece of information: he weighs 80 pounds. You would guess that he is probably a child. The reason for this decision is that it would be highly unlikely for him to weigh so little if he were an adult; it seems much more likely that he is a child. This decision was not difficult to make because from experience we are familiar with the random variable "weight" (of an adult male). We have at least a rough idea of which values are likely and which are unlikely for adult males, and 80 pounds is very unlikely (though not impossible).

The above type of decision making is not uncommon in everyday situations. It is often used without giving it a second thought. It is also used in statistics. However, in statistics we often deal with quantities (random variables) that are not part of our everyday experience. Hence we cannot rely on our intuition to make a decision. Since we often do not know from experience which values of a random variable are likely and which are not, we need some kind of mathematical device that tells us this. And we have seen that such a device is called a "probability distribution."

Now let us investigate how statistics is used in decision making. A drug company was marketing a weight-reducing medication, and over the years it was determined that people using this drug for a period of 3 months have a mean weight loss of 25 pounds with a standard deviation of 12 pounds. The company was satisfied with the performance of the drug, except that in some cases it caused a nervous reaction. The manufacturer claimed that he could add an ingredient that would block this effect while theoretically not changing the average weight loss for the new drug.

However, a researcher at the company suspected that this ingredient might change the weight-reducing characteristics. The researcher wished to study the population consisting of future users of the new drug. Let μ = the mean weight loss for this population. The researcher wanted to choose between two statements, also called **hypotheses,** concerning this population: the manufacturer's claim that the mean weight loss for the new drug is 25 pounds is true or it is not true. In other words, she will choose between

$$\mu = 25 \text{ pounds or } \mu \neq 25 \text{ pounds}$$

(Note that $\mu \neq 25$ includes the possibility of a mean weight loss of less than 25 pounds, which would be unacceptable, or a mean loss of more than 25 pounds, which may constitute a health hazard.)

The drug manufacturer felt the new ingredient would have *no effect* on the drug's performance; there would be *no change* in the mean. So he believes $\mu = 25$. This is called the **null hypothesis.**

Definition The *null hypothesis* is a statement asserting no effect, or no change, or no difference. It usually takes the form of a statement about a population parameter (or parameters) containing an equal sign. The null hypothesis is labeled H_0.

The researcher felt there might be a change of some sort ($\mu \neq 25$ lb). This is called the **alternate hypothesis.**

Definition The *alternate hypothesis* is a statement that a researcher believes might be true instead of the null hypothesis. It usually contains the symbol $\neq$, $>$, or $<$. The alternate hypothesis is labeled H_a.

The procedure for choosing between hypotheses is called **hypothesis testing.** In hypothesis testing, the idea is to give the benefit of the doubt to the null hypothesis. The null hypothesis will only be rejected (and the alternate hypothesis accepted), if the sample data suggest beyond a reasonable doubt that the null hypothesis is false. We will have more to say about this later.

Thus the hypotheses the company researcher wishes to investigate are

$$H_0: \mu = 25$$
$$H_a: \mu \neq 25$$

Note that the population in question (users of the new drug) does not exist as of yet because the drug has not been marketed. But the researcher can still obtain a sample from this population. She could obtain some volunteers to try the drug and then record their weight losses. The company has a limited supply of the experimental drug, so let us suppose that the researcher selects only 36 people to try the drug. Also, in what follows, we will assume that while the mean μ for the new drug is not known for certain, there is every reason to believe that the standard deviation for the new drug is the same as that of the old drug. Hence $\sigma = 12$ pounds.

To investigate μ, it would seem natural to look at the sample mean weight loss $\bar{x}$ of the 36 people selected to try the drug. Although we do not expect the value of $\bar{x}$ to be exactly equal to μ, there is some relationship between the two. It is likely that $\bar{x}$ will not be too far from μ.

Now the procedure in hypothesis testing is to suppose for the moment that the null hypothesis H_0 is true. Therefore, let us assume for the moment that $\mu = 25$ pounds. Then it is likely that the sample mean $\bar{x}$ will not be too far from 25 pounds. For example, if $\bar{x} = 24$ pounds, the researcher would probably not reject the manufacturer's claim that $\mu = 25$. After all, we can expect some variation in $\bar{x}$ due to chance. However, if the value of $\bar{x}$ were only 2 pounds, the researcher would be inclined to reject the manufacturer's claim that $\mu = 25$. The reason is that it seems unlikely that we would get a value of $\bar{x}$ so far away from 25 pounds if the manufacturer's claim were really correct.

We cannot always rely on intuition to tell which values of $\bar{x}$ are consistent with the claim that $\mu = 25$ and which are not. For example, suppose that we got a value $\bar{x} = 18$ pounds. Should the manufacturer's claim be rejected or not?

What we need to be able to do is to determine those values of $\bar{x}$ that are so far away from 25 pounds, that such values would be very unlikely to occur if μ were really 25 pounds. If it then happens that we do get one of these values for our sample mean, the manufacturer's claim would be rejected; otherwise it would not be rejected.

The way we determine which values of $\bar{x}$ are likely and which are unlikely (if μ were 25) is to use the probability distribution of $\bar{x}$. The Central Limit Theorem (Section 6.7) tells us that when $\bar{x}$ is obtained from a random sample of size n, then if n is sufficiently large (at least 30), $\bar{x}$ will be approximately normal. In addition,

$$\mu_{\bar{x}} = \mu \quad \text{and} \quad \sigma_{\bar{x}} = \frac{\sigma}{\sqrt{n}}$$

Now we have seen that $\sigma = 12$. If the manufacturer's claim were true, $\bar{x}$ is approximately normal with

$$\mu_{\bar{x}} = 25 \text{ pounds} \quad \text{and} \quad \sigma_{\bar{x}} = \frac{12 \text{ pounds}}{\sqrt{36}} = 2 \text{ pounds}$$

We wish to find those values of $\bar{x}$ that are so far away from 25 as to be unlikely. By "unlikely" we mean "having very low probability." It is up to the researcher to define what is meant by "low." A value that is often selected is .05. We will use this value in this discussion. We call it the **level of significance** of the test. (The level of significance is denoted by the symbol α and can change from one test to the next.)

We will find it convenient to work with the standard score (z score) for $\bar{x}$ rather than $\bar{x}$ itself. This is just a different unit of measurement for the sample mean weight loss, namely, the number of standard deviations above or below the mean, instead of pounds. If we assume that the claim $\mu = 25$ is true,

$$z = \frac{\bar{x} - \mu_{\bar{x}}}{\sigma_{\bar{x}}} = \frac{\bar{x} - \mu}{\sigma/\sqrt{n}} = \frac{\bar{x} - 25}{2}$$

and this z score is approximately standard normal. When $\bar{x} = 25$, $z = 0$. Therefore, values of $\bar{x}$ far to the right or left of 25 correspond to values of z far to the right or left of 0. We will use the value of z to make a decision concerning hypotheses H_0 and H_a. Such an expression is called a **test statistic.**

> **Definition** A *test statistic* is a quantity that is used to make a decision in a test of hypotheses.

Notice from Figure 7.7a that the probability that a value of z will be either ≥ 1.96 or ≤ -1.96 is $.025 + .025 = .05$. [We used Appendix Table B.3 to find the value 1.96 by looking up the value of z corresponding to the area .4750. Observe that $.4750 = .5 - .025$ and $z(.025) = 1.96$.]

Keep in mind that

$$z = \frac{\bar{x} - 25}{2}$$

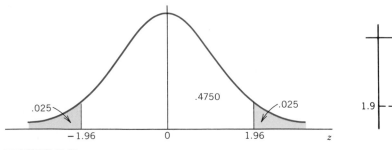

FIGURE 7.7a
$P(z \geq 1.96 \text{ or } z \leq -1.96) = .05$

A value of z that is ≥ 1.96 or ≤ -1.96 will correspond to a value of $\bar{x}$ that is so far from 25 (to the right or left) that the probability of getting a value that far away from 25 would be very small (only .05) if μ were 25. (See Figure 7.7b.) The researcher would now obtain the value of $\bar{x}$ from the sample and calculate its z score. If the z score is ≥ 1.96 or ≤ -1.96, this would correspond to a value of $\bar{x}$ that would be very unlikely (only a 5% chance) if the claim that $\mu = 25$ were correct. So if a z score that is ≥ 1.96 or ≤ -1.96 does occur, the researcher would reject the manufacturer's claim that $\mu = 25$. That is, we would reject the null hypothesis H_0 in favor of the alternate hypothesis H_a.

For example, suppose that we observed from our sample a mean weight loss of $\bar{x} = 18$ pounds. Then the observed value of z would be

$$z = \frac{\bar{x} - 25}{2} = \frac{18 - 25}{2} = -3.5$$

Since $-3.5 < -1.96$, the claim H_0 ($\mu = 25$ pounds) would be rejected in favor of H_a ($\mu \neq 25$ pounds). In fact, it would appear that $\mu < 25$ pounds. (See Figure 7.7b.)

Therefore, values of $z \geq 1.96$ or ≤ -1.96 provide strong evidence in favor of the alternate hypothesis H_a. These values of z constitute what is called the **critical region**. The values ± 1.96 are called **critical values**.

The **critical region** consists of those values of the test statistic that provide strong evidence in favor of the alternate hypothesis. Hence a value in the critical region leads to rejection of the null hypothesis.

Notice that in the above procedure it was not necessary to find the values of a and b on the $\bar{x}$ axis in Figure 7.7b. However, these values may easily be

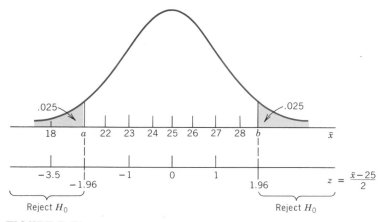

FIGURE 7.7b

found and would enable us to describe the critical region in terms of $\bar{x}$ values. The value a on the $\bar{x}$ axis corresponds to $z = -1.96$. Similarly, b corresponds to $z = 1.96$ so

$$-1.96 = \frac{a - 25}{2} \qquad 1.96 = \frac{b - 25}{2}$$

$$a = 21.08 \qquad\qquad b = 28.92$$

Therefore, a value of $\bar{x}$ that is ≥ 28.92 pounds or ≤ 21.08 pounds will lead to rejection of the claim that $\mu = 25$ pounds.

Errors

In the above discussion we have seen that it would be very unlikely that the value of the test statistic would fall in the critical region if the null hypothesis were true. (That is why we reject H_0 when this happens.) The probability was only .05. We called .05 the level of significance of the test; it is denoted by α (alpha).

> The probability that the test statistic will fall in the critical region if the null hypothesis is true is equal to the **level of significance** of the test α.

Although it is unlikely that the test statistic will fall in the critical region when H_0 is true, it is still possible. In this case, we will reject H_0 and make an error in doing so. This is called **type I error**—rejecting the null hypothesis when it is true. So a type I error will occur if the test statistic falls in the critical region when H_0 is actually true. We saw that the probability of this happening is α. Thus

$$\alpha = P(\text{type I error})$$

The other kind of error that could be made is to fail to reject H_0 when it is false. This is called a **type II error**. A type II error will occur if the test statistic does not fall in the critical region when H_0 is in fact false. We represent the probability of a type II error by the symbol β (beta).

TABLE 7.1
Possible Decisions in a Hypothesis Test

Null Hypothesis	Decision	
	Reject H_0	Fail to Reject H_0
True	Type I error	Correct decision
False	Correct decision	Type II error

$$\beta = P(\text{type II error})$$

Whereas the value of α is always known (in fact, the researcher chooses this value), the value of β is usually unknown. To understand why this is so, recall that in our preceding example β is the probability that the test statistic

$$\frac{\bar{x} - 25}{2}$$

falls outside the critical region when H_0 is false; *but* when H_0 is false, that is, when $\mu \neq 25$, we can no longer say that the above expression has a standard normal distribution. Therefore, in the absence of more information, we cannot find the probability that its value will fall outside the critical region.

The following procedure will frequently be used in the remainder of this book to investigate hypotheses concerning a population. It need not be restricted to tests concerning a population mean.

Summary of the Steps in Hypothesis Testing

1. **Identify the null hypothesis H_0 and the alternate hypothesis H_a.** These will often be conjectures (or suspicions or beliefs) concerning the value of one or more population parameters. As a rule of thumb, the null hypothesis will usually contain an equal sign. The alternate hypothesis will usually contain the symbol $>$, $<$, or $\neq$.

2. **Choose α, the level of significance.** The value of α should be small, usually $\leq .10$. But since α is the probability of a type I error, the actual value of α chosen depends on how serious a type I error is in a given situation. The more serious the type I error, the smaller we should choose α.

3. **Select the test statistic and determine its value from the sample data.** This value is called the *observed value* (O.V.) of the test statistic. The test statistic will be a quantity that has some relationship to the parameter in question. For example, when investigating μ we could look at $\bar{x}$. Actually (when n is large), we look at the standard score for $\bar{x}$, namely, $(\bar{x} - \mu)/(\sigma/\sqrt{n})$. In this expression we use the value of μ given in the null hypothesis H_0, because H_0 is assumed true until there is evidence to the contrary.

4. **Determine the critical region.** The critical region consists of those values of the test statistic that strongly favor the alternate hypothesis H_a. The actual size of the critical region depends on the level of significance α. This is because the critical region is chosen in such a way that the probability will be α that the test statistic will fall in the critical region (if H_0 were true).

5. **Make your decision.** If the test statistic falls in the critical region, reject H_0 in favor of H_a. When this occurs, some statisticians say the results are *statistically significant*, also giving the α level. If the test statistic does not fall in the critical region, we fail to reject H_0. That is, we conclude that there is not enough evidence to reject H_0. You should interpret your decision in ordinary, nontechnical language.

Choosing H_0 and H_a

Often the statement of the problem will involve only one hypothesis, which will take the form of a claim, a belief, or a suspicion about the population. It is up to you to identify this as H_0 or H_a, and then write the other hypothesis.

For example, a nutritionist claimed that a food company's cans of soup contained more than 900 milligrams of sodium on the average.

- Write the statement in symbolic form:

$$\mu > 900 \leftarrow \text{Claim}$$

- Now write the statement with an $=$ sign instead of $>$:

$$\mu = 900 \leftarrow \text{Contains } = . \text{ It is the}$$
$$\text{null hypothesis}$$

- Thus we test the hypotheses

$$H_0: \mu = 900$$
$$H_a: \mu > 900 \leftarrow \text{Claim}$$

The preceding claim was the alternate hypothesis. Sometimes the claim is the null hypothesis. For example, an auto company claimed that the mean weight of its pickup trucks was 2 tons:

$$\mu = 2$$

This is a statement of no difference; it contains an $=$ sign. So it is H_0. What about H_a? If H_0 is false, then μ could be either more than 2, or less than 2. If we have no *prior* reason to believe which of these might be true, we should allow for either possibility and use $\mu \neq 2$ as the alternate hypothesis. So we would test:

$$H_0: \mu = 2 \leftarrow \text{Claim}$$
$$H_a: \mu \neq 2$$

The following table gives the hypotheses tests to be performed when just one statement about a population is given:

Summary of Types of Hypothesis Tests

Statement	μ is	more than 100	less than 100	different from 100	equal to 100
Null Hypothesis	H_0	$\mu = 100$	$\mu = 100$	$\mu = 100$	$\mu = 100$
Alternate Hypothesis	H_a	$\mu > 100$	$\mu < 100$	$\mu \neq 100$	$\mu \neq 100$

Example 7.7

A sports biologist claimed that female distance runners tend to be taller on the average than women in general, who have an average height of 64 inches. To study this she obtained a random sample of 40 female distance runners and their heights were recorded, giving $\bar{x} = 65.6$ inches and $s = 3.3$ inches. Using these results, test the claim at the 5% level of significance, that is, use $\alpha = .05$. (*Note:* Although we will need the value of σ, it is not given. However, since the sample size is large, we can use s as an estimate. Hence $\sigma \doteq 3.3$ inches.)

Solution

We will go through the five steps outlined above.

1. Hypotheses: The claim is that $\mu > 64$. Thus our hypotheses are

$$H_0: \mu = 64$$
$$H_a: \mu > 64$$

2. Level of significance: This is given as $\alpha = .05$.

3. Test statistic and its observed value: It is natural to look at the value of $\bar{x}$, or more precisely, its standard score:

$$z = \frac{\bar{x} - \mu}{\sigma/\sqrt{n}} = \frac{65.6 - 64}{3.3/\sqrt{40}} = \frac{1.6}{.522} = 3.07$$

Recall that we are assuming for the moment that H_0 is true. Therefore, for μ we substituted 64, the value given in H_0.

4. Critical region (favors H_a): Large values of $\bar{x}$ would favor the alternate hypothesis H_a. Large values of $\bar{x}$ correspond to large values of z. How large should z be to convince us to reject H_0 in favor of H_a? The answer is, so large that such a value would be unlikely (if H_0 were true). How unlikely? So unlikely that the probability is only $\alpha = .05$. (See Figure 7.8.)

From Appendix Table B.3, we see that $P(0 < z < 1.65) = .45$. So $P(z \geq 1.65) = .5 - .45 = .05$. Thus our critical value is $z(.05) = 1.65$. [Recall from Section 6.4, $z(.05)$ is the value of z such that the area to the right of it is .05.] The critical region consists of values of $z \geq 1.65$.

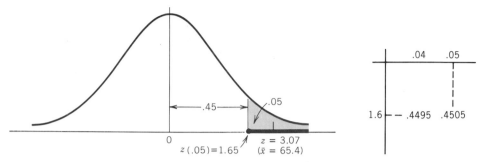

FIGURE 7.8

5. Decision: Our observed value is $z = 3.07$. Since $3.07 > 1.65$, it is in the critical region. Hence we reject H_0 in favor of H_a. Therefore, it does appear that female distance runners tend to be taller on the average than women in general.

Example 7.8
A high-school mathematics department made a change in its curriculum for college-bound students that would allow the substitution of Computer Mathematics for Algebra II. One of the instructors in the department felt that this would result in students having lower math skills. He administered a math skills test to 49 students who elected the new option. Their mean score was 67.5. Past performance of students (under the old option) showed a mean of 70 and standard deviation of 10 points. Is there evidence to conclude that students electing the new option have lower math skills? Use the 2% level of significance, and assume that the standard deviation for all students who elect the new option is still 10.

Solution
The population consists of all students now or in the future who will elect the new option. Let μ = the mean score for this population.

1. Hypotheses: Obviously, the instructor feels that $\mu < 70$. The hypotheses are

$$H_0\text{: } \mu = 70$$
$$H_a\text{: } \mu < 70$$

2. Level of significance: $\alpha = .02$.

3. Test statistic and its observed value: Here again we use the standard score of z. Remember that the value of the parameter μ is always obtained from the null hypothesis H_0.

$$z = \frac{\bar{x} - \mu}{\sigma/\sqrt{n}} = \frac{67.5 - 70}{10/\sqrt{49}} = -1.75$$

4. Critical region: This is always obtained by examining the alternate hypothesis H_a and using α. Values of $\bar{x}$ that favor H_a are those that are far to the left of 70. Since

$$z = \frac{\bar{x} - \mu}{\sigma/\sqrt{n}}$$

where $\mu = 70$, values of $\bar{x}$ far to the left of 70 correspond to values of z far to the left of 0. Values of z that lead to rejection of H_0 are those so far to the left of 0 that the probability is only $\alpha = .02$ that such values would occur if H_0 were true. (See Figure 7.9.) Now keep in mind the symmetry of the standard normal distribution. Using Appendix Table B.3, look up the z value corresponding to areas as close to .48 as possible. This value is 2.05 [symbolically written z(.02)]. Therefore, by symmetry the desired critical value is $-z(.02) = -2.05$. Thus the critical region consists of values of z such that $z \leq -2.05$.

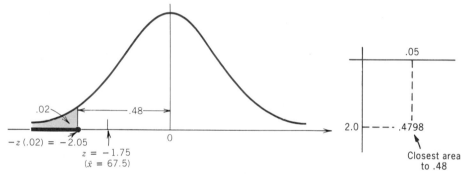

FIGURE 7.9

5. Decision: From step 3, $z = -1.75$. This is not in the critical region. Therefore, we fail to reject H_0. There is not enough evidence to justify the instructor's suspicion that the new mathematics option will result in lower scores (on the average).

In the preceding pages we have seen examples of three basic types of tests with their corresponding critical regions. A review of these tests follows.

1. In Example 7.7 we tested

$H_0: \mu = 64$

$H_a: \mu > 64$

$\alpha = .05$

The critical value is

$z(\alpha) = z(.05) = 1.65$

See Figure 7.10.

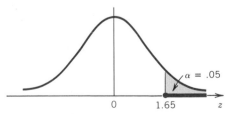

FIGURE 7.10
Right-Tailed Critical Region

Because of the location of the critical region, this is called a **right-tailed test.**

2.   In Example 7.8 we tested

$H_0: \mu = 70$

$H_a: \mu < 70$

$\alpha = .02$

The critical value is

$-z(\alpha) = -z(.02) = -2.05$

See Figure 7.11.
This is called a **left-tailed test.**

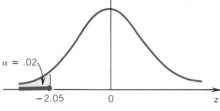

FIGURE 7.11
Left-Tailed Critical Region

3.   In the introductory example of this section we tested

$H_0: \mu = 25$

$H_a: \mu \neq 25$

$\alpha = .05$

The critical values are

$\pm z\left(\dfrac{\alpha}{2}\right) = \pm z(.025) = \pm 1.96$

See Figure 7.12.
This is called a **two-tailed test.**

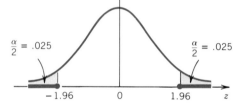

FIGURE 7.12
Two-Tailed Critical Region

All the above tests used a test statistic z that had a standard normal distribution. A statistical test that uses a standard normal distribution is called a **z test.** Now we summarize the essential features of a z test for a population mean.

The z Test for a Population Mean (Based on a Large Sample)   To test hypotheses concerning μ (when sample size is ≥ 30), we use the test statistic

$$z = \frac{\bar{x} - \mu}{\sigma/\sqrt{n}}$$

which has the standard normal distribution (approximately). For μ we substitute the value given in the null hypothesis. The observed value of z is computed by substituting $\bar{x}$ and n obtained from the sample data. (If σ is unknown, we use s.) If z falls in the critical region, we reject the null hypothesis H_0. Assume that α is the level of significance of the test. Critical values of z are obtained in Appendix Table B.3. The possible critical regions are described as follows:

(a)   If the alternate hypothesis contains the symbol $>$, we conduct a right-tailed test. The critical region is displayed in Figure 7.13.

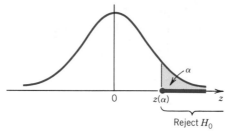

FIGURE 7.13
Right-Tailed Test

(b) If the alternate hypothesis contains the symbol $<$, we conduct a left-tailed test. The critical region is displayed in Figure 7.14.

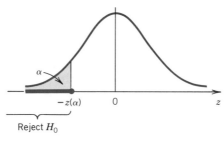

FIGURE 7.14
Left-Tailed Test

(c) If the alternate hypothesis contains the symbol $\neq$, we conduct a two-tailed test. The critical region is displayed in Figure 7.15.

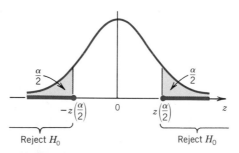

FIGURE 7.15
Two-Tailed Test

Note that we can tell which type of test (i.e., which type of critical region) is to be used by examining the alternate hypothesis H_a because those values of the test statistic that strongly favor H_a constitute the critical region. We summarize this in Table 7.2. The actual critical values will vary from one situation to the

TABLE 7.2
Hypothesis Tests Concerning μ

If H_a contains	$>$	$<$	$\neq$
Perform	Right-tailed test	Left-tailed test	Two-tailed test

next. These can be found, once the level of significance α is specified. *Note:* It is customary to write our hypotheses in such a way that the symbol for the parameter (such as μ) appears first. For example, we write

$$\mu < 70 \text{ rather than } 70 > \mu$$

We will follow this convention throughout the book. Notice that if we did not always follow this practice, Table 7.2 would not always be correct.

Which Hypothesis Are We Trying to "Prove"?

Our method enables us to decide if there is sufficient evidence to "prove" the alternate hypothesis H_a but not the null hypothesis H_0. If we reject H_0, we accept H_a. In this case we know the risk (chance) that we made an error (type I error)—it is α, which is always known and always chosen small. However, if we fail to reject H_0, we have seen that the chance of an error (type II error) is β, which is usually unknown. Sometimes β can be quite large. Therefore if we fail to reject H_0, we are not prepared to say that the null hypothesis has been established, because we don't know our chance of being wrong. This is why we avoid strong language like "accept H_0," settling instead for the much weaker "fail to reject H_0."

Sometimes an analogy is drawn between hypothesis testing and a criminal trial. The defendant is presumed innocent unless there is evidence of his guilt beyond a reasonable doubt. In hypothesis testing the null hypothesis is presumed to be true unless there is sufficient evidence to the contrary. This evidence consists of the test statistic falling in the critical region—something that would be unlikely to occur if H_0 were true. In a criminal trial a verdict of "not guilty" does not necessarily mean the defendant did not commit the crime; it just means there was not sufficient evidence of guilt. In hypothesis testing, if we fail to reject H_0, we don't assert that it is definitely true. The statistical evidence was just not strong enough to reject it.

The lesson in all of this for researchers is: formulate the statement you wish to prove as the alternate hypothesis H_a. Then hope your test will lead to rejection of H_0. In this case you can place some faith in the truth of H_a, because you know the chance that you made an error (α), and this chance is small.

Controlling Errors

Since α is the probability of a type I error, it measures the risk of an error when we reject the null hypothesis. In general α is chosen to be small; however, the more serious a type I error, the smaller we would want the value of α. Determining the seriousness of a type I error in a given situation is a subjective judgment that must be made by the researcher. In any case, *the smaller the value of α, the more confident we may feel about our decision if we reject H_0.* Some of the commonly chosen values for α are .01, .02, .025, .05, and .10. Since α measures the risk associated with a type I error, the reader might wonder why we would settle for a value of .05 for α when we could choose .01 or something even smaller. The answer lies in the fact that when we reduce the value of α, there will be an increase in the value of β, the probability of a type II error (failing to reject H_0 when it is false). (To get an idea of why this is true, see Exercise 7.119.)

Therefore, even though we don't usually know the value of β, we can have an influence on its size by our choice of α. For a fixed sample size, the smaller the value of α, the larger the value of β. The larger the value of α, the smaller the value of β. If a type I error is very serious when compared to a type II error, we should choose a rather small value of α, say, .01 or .02. However, if we are very concerned about a type II error, we may choose a larger value for α, say, .10.

We should also mention that a technique that can often be used to decrease the size of β is to increase the sample size.

Exercises

7.41 For the claims in parts (a) through (g), **(i)** find H_0 and H_a, **(ii)** give the type of critical region (right-tailed, left-tailed, or two-tailed), and **(iii)** explain the meaning of a type I and type II error.

(a) The mean amount μ of rainfall per year is more than 72 inches.

(b) The mean number μ of books borrowed per day from a library is 250.

(c) The mean temperature μ, taken at 1 P.M., in a coastal town for the month of July is less than 78°F.

(d) The mean age of professors at a large university is more than 36.

(e) The mean salary of employees at an industrial plant is less than $29,500.

(f) The mean number of families per month that are below poverty level in a state is 18,000.

(g) The mean grade-point average of graduating seniors at a university is greater than 2.3.

7.42 Suppose you are testing the claim that the mean μ of a population is greater than 90. Assume that the sample size is large.

(a) Give H_0 and H_a.

(b) Specify the critical region in terms of the standard normal z for the given level of significance in each of the following:

    (i) $\alpha = .01$.   **(ii)** $\alpha = .05$.   **(iii)** $\alpha = .10$.

7.43 Repeat Exercise 7.42, assuming that you are to test the claim that the population mean μ is 90.

7.44 Repeat Exercise 7.42, assuming that you are to test the claim that the population mean μ is less than 90.

7.45   The claim to be tested is that the mean μ of a population is smaller than 30. Suppose that $\bar{x} = 24$, and the population standard deviation $\sigma = 40$. For each sample size n, determine whether H_0 would be rejected at each significance level (i) .01, (ii) .05, and (iii) .10.

   (a) $n = 100$.   **(b)** $n = 225$.   **(c)** $n = 400$.

7.46   Sometimes a critical region is given in terms of $\bar{x}$. Assume that you are to test the claim that the mean μ of a particular population is greater than 30. Assume that $\sigma = 40$ and you decide to use $n = 64$ random observations. Determine the level of significance α in each of the following, where the critical region is given in terms of $\bar{x}$.

   (a) $\bar{x} \geq 36$.   **(b)** $\bar{x} \geq 42$.   **(c)** $\bar{x} \geq 38$.   **(d)** $\bar{x} \geq 40$.

7.47   A bus company advertised a mean time of 150 minutes for a trip between two cities. A consumer group had reason to believe that the mean time was more than 150 minutes. A sample of 40 trips showed a mean of 153 minutes and a standard deviation of 7.5 minutes. Using a 5% level of significance, is there sufficient evidence to support the consumer group's contention? What type of error has possibly been committed? Explain the error in ordinary language.

7.48   Under laboratory test conditions a handball should bounce 48 inches. The manufacturer was receiving complaints that the balls were not as lively as they should be. The manufacturer randomly sampled 50 handballs and found a mean bounce of 46.5 inches and a standard deviation of 5 inches. At the 5% level of significance, is there sufficient evidence to suggest the manufacturer's handballs are bouncing less than 48 inches on average? What type of error has possibly been committed? Explain the error in ordinary language.

7.49   To confirm her belief that abused children would show elevated levels of depression, a psychologist gave a test called the Profile of Mood States (POMS) to a sample of 50 abused children. The results showed a mean depression score of 17.3 with a standard deviation of 5.4. At the 5% level, can she conclude that abused children in general have a mean depression level of more than 15 (the mean for college students)?

7.50   Prior to expanding the facilities of a town library it was determined that 385 books per day were loaned out. It was believed that a new addition to the building would increase the mean number of books loaned out per day. After completion of the new addition, a random sample of 35 days showed a mean of 395 and a standard deviation of 26 books loaned out per day. At the 10% level of significance, is there sufficient evidence to indicate that more books are being loaned out per day, on the average?

7.51   The owner of an artesian well-drilling company suspected that for a large tract of development land, the average depth of water below the surface was less than 500 feet. A sample of 32 wells gave a mean depth of $\bar{x} = 486$ feet with a standard deviation of $s = 53$ feet. At the 1% level of significance, is the owner's suspicion justified?

7.52   A plastic has a mean breaking strength of 27 and a standard deviation of 6 pounds per square inch. A new process is developed and will replace the old one, provided that there is substantial evidence that it improves the strength of the product. A random sample of 40 pieces made with the new process gives a sample mean of 30 pounds per square inch. Assuming that the variability is unchanged (i.e., $\sigma = 6$), is there sufficient evidence to suggest that the strength of the product has increased at the 1% level of significance?

7.53 Repeat Exercise 7.52 using $\sigma = 15$. Comment on the relationship between the ability to detect a change and the amount of spread of data values about the mean.

7.54 A cereal is packaged to contain 16 ounces, on the average. A consumer agency has received many complaints claiming that packages of the cereal contain less than 16 ounces. To test the claim that the mean content μ is less than 16 ounces, the consumer agency randomly selects 100 packages and finds that $\bar{x} = 15.1$ with $s = 3$.

(a) Complete the test at the 1% level of significance.

(b) Comment on why it seems appropriate for the consumer agency to use a small level of significance.

(c) If the mean content μ really were 16 ounces, how many times in 10,000, on the average, would a sample mean result in a value of 15.1 or less? (Use $\sigma = 3$.)

(d) Based on your answer in part (c), does the evidence appear substantial that the population mean μ is less than 16?

7.55 A manufacturer of floor mops would like his product to last 700 hours, on the average. He hopes the mean number of hours is not a lot less than 700 (people will not continue to buy his product) nor a lot more than 700 (people would seldom have to buy the product). However, the manufacturer has reason to believe the mean may have changed. A random sample of 48 items shows that $\bar{x} = 675$ and $s = 77$ hours.

(a) Complete the test at the 5% level of significance.

(b) What type of error may have been made?

(c) What is the probability of committing this error?

7.56 A typing instructor wondered whether a new method of instruction would result in a change in the mean typing speed of students. The old method produced a mean of 64 words per minute. The results (words per minute) of 38 students using the new method were as follows:

56	60	60	76	60	59	41	66	67	33	55	43	61
71	73	60	56	77	43	58	65	67	71	56	48	90
82	46	68	36	27	52	64	54	49	69	46	68	

At the 5% level of significance, can the instructor conclude that the new mean is different from 64?

7.5 *P* VALUES

The method of hypothesis testing discussed in the previous section is called the **classical approach** to hypothesis testing, and is the method that will be emphasized in this book. However, some statisticians use a different approach.

They omit the step of finding the critical region. Instead, they find the "P value" associated with the value of the test statistic. Before giving a formal definition of this term, let us look at Example 7.7 again.

In this example, we investigated μ, the mean height of female distance runners. We tested

$$H_0: \mu = 64 \text{ inches}$$

$$H_a: \mu > 64 \text{ inches}$$

We used $\alpha = .05$. The critical region consisted of all $z \geq 1.65$, since $z(.05) = 1.65$. The average height for the sample was $\bar{x} = 65.6$ inches. Its z score (the observed value of the test statistic) was

$$z = \frac{\bar{x} - \mu}{\sigma/\sqrt{n}} = \frac{65.6 - 64}{3.3/\sqrt{40}} = 3.07$$

(*Note:* We used 64 for μ. This is the value of μ if H_0 is true.) The value $z = 3.07$ is quite far into the critical region. Now the larger the value of $\bar{x}$ (or equivalently the z score) the stronger the evidence in support of the alternate hypothesis H_a: $\mu > 64$. The probability of getting a value of the test statistic z as favorable or more favorable to H_a than $z = 3.07$ (if H_0 were true) is (from Appendix Table B.3)

$$P(z \geq 3.07) = .5 - .4989 = .0011$$

Hence if H_0 were true, the chances of getting a value of $z \geq 3.07$ is only about 1 in 1000. In other words, it would be very unlikely that we would get a value of z so large if H_0 were really true. Therefore, since we did get a value this large (i.e., so favorable to H_a), we are inclined to reject H_0 in favor of H_a. The probability $P(z \geq 3.07) = .0011$ is called the **P value** associated with $z = 3.07$. (See Figure 7.16.)

Definition   Suppose that we are conducting a test of hypotheses and we calculate the observed value of the test statistic. The *P value* is the probability of getting a value of the test statistic as favorable or more favorable to the alternate hypothesis than the observed value (if H_0 were true).

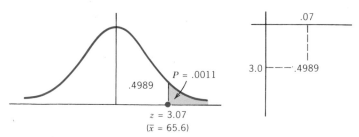

FIGURE 7.16
P value $= P(z \geq 3.07) = .5 - .4989 = .0011$

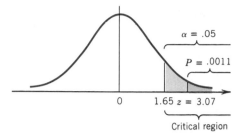

FIGURE 7.17

The importance of the P value is that it enables us to judge just how strong (or weak) the evidence is in support of the alternate hypothesis H_a. *The smaller the P value, the stronger the evidence is in support of the alternate hypothesis.*

How small should the P value be in order for us to reject H_0 in favor of H_a? This is a subjective judgment. Some journals will not accept results as statistically significant (i.e., justifying a rejection of the null hypothesis), unless P is less than or equal to .05.

What this amounts to is using the P value in conjunction with a predetermined level of significance. To see how this works, recall that the hypotheses discussed above were tested in Example 7.7 using a level of significance $\alpha =$ .05. The critical value was 1.65. In Figure 7.17 we show the relationship between α and the P value associated with the observed value of $z = 3.07$.

The fact that the observed value $z = 3.07$ is in the critical region corresponds to the fact that $P = .0011$ is less than $\alpha = .05$. We could have tested the above hypotheses at the 5% level of significance *without finding the critical region.* Since P is less than .05, the observed value must be in the critical region; therefore we reject the null hypothesis H_0.

This leads to a slightly different approach to testing hypotheses, summarized as follows:

> **The P Value Approach to Testing Hypotheses**   When testing hypotheses at a level of significance α, we reject the null hypothesis if
>
> $$P \leq \alpha$$
>
> We would fail to reject H_0 if
>
> $$P > \alpha$$

The following example shows how to compute the P value for a two-tailed test.

Example 7.9
At the beginning of this section we discussed the procedure for testing a drug manufacturer's claim that his new weight-reducing medication would have a

mean weight loss μ of 25 pounds. The hypotheses were

$$H_0: \mu = 25$$
$$H_a: \mu \neq 25$$

The sample size was $n = 36$ and $\sigma = 12$ pounds. If the sample mean weight loss is $\bar{x} = 18$, find the P value and interpret this value. Would the null hypothesis be rejected if the level of significance were $\alpha = .005$?

Solution
The observed value of the test statistic is

$$z = \frac{\bar{x} - \mu}{\sigma/\sqrt{n}} = \frac{18 - 25}{12/\sqrt{36}} = \frac{-7}{2} = -3.5$$

The values of the test statistic z that are as favorable or more favorable to H_a than $z = -3.5$ are those values of z which are ≤ -3.5 or ≥ 3.5. (Remember that values of $\bar{x}$ that favor H_a in this problem are those that are much larger than or much less than 25. These $\bar{x}$ values correspond to values of z that are much larger than or much less than 0.) Therefore, the P value associated with $z = -3.5$ is the probability

$$P(z \leq -3.5 \text{ or } z \geq 3.5)$$
$$= 2 \cdot P(z \geq 3.5)$$
$$= 2[.5 - P(0 \leq z \leq 3.5)]$$
$$= 2(.5 - .4998) = 2(.0002) = .0004$$

(See Figure 7.18.)
It is very unlikely (the probability is only .0004) that we would observe a value of z as favorable or more favorable to the alternate hypothesis H_a than $z = -3.5$ if the null hypothesis H_0 were really correct. Now suppose that the level of significance is $\alpha = .005$. Since the P value is less than .005, we would reject the null hypothesis.

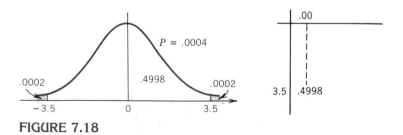

FIGURE 7.18

The following is a graphic display of P values:

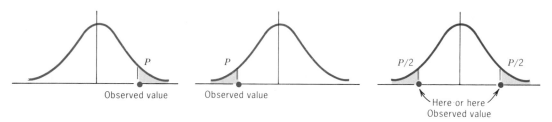

| For a right-tailed test | For a left-tailed test | For a two-tailed test |

FIGURE 7.19
P **Values**

Exercises

In Problems 7.57–7.62 find the *P* value and decide whether to reject or fail to reject the null hypothesis. Assume $\alpha = .05$.

7.57 $H_0: \mu = 80; H_a: \mu > 80.$ Observed value: $z = 1.75.$

7.58 $H_0: \mu = 23.5; H_a: \mu > 23.5.$ Observed value: $z = 1.2.$

7.59 $H_0: \mu = 15; H_a: \mu < 15.$ Observed value: $z = -.83.$

7.60 $H_0: \mu = 3.7; H_a: \mu < 3.7.$ Observed value: $z = -2.95.$

7.61 $H_0: \mu = 80; H_a: \mu \neq 80.$ Observed value: $z = 1.75.$

7.62 $H_0: \mu = 125.5; H_a: \mu \neq 125.5.$ Observed value: $z = 3.5.$

7.63   In parts (a) through (c), test the hypothesis that a population mean μ is more than 30 . Assume $\sigma = 40$ and the sample size is $n - 64$. Determine the *P* value for each of the following three values of $\bar{x}$:

    (a) $\bar{x} = 35.$ **(b)** $\bar{x} = 37.$ **(c)** $\bar{x} = 39.$

    (d)   Determine the *P* value in parts (a) through (c) for testing the hypothesis that μ is not equal to 30.

7.64   Repeat Exercise 7.63, replacing $n = 64$ with $n = 100$. Compare your results with those obtained in Exercise 7.63.

7.65   A consumers' group felt that an insecticide was effective for less than the advertised 12 days (on the average). The group used a random sample of size 36 and obtained a mean of 11.25 with a standard deviation of 3 days.

    (a)   Find the *P* value.

    (b)   For which of the following levels of significance would H_0 be rejected:

        (i) $\alpha = .10?$ **(ii)** $\alpha = .05?$ **(iii)** $\alpha = .01?$

    [*Hint:* These questions can be answered using your answer in part (a).]

    (c)   For each case in part (b), what type of error has possibly been committed?

7.66   A city assessor suspected that the mean income per household had increased over the old mean of $25,200. A random sample of 40 households in the city showed a mean income of $27,400 with a standard deviation of $6325.

    (a)   Find the *P* value.

(b) Would H_0 be rejected at level of significance

    (i) $\alpha = .05$?   **(ii)** $\alpha = .01$?

7.67 In Exercise 7.49 we were interested in whether abused children had a mean depression score of more than 15 on a psychological test. The sample data were $n = 50$, $\bar{x} = 17.3$, and $s = 5.4$. The level of significance was 5%. Complete the hypothesis test using the P value approach.

7.68 A vendor was concerned that a soft drink machine was not dispensing 6 ounces per cup, on average. A sample of size 40 gave a mean amount per cup of 5.95 ounces with a standard deviation of .15.

(a) Find the P value.

(b) For which of the following levels of significance would H_0 be rejected?

    (i) $\alpha = .10$?   **(ii)** $\alpha = .05$?   **(iii)** $\alpha = .01$?

[*Hint:* These can be answered using your answer in part (a).]

(c) For each case in part (b), what type of error has possibly been committed?

7.69 In Exercise 7.51 we discussed a conjecture to the effect that the mean depth of water below the surface in a large development tract was less than 500 feet. The sample data were $n = 32$ test holes, $\bar{x} = 486$ feet, and $s = 53$ feet. Complete the test using the P value approach and the 1% level of significance.

7.70 A tire gauge, when tested on a machine, is designed to measure 30 pounds per square inch (psi). The manufacturer was concerned that the gauges weren't meeting specifications. A sample of 50 readings gave a sample mean of 30.1 psi and a standard deviation of .25.

(a) Find the P value.

(b) For which of the following levels of significance would H_0 be rejected:

    (i) $\alpha = .10$?   **(ii)** $\alpha = .05$?   **(iii)** $\alpha = .01$?

[*Hint:* These can be answered using your answer in part (a).]

(c) For each case in part (b), what type of error has possibly been committed?

7.71 In Exercise 7.56 we studied the question of whether a typing instructor could conclude that a new method of instruction gave a mean typing speed different from 64 words per minute. Below is a partial Minitab printout.

TEST OF MU $= 64.00$ VS MU N.E. 64.00

N	MEAN	STDEV	Z	P VALUE
38	58.76	13.59	-2.38	

(*Note:* N.E. $=$ not equal.)

(a) Find the P value.

(b) Give the levels of significance for which the null hypothesis would be rejected.

7.72 Suppose that the outcome of a test of hypotheses gives a .07 P value. For what levels of significance would the null hypothesis be rejected?

7.6

DECISION MAKING CONCERNING A POPULATION PROPORTION

We are often interested in studying the (unknown) proportion p of all elements in a population that possess some characteristic. We have seen that this is called a *population proportion.* For instance, we might investigate the proportion p of all voters who favor nuclear power for production of consumer energy. To study this, suppose that we interview a random sample of n voters and record the number (x) who favor nuclear power. This is a binomial experiment. A trial consists of interviewing a voter. A success would occur if the voter favors nuclear power. It is important to note that *the probability of success is equal to the proportion p of all voters in the population who favor nuclear power:* $P(S) = p$.

If we intend to test hypotheses concerning the proportion p, the number of voters x who favor nuclear power out of a sample of n voters will have a bearing on our decision. The quantity x can vary from one sample to the next and, in fact, is a binomial random variable. In Section 6.6 we saw that x has an approximately normal distribution with mean np and standard deviation $\sqrt{npq}$ when n is sufficiently large. So when n is sufficiently large, the standard score

$$z = \frac{x - np}{\sqrt{npq}}$$

is approximately standard normal. When this is the case, we use this expression as a test statistic for testing hypotheses concerning p. How large must the sample size be in order for z to be approximately standard normal? In Section 6.6 we saw that n would be large enough if both np and nq were at least 5 (where $q = 1 - p$). We are assuming that the value of p is not known for sure, but when testing hypotheses concerning p there will be some particular value of p under consideration. For example, suppose that we were testing

$$H_0: p = .20$$
$$H_a: p \neq .20$$

Then we would use .20 for p and .80 for q, and we would require the sample size n to be large enough so that both $(n)(.20)$ and $(n)(.80)$ are at least 5.

When testing hypotheses concerning p, we will use the same five-step procedure we discussed (in Section 7.4) when we were testing hypotheses concerning μ. The only change will be the formula for calculating z. The critical regions can be determined in the same manner as in Section 7.4 (see Table 7.2). In fact, we should point out that when testing *any* hypotheses concerning a population, the same five steps outlined in Section 7.4 may be used. The major change from one test to another will be the test statistic and critical region.

We should keep in mind that the methods of this section will work when the sample size n is sufficiently large. For the examples we consider in this section, this will be the case.

Example 7.10

A sociologist wanted to conduct a test to see if a majority of voters in a voting district favored an Equal Rights Amendment (E.R.A.) to the U.S. Constitution. A random sample of 100 voters was obtained and 56 said they favored E.R.A. Is there sufficient evidence to safely conclude that a majority of voters in the town favor E.R.A.? Use the 5% level of significance.

Solution

Let p = the proportion of all voters in the town who favor E.R.A.

1. Hypotheses: A majority favor E.R.A. if $p > .5$. Therefore we will test the hypotheses:

$$H_0: p = .5$$
$$H_a: p > .5$$

   (In a problem like this some statisticians would prefer to state the null hypothesis as $p \leq .5$. But the test is completed in exactly the same way.)

2. Level of significance: $\alpha = .05$.

3. Test statistic: We use the statistic

$$z = \frac{x - np}{\sqrt{npq}}$$

   The sample size is $n = 100$. For p we use the value appearing in the null hypothesis, so $p = .5$ and $q = 1 - p = .5$. Also, x is the number of voters favoring E.R.A., which in this case is 56. Hence the observed value of the test statistic is

$$z = \frac{56 - 100(.5)}{\sqrt{(100)(.5)(.5)}} = \frac{6}{\sqrt{25}} = \frac{6}{5} = 1.2$$

4. Critical region (favors H_a): The alternate hypothesis contains the symbol $>$. We use a right-tailed test with $\alpha = .05$. To convince yourself of this, note that large values of x favor H_a. Large values of x yield large values of z. How large should z be for us to reject H_0? Large enough so that it would be unlikely that we would observe such a large value if H_0 were true. Since $\alpha = .05$, "unlikely" in this case will mean only a 5% chance. So indeed we would choose our critical region to be all $z \geq z(.05) = 1.65$, since there is only a 5% chance of getting a z value in this region. (See Figure 7.20.)

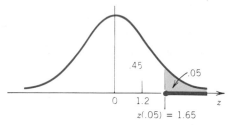 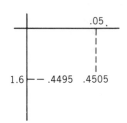

FIGURE 7.20

5. Decision: Since $z = 1.2$, it is not in the critical region. Thus we fail to reject H_0. This means there is insufficient evidence to conclude that a majority of the voters in the town favor an Equal Rights Amendment.

Example 7.11
An official of the Civil Rights Division of the U.S. Department of Justice suspected that a large manufacturing concern practiced discrimination against minorities in its hiring policies. To investigate this, she examined the files of 80 randomly selected employees of the firm and found that 7 of them were minority members. If employees of this company are obtained from a labor pool of which 15% are minorities, is the official justified in her suspicions? Use the 5% level of significance.

Solution

1. Hypotheses: Let p be the proportion of employees of the company who are minorities. Since 15% of the labor pool are minorities, a policy of discrimination in hiring would result in a value of p less than .15. Thus our hypotheses are

$$H_0: p = .15$$
$$H_a: p < .15$$

2. Level of significance: $\alpha = .05$.

3. Test statistic and its observed value:

$$z = \frac{x - np}{\sqrt{npq}} = \frac{7 - (80)(.15)}{\sqrt{(80)(.15)(.85)}} = \frac{-5}{\sqrt{10.2}} = \frac{-5}{3.194} \doteq -1.57$$

4. Critical region: Since H_a involves $<$, we use a left-tailed test. Since $\alpha = .05$, the critical region consists of values of $z \le -z(.05) = -1.65$. (See Figure 7.21.)

5. Decision: Since $z = -1.57$ is not in the critical region, we fail to reject H_0. Hence there is not sufficient evidence to conclude that the company discriminates against minorities.

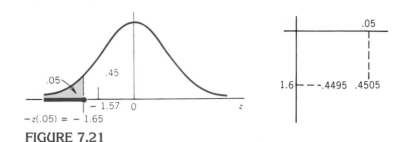

FIGURE 7.21

The essential features of a large-sample test concerning a population proportion follow.

The z Test for a Population Proportion: This test is based on a sufficiently large sample. To test hypotheses concerning p, use the test statistic

$$z = \frac{x - np}{\sqrt{npq}}$$

(where $q = 1 - p$). This has the standard normal distribution (approximately) when np and nq are at least 5. For p we use the value given in the null hypothesis. The observed value of z is computed from the sample data. If z falls in the critical region, reject the null hypothesis H_0. Suppose that α is the level of significance of the test. Critical values of z are obtained in Appendix Table B.3. The possible critical regions are described as follows:

(a) If the alternate hypothesis H_a contains the symbol $>$, we conduct a right-tailed test. The critical region is shown in Figure 7.22.

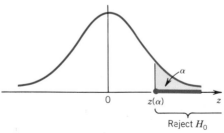

FIGURE 7.22
Right-Tailed Test

(b) If the alternate hypothesis H_a contains the symbol $<$, we conduct a left-tailed test. The critical region is shown in Figure 7.23.

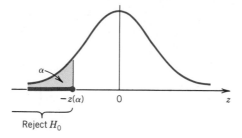

FIGURE 7.23
Left-Tailed Test

(c) If the alternate hypothesis H_a contains contains the symbol $\neq$, conduct a two-tailed test. The critical region is shown in Figure 7.24.

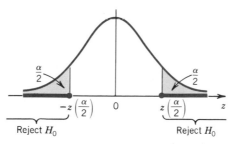

FIGURE 7.24
Two-Tailed Test

Exercises

7.73 For the claims in parts (a) through (g), **(i)** find H_0 and H_a, **(ii)** give the type of critical region (right-tailed, left-tailed, or two-tailed), and **(iii)** explain the meaning of a type I and type II error.

(a) The proportion p of voters in a state who favor the death penalty under prescribed conditions is .53.

(b) The proportion p of a product that breaks down before the guarantee expires is less than .06.

(c) The proportion p of full-time college students younger than 23 years of age is less than .90.

(d) The proportion p of minority students at a large university is more than .07.

(e) The proportion p of persons surviving a cancer for 5 years is .40.

(f) The proportion p of defective items from a large batch is less than .05.

(g) The proportion p of high school graduates in 1990 seeking full-time employment was more than .30.

7.74   In parts (a), (b), and (c), specify H_0 and H_a. Also give the critical region in terms of the standard normal z for $\alpha = .10$, $\alpha = .05$, and $\alpha = .01$. Assume the claim is that the proportion p of children in a large city that watch television at least 4 hours per day is

(a)   Less than .8.   (b)   .8.   (c)   More than .8.

7.75   Let p be the proportion of voters in a large city who favor a restructuring of the police department. Consider the hypothesis that less than 70% of the voters are in support of the idea. In each of the following, **(i)** give the critical region, **(ii)** determine whether you would reject H_0, and **(iii)** give the type of error (I or II) that has possibly been committed. Note that n refers to the sample size, α the level of significance, and x the number of voters in the sample who favor the restructuring.

(a) $n = 100$, $\alpha = .10$, $x = 65$.   (b) $n = 100$, $\alpha = .05$, $x = 60$.

(c) $n = 400$, $\alpha = .01$, $x = 260$.   (d) $n = 1600$, $\alpha = .01$, $x = 1040$.

7.76   The manager of the circulation department of a newspaper claimed that 75% of subscribers would renew their subscriptions. A sample of 100 subscribers showed that 63 renewed their subscriptions. At the 1% level of significance, is there sufficient evidence to conclude that the manager is wrong?

(a)   Use the classical approach.   (b)   Use the P value approach.

7.77   A politician believed that more than half of the voters in her district support her stand on a controversial issue. Suppose that 105 registered voters in a sample of size 200 support her on this issue. Is there sufficient evidence to support the politician's belief at the 10% level of significance?

7.78   It was conjectured that a majority of recovering alcoholics exhibit abnormal levels of depression. A psychologist gave a test called the Depression Adjective Checklist (DACL) to a sample of 45 recovering alcoholics and found that 25 had above-average depression levels. At the 5% level of significance, does this result support the conjecture?

(a)   Use the classical approach.   (b)   Use the P value approach.

7.79   An official at a large university suspected that the proportion p of their football players who have failed to receive a diploma within 4 years is more than .20. A scholarship committee obtained a random sample of 225 former players. The sample showed that 162 had received a diploma within 4 years. Using the P value approach, is the spokesperson's suspicion justified at the 1% level of significance?

7.80   A retailer has received a large shipment of VCR's. He decided to accept the shipment provided there is no evidence to suggest that the shipment contains more than 5% nonacceptable items. The retailer found 14 nonacceptable items in a random selection of 160 items. Using a 5% level of significance, what is the retailer's decision?

7.81   The chamber of commerce of an island resort claimed that the proportion p of all automobiles transported to the island during the summer that are from out-of-state is different from last year's 20%. A random sample of size 40 gives 11 automobiles from out-of-state. At the 5% level of significance, test the claim using

(a)   The classical approach.   (b)   The P value approach.

7.82   A politician claimed she has gained support since the last election. In that election she received 52% of the vote.

(a)   A poll of 100 randomly selected voters showed that 57 would vote for her at this time. Test her claim using a 5% significance level.

(b)   Suppose a random sample of size $n = 4(100) = 400$ shows $4(57) = 228$ voters indicating they would vote for her at this time. Test her claim using a 5% significance level.

(c)   Compare your answers in parts (a) and (b). Note that the proportion of voters who would vote for her at this time is the same in both parts. Comment.

7.83   Over the past year, 90% of customers at a clothing store spent more than $45. The owner sampled 36 receipts from customers serviced by one of the clerks. The amounts (rounded to the nearest dollar) are

47	46	56	70	52	58	48	57	49	61	52	40
60	22	74	59	60	30	61	44	62	41	53	57
50	52	57	59	69	51	58	56	44	36	47	51

Using a 1% level of significance, can the owner conclude that the clerk is not living up to expectations?

7.84   A pain killer is known to be effective in 60% of the patients. Suppose you are the manufacturer of a new pill designed to kill the pain. Further suppose that you will not market the pill unless there is very strong evidence that the effective rate is more than 60%. Which one of the following levels of significance would you use? Justify your decision.

(a) $\alpha = .10$.   (b) $\alpha = .05$.   (c) $\alpha = .01$.

7.85   In Exercises 5.61 and 6.36 we discussed the Hazelwood case. Out of 405 teachers hired in 2 years in Hazelwood, St. Louis County, Mo., 15 (or 3.7%) were black. The Equal Employment Opportunity Commission (EEOC) sued Hazelwood for discrimination under Title VII of the Civil Rights Act of 1964.

(a)   Let p = the proportion of black teachers in the population of teachers from which Hazelwood recruits. (Think of the 405 teachers hired as a sample from the population.) EEOC noted that the percentage of black teachers in St. Louis County, plus the nearby city of St. Louis, was 15.4%. Is there evidence that p is less than .154 (in which case there may be a problem)? Use the 2.5% level of significance, a level favored by some statisticians in a test of this type. (*Note:* The EEOC won in the court of appeals.)

(b)   On appeal to the Supreme Court, the decision of the court of appeals was vacated. The Supreme Court noted that the relevant job market for comparison might well be St. Louis County alone (not including the city of St. Louis). Here the percentage of black teachers was only 5.7%. Redo part (a) using 5.7% instead of 15.4%.

7.7 USING MINITAB (OPTIONAL)

In the Minitab computer output below are the results of a z test on the aggressiveness scores of 30 males discussed in Section 2.7. The psychologist who developed the test suspected that the data came from a population with mean 120. So the test is

$$H_0: \mu = 120$$
$$H_a: \mu \neq 120$$

In addition to typing the usual SET command and the data, we typed the following command:

ZTEST OF MU = 120, SIGMA = 15, DATA IN C1

Some of the words in this command are optional and are inserted for clarity. (Minitab only needs the command name and essential constants or columns.) The following command would accomplish the same result:

ZTEST 120 15 C1

The population value 15 was used for σ because the psychologist felt fairly sure about this value.

```
MTB > SET THE FOLLOWING DATA IN C1
DATA> 136 133 122 105 141 118 148 113 128 125 120 146
DATA> 137 135 98 126 108 93 140 110 134 131 102 115
DATA> 106 110 118 147 136 146
DATA> END
MTB > ZTEST OF MU = 120, SIGMA = 15, DATA IN C1

TEST OF MU = 120.00 VS MU N.E. 120.00
THE ASSUMED SIGMA = 15.0
```

	N	MEAN	STDEV	SE MEAN	Z	P VALUE
C1	30	124.23	15.75	2.74	1.55	0.12

Notice that the computer prints out the sample size (30), the sample mean (124.23), the sample standard deviation (15.75), the standard error of the mean ($\sigma/\sqrt{n} = 2.74$), the observed value of the test statistic (1.55), and the P value (.12). The observed value is, of course, computed from

$$z = \frac{\bar{x} - \mu}{\sigma/\sqrt{n}} = \frac{124.23 - 120}{2.74} \doteq 1.55$$

The null hypothesis ($\mu = 120$) would be rejected if the value you have in mind for α is greater than or equal to the P value, .12. If $\alpha = .05$, for example, we fail to reject the null hypothesis.

A two-tailed test is automatically done unless the ALTERNATIVE subcommand is used. To do a right-tailed test:

$$H_0: \mu = 120$$
$$H_a: \mu > 120$$

type

$$\text{ZTEST OF MU } = 120, \text{SIGMA } = 15, \text{DATA IN C1};$$
$$\text{ALTERNATIVE } = 1.$$

The 1 in the subcommand is the code for a right-tailed test. (If we had used -1 instead, a left-tailed test would have been done.) The only difference in output between a left-tailed, right-tailed, or two-tailed test is the P value. The two-tailed P value in the printout above is .12. If we had done a right-tailed test, the P value would be .06.

In the above tests, we assumed $\sigma = 15$. If we had had no idea before-hand about the value of σ, we could have estimated σ by s (since the sample size is large). This could be calculated for the data in C1 using the command STANDARD DEVIATION C1. The computer would return the value 15.75, and we could use this for σ in the ZTEST command. This can be accomplished more easily by use of a t test. For a two-tailed t test type

$$\text{TTEST OF MU } = 120, \text{DATA IN C1}$$

or in abbreviated form

$$\text{TTEST } 120 \text{ C1}$$

We will have more to say about t tests in Chapter 8. The point here is that when the sample size is large, a t test gives approximately the same results as a z test with σ estimated by s.

The following command produces a 95% confidence interval for μ based on a standard normal distribution.

$$\text{ZINTERVAL } 95 \text{ PERCENT CONFIDENCE, SIGMA } = 15, \text{DATA IN C1}$$

(Of course, we could have chosen any level of confidence.)

```
MTB > ZINTERVAL 95 PERCENT CONFIDENCE, SIGMA = 15, DATA IN C1
THE ASSUMED SIGMA =15.0

       N     MEAN   STDEV  SE MEAN   95.0 PERCENT C.I.
C1    30   124.23   15.75    2.74   ( 118.86, 129.61)
```

Here again, if we chose s as an estimate for σ, we could obtain our confidence interval by use of a command to be discussed in Chapter 8:

$$\text{TINTERVAL } 95 \text{ PERCENT CONFIDENCE, DATA IN C1}$$

Finding P Values

We have seen that Minitab does a statistical test by computing the P value. Minitab commands work on the original data values. However, there are times when we only have summary data—the sample size, sample mean, and sample standard deviation. A Minitab command, such as ZTEST, won't work on this information. Nevertheless, we can use Minitab to find the P value.

Recall that in Section 7.5 we found the P value for a test concerning the mean height of female distance runners:

$$H_0: \mu = 64 \text{ in.}$$

$$H_a: \mu > 64 \text{ in.}$$

Our sample data was $\bar{x} = 65.6$, $s = 3.3$, $n = 40$. Using 3.3 as an estimate for σ, we computed the observed value of our test statistic to be

$$z = \frac{\bar{x} - \mu}{\sigma/\sqrt{n}} = \frac{65.6 - 64}{3.3/\sqrt{40}} \doteq 3.07$$

We saw that the P value was

$$P = P(z \geq 3.07) = .0011$$

We can use Minitab to compute the observed value and then use the CDF command to find the P value. Note that CDF 3.07 would give the probability $P(z \leq 3.07)$, and hence the P value is $1 - $ CDF 3.07. (See Figure 7.16 in Section 7.5.) The program to compute the P value is as follows:

```
MTB > LET K1 = (65.6 - 64)/(3.3/SQRT(40))
MTB > CDF K1, STORE IN K2;
SUBC>   NORMAL 0, 1.
MTB > LET K3 = 1 - K2 #THIS IS THE P VALUE
MTB > PRINT K3
K3       0.00108314
```

In the above program, K1 = 3.07. CDF 3.07 is computed and stored as K2. K3 is the P value. Observe also that we used the square root function (SQRT) to compute $\sqrt{40}$.

For the above right-tailed test, the P value was the area of the right tail of the standard normal distribution. For a left-tailed test, the P value is the area of a left tail (to the left of the observed value). The CDF command gives the area of a left tail directly. For a right tail, we must subtract the CDF value from 1, as in the above printout. For a two-tailed test, we find the area of a tail and double it. The tail we find will be a right tail if the observed value is positive, and a left tail if the observed value is negative. For example, if we had wanted to do a two-tailed test instead of a right-tailed test in the above program, the LET command used to compute the P value would be

$$\text{LET K3} = 2 * (1 - K2)$$

One might argue that it's just as easy to find the P value using Appendix Table B.3. However, we shall see that tables for other statistical tests (such as the t test) are not as convenient for finding P values. The CDF command can then be quite useful.

Exercises

Suggested exercises for use with Minitab are 7.21, 7.23, 7.56, 7.112, and 7.113. Find the P values for the following problems: 7.76 (b), 7.106 (b), 7.107 (b), and 7.111 (a).

7.8 SUMMARY

In this chapter the reader was introduced to the two types of statistical inference: estimation and decision making (hypothesis testing). We restricted our attention to large-sample methods. We discussed inference concerning a population mean μ and a population proportion p. We obtained point estimates and confidence intervals for these parameters and discussed how one uses the appropriate test statistics to test hypotheses concerning these parameters. These results are summarized in the following table:

Parameter	Point Estimate	$1 - \alpha$ Confidence Interval	Test Statistic for a Hypothesis Test
μ	$\bar{x}$	$\bar{x} \pm z\left(\dfrac{\alpha}{2}\right)\dfrac{\sigma}{\sqrt{n}}$	$z = \dfrac{\bar{x} - \mu}{\sigma/\sqrt{n}}$ (if σ unknown, use s)
p	$\hat{p} = \dfrac{x}{n}$	$\hat{p} \pm z\left(\dfrac{\alpha}{2}\right)\sqrt{\dfrac{\hat{p}\hat{q}}{n}}$	$z = \dfrac{x - np}{\sqrt{npq}}$

In order for our point estimate for a parameter to have the desired accuracy (maximum error of estimate to be E with confidence $1 - \alpha$), the sample size should be at least:

Parameter	Sample Size
μ	$n = \left[\dfrac{z(\alpha/2) \cdot \sigma}{E}\right]^2$
p	$n = \left[\dfrac{z(\alpha/2)}{E}\right]^2 \cdot \dfrac{1}{4}$

The five steps in a test of hypotheses are

1. Determine the null hypotheses H_0 and the alternate hypothesis H_a. The null hypothesis will usually contain the symbol $=$. The alternate hypothesis will usually contain the symbol $>$, $<$, or $\neq$.
2. Determine the level of significance α.

3. Select the test statistic and compute its observed value from the sample data. When the parameter being tested occurs in the formula for the test statistic, we substitute the value appearing in the null hypothesis.

4. Find the critical region. This consists of values of the test statistic that strongly favor the alternate hypothesis.

5. Make your decision. If the observed value of the test statistic falls in the critical region, reject the null hypothesis H_0 in favor of the alternate hypothesis H_a. Otherwise, we fail to reject H_0.

The **P value** is the probability of obtaining a value of the test statistic as favorable or more favorable to the alternate hypothesis than the observed value. The smaller the P value, the stronger the evidence in favor of the alternate hypothesis. Using the P value, we can test hypotheses without finding the critical region. We reject H_0 if $P \leq \alpha$ and fail to reject H_0 if $P > \alpha$.

Review Exercises

7.86 Suppose that 100 95% confidence intervals for a population mean μ were obtained. Approximately how many of these intervals should contain the value μ? Approximately how many 99% confidence intervals should contain μ if 1000 such intervals were constructed?

7.87 A golf ball was tested under laboratory conditions to estimate the mean distance μ the ball travels when subjected to a particular force. A random sample of 60 golf balls showed that they moved a mean distance of 210 with a standard deviation of 6 yards. Find **(a)** a 95% confidence interval for μ, and **(b)** the maximum error of estimate.

7.88 An automobile was designed to weigh, on the average, 1850 pounds. The production manager suspected that the mean weight had changed and wanted to get an estimate of it. A sample of 35 automobiles showed a mean weight of 1845 with a standard deviation of 21 pounds. Find **(a)** a 90% confidence interval for μ, and **(b)** the maximum error of estimate.

7.89 A city assessor needed an estimate of the mean income per household. A random sample of 40 households in the city showed a mean income of $29,400 with a standard deviation of $6325. Find a 99% confidence interval for μ. At the 99% confidence level, can the assessor conclude that the mean income per household has increased over last year's figure of $25,100?

7.90 A graduate dean felt that the mean verbal GMAT (Graduate Management Aptitude Test) of 88 students applying to his school was low. He believed that the mean GMAT score for applicants should be about 35. The mean and standard deviation for the 88 students were 34.08 and 7.04, respectively. Realizing that scores vary from one time to the next, he wished to get a range of possible mean scores. Find a 95% confidence interval for the population mean μ. Can he conclude that the mean score is below 35 (at the 95% level of confidence)?

7.91 Thirty-two data values selected from a population gave $\Sigma x = 2284$ and $\Sigma x^2 = 163,232$. Find a 99% confidence interval for the mean μ of the population from which the data were selected.

7.92 On a stretch of an interstate highway with a speed limit of 55 miles per hour, an unusual number of accidents were being reported. Forty cars were randomly checked and clocked for speed by the state police. The speeds were

66	79	58	65	64	71	70	67	55	67	70	60	66	66
60	63	75	60	57	63	72	61	70	72	72	67	68	
59	71	64	69	77	61	74	59	51	56	75	58	66	

(a) Estimate the mean speed of all cars with a 95% confidence interval.

(b) If anything above 62 miles per hour is considered excessive, can the police conclude that there is excessive speeding (on the average)?

7.93 An economist estimated the mean salary μ of a household head as between $27,170 and $27,830. The sample size was 225 with sample standard deviation $3000.

(a) What is the value of the sample mean?

(b) What level of confidence is being used?

7.94 An economist estimated the mean salary μ of a household head as between $28,010.70 and $28,989.30 with 98% confidence. If she reported the sample size as 225 but did not give the sample standard deviation s, what is the value of s?

7.95 An economist estimated the mean salary μ of a household head as between $27,510.70 and $28,489.30 with 98% confidence. If he reported the sample standard deviation as $s = $3570 but did not give the sample size, what is the size of the sample?

7.96 Researchers in a large metropolitan area were studying systolic blood pressures of male runners, and they wished to obtain an estimate of the mean μ. Assume that $\sigma = 17$. How large a random sample is needed to estimate μ to within 2 units with 95% confidence?

7.97 A company was working on a new liquid diet, and an official wanted to estimate the mean weight loss μ. How large a random sample is required to estimate μ to within .5 pounds with 98% confidence? Use $\sigma = 3$.

7.98 Consider a population with unknown mean μ and standard deviation $\sigma = 10$. How large a random sample is required to estimate μ to within 2 units with

(a) 90% confidence?   (b) 99% confidence?

7.99 Refer to Exercise 7.98. Suppose you were to find the required sample size to estimate the population mean μ to within 2 units with 95% confidence. Assume that $\sigma = 10$.

(a) Without finding the sample size, would the required sample size be larger or smaller than that needed for

(i) 90% confidence?   (ii) 99% confidence?

(b) Now find the required sample size and compare your answers with Exercise 7.98.

7.100 Refer to Exercise 7.98. Suppose you were to find the required sample size to estimate the population mean μ to within 1 unit with 90% confidence. Assume that $\sigma = 10$.

(a) Without finding the sample size, would the required sample size be larger or smaller than that required to estimate μ to within 2 units with 90% confidence?

(b) Now find the required sample size and compare your answer with Exercise 7.98 (a).

7.101 Fill in the blanks and construct a confidence interval for a population proportion p in each of the following. Note that n refers to the sample size and x the number of successes.

	$\hat{p}$	x	n	Percentage Confidence Interval
(a)	.50		100	95
(b)		50	125	90
(c)	.72		400	99
(d)		45	225	95
(e)	.60		60	90
(f)		30	50	99

7.102 There was concern among health officials in a community that an unusually large percentage of babies with abnormally low birth weight were being born. Abnormal low birth weight here is defined as less than 88 ounces. A sample of 180 births showed 18 babies with abnormally low birth weight.

(a) Find a 95% confidence interval for the true proportion of babies with abnormally low birth weight in the community.

(b) Suppose the true proportion is 5%. At the 95% confidence level, would the health officials conclude that the percentage of abnormally low birth weights in the community is larger than 5%?

7.103 In a poll of 300 registered voters in a precinct, 126 voters said they knew "very little or nothing" about the politics in Central America.

(a) Find a 90% confidence interval for the proportion p of all registered voters in the city who know "very little or nothing" about the politics in Central America.

(b) What is the maximum error of the estimate for p?

7.104 A poll of 450 residents in a large city was conducted to estimate the proportion p of residents who believed "the federal government was not doing enough about the air pollution problem." Of those polled, 306 residents agreed that the federal government was not doing enough.

(a) Estimate p with a 95% confidence interval.

(b) What is the maximum error of estimate for p?

7.105 How large a random sample is required to estimate the proportion p of women employed in a large city with 95% confidence to within

(a) Five percentage points?

(b) Three percentage points?

(c) One percentage point?

7.106   A sample of 40 recovering alcoholics was given the State–Trait Anxiety Inventory test (STAI). The state anxiety score showed a mean of 38 with a standard deviation of 7. A psychologist suspected that recovering alcoholics in general had a higher mean state anxiety than the norm of 35. Do the sample data justify the suspicion? Complete the test with a 5% significance level.

   (a)   Use the classical approach.   **(b)**   Use the P value approach.

7.107   A manufacturer of transparent tape was worried that the mean length of the tape was not 450 inches. The price and length of the tape were designed to yield a maximum profit. A random sample of 43 rolls of tape gave a sample mean length of 450.4 inches and a sample standard deviation of .95 inches. Complete the test at the 1% significance level.

   (a)   Use the classical approach.   **(b)**   Use the P value approach.

7.108   An artist claimed that a framer was cutting mats that were longer on a side than they should be. In particular, the desired length was 3.25 inches. A sample of 31 paintings framed by the framer showed a mean length of 3.35 and a standard deviation of .40 inches. Test the claim at the 5% level.

7.109   The manager of a bicycle factory believed he was being supplied spokes that were less than the desired 12 inches in length, on the average. A random sample of 38 spokes gave a mean length of 11.85 and a standard deviation of .65 inches. Test the manager's claim with a 5% level of significance.

7.110   A manufacturer of flashlight bulbs claimed the mean life μ of his product is more than 500 hours. A random sample of 100 items gave a mean of 504 with a standard deviation of 20.

   (a)   Find the P value.

   (b)   Is the test statistically significant at the 10% level of significance? At the 1% level of significance?

7.111   Cans of motorcycle oil are supposed to contain 16 ounces of oil (on the average). To test this, 36 cans were obtained and the results showed a mean of 15.5 and a standard deviation of 1 ounce.

   (a)   Find the P value.

   (b)   What conclusion is reached at the 10% level of significance? At the 1% level of significance?

7.112   Health clinic officials believed that, on average, patients were waiting more than 1 year between physical examinations. Below are the times (in years) between physical exams for 32 patients. Can the officials be confident of their conclusion at the 5% level? The data are

0.6	0.7	0.8	0.8	0.8	0.9	1.1	1.2	1.2	1.3	1.3
1.3	1.3	1.3	1.4	1.4	1.5	1.5	1.5	1.5	1.6	1.6
1.7	1.7	1.8	1.8	1.8	1.8	1.9	1.9	1.9	2.0	

7.113   The director of a computing center claimed that the mean down time per week (when computer is not operating) is less than 60 minutes. The following down times (in minutes) for 32 weeks were recorded.

61	50	53	55	52	42	53	56	49	53	51
61	55	65	61	70	61	64	51	56	46	53
55	54	60	53	56	47	52	66	49	57	

   Test the claim at the 5% level of significance.

7.114   The manufacturer of an automobile battery claimed an average battery life of 5 years. Thirty-five of these batteries gave the following years of service:

4.8	5.0	4.9	4.7	5.2	4.9	4.6	4.7	5.2	5.2	4.9	5.0
4.9	4.4	5.0	5.0	5.0	5.1	4.8	5.0	4.7	5.2	4.5	4.9
5.0	4.7	4.6	5.1	4.6	4.5	5.7	4.9	4.8	4.6	4.8	

Test the claim at the 1% level of significance.

7.115   An executive of a large business believed that more than 20% of the unpaid bills are overdue. A random sample of 50 bills shows that 15 are overdue. At the 5% level of significance, is there sufficient evidence to support the executive's belief that the proportion p of overdue bills is more than .20?

7.116   A regular participant in a state lottery game suspected that the number 8 appears too infrequently. (The number 8 should appear about $\frac{1}{10}$ of the time in the long run.) A random selection of 100 numbers shows that the number 8 appeared 4 times. At the 1% level of significance, is there sufficient evidence to support the participant's suspicion? At the 10% level?

7.117   Megabucks, a lottery game conducted by the Massachusetts State Lottery Commission, consists of selecting 6 numbers (no repetitions) from a collection of 36 numbers, {1, 2, 3, 4, ..., 35, 36}. A regular player felt that too many single-digit numbers were being drawn by the commission. The player looked at 360 numbers drawn from September 30, 1987–April 23, 1988. There were 103 single-digit numbers. Was the player justified in his assertion at the 5% level of significance?

7.118   A city planning board denied a developer's request to increase the size of an existing mall. The board members claimed that they had the support of more than 60% of the voters. In a random sample of 300 city voters, 192 supported the board's position. Test the board's claim by finding a P value. At what levels of significance is the board justified in claiming that they have the support of more than 60% of the voters?

7.119   Assume that you are testing the hypotheses H_0: $\mu = 150$ and H_a: $\mu = 146$. Further assume that $\sigma = 20$ and use a left-tailed test.

(a)   For level of significance $\alpha = .05$ and sample size $n = 100$, calculate the probability of a type II error β. (*Hint:* Express the critical region in terms of $\bar{x}$. From the definition of β, it follows that $\beta =$ the probability that $\bar{x}$ does *not* fall in the critical region when the alternate hypothesis is true.)

(b)   For $\alpha = .10$ and $n = 100$, calculate β.

(c)   Comment on the relationship between the magnitudes of α and β, assuming the same sample size.

(d)   Calculate β for level of significance $\alpha = .01$ and sample size $n = 100$.

(e)   Calculate β for $\alpha = .01$ and $n = 400$.

(f)   Comment on the relationship between the sample size and β, assuming a fixed level of significance.

7.120   A large box is filled with red and blue marbles. Let p be the proportion of red marbles and consider the hypotheses H_0: $p = .4$ and H_a: $p = .7$. A sample of size 5 is to be selected. Let x be the number of red marbles obtained and consider

x as a binomial random variable. Suppose that H_0 is to be rejected if the number x of red marbles selected is 4 or 5.

(a) Find the probability of a type I error α. (*Hint:* Use Appendix Table B.2.)

(b) Find the probability of a type II error β. (*Hint:* β = probability that x is not in the critical region when H_a is true.)

7.121 Consider Exercise 7.120, replacing the critical region {4, 5} with {3, 4, 5.}

(a) Without calculating α, will it be larger, smaller, or the same as the α in Exercise 7.120 (a)? Now calculate α and compare it with the α in Exercise 7.120 (a).

(b) Without calculating β, will it be larger, smaller, or the same as the β in Exercise 7.120 (b)? Now calculate β and compare it with the β in Exercise 7.120 (b).

Notes

Haskell, W. L., S. Lewis, C. Perry, P. Stern, and P. D. Wood, "Plasma Lipoprotein Distributions in Male and Female Runners," in P. Milvy, ed., *The Long Distance Runner* (New York: Urizen Books, 1977). Originally published as part of Volume 301 of the *Annals of the New York Academy of Sciences* under the title: "The Marathon: Physiological, Medical, Epidemiological and Psychological Studies".

Keys, A., ed., "Coronary Heart Disease in Seven Countries," *Circulation* 41, (Suppl. 1): 1 (1970).

Perron, F. E., "Growth, Fecundity and Mortality of Conus Pennaceus in Hawaii," *Ecology* 64(1)(1983).

Vaillant, G., *The Natural History of Alcoholism: Causes, Patterns and Paths to Recovery* (Cambridge: Harvard University Press, 1983).

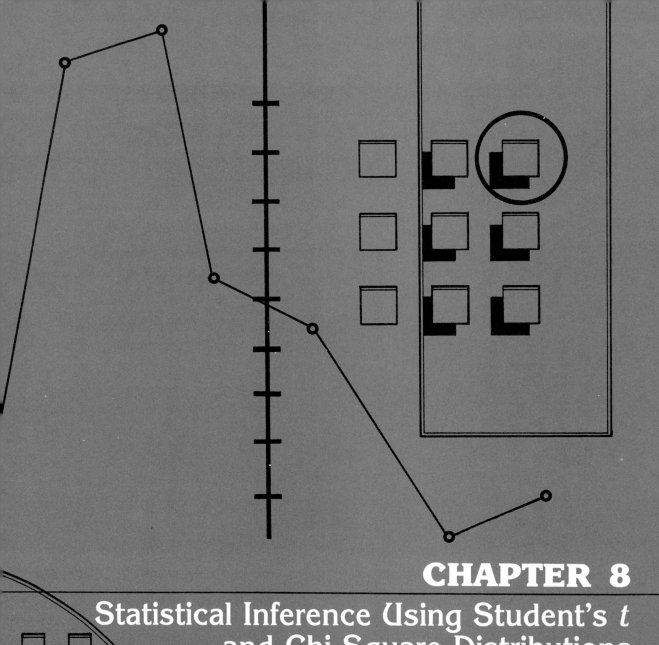

CHAPTER 8
Statistical Inference Using Student's *t*
and Chi-Square Distributions

8.1 INTRODUCTION

8.2 STUDENT'S t DISTRIBUTION

8.3 INFERENCE CONCERNING A MEAN BASED ON A SMALL SAMPLE

8.4 THE CHI-SQUARE DISTRIBUTION

8.5 INFERENCE CONCERNING A POPULATION VARIANCE

8.6 USING MINITAB (OPTIONAL)

8.7 SUMMARY
REVIEW EXERCISES
NOTES

8.1 INTRODUCTION

In Chapter 7 we formally introduced the subject of statistical inference (making judgments about a population based on information contained in a sample from the population). We discussed the estimation of certain population parameters and hypothesis testing of statements about a population. However, when investigating a population mean in Chapter 7, we restricted our attention to those cases where the sample size was large ($n \geq 30$).

In this chapter we will see how this restriction may be removed under certain conditions. We will develop methods for carrying out statistical inference concerning a population mean even when the sample size is small ($n < 30$). For this purpose we will introduce Student's t distribution. We will also develop methods for studying a population variance, and these methods, which are based on the chi-square distribution, can be used whether the sample size is large or small.

In that the methods of this chapter are applicable when the sample size is small, they can be extremely helpful because in some situations it may not be feasible to obtain a large sample for one reason or another. For example, suppose that we wish to test the effectiveness of an experimental drug for the treatment of hypertension. This drug may be very expensive to produce on an experimental basis, so the amount available may be very limited. Hence the number of subjects that can be tested would necessarily be small.

8.2 STUDENT'S *t* DISTRIBUTION

William S. Gosset, who wrote under the pen name of Student, was interested in statistical inference based on small samples. In a paper published in 1908 he developed some small sample methods for statistical inference.* In that paper he discussed some data obtained by scientists A. Cushny and A. Peebles, who were studying the effects of optical isomers of hyoscyamine hydrobromide in producing sleep. The following data give additional hours of sleep per night (x) induced in 10 patients treated with (L)-hyoscine. See Table 8.1.

TABLE 8.1
Additional Hours of Sleep Produced by (L)-Hyoscine

Patient	Additional Hours of Sleep
1	1.9
2	.8
3	1.1
4	.1
5	−.1 (lost sleep)
6	4.4
7	5.5
8	1.6
9	4.6
10	3.4
	$\bar{x} = 2.33$
	$s = 2.00$

---

*Student, 1908, pp. 1–25.

The drug would be considered to increase sleep if the mean additional hours of sleep for all possible patients (μ) were greater than 0. Therefore, it would be of interest to test the hypotheses

$$H_0: \mu = 0$$

$$H_a: \mu > 0$$

We developed a general method for testing such hypotheses in Chapter 7 but only for large samples, that is, when the sample size $n \geq 30$. But in the present case our sample size is $n = 10$, so the method of Chapter 7 does not apply. To understand why this is true, let us see where we would go wrong if we tried to apply the method developed in Chapter 7. In that chapter we used a test statistic to test such hypotheses that was based on the Central Limit Theorem (see Section 6.7). This theorem states that for a sample mean $\bar{x}$ based on random samples of size n,

1. $\mu_{\bar{x}} = \mu$ (the mean of the population).
2. $\sigma_{\bar{x}} = \sigma/\sqrt{n}$.
3. When $n \geq 30$, $\bar{x}$ is approximately normal.

It follows from this that the standard score of $\bar{x}$

$$\frac{\bar{x} - \mu_{\bar{x}}}{\sigma_{\bar{x}}} = \frac{\bar{x} - \mu}{\sigma/\sqrt{n}}$$

is approximately standard normal (when $n \geq 30$). This was the test statistic that we used to test hypotheses concerning a population mean. When σ was unknown (and this is usually the case), we estimated this by s, so in these cases we are really using the test statistic

$$\frac{\bar{x} - \mu}{s/\sqrt{n}}$$

If we tried to apply the above statistic to the hypotheses discussed above, we would obtain its observed value by substituting the appropriate values: $\bar{x} = 2.33$, $s = 2.00$, $n = 10$, and $\mu = 0$ (the value appearing in the null hypothesis). This gives

$$\frac{2.33 - 0}{2.00/\sqrt{10}} \doteq 3.68$$

We would then determine if this value was in the critical region (step 4) for some particular level of significance, say, .05. If we tried to use the critical value for a right-tailed test obtained from Appendix Table B.3, namely, $z(.05) = 1.65$, *this is where we would go wrong*. The reason is that Appendix Table B.3 would apply only if the statistic used had the standard normal distribution (at least approximately). But since the sample size ($n = 10$) is small, we can no longer assume that this is true.

If the probability distribution of the expression

$$\frac{\bar{x} - \mu}{s/\sqrt{n}}$$

were known for small values of n, a table could be developed that would enable us to find the correct critical values. We could then test hypotheses concerning the population mean μ based on small samples using the same five steps we used in Chapter 7.

Gosset found the probability distribution for the above expression. From this probability distribution the necessary table was developed (Appendix Table B.4) to find critical values. We shall see how to use this table shortly. It is interesting to note that Gosset was an employee of the Guinness Brewery in Dublin involved in statistical analysis based on small samples obtained in the brewing process. The samples were necessarily small because of the unwanted variability in temperature and ingredients that would occur in large samples. The Guinness Company was opposed to employees publishing research results, so Gosset published under the pseudonym of Student. He called the above expression t:

$$t = \frac{\bar{x} - \mu}{s/\sqrt{n}}$$

Its probability distribution has become known as **Student's *t* distribution.** However, in order for the above expression to have Student's *t* distribution, it is necessary that the population (of x values) be normal or approximately normal. (No such assumption was needed for the large-sample method of Chapter 7.)

Properties of Student's *t* Distribution

For random samples of size n obtained from an approximately normal population* with mean μ, the random variable

$$t = \frac{\bar{x} - \mu}{s/\sqrt{n}}$$

---

*Actually, the mathematical derivation of Student's *t* distribution assumes that the population (of x values) is normal. However, it has been found that this assumption is not necessary as long as the x distribution is roughly mound shaped. This is why we use the term *approximately normal*.

has Student's t distribution, which has the following properties:

1. There is not just one t distribution but, in fact, an infinite number of them. Each one has a number associated with it called its **degrees of freedom,** df. For the expression

$$t = \frac{\bar{x} - \mu}{s/\sqrt{n}}$$

the degrees of freedom, df, is $n - 1$. We will not go into the mathematical fine points involved in the notion of degrees of freedom. We will simply follow the custom of using this to identify the particular t distribution with which we are working.

2. A t distribution resembles the standard normal distribution in shape. Its curve is symmetric with respect to a vertical line through 0, and the curve extends indefinitely in the positive and negative directions. The expected value of t is 0. However, it is more spread out than the standard normal curve. (It has a larger standard deviation.) As with all continuous probability distributions, the total area under a t curve is 1. (See Figure 8.1.)

3. As the value of n gets larger and larger, Student's t curves get closer and closer to the standard normal curve. In fact, for $n \geq 30$ a t curve is approximately standard normal. This means that for $n \geq 30$

$$t = \frac{\bar{x} - \mu}{s/\sqrt{n}}$$

is approximately standard normal.

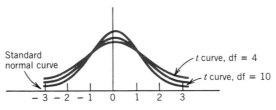

FIGURE 8.1

Appendix Table B.4 relates probabilities to various values of t that we will need. Specifically, for a particular t distribution it gives the value of t such that the area under the t curve to the right of this value is equal to some desired probability. For example, suppose that we are concerned with the t distribution with df $= 15$, and we want the value of t such that the area under the curve to the right of it is .05. We denote this t value by $t(.05)$. Locate the value of df in the left-hand column (df $= 15$). Now look at the column under $t(.05)$. The intersection of this column with the row corresponding to df $= 15$ contains the desired value of t, namely, 1.753. Therefore, $t(.05) = 1.753$. This means that $P(t \geq 1.753) = .05$. (See Figure 8.2.)

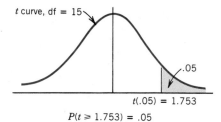

df	t (.005)		t(.01)	t(.025)		t(.05)		t(.10)	•••
•	•		•	•		•		•	
•	•		•	•		•		•	
•	•		•	•		•		•	
•	•		•	•		•		•	
14	2.977	•••	2.624	2.145		1.761		1.345	•••
15	2.947	•••	2.602	2.131		1.753		1.341	•••
16	2.921	•••	2.583	2.120		1.746		1.337	•••
•	•		•	•		•		•	
•	•		•	•		•		•	

$t(.05) = 1.753$

$P(t \geq 1.753) = .05$

FIGURE 8.2

Remark   For values of df $>$ 29 we use the last row in Appendix Table B.4. These are the same values we would get using Appendix Table B.3, since for df $>$ 29, the *t* distributions are approximately standard normal.

Example 8.1
For Student's *t* distribution with df $=$ 14, find the value of *t* such that the area (probability) under the curve to the left of this value is .10.

Solution
By the symmetry of the *t* distribution the desired value of *t* is $-t(.10)$. (See Figure 8.3.) From Appendix Table B.4 we see that $t(.10) = 1.345$, so the *t* value with the desired property is $-t(.10) = -1.345$.

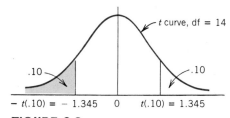

FIGURE 8.3
$P(t < -1.345) = .10$

Example 8.2
For Student's *t* distribution with df $=$ 16, find the *t* value *b* such that the area under the curve between $t = -b$ and $t = b$ is .95. Note that this means that $P(-b < t < b) = .95$.

Solution
The reader should keep in mind that the total area under the curve is 1, so the area in each tail of the curve in Figure 8.4 is .025. This means that $b = t(.025) = 2.120$.

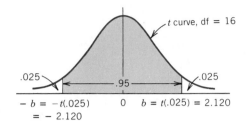 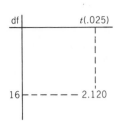

FIGURE 8.4
$P(-2.120 < t < 2.120) = .95$

Exercises

8.1 In each part of the following, find **(i)** $t(.005)$, **(ii)** $t(.01)$, **(iii)** $t(.025)$, **(iv)** $t(.05)$, and **(v)** $t(.10)$.

    (a) Assume a *t* distribution with 7 degrees of freedom.

    (b) Assume a *t* distribution with 12 degrees of freedom.

    (c) Assume a *t* distribution with 25 degrees of freedom.

8.2 Complete each of the following underlined parts.

(a)	$t(.05) = 1.833$	df = ___ .	**(b)**	$t(\underline{}) = 2.602$	df = 15.
(c)	$t(.10) =$ ___	df = 21.	**(d)**	$t(.025) = 2.306$	df = ___ .
(e)	$t(\underline{}) = 2.797$	df = 24.	**(f)**	$t(.995) =$ ___	df = 4.
(g)	$t(.01) = 2.567$	df = ___ .	**(h)**	$t(\underline{}) = -1.319$	df = 23.
(i)	$t(.005) =$ ___	df = 3.	**(j)**	$t(.90) = -1.328$	df = ___ .
(k)	$t(\underline{}) = -2.179$	df = 12.	**(l)**	$t(.95) =$ ___	df = 18.

8.3 Assuming a *t* distribution with 22 degrees of freedom, find each of the following probabilities. (*Hint:* A diagram of the required area might help.)

    (a) $P(-2.074 < t < 2.074)$.   **(b)** $P(t > 1.717)$.

    (c) $P(t < 2.508)$.   **(d)** $P(t > -1.321)$.

    (e) $P(1.321 < t < 2.074)$.   **(f)** $P(-2.508 < t < 1.717)$.

8.4 Below is the *t* curve with 5 degrees of freedom. Find the area A and the *t* value a.

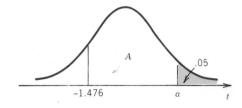

INFERENCE CONCERNING A MEAN BASED ON A SMALL SAMPLE

In this section we will see how to conduct inference concerning the mean when the sample size is small ($n < 30$). We will use the expression

$$t = \frac{\bar{x} - \mu}{s/\sqrt{n}}$$

which we saw in the last section has Student's *t* distribution with df $= n - 1$ if *x* is approximately normal (roughly mound shaped). We use the above expression as a test statistic. To carry out the test, we use the same five-step procedure we developed in Chapter 7; the only difference is that we use Appendix Table B.4 instead of Appendix Table B.3 to find critical values.

When a statistic that has a Student's *t* distribution is used in a statistical test, we call it a **t test.** In future applications of the *t* statistic in this book (in both examples and exercises), you may assume that the requirement of approximate normality is satisfied.

Example 8.3

For the data in Table 8.1 discussed at the beginning of Section 8.2, test the claim that the drug increases sleep (on the average). Use the 5% level of significance.

Solution

1. Hypotheses: The drug increases sleep if the average additional hours of sleep μ is greater than 0; therefore we will test

    $$H_0: \mu = 0$$
    $$H_a: \mu > 0$$

2. Level of significance: $\alpha = .05$.

3. Test statistic and observed value:

    $$t = \frac{\bar{x} - \mu}{s/\sqrt{n}} = \frac{2.33 - 0}{2.00/\sqrt{10}} \doteq 3.68$$

    (Note that we substitute the value 0 appearing in H_0 for μ in the test statistic.) The degrees of freedom df $= n - 1 = 9$.

4. Critical region: Values of $\bar{x}$ far above 0 favor H_a. These correspond to values of *t* far to the right of 0. Therefore, we perform a right-tailed test. When df $= 9$, the critical value from Appendix Table B.4 is $t(\alpha) = t(.05) = 1.833$. Thus the critical region consists of values of $t \geq 1.833$. (See Figure 8.5.)

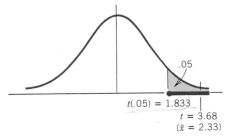 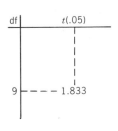

FIGURE 8.5

5. Decision: The observed value of 3.68 is in the critical region, so we reject H_0. This means that the evidence suggests that $\mu > 0$; therefore, the drug appears to increase sleep.

Example 8.4

In order to pass inspection in Massachusetts, an automobile may not show a hydrocarbon level of more than 220 parts per million (ppm). Officials of a large car rental company felt that their cars would average less than 200 ppm. To investigate this, 16 cars were tested, giving a mean of 180 ppm with a standard deviation of 30 ppm. At the 10% level of significance, do the officials appear to be correct?

Solution

1. Hypotheses: The officials believed that the mean amount of hydrocarbons μ for their cars would be less than 200 ($\mu < 200$). The hypotheses are

$$H_0: \mu = 200$$
$$H_a: \mu < 200$$

2. Level of significance: $\alpha = .10$

3. Test statistic and value: Since the sample size is small, we use the t statistic

$$t = \frac{\bar{x} - \mu}{s/\sqrt{n}} = \frac{180 - 200}{30/\sqrt{16}} \doteq -2.67 \qquad df = n - 1 = 15$$

4. Critical region: Since H_a involves the symbol $<$, we perform a left-tailed test. From Appendix Table B.4 we see that $t(\alpha) = t(.10) = 1.341$. Hence $-t(.10) = -1.341$. The critical region consists of values of $t \le -1.341$. (See Figure 8.6.)

5. Decision: Since the value $t = -2.67$ is in the critical region, we reject H_0 in favor of H_a. This means that the officials' belief ($\mu < 200$) appears to be correct.

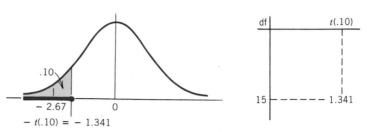

FIGURE 8.6

Example 8.5

An ice cream company claimed that its product contained 500 calories per pint (on the average). To test this claim, 25 one-pint containers were analyzed, giving $\bar{x} = 507$ calories and $s = 22$ calories. Test the claim at the 2% level of significance, using

(a)   The classical approach.   **(b)**   The P value approach.

Solution

(a)   The classical approach:

1.   Hypotheses: The claim is that $\mu = 500$. Thus we will test

$$H_0\text{: } \mu = 500$$
$$H_a\text{: } \mu \neq 500$$

2.   Level of significance: $\alpha = .02$.

3.   Test statistic and its observed value: The sample size is small (<30). Therefore we do a t test:

$$t = \frac{\bar{x} - \mu}{s/\sqrt{n}} = \frac{507 - 500}{22/\sqrt{25}} \doteq 1.59 \qquad df = 24$$

4.   Critical region: Since H_a involves $\neq$, we use a two-tailed test. From Appendix Table B.4, $t(\alpha/2) = t(.01) = 2.492$. Thus the critical region consists of values of t that are ≥ 2.492 or ≤ -2.492. (See Figure 8.7.)

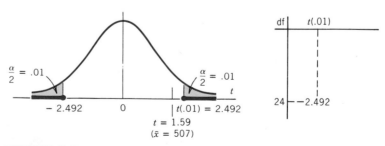

FIGURE 8.7

5. Decision: The observed value $t = 1.59$ is not in the critical region; hence we fail to reject H_0. This means that there is insufficient evidence to reject the company's claim at the 2% level of significance.

(b) The P value approach: Notice that Appendix Table B.4 does not lend itself to finding exact P values. However, using this table we can find an upper and/or lower bound for P. This amounts to finding an interval estimate for P. Recall that the P value is the probability of obtaining a value of the test statistic as favorable or more favorable to H_a than the observed value $t = 1.59$. This is displayed in Figure 8.8.

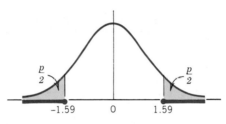

FIGURE 8.8
***P* Value**

Look at the row for df $= 24$ in Appendix Table B.4. Notice that 1.59 is between the table values 1.711 and 1.318. Now $t(.05) = 1.711$ and $t(.10) = 1.318$. From Figure 8.9 we can see that $.05 < P/2 < .10$. Hence,

$$.10 < P < .20$$

We are testing the claim using $\alpha = .02$. Clearly $P > .02$. Hence, we fail to reject the claim H_0.

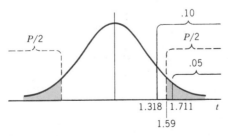

	Area of right tail	.05	.10
	t value	1.711	1.318

FIGURE 8.9
.10 < *P* < .20

We summarize the essential features of a *t* test concerning a population mean as follows:

The *t* Test for a Population Mean (Based on a Small Sample) To test hypotheses concerning μ when the sample size n is less than 30, use the test statistic

$$t = \frac{\bar{x} - \mu}{s/\sqrt{n}}$$

This has Student's *t* distribution with df $= n - 1$. For μ we substitute the value given in the null hypothesis. The observed value of *t* is computed using the sample data; if it falls in the critical region, reject the null hypothesis H_0. Suppose that α is the level of significance of the test. Critical values of *t* are obtained in Appendix Table B.4. The possible critical regions are described as follows:

(a) If the alternate hypothesis contains the symbol >, we conduct a right-tailed test. The critical region is shown in Figure 8.10.

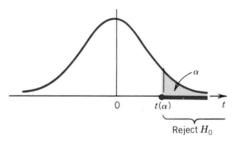

FIGURE 8.10
Right-Tailed Test

(b) If the alternate hypothesis contains the symbol <, we conduct a left-tailed test. The critical region is shown in Figure 8.11.

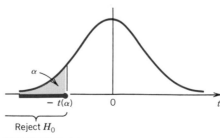

FIGURE 8.11
Left-Tailed Test

(c) If the alternate hypothesis contains the symbol $\neq$, we conduct a two-tailed test. The critical region is shown in Figure 8.12.

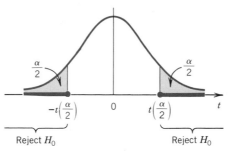

FIGURE 8.12
Two-Tailed Test

Assumption for the test: The population is normal or approximately normal (i.e., it has a distribution that is roughly mound shaped).

Remark   We have said that the t test is applied when the sample size is less than 30 and the population (of x values) is normal or approximately so. We have also tacitly assumed that the population standard deviation σ is unknown, because this is usually the case. On those rare occasions when σ is known, we should conduct a z test, even when the sample size is small, as long as x is (approximately) normal. That is, we should use the test statistic

$$z = \frac{\bar{x} - \mu}{\sigma / \sqrt{n}}$$

The reason for this is that the Central Limit Theorem assures us that this expression still has a standard normal distribution (approximately) even if the sample size is less than 30, provided that x is (approximately) normal. (See the remarks following the Central Limit Theorem in Section 6.7.)

Confidence Intervals Based on Small Samples

In Example 8.4 we discussed the mean level of hydrocarbon pollutants for a fleet of rental cars. We saw that a sample of 16 of these cars showed a sample mean level of 180 parts per million (ppm) with $s = 30$ ppm. Suppose that we wanted to find a 90% confidence interval for μ. In Section 7.2 we showed how to find a confidence interval for a population mean when the sample size is $\geq$ 30. We used the standard normal expression

$$z = \frac{\bar{x} - \mu}{\sigma / \sqrt{n}}$$

to develop the $1 - \alpha$ confidence interval with endpoints

$$\bar{x} \pm z\left(\frac{\alpha}{2}\right) \cdot \frac{\sigma}{\sqrt{n}}$$

But for the problem at hand, $n = 16$. So the above procedure for finding a confidence interval would not be appropriate.

We have seen that when the sample size is small (and x is approximately normal), the expression

$$t = \frac{\bar{x} - \mu}{s/\sqrt{n}}$$

has Student's t distribution with degrees of freedom, df $= n - 1$. Proceeding by analogy, it seems reasonable that a $1 - \alpha$ confidence interval for μ (when the sample size is small) would be described by the endpoints

$$\bar{x} \pm t\left(\frac{\alpha}{2}\right) \cdot \frac{s}{\sqrt{n}}$$

When estimating μ by $\bar{x}$, the maximum error of the estimate with confidence $1 - \alpha$ would be

$$E = t\left(\frac{\alpha}{2}\right) \cdot \frac{s}{\sqrt{n}}$$

For example, to find our 90% confidence interval for the mean level of hydrocarbon pollutants μ discussed above, df $= n - 1 = 16 - 1 = 15$.

$$1 - \alpha = .90 \qquad \alpha = .10$$

$$t\left(\frac{\alpha}{2}\right) = t(.05) = 1.753 \qquad \text{(from Appendix Table B.4)}$$

It follows that a 90% confidence interval is given by

$$180 \pm (1.753)\left(\frac{30}{\sqrt{16}}\right)$$

$$180 \pm 13.15$$

Therefore, our 90% confidence interval is

$$166.85 < \mu < 193.15$$

We summarize the procedure for finding a 1 - α confidence interval based on a small sample as follows:

For an approximately normal population, a $1 - \alpha$ confidence interval for the population mean μ (when the sample size is <30) is given by the endpoints

$$\bar{x} \pm t\left(\frac{\alpha}{2}\right) \cdot \frac{s}{\sqrt{n}}$$

When estimating μ by $\bar{x}$, the maximum error of the estimate with confidence $1 - \alpha$ is

$$E = t\left(\frac{\alpha}{2}\right) \cdot \frac{s}{\sqrt{n}}$$

Example 8.6

In a study of college dropouts, a psychologist administered a questionnaire, designed to measure anxiety level, to 20 dropouts. Scores on the questionnaire can range from 25 (lowest anxiety level) to 100 (highest anxiety level). College students in general show a mean of about 60. The 20 scores for the dropouts gave $\bar{x}$ = 65.2 and s = 7.5. Find a 95% confidence interval for the mean score μ for the population of dropouts from which the sample was obtained. If μ is estimated by $\bar{x}$, find the maximum error of the estimate with 95% confidence.

Solution

Since $1 - \alpha$ = .95, α = .05; df = $n - 1$ = 19. Using Appendix Table B.4, we see that $t(\alpha/2)$ = $t(.025)$ = 2.093. A $1 - \alpha$ confidence interval is given by

$$\bar{x} \pm t\left(\frac{\alpha}{2}\right) \cdot \frac{s}{\sqrt{n}}$$

In this case we get

$$65.2 \pm (2.093) \cdot \frac{7.5}{\sqrt{20}}$$

$$65.2 \pm 3.51$$

Therefore our 95% confidence interval for μ is

$$61.69 < \mu < 68.71$$

Therefore we are 95% confident that μ is between 61.69 and 68.71. Since the confidence interval was computed to be 65.2 ± 3.51, it follows that the maximum error of the estimate is, $E = 3.51$.

Exercises

8.5   In each of the following parts, **(i)** test the claim, and **(ii)** indicate the possible type of error committed (I or II). Note that n refers to the sample size.

   (a)   Test the claim that the mean μ is more than 16.

$$n = 14, \qquad \alpha = .10, \qquad \bar{x} = 18, \qquad s = 4$$

   (b)   Test the claim that the mean μ is less than 27.

$$n = 9, \qquad \alpha = .05, \qquad \bar{x} = 23, \qquad s = 7$$

   (c)   Test the claim that the mean μ is not 30.

$$n = 6, \qquad \alpha = .01, \qquad \bar{x} = 25, \qquad s = 4$$

   (d)   Test the claim that the mean μ is greater than 125.

$$n = 40, \qquad \alpha = .05, \qquad \bar{x} = 128, \qquad s = 18$$

   (e)   Test the claim that the mean μ is less than 50.

$$n = 20, \qquad \alpha = .025, \qquad \bar{x} = 45, \qquad s = 10$$

   (f)   Test the claim that the mean μ is 60.

$$n = 8, \qquad \alpha = .10, \qquad \bar{x} = 70, \qquad s = 16$$

8.6   Verify the conclusions reached in Exercise 8.5 by using the P value approach. (*Hint:* Estimate P by giving an interval containing P such as $.025 < P < .05$ or $P < .005$.)

8.7   At a large university, a department chairperson believed the mean salary μ of assistant professors throughout the university was higher than $30,750 (which was the mean salary of assistant professors in the chairperson's department). A random sample of nine other assistant professors gave a sample mean salary of $31,100 with standard deviation $420. Is there sufficient evidence to justify the chairperson's contention at the 5% level of significance? At the 1% level of significance?

   (a)   Use the classical approach.

   (b)   Use the P value approach. (*Hint:* Estimate P by giving an interval containing P such as $.025 < P < .05$ or $P < .005$.)

8.8   A federal prison warden claimed that first-time released prisoners from a federal institution were imprisoned less than 22 months, on the average. A sample of 24 first-time released prisoners showed that they had served a mean time of 20.5 months with a standard deviation of 2.5 months. At the 1% significance level, do the data support the warden's claim?

8.9   A director of admissions at a university claimed that families with an income of $30,000 a year contributed an average of $6000 per family toward a child's education. A sample of 20 such families whose children attended the university revealed a mean contribution of $6200 with a standard deviation of $300. Test the claim with a 5% level of significance.

8.10   The mean amount of money spent per customer in a department of a retail store is $30. The department manager claimed he could increase that figure by stocking a new product. The new product was put on the shelves. A sample of 25 customers indicated a mean of $34 spent with a standard deviation of $6. Test the manager's claim with a 1% significance level.

8.11   A production manager noticed that the mean time to complete a job was 160 minutes. The manager made some changes in the production process in an attempt to reduce the mean time to finish the job. A sample of 8 jobs gave a mean time of 148 and a standard deviation of 9.5 minutes. Give an interval estimate for the P value. What values of α can be determined from your answer for which the null hypothesis would be rejected?

8.12   An alkaline battery for an AM-FM stereo cassette radio was designed to last 30 hours, on average, for FM play. There were consumer complaints that the batteries were lasting less than 30 hours. The manufacturer randomly sampled 38 batteries. The mean life was 29.3 hours with a standard deviation of 2.95 hours. Is there sufficient evidence, at the 5% level of significance, to indicate the mean battery life is less than 30 hours?

8.13   A partner of a walk-in dental clinic suspected that the mean waiting time for patients had increased from the past average of 10 minutes. A sample of 18 patients showed a mean waiting time of 12 minutes with a standard deviation of 5 minutes. Give an interval estimate for the P value using Appendix Table B.4. What values of α can be determined from your answer for which the null hypothesis would be rejected?

8.14   A 6th grade geography teacher in a city school system gave a standardized geography test to a sample of 23 students. The mean was 68.3 with a standard deviation of 12.2. Using a 5% level of significance, can the teacher conclude the city's 6th graders' performance differs from the national mean of 72?

8.15   In a study of anxiety levels in athletes, 27 runners were given a test called the State-Trait Anxiety Inventory (STAI). The sample showed a mean trait anxiety level of 31.68 with a standard deviation of 9.27 (*Source:* Morgan, W. P., and M. L. Pollock, "Psychological Characterization of the Elite Distance Runner," in *The Long Distance Runner*, ed. P. Milvy (New York: Urizen Books, 1977). (Trait anxiety is a measure of enduring or underlying anxiety level.) Do these results indicate that runners in general have a mean trait anxiety level different from 36 (the norm for college students)? Use the 5% level of significance.

8.16   Experiments conducted at the University of Michigan showed that a standing rider on a transit vehicle can withstand acceleration of up to 4.75 miles per hour per second (mphps) without experiencing discomfort, provided the acceleration is smooth. A transit official specified that a large fleet of new transit vehicles should be able to reach an acceleration rate of 4.75 mphps. An engineer felt that due to motor design, the vehicles' acceleration might be less than 4.75 mphps (on the average). A sample of 10 vehicles showed a mean acceleration rate of 4.58 mphps with a standard deviation of .25 mphps. At the 1% level of significance, is there evidence to justify the engineer's suspicion?

8.17   The Speedy Oil Change Company advertised a 15-minute wait for an oil change. A sample of 23 oil changes showed a mean time of 16.5 minutes with a standard deviation of 4.2 minutes. At the 5% level of significance, is there evidence that the mean time for an oil change is different from 15 minutes?

8.18 A medical doctor sensed that smokers in the 40–45 age group with a rare disease had smoked on average more than 20 years. A sample of 10 patients gave the following years of smoking:

22.0   21.3   19.6   19.6   21.4   24.0   25.9   19.7   25.5   25.1

Using a 1% significance level, is there sufficient evidence to justify the doctor's belief?

8.19 A product was declared acceptable if it did not require repair for at least 12 months. A manufacturer claimed that her company's product would not require repair for more than 13 months on the average. A sample of 12 customers who had purchased her product provided the following information on how many months elapsed before repair was needed on their purchases:

13.1   13.3   14.5   11.7   14.0   12.9
15.4   12.3   12.9   12.6   14.9   13.1

Using a 5% level of significance, test the manufacturer's claim.

8.20 Refer to Exercise 8.17. Find a 95% confidence interval for the mean time for an oil change.

8.21 A psychologist wanted to estimate the mean self-esteem level μ of his patients. Fourteen patients were given a test designed to measure self-esteem. The sample mean and standard deviation were 25.3 and 5.3, respectively.

(a) Construct a 98% confidence interval for μ.

(b) Using the confidence interval, could you conclude that μ is smaller than the norm of 28.5?

(c) Find the maximum error of estimate for μ.

8.22 A study of 28 men's diets in Crevalcore, Greece, gave a sample mean of 28.7 with a sample standard deviation of 5.7% calories from fats. (*Source*: Keys, 1970, p. I–166.)

(a) Find a 95% confidence interval for the mean μ of the population from which the men were selected.

(b) Using the confidence interval, could you conclude that μ is different from the mean of 36.2 for American males?

(c) Find the maximum error of estimate for μ.

8.23 A group of physical education majors was discussing the heights of female runners and whether or not female runners tended to be tall, on the average. They decided to estimate the mean height of female runners. A sample of 12 runners showed a sample mean height of 65.80 inches and a sample standard deviation of 1.95 inches.

(a) Find a 95% confidence interval for the mean μ of the population of female runners from which the 12 runners were selected.

(b) Find the maximum error of estimate for μ.

8.24 Find a 99% confidence interval for μ using the data in Exercise 8.18.

8.25 Find a 90% confidence interval for μ using the data in Exercise 8.19.

THE CHI-SQUARE DISTRIBUTION

In many situations we are concerned with making judgments about a population variance σ^2 (or standard deviation σ). For example, the variance of scores on an examination is very important. If the variance is too small, it will be difficult to differentiate between scores.

A college mathematics professor developed a mathematics examination to screen incoming students for purposes of placing them in the appropriate mathematics course. The new examination is to be used in place of a previously used examination that showed a variance of 100 points. The professor wants to compare the variance of the new examination with the old variance. He felt that the new examination would show a larger variance than the old examination ($\sigma^2 > 100$). Therefore he wished to test the hypotheses

$$H_0: \sigma^2 = 100$$

$$H_a: \sigma^2 > 100$$

To investigate this, the professor decided to give the examination to a sample of 20 students and examine the sample variance s^2. The value of s^2 will influence the decision in the hypothesis test. After all, s^2 can be thought of as a rough estimate of σ^2. The 20 scores are given in Table 8.2.

If the observed value of s^2 is too large to be consistent with the hypothesis H_0, then H_0 will be rejected and H_a accepted. "Too large" means that the value of s^2 is so much larger than 100 that it would be unlikely that we would observe such a large value if H_0 were true. "Unlikely" means having a low order of probability. The researcher selects this value. As before, it is called the *level of significance* of the test and is denoted by the symbol α.

TABLE 8.2
Scores on a Mathematics Screening Exam

Student	Score	Student	Score
1	70	11	88
2	80	12	50
3	73	13	70
4	64	14	65
5	91	15	75
6	52	16	50
7	48	17	90
8	92	18	53
9	75	19	68
10	67	20	89

$$s^2 = 217.63$$

Instead of working with the quantity s^2, statisticians prefer to use a standardized form of s^2 when investigating σ^2. (This is analogous to using the standard score of $\bar{x}$ when investigating μ.) The quantity statisticians use is called **chi-square,** denoted by χ^2:

$$\chi^2 = \frac{(n-1)s^2}{\sigma^2}$$

where n is the sample size. If we assume for the moment that H_0 is true ($\sigma^2 = 100$), then

$$\chi^2 = \frac{(20-1)s^2}{100} = .19s^2$$

Large values of s^2 correspond to large values of χ^2. Therefore, we will reject H_0 if the value of χ^2 is too large—so large that it would be unlikely that we would observe so large a value if H_0 were true.

To determine the values of χ^2 that are so large as to be unlikely, we need to know something about the probability distribution of χ^2. Under appropriate conditions the quantity χ^2 has a probability distribution known as a **chi-square distribution.**

Properties of a Chi-Square Distribution

For random samples of size n from a normal population with variance σ^2, the probability distribution of the quantity

$$\chi^2 = \frac{(n-1)s^2}{\sigma^2}$$

is a chi-square distribution that has the following properties:

1. There is not just one chi-square distribution but, in fact, an infinite number of them. Each one has a number associated with it called its degrees of freedom, df. For the above expression, df $= n - 1$. We use the degrees of freedom to specify which chi-square distribution we are using.

2. The shape of a chi-square curve is not symmetric but is skewed to the right. It begins at 0 and extends indefinitely in a positive direction. The total area under the curve is 1. (See Figure 8.13.)

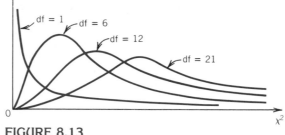

FIGURE 8.13
Chi-Square Curves

3.   The expected value of χ^2 is the degrees of freedom, df.

Appendix Table B.5 relates probabilities to various values of χ^2 that we will need. For a particular chi-square distribution, it gives the value of χ^2 such that the area under the chi-square curve to the right of this value is equal to some desired probability. For example, suppose that we are concerned with the chi-square distribution with df = 10, and we want the value of χ^2 such that the area under the curve to the right of it is .05. We denote this value by $\chi^2(.05)$. Locate the value of the degrees of freedom in the left-hand column of the table (df = 10). Now look at the column under $\chi^2(.05)$. The intersection of this column with the row corresponding to df = 10 contains the desired value of χ^2, namely, 18.307. Thus $\chi^2(.05) = 18.307$. This means that $P(\chi^2 \geq 18.307) = .05$. (S e Figure 8.14.)

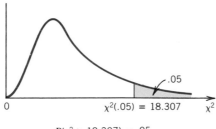

$P(\chi^2 \geq 18.307) = .05$

FIGURE 8.14
A Portion of Appendix Table B.5

df	$\chi^2(.995)$	•••	$\chi^2(.10)$	$\chi^2(.05)$	$\chi^2(.025)$ •••
•	•		•	•	•
•	•		•	•	•
•	•		•	•	•
•	•		•	•	•
•	•		•	•	•
9	1.735	•••	14.684	16.919	19.023 •••
10	2.156	•••	15.987	18.307	20.483 •••
11	2.603	•••	17.275	19.675	21.920 •••
•	•		•	•	•
•	•		•	•	•

Example 8.7
For the chi-square distribution with df = 11, find the value of χ^2 such that the area under the chi-square curve to the left of this value is .005.

Solution
The total area under the curve is 1. So the area to the right of the desired value is $1 - .005 = .995$. This means that the desired value will be $\chi^2(.995)$ with

df = 11. Looking up this value in Appendix Table B.5, we see that $\chi^2(.995) = 2.603$. This means that $P(\chi^2 < 2.603) = .005$. (See Figure 8.15.)

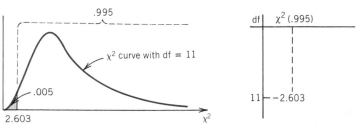

FIGURE 8.15
$P(\chi^2 < 2.603) = .005$

Before turning to applications of the chi-square distribution, we should observe that there are no requirements on the sample size n in the expression

$$\chi^2 = \frac{(n-1)s^2}{\sigma^2}$$

Therefore, our methods will apply in small-sample or large-sample situations. However, in order for the above expression to have a chi-square distribution, we require that the population (of x values) be normal. Contrast this with the requirement on the population in applications of Student's t distribution in Section 8.3, where we said that the population need only be approximately normal—actually its distribution need only be roughly mound shaped. The requirement of normality when applying the chi-square distribution in the study of variance is a severe limitation that the reader should keep in mind.

In examples and exercises of this chapter where the chi-square distribution is used, we will assume that the populations involved are normal.

Exercises

8.26 In each of the following parts, find (i) $\chi^2(.01)$, (ii) $\chi^2(.025)$, (iii) $\chi^2(.05)$, (iv) $\chi^2(.95)$, and (v) $\chi^2(.99)$.

(a) Assume a chi-square distribution with 9 degrees of freedom.

(b) Assume a chi-square distribution with 15 degrees of freedom.

(c) Assume a chi-square distribution with 25 degrees of freedom.

8.27 Assume a chi-square distribution with 6 degrees of freedom. Fill in the blanks.

(a) $P(\chi^2 \geq \underline{\quad}) = .01$.   **(b)** $P(\chi^2 \geq \underline{\quad}) = .99$.

(c) $P(\chi^2 > \underline{\quad}) = .05$.   **(d)** $P(\chi^2 < \underline{\quad}) = .90$.

(e) $P(\chi^2 \leq \underline{\quad}) = .01$.   **(f)** $P(\chi^2 < \underline{\quad}) = .05$.

8.28 Complete each of the following underlined parts.

(a) $\chi^2(.10) = 17.275$ df = $\underline{\quad}$.   **(b)** $\chi^2(\underline{\quad}) = 2.167$ df = 7.

(c) $\chi^2(.95) = \underline{\quad}$ df = 10.   **(d)** $\chi^2(.01) = 44.314$ df = $\underline{\quad}$.

(e) $\chi^2(\underline{\quad}) = 18.549$ df = 12.   (f) $\chi^2(.05) = \underline{\quad\quad}$ df = 17.

(g) $\chi^2(.90) = \underline{\quad\quad}$ df = 20.   (h) $\chi^2(.99) = \underline{\quad\quad}$ df = 11.

8.29 Assuming a χ^2 distribution with 18 degrees of freedom, find each of the following probabilities:

(a) $P(\chi^2 > 25.989)$.   (b) $P(\chi^2 < 7.015)$.

(c) $P(7.015 < \chi^2 < 9.390)$.   (d) $P(10.865 < \chi^2 < 28.869)$.

8.30 Below is the χ^2 curve with 14 degrees of freedom. Find the area A and the chi-square value a.

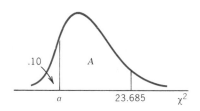

8.5 INFERENCE CONCERNING A POPULATION VARIANCE

In Section 8.4 we noted that inference concerning a population variance may be conducted using the quantity

$$\chi^2 = \frac{(n-1)s^2}{\sigma^2}$$

which has a chi-square distribution with df $= n - 1$ when the population is normal. In hypothesis testing concerning σ^2 we use this expression as a test statistic. The value we use for σ^2 in this quantity is the value appearing in the null hypothesis.

When a statistic that has a chi-square distribution is used in a statistical test, we call it a **chi-square test.**

Example 8.8

For the data in Table 8.2 discussed at the beginning of Section 8.4, we saw that a sample of size $n = 20$ gave a sample variance of $s^2 = 217.63$. Test the claim that the variance σ^2 for the new examination is larger than the variance for the old examination (which was 100). Use the 1% level of significance.

Solution

1. Hypotheses: We have seen that the hypotheses are
$$H_0: \sigma^2 = 100$$
$$H_a: \sigma^2 > 100$$

2. Level of significance: $\alpha = .01$

3. Test statistic and observed value:
$$\chi^2 = \frac{(n-1)s^2}{\sigma^2} = \frac{(20-1)(217.63)}{100} \doteq 41.35$$
$$df = n - 1 = 20 - 1 = 19$$

4. Critical region: Very large values of s^2 favor H_a. These correspond to very large values of χ^2, so large that it would be unlikely (only 1% chance) that we would observe such large values if H_0 were true. Hence we perform a right-tailed test. The right critical value obtained from Appendix Table B.5 is $\chi^2(\alpha) = \chi^2(.01) = 36.191$. The critical region shown in Figure 8.16 consists of values of $\chi^2 \geq 36.191$.

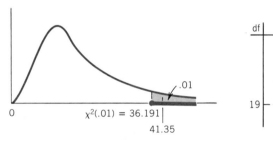

FIGURE 8.16

5. Decision: The observed value of 41.35 is in the critical region, so we reject H_0. This means that the new test does appear to have a larger variance than the old one.

Example 8.9
A machine is designed to fill 32-ounce milk containers. The actual measurements of milk will vary somewhat from 32 ounces; it is said that for this machine the standard deviation is .1 ounce. To test this claim, 28 containers are randomly selected and the actual amount of milk in each is measured. This results in a sample standard deviation of $s = .13$ ounce. Complete the test of the claim at the 5% level of significance using

(a) The classical approach.   (b) The P value approach.

Solution

(a) The classical approach:

1. Hypotheses: It is claimed that $\sigma = .1$ ounce or equivalently $\sigma^2 = (.1)^2 = .01$. Therefore, we will test

$$H_0: \sigma^2 = .01$$
$$H_a: \sigma^2 \neq .01$$

2. Level of significance: $\alpha = .05$.
3. Test statistic and its observed value:

$$\chi^2 = \frac{(n-1)s^2}{\sigma^2} = \frac{(27)(.13)^2}{.01} \doteq 45.63 \qquad \text{df} = 27$$

4. Critical region: The fact that H_a contains the symbol $\neq$ implies that this will be a two-tailed critical region. From Appendix Table B.5 we find that the right critical value is

$$\chi^2\left(\frac{\alpha}{2}\right) = \chi^2(.025) = 43.194$$

Note that the area under the curve to the right of the left critical value is $1 - \alpha/2 = 1 - .025 = .975$. Thus the left critical value is $\chi^2(1 - \alpha/2) = \chi^2(.975) = 14.573$. The critical region is displayed in Figure 8.17a. The critical region consists of values of $\chi^2 \leq 14.573$ or ≥ 43.194.

5. Decision: Since the observed value $\chi^2 = 45.63$ is in the critical region, we reject H_0. Therefore, it would seem that the claim is untrue.

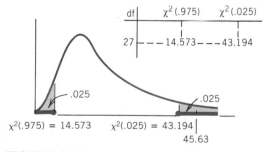

FIGURE 8.17a
Critical Region

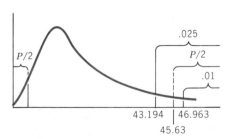

FIGURE 8.17b
P Value .02 < P < .05

(b) The P value approach:
We are using a two-tailed test. Thus the P value is twice the area under the curve to the right of the observed value 45.63. See Figure 8.17b. Ex-

amining the df = 27 row of Appendix Table B.5, we see that 45.63 is between 43.194 and 46.963.

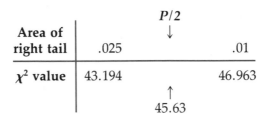

Thus $.01 < P/2 < .025$. Hence, $.02 < P < .05$. Now $\alpha = .05$ and since $P < .05$, we reject the null hypothesis.

Example 8.10

The systolic blood pressure readings of males between the ages of 35 and 59 show a standard deviation of about 17 millimeters. A sample of 41 male runners (age 35–59) showed a (sample) standard deviation of 15 millimeters (Haskell, et al., 1977, p. 149). Test the claim that runners in this age group show less variability in their systolic blood pressures. Use the 5% level of significance. What can be said about the P value?

Solution

1. Hypotheses: Let σ = the standard deviation of systolic blood pressures of male runners aged 35 to 59. The claim is $\sigma < 17$ (or $\sigma^2 < 289$). The hypotheses are

$$H_0: \sigma^2 = 289$$
$$H_a: \sigma^2 < 289$$

2. Level of significance: $\alpha = .05$.

3. Test statistic and its observed value:

$$\chi^2 = \frac{(n-1)s^2}{\sigma^2} = \frac{(40)(15)^2}{289} = 31.14 \qquad df = n - 1 = 40$$

4. Critical region: Since H_a says that $\sigma^2 < 289$, values of s^2 much smaller than 289 strongly favor H_a. These correspond to small values of χ^2, so we use a left-tailed test. From Appendix Table B.5 we find $\chi^2(1 - \alpha) = \chi^2(1 - .05) = \chi^2(.95) = 26.509$. Hence the critical region consists of values of $\chi^2 \leq 26.509$. (See Figure 8.18a.)

5. Decision: The observed value $\chi^2 = 31.14$ is not in the critical region; hence we fail to reject H_0. This means that there is not enough evidence to conclude (at the 5% level) that there is less variability in systolic blood pressures for the population.

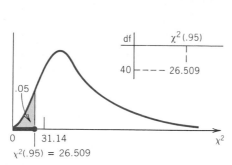

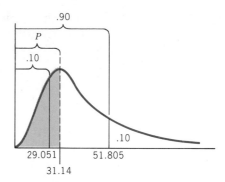

FIGURE 8.18*a*
Critical Region

FIGURE 8.18*b*
P Value: .10 < *P* < .90

P value: Since we are dealing with a left-tailed test, the *P* value is the area under the curve to the *left* of 31.14. In Appendix Table B.5 locate the row for df = 40. Notice that 31.14 is between the values 29.051 and 51.805. The table relates χ^2 values to *right* tail areas. For example, $\chi^2(.90) = 29.051$. So the area to the right of 29.051 is .90. Thus the area to the left is $1 - .90 = .10$. Also, $\chi^2(.10) = 51.805$, so the area to the *left* of 51.805 is .90. See Figure 8.18*b*.

		P	
Area of left tail	.10	↓	.90
χ^2 **value**	29.051		51.805
		↑ 31.14	

Thus .10 < *P* < .90. Therefore we cannot reject H_0 when $\alpha = .05$ since $P > .05$.

We summarize the essential features of a chi-square test concerning a population variance as follows:

Chi-Square Test for a Population Variance   To test hypotheses concerning σ^2, the test statistic is

$$\chi^2 = \frac{(n-1)s^2}{\sigma^2}$$

This has a chi-square distribution with df = $n - 1$. For σ^2 we substitute the value given in the null hypothesis. The observed value of χ^2 is computed using the sample data; if it falls in the critical region, we reject the null hypothesis H_0. Suppose that α is the level of significance of the test. Critical values of χ^2 are found in Appendix Table B.5. The possible critical regions are described as follows:

(a) If the alternate hypothesis contains the symbol $>$, we conduct a right-tailed test. The critical region is shown in Figure 8.19.

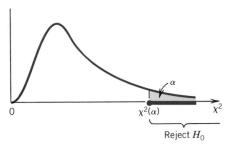

FIGURE 8.19
Right-Tailed Test

(b) If the alternate hypothesis contains the symbol $<$, we conduct a left-tailed test. The critical region is shown in Figure 8.20.

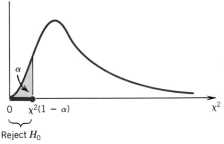

FIGURE 8.20
Left-Tailed Test

(c) If the alternate hypothesis contains the symbol $\neq$, we conduct a two-tailed test. The critical region is shown in Figure 8.21.

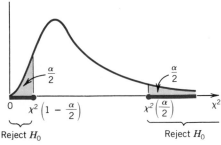

FIGURE 8.21
Two-Tailed Test

Assumption for the test: The population is normal.

Confidence Intervals (Optional)

Let us now turn to the subject of estimating a population variance σ^2. We will omit the derivation of the formula for a $1 - \alpha$ confidence interval for σ^2 and give only the final result.

- We use the sample variance s^2 as a point estimate for σ^2.
- A $1 - \alpha$ confidence interval for σ^2 for a normal population is

$$\frac{(n - 1)s^2}{\chi^2(\alpha/2)} < \sigma^2 < \frac{(n - 1)s^2}{\chi^2(1 - \alpha/2)}$$

The values appearing in the denominators of the above inequality are shown in Figure 8.22.

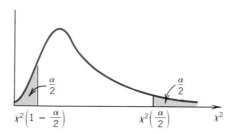

FIGURE 8.22

Example 8.11

In Example 8.8 we discussed a new mathematics screening examination. The test scores of a sample of 20 students showed a sample variance of $s^2 = 217.63$. Find a 95% confidence interval for the (population) variance σ^2 for all future test scores on the exam.

Solution

$$1 - \alpha = .95 \qquad \alpha = .05$$

Now df $= n - 1 = 20 - 1 = 19$. Using Appendix Table B.5, we see that

$$\chi^2\left(\frac{\alpha}{2}\right) = \chi^2(.025) = 32.852$$

$$\chi^2\left(1 - \frac{\alpha}{2}\right) = \chi^2(.975) = 8.907$$

Therefore, a 95% confidence interval for σ^2 is

$$\frac{(19)(217.63)}{32.852} < \sigma^2 < \frac{(19)(217.63)}{8.907}$$

or

$$125.87 < \sigma^2 < 464.24$$

If we want a 95% confidence interval for the standard deviation σ, we just take the square root of each term, getting

$$11.22 < \sigma < 21.55$$

Therefore we are 95% confident that σ is between 11.22 and 21.55.

Exercises

8.31   In each of the following parts, **(i)** test the claim, and **(ii)** indicate the possible type of error committed (I or II). Note that n refers to the sample size.

    (a)   Test the claim that the variance σ^2 is more than 100.

        $n = 16,$ $\alpha = .05,$ $s^2 = 173$

    (b)   Test the claim that the variance σ^2 is less than 70.

        $n = 12,$ $\alpha = .01,$ $s^2 = 12.7$

    (c)   Test the claim that the variance σ^2 is not 50.

        $n = 19,$ $\alpha = .10,$ $s^2 = 33.3$

    (d)   Test the claim that the variance σ^2 is greater than 30.

        $n = 9,$ $\alpha = .10,$ $s^2 = 45$

    (e)   Test the claim that the variance σ^2 is less than 65.

        $n = 22,$ $\alpha = .05,$ $s^2 = 40.2$

    (f)   Test the claim that the variance σ^2 is 20.

        $n = 26,$ $\alpha = .01,$ $s^2 = 38.4$

8.32   Verify the conclusions reached in Exercise 8.31 by using the P value approach. (*Hint*: Estimate the P value by giving an interval such as $.025 < P < .05$ or $P < .005$.)

8.33   A dispenser of soft drinks is designed to fill containers with a mean μ of 16 ounces and a variance σ^2 of .0001. The manufacturer is concerned that the variability of the amount being dispensed has increased. A random sample of size 15 gives a sample variance of .0002. Is there sufficient evidence to indicate that the variability has increased? Use a 1% level of significance.

    (a)   Use the classical approach.   **(b)**   Use the P value approach.

8.34   A tablet is supposed to have a mean weight μ of 325 milligrams and a standard deviation σ of .70 milligrams. A physician suspected that the standard deviation was larger than .70 milligrams. A sample of 20 tablets showed a sample standard deviation of .94 milligrams. At the 5% significance level, is there sufficient evidence to justify the physician's suspicion?

8.35   Under test conditions one brand of golf ball, manufactured by the Avex company, moves a mean distance μ of 200 yards with a variance σ^2 of 4. The company decided to modify the golf ball with a tougher cover. It was felt that the variance will remain the same. Twenty balls with the tougher cover are tested and found to have a sample variance of 6. Is there sufficient evidence to suggest the variance has changed? Use a 5% level of significance.

(a) Use the classical approach.   **(b)** Use the P value approach.

8.36   A company has a policy of not marketing a speedometer if it believes the standard deviation σ of measurements is greater than 3 miles per hour. Twelve speedometers were tested and the sample variance was 23. Using a 1% level of significance, will the company decide to market the speedometer?

8.37   An automobile company specified that replacement of a defective head gasket should take 2 hours on the average, with a standard deviation of 20 minutes. On warranty work, the company stated that it would pay a dealership for labor up to one standard deviation above the mean (up to 2 hours and 20 minutes). One dealer felt that the standard deviation σ was more than 20 minutes. A sample of 20 jobs gave a sample standard deviation of 32 minutes. At the 1% level of significance, is the dealer's suspicion justified?

8.38   In the Crevalcore area of Greece, a study of 28 men's diets showed a sample standard deviation of 1.6% of calories from proteins (*Source*: Keys, 1970, p. I–166). Find a 90% confidence interval for the variance σ^2 of the population from which the men were selected.

8.39   A medical doctor claimed the variance σ^2 is less than 156 for diastolic blood pressure of males aged 60–64 in the Framingham Heart Study. A sample of 9 such males gave the following readings:

$$80 \quad 82 \quad 102 \quad 84 \quad 88 \quad 86 \quad 88 \quad 98 \quad 106$$

Test the claim using a 1% level of significance.

8.40   The data below represent the length of time (in minutes) of six randomly selected minor league baseball games. The times are

$$153 \quad 137 \quad 127 \quad 143 \quad 152 \quad 148$$

Find a 95% confidence interval for the variance σ^2 of the population from which the times were selected.

8.41   **(a)** Using the data in Exercise 8.33, find a 98% confidence interval for the population variance σ^2.

   (b) Using the data in Exercise 8.34, find a 90% confidence interval for the population variance σ^2.

   (c) Using the data in Exercise 8.36, find a 99% confidence interval for the population variance σ^2.

8.6 USING MINITAB (OPTIONAL)

In the Minitab printout below are the results of a t test on a sample of 15 data values. The data represent daily production of rubber gaskets (in pounds) for

15 selected days using a new machine being tested by a company. The company wanted to test the following hypotheses:

$$H_0: \mu = 200 \text{ lbs.}$$

$$H_a: \mu > 200 \text{ lbs.}$$

We typed

TTEST OF MU = 200, DATA IN C1;

ALTERNATIVE = 1.

As noted in Chapter 7, the 1 in the subcommand indicates a right-tailed test; −1 would indicate a left-tailed test. If you don't use a subcommand, Minitab does a two-tailed test.

```
MTB > SET THE FOLLOWING DATA IN C1
DATA> 195 210 211 197 215 187 220 203 198 207 218 221 193 206 214
DATA> END
MTB > TTEST OF MU = 200, DATA IN C1;
SUBC>   ALTERNATIVE = 1.

TEST OF MU = 200.00 VS MU G.T. 200.00

                 N     MEAN    STDEV    SE MEAN       T    P VALUE
C1              15   206.33    10.54       2.72    2.33      0.018
```

Since the *P* value is .018, we would reject the null hypothesis for any $\alpha \geq .018$.

The following command produces a confidence interval for μ using the data in column 1. We have chosen the 95% level of confidence.

```
MTB > TINTERVAL 95 PERCENT CONFIDENCE, DATA IN C1

            N    MEAN   STDEV   SE MEAN   95.0 PERCENT C.I.
C1         15  206.33   10.54      2.72  (  200.49,   212.17)
```

Therefore, the confidence interval is from 200.49 to 212.17.

Minitab does not provide a chi-square test for variance.

Finding *P* Values

As noted in Section 7.8, Minitab statistical tests usually return a *P* value. However, Minitab commands, such as TTEST, work on original data values. But often there is only summary data ($\bar{x}$, *s*, *n*) available to us. This is often the case, for example, in journal articles. In this chapter we have seen how to use Appendix Table B.4 to get an *interval estimate* of a *P* value. We can write a short Minitab program to compute the *exact P* value.

Recall that in Example 8.5 we investigated hypotheses concerning the mean calories per pint of ice cream:

$$H_0: \mu = 500$$

$$H_a: \mu \neq 500$$

The summary data for the sample was: $\bar{x} = 507$, $s = 22$, $n = 25$. The following program computes the P value for this test. Note that the subcommand specifies a t distribution with 24 degrees of freedom.

```
MTB > LET K1 = (507 - 500)/(22/SQRT(25)) # THE O.V.
MTB > CDF K1, STORE IN K2;
SUBC>    T DIST, DF = 24.
MTB > LET K3 = 2*(1 - K2) # THE P VALUE
MTB > PRINT K3
K3        0.124715
```

In the above printout K2 was the area to the left of the observed value (O.V.), and $1 - K2$ was the area of the right tail. Since we have a two-tailed test, it was necessary to double this to get the P value. See Figure 8.23.

When computing the P value, we must be careful about the sign of the observed value. In the above case, $\bar{x} = 507$, so the observed value is positive. If $\bar{x}$ were, say, 493, the observed value would be negative. Hence CDF K1 would give us the area of the left tail (stored as K2). To get the P value we would just double K2.

Although Minitab does not provide a command for a chi-square test on a data set, it is still possible to use summary data (n, s^2) to find the P value for a test concerning σ^2. In Example 8.9 we conducted a two-tailed chi-square test concerning variability in the amount of milk in milk bottles. We tested

$$H_0: \sigma^2 = .01$$

$$H_a: \sigma^2 \neq .01$$

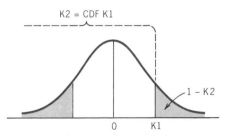

FIGURE 8.23
P **Value = 2 * (1 − K2)**

The summary data were $s = .13$ oz and $n = 28$. The observed value and degrees of freedom were

$$\chi^2 = \frac{(n-1)s^2}{\sigma^2} = \frac{(27)(.13)^2}{.01} = 45.63, \qquad df = 27$$

When deciding whether our *P* value will be obtained by doubling the area of a right tail or a left tail, keep in mind that the expected value of χ^2 equals the degrees of freedom. Thus the observed value of 45.63 is well over to the right. Hence, we should find the area of the right tail, and double it. The following program computes the observed value and then the *P* value.

```
MTB > LET K1 = (27*.13**2)/.01 # THE O.V.
MTB > CDF K1, STORE IN K2;
SUBC>   CHISQUARE, DF = 27.
MTB > LET K3 = 2*(1 - K2) # THE P VALUE
MTB > PRINT K3
K3      0.0278752
```

So the *P* value is about .03. *Note*: Minitab treats "chisquare" as one word. So don't put a dash between i and s.

One-tailed tests are fairly straightforward. For a right-tailed test, the *P* value is the area to the right of the observed value. For a left-tailed test, the *P* value is the area to the left of the O.V.

However, finding the *P* value for a two-tailed chi-square test can be tricky at times. It may not be obvious whether to double the area of a right tail or a left tail. When in doubt, you can compute CDF of the observed value and store it as K2. Then print the value of K2. If K2 < .5, the *P* value (for a two-tailed test) is 2*K2. If K2 > .5, the *P* value is 2*(1 - K2).

We noted earlier that Minitab does not provide a chi-square test for variance (on a data set). If you have raw data stored in a column, you can use Minitab commands to compute the observed value and proceed as outlined above. For example, if the data for the above chi-square test were in column 1, we could replace the first command with

```
LET K1 = (COUNT(C1) - 1)*STDEV(C1)**2/.01
```

When doing a two-tailed test, you may want to print the value of K1 to see if you should double the area of a right tail or a left tail.

Exercises

Suggested exercises for use with Minitab are 8.18, 8.19, 8.24, 8.25, and 8.53. Find the *P* value for the following exercises: 8.10, 8.12, 8.15, and 8.31.

8.7 SUMMARY

In this chapter we introduced **Student's t** and **chi-square distributions** and studied their use in statistical inference.

Student's t distribution is used to study the mean μ of an approximately normal population when the sample size is small ($n < 30$). In hypothesis testing concerning μ (based on a small sample), we use the test statistic

$$t = \frac{\bar{x} - \mu}{s/\sqrt{n}}$$

which has a Student's t distribution with df $= n - 1$. Appendix Table B.4 contains critical values of t. A $1 - \alpha$ confidence interval for μ based on a small sample is given by the endpoints

$$\bar{x} \pm t\left(\frac{\alpha}{2}\right) \cdot \frac{s}{\sqrt{n}}$$

A chi-square distribution can be used to study the variance σ^2 of a normal population. In hypothesis testing concerning σ^2 we use the test statistic

$$\chi^2 = \frac{(n - 1)s^2}{\sigma^2}$$

which has a chi-square distribution with df $= n - 1$. Appendix Table B.5 contains critical values of χ^2. A $1 - \alpha$ confidence interval for σ^2 is

$$\frac{(n - 1)s^2}{\chi^2(\alpha/2)} < \sigma^2 < \frac{(n - 1)s^2}{\chi^2(1 - \alpha/2)}$$

Review Exercises

8.42 Find the following values:
  (a) $t(.01)$ when df $= 11$.   **(b)** $t(.05)$ when df $= 23$.
  (c) $t(.10)$ when df $= 14$.   **(d)** $t(.975)$ when df $= 21$.

8.43 Assuming a t distribution with 16 degrees of freedom, find each of the following probabilities. (*Hint*: A diagram of the required area might help.)
  (a) $P(-1.337 < t < 1.337)$.   **(b)** $P(1.746 < t < 2.583)$.
  (c) $P(t < 2.12)$.   **(d)** $P(-1.337 < t < 1.746)$.

8.44 Below is the t curve with 19 degrees of freedom. Find the areas A, B, C, and D.

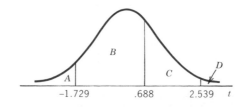

8.45   The mean number of calories μ per 6-ounce can of tomato paste is supposed to be 150 calories. A weight watchers group thought that μ was larger than 150. A random sample of 14 cans yielded a sample mean of 151.2 and a sample standard deviation of 2.4 calories. Do the sample results support the weight watchers' contention at the 5% significance level?

    (a)   Use the classical approach.   **(b)**   Use the P value approach.

8.46   The mean percentage of sodium carbonate μ per 21-ounce can of a cleanser is supposed to be 11%. The manufacturer thinks that the mean percentage may have changed. A random sample of 20 cans showed a mean percentage of sodium carbonate as 11.5 with a variance of 2.25%. Using a 10% level of significance, is there sufficient evidence indicating that the mean percentage of sodium carbonate μ is different from 11?

8.47   A method of manufacturing a ball bearing produced a mean diameter μ of .120 and a standard deviation σ of .015 inches. A new process to manufacture the ball bearing was developed, and it was believed that there would be no change in the variance. To test the hypothesis of no change in the mean μ, 18 ball bearings produced by the new process were selected and the mean diameter was .114 inches. Complete the test at the 10% significance level.

8.48   Applicants to a graduate program in business in the past had a mean high school grade-point average μ of 3.23. The admissions committee believed the current applicants had higher averages. A quick check of 25 current applicants showed a mean of 3.3 with standard deviation .375. With a 5% significance level, do the sample data support the committee's belief?

8.49   An economist claimed that the mean tax deduction for charities was less than $600 for families with an income of about $35,000. A sample of 15 tax returns of families with an income of $35,000 showed a mean deduction of $450 with a standard deviation of $300. Test the economist's claim using a 5% level of significance.

    (a)   Use the classical approach.   **(b)**   Use the P value approach.

8.50   The chief of police in a small town received complaints that cars were speeding on a street in the residential area. The posted speed was 25 miles per hour. The police chief decided to gather some information. A sample of 14 cars showed a mean speed of 36.1 miles per hour with a standard deviation of 10.4. Discuss the chief's findings using a P value approach.

8.51   Twenty-seven runners were given a psychological test called the Profile of Mood States (POMS). The sample showed a mean tension score of 10.46 with a standard deviation of 5.57 (*Source*: Morgan, W. P., and M. L. Pollock, "Psychological Characterization of the Elite Distance Runner," in *The Long Distance Runner*, ed. P. Milvy (New York: Urizen Books, 1977). Do these results indicate that runners in general have a mean tension level different from 13 (the norm for college students)? Use the 5% level of significance.

8.52   A transit authority believed that introduction of its new rail transit vehicles would increase patronage. Past records showed an average of 2200 passengers per day on the number 10 transit line. With the new transit vehicles, a sample of 20 days showed an average of 2315 passengers per day with a standard deviation of 115. At the 5% level of significance, does it appear that there has been an increase in mean daily patronage for this route?

    (a)   Use the classical approach.   **(b)**   Use the P value approach.

8.53 The following data represent lactate readings for 15 randomly selected females from the Framingham Heart Study:

$$180 \quad 145 \quad 112 \quad 159 \quad 189 \quad 173 \quad 108 \quad 163$$
$$136 \quad 191 \quad 124 \quad 131 \quad 134 \quad 184 \quad 140$$

Test the hypothesis that the population mean μ of such females is more than 130. Use a 5% level of significance.

8.54 A trucking company interested in the mean travel time for a new route obtained the following travel times for seven trips (in minutes):

$$152 \quad 160 \quad 168 \quad 130 \quad 162 \quad 146 \quad 178$$

(a) Find a 90% confidence interval for the mean travel time μ for the new route.

(b) Find the maximum error of the estimate for μ with 90% level of confidence.

8.55 **(a)** Consider the data in Exercise 8.45. Find a 90% confidence interval for the population mean μ.

(b) Consider the data in Exercise 8.46. Find a 98% confidence interval for the population mean μ.

(c) Consider the data in Exercise 8.47. Find a 99% confidence interval for the population mean μ.

8.56 Assuming a χ^2 distribution with 10 degrees of freedom, find each of the following probabilities:

(a) $P(\chi^2 > 2.558)$. **(b)** $P(\chi^2 < 3.94)$.

(c) $P(15.987 < \chi^2 < 23.209)$. **(d)** $P(4.865 < \chi^2 < 18.307)$.

8.57 Find the following values:

(a) $\chi^2(.01)$ when df = 12. **(b)** $\chi^2(.05)$ when df = 19.

(c) $\chi^2(.975)$ when df = 30. **(d)** $\chi^2(.95)$ when df = 90.

8.58 Below is the χ^2 curve with 8 degrees of freedom. Find the areas A, B, and C.

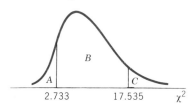

8.59 A company spokesman maintained the standard deviation was less than 3 calories for 6-oz cans of its brand of apple sauce. The company's laboratory sampled 14 six-ounce cans, and the sample variance was 5.76. Is there sufficient evidence to justify the spokesman's claim? Use a 1% level of significance.

(a) Use the classical approach. **(b)** Use the P value approach.

8.60 It was claimed that the variance σ^2 of the breaking strength of a yarn was more than 5.1. A random sample of 10 pieces of yarn produced a standard deviation of 2.58 pounds in the breaking strength. At the 10% level of significance, is there sufficient evidence to indicate that the claim is true?

8.61 The turkey bologna produced by a meat-processing company had a mean fat content of 10% with a standard deviation of .75%. The quality control department at the company tested a sample of size 25 and found a sample mean fat content of 10.3% with a standard deviation of .83%. At the 5% level of significance, test the claim that

(a) The mean μ is no longer 10%.

(b) The standard deviation σ is no longer .75%.

8.62 **(a)** Using the data in Exercise 8.59, find a 99% confidence interval for the variance σ^2 of the population from which the cans were selected.

(b) Using the data in Exercise 8.60, find a 90% confidence interval for the population variance σ^2.

8.63 An insurance company believed that ordinarily the cost of straightening a bumper should be under $450. A sample of 8 jobs at an auto body shop gave a sample mean of $420 with a standard deviation of $40.50. Is there sufficient evidence to conclude that the mean charge for the job at the shop is less than $450?

(a) Use a 5% level of significance.

(b) Use a 1% level of significance.

8.64 In Exercise 8.63, suppose that the sample standard deviation were $60 instead of $40.50.

(a) Without any computations, which (if any) of the conclusions reached in Exercise 8.63 might possibly be different?

(b) Now test the hypothesis of Exercise 8.63 with $s = $60.

    (i) Use a 5% level of significance.

    (ii) Use a 1% level of significance.

    What are your conclusions?

8.65 In Exercise 8.63, suppose that the sample standard deviation were $27 instead of $40.50.

(a) Without any computations, which (if any) of the conclusions reached in Exercise 8.63 might possibly be different?

(b) Now test the hypothesis of Exercise 8.63 with $s = $27.
    (i) Use a 5% level of significance.

    (ii) Use a 1% level of significance.

    What are your conclusions?

8.66 In Exercise 8.63, suppose that the sample mean were $402 instead of $420. (Use $s = $40.50.)

(a) Without any computations, which (if any) of the conclusions reached in Exercise 8.63 might possibly be different?

(b) Now test the hypothesis of Exercise 8.63 with the sample mean of $402.
    (i) Use a 5% level of significance.

    (ii) Use a 1% level of significance.

    What are your conclusions?

8.67 In Exercise 8.63, suppose that the sample mean were $432 instead of $420. (Use $s = $40.50.)

(a) Without any computations, which (if any) of the conclusions reached in Exercise 8.63 might possibly be different?

(b) Now test the hypothesis of Exercise 8.63 with the sample mean of $432.

    (i) Use a 5% level of significance.

    (ii) Use a 1% level of significance.

    What are your conclusions?

8.68 **(a)** Use the sample results of Exercise 8.63 ($\bar{x}$ = $420, s = $40.50, and n = 8) to obtain a 95% confidence interval for the mean cost μ.

    (b) In parts (i)–(iv), will the length of the 95% confidence interval for μ be longer, shorter, or the same as in part (a)? Do not calculate.

        (i) Replace s = $40.50 with s = $60.

        (ii) Replace s = $40.50 with s = $27.

        (iii) Replace $\bar{x}$ = $420 with $\bar{x}$ = $402.

        (iv) Replace $\bar{x}$ = $420 with $\bar{x}$ = $432.

8.69 Look at Appendix Table B.4 and note the relationship between sizes for degrees of freedom and critical values. Now, without using Appendix Table B.4, match the degrees of freedom with the appropriate entry in the left-hand column.

	df
$P(t > 1.812)$ = .05	10
$P(t > 1.753)$ = .05	20
$P(t > 1.725)$ = .05	25
$P(t > 1.708)$ = .05	15

8.70 The following table represents some probabilities of a chi-square distribution with one degree of freedom:

b	$P(\chi^2 < b)$
.455	.50
1.638	.80
2.706	.90
3.841	.95
6.635	.99

    (a) Complete the following table where z is the standard normal variable (round to two decimal places):

b	$P(z^2 < b)$
.455	
1.638	
2.706	
3.841	
6.635	

    [*Hint*: Note that $P(z^2 < b) = P(-\sqrt{b} < z < \sqrt{b})$. Now use Appendix Table B.3 to evaluate the probabilities.]

    (b) Compare the probabilities in the z^2 table with the probabilities in the χ^2 table. Conjecture a relationship between the chi-square distribution with 1 degree of freedom and the z^2 distribution.

Notes

Student, "The Probable Error of a Mean," *Biometrika* 6 (1908).

Keys, A., ed., "Coronary Heart Disease in Seven Countries," *Circulation* 41, (Suppl. 1): 1 (1970).

Morgan, W. P., and M. L. Pollock, "Psychological Characterization of the Elite Distance Runner," in Paul Milvy, ed. *The Long Distance Runner* (New York: Urizen Books, 1977).

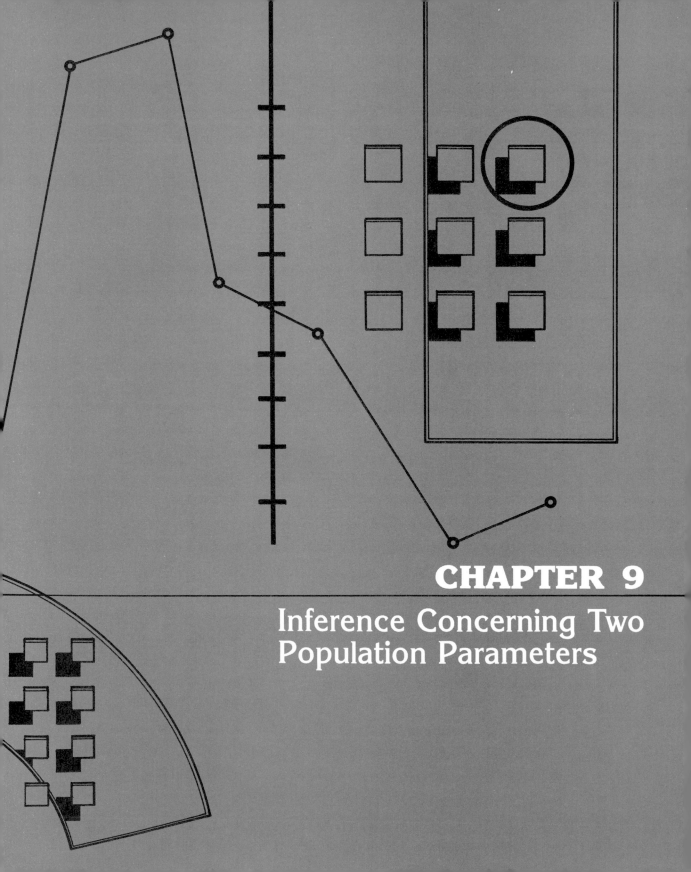

CHAPTER 9

Inference Concerning Two
Population Parameters

INTRODUCTION

In many situations in statistics, the primary objective is to study how one parameter compares with another parameter rather than to study the particular value of some parameter. For example, suppose there are two medications used in the treatment of the same illness, and we let p_1 and p_2 represent the proportion of people who will be cured with medication 1 and medication 2, respectively. Then we would want to know which proportion is larger. Or we may wish to find out which of two types of automobiles has the greater mean gas mileage.

Or we may wish to compare the variances for scores on two examinations to see which is larger.

In this chapter we will develop methods for comparing two means, two variances, or two proportions for a pair of populations.

9.2   THE F DISTRIBUTION

Suppose that we wish to compare two (unknown) population variances σ_1^2 and σ_2^2. It would seem natural to obtain samples from each population and examine the sample variances s_1^2 and s_2^2. For example, two fertilizer companies that marketed a fertilizer for wheat merged to form a new company that would market only one of the fertilizers. Both fertilizers result in about the same yield (in bushels of wheat per acre), on the average. But the variability is unknown. It is, of course, desirable for the variability of such a product to be at a minimum. Hence the company wishes to compare the population variances σ_1^2 and σ_2^2 for the two fertilizers. (The population for a given fertilizer can be viewed as the collection of data representing the yield in bushels per acre of wheat for each farm that has used or will use the fertilizer.)

To study this situation the company obtained the yields of some farms that used the fertilizers. Eleven farms used fertilizer A and 11 farms used fertilizer B. The resulting data are given in Table 9.1.

TABLE 9.1
Wheat Yields for Two
Types of Fertilizer (Bushels
per Acre)

Fertilizer A	Fertilizer B
74	70
61	69
74	71
80	70
70	69
71	71
60	70
79	69
64	68
70	73
67	70
$\bar{x}_1 = 70.0$	$\bar{x}_2 = 70.0$
$s_1^2 = 44.0$	$s_2^2 = 1.8$

You probably feel that the sample variances shown are *significantly different*; that is, they are far enough apart to indicate that the population variances are different. After all, s_1^2 is almost 25 times larger than s_2^2 , since $s_1^2/s_2^2 = 24.44$. Often, however, it is not so easy to decide whether $\sigma_1^2 = \sigma_2^2$ or not by simply looking at s_1^2 and s_2^2. Hence we need some method other than intuition to investigate these issues.

In general, suppose that we are investigating two population variances σ_1^2 and σ_2^2 and we wish to test

$$H_0: \sigma_1^2 = \sigma_2^2$$

$$H_a: \sigma_1^2 \neq \sigma_2^2$$

Note that if we divide by σ_2^2, we can rewrite these hypotheses as

$$H_0: \frac{\sigma_1^2}{\sigma_2^2} = 1$$

$$H_a: \frac{\sigma_1^2}{\sigma_2^2} \neq 1$$

Now if we obtain samples from each population, we may think of the sample variances s_1^2 and s_2^2 as rough estimates of σ_1^2 and σ_2^2, respectively. Therefore, if H_0 is true ($\sigma_1^2 = \sigma_2^2$), the values of s_1^2 and s_2^2 should not be too far apart. This means that the ratio s_1^2/s_2^2 should not be too far from 1. Now the smallest possible value such a ratio can assume is 0 (if $s_1^2 = 0$), and there is no limit to how large the ratio can be. If the ratio s_1^2/s_2^2 is so far from 1 (i.e., so large or so small) that it would be unlikely that we would observe such a value if σ_1^2 and σ_2^2 were equal, we would be inclined to reject H_0 in favor of H_a.

If we are to know what values of s_1^2/s_2^2 are unlikely, we should know something about the probability distribution of this expression. We use the symbol *F* to represent this ratio.

$$F = \frac{s_1^2}{s_2^2}$$

Under suitable conditions this ratio has a probability distribution known as an **F distribution.**

Properties of an *F* Distribution

Suppose that we obtain random samples from two populations. We will use the notation in Table 9.2.

TABLE 9.2

Population	Sample Size	Sample Variance	Population Variance
1	n_1	s_1^2	σ_1^2
2	n_2	s_2^2	σ_2^2

We assume that the populations are normal. Further, we assume that the samples are obtained *independently* of one another. That is, the individual data values we get in one sample are not related to any of the values in the other sample.* Then if $\sigma_1^2 = \sigma_2^2$, the ratio

$$F = \frac{s_1^2}{s_2^2}$$

has an F distribution. An F distribution has the following properties:

1. For an F distribution we have a pair of degrees of freedom: the degrees of freedom of the numerator, df_1, and the degrees of freedom of the denominator, df_2. For the ratio s_1^2/s_2^2, $df_1 = n_1 - 1$ and $df_2 = n_2 - 1$. We often express the degrees of freedom as an ordered pair of numbers, writing

$$df = (n_1 - 1, n_2 - 1)$$

There are an infinite number of F distributions, one for each possible pair of degrees of freedom.

2. The graph of an F distribution (an F curve) starts at 0 and extends indefinitely to the right. It is skewed to the right. Of course, the total area under the curve is 1. (See Figure 9.1.)

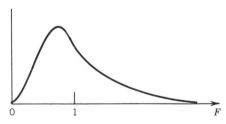

FIGURE 9.1
An F Curve

---

*For example, we have independent samples if a group of farmers tries fertilizer A and a *different* group tries fertilizer B. If only one group of farmers is selected, and each farmer tries both fertilizers, the samples would be *dependent*. We will have more to say about this in Section 9.4.

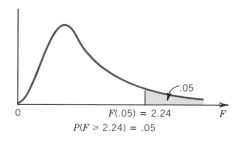

Degrees of freedom of numerator

			20	24	30
15	· · · · · · · ·	2.33	2.29	2.25 · · · · ·
16	· · · · · · · ·	2.28	2.24	2.19 · · · · ·
17	· · · · · · · ·	2.23	2.19	2.15 · · · · ·

$F(.05) = 2.24$

$P(F \geq 2.24) = .05$

FIGURE 9.2

Appendix Table B.6 relates probabilities to various values of F that we will need. The table enables us to find the value of F such that the area under the curve to the right of this value is some desired value (probability). Note that Appendix Table B.6 actually consists of three different tables corresponding to three such probabilities (.01, .025, and .05). Suppose that we are interested in the F distribution with degrees of freedom df $= (24, 16)$, and we want the value of F such that the area under the F curve to the right of it is .05. We denote this value by $F(.05)$. Locate the appropriate table for .05. In the row across the top find the degrees of freedom for the numerator, 24. In the column on the left locate the degrees of freedom of the denominator, 16. The intersection of the column under 24 and the row next to 16 contains the desired value of F, namely, 2.24. Thus $F(.05) = 2.24$. This means that $P(F \geq 2.24) = .05$. In Figure 9.2 we have included a portion of the F table (with right tail area .05).

Before turning to applications of the F distribution, we should note the following concerning the quantity

$$F = \frac{s_1^2}{s_2^2}$$

The sample sizes n_1 and n_2 need not be large for this expression to have the F distribution; it has an F distribution regardless of the sample sizes. However, we do require the populations to be normal. This rather severely limits the use of this expression in the study of variances. In the examples and exercises of this chapter where the F distribution is used, we will assume that the normality condition is met.

Exercises

9.1 In parts (a)–(d) find **(i)** $F(.01)$, **(ii)** $F(.025)$, and **(iii)** $F(.05)$.

    (a) Assume an F distribution with (5, 8) degrees of freedom.

    (b) Assume an F distribution with (8, 5) degrees of freedom.

    (c) Assume an F distribution with (20, 10) degrees of freedom.

    (d) Assume an F distribution with (1, 7) degrees of freedom.

9.2 Assume an F distribution. Fill in the blanks.

    (a) $P(F > \underline{\quad}) = .01$ df $= (15, 12)$.

    (b) $P(F \leq \underline{\hphantom{xxx}}) = .95$ df = (9, 5).

    (c) $P(F \geq \underline{\hphantom{xxx}}) = .025$ df = (6, 20).

    (d) $P(F > \underline{\hphantom{xxx}}) = .05$ df = (12, 9).

9.3 INFERENCE CONCERNING TWO POPULATION VARIANCES

In the last section we said that we could compare two (unknown) population variances using the quantity

$$F = \frac{s_1^2}{s_2^2}$$

where s_1^2 and s_2^2 were sample variances for independent random samples of size n_1 and n_2, respectively, obtained from two normal populations. If the variances of the populations are equal ($\sigma_1^2 = \sigma_2^2$), then the expression F has an F distribution with degrees of freedom df = ($n_1 - 1, n_2 - 1$).

    To test hypotheses concerning variances, we can use the above expression as a test statistic. We call such a test an **F test.** Critical values of F may be found in Appendix Table B.6. This table enables us to find the values of $F(.01)$, $F(.025)$, or $F(.05)$. Notice that these are right critical values. In testing hypotheses, we will see that it will not be necessary to find left critical values if we use the following rule: *labeling will be done so that the numerator s_1^2 represents the larger of the two sample variances.* However, a word of caution about this: if your alternate hypothesis is H_a: $\sigma_1^2 < \sigma_2^2$, and yet for the sample data $s_1^2 > s_2^2$, then don't do an F test, because there is clearly no evidence to support the alternate hypothesis.

Remark   If you don't care to follow the labeling convention discussed above, this is all right. Just remember to put the larger sample variance in the numerator.

Example 9.1
A manufacturer of 6-volt batteries is considering two different production processes: process A and process B. In either process we can expect some degree of variability of individual batteries from the advertised 6 volts. An engineer claims that the variance in voltages of all batteries produced by process B is

greater than that of process A. To investigate this claim, samples of voltages from batteries produced by each process are obtained with the following results:

	n	s^2
A	21	.15
B	25	.33

Complete the test at the 5% level of significance. What can be said about the P value?

Solution
By our labeling convention, s_1^2 will denote the larger sample variance, so $s_1^2 = .33$ and $s_2^2 = .15$. We now proceed with the usual five steps.

1. Hypotheses: It is claimed that the population variance for process B is greater than the population variance for process A. Because the sample variance s_1^2 refers to process B, we denote the population variance for process B by σ_1^2. Therefore, the claim is that $\sigma_1^2 > \sigma_2^2$. This is the alternate hypothesis. Hence the hypotheses are

$$H_0: \sigma_1^2 = \sigma_2^2$$
$$H_a: \sigma_1^2 > \sigma_2^2$$

2. Level of significance: $\alpha = .05$.

3. Test statistic and its observed value:

$$F = \frac{s_1^2}{s_2^2} = \frac{.33}{.15} = 2.20$$
$$\text{df} = (n_1 - 1, n_2 - 1) = (24, 20)$$

4. Critical region: Values of s_1^2 much larger than s_2^2 favor H_a. But when s_1^2 is much larger than s_2^2, s_1^2/s_2^2 will be very large. Therefore, large values of F favor H_a. Thus we will conduct a right-tailed test. We will find the appropriate right critical value from Appendix Table B.6. (Notice how our labeling convention has led to a right-tailed test, making it unnecessary to find a left critical value.) From Appendix Table B.6 $F(.05) = 2.08$ when df = (24, 20). The critical region, shown in Figure 9.3a, consists of values of $F \geq 2.08$.

5. Decision: The observed value $F = 2.20$ lies in the critical region, so we reject H_0. Thus it does appear that process B shows more variability than process A.

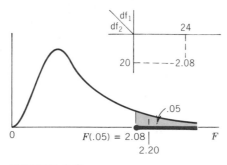

FIGURE 9.3a
Critical Region

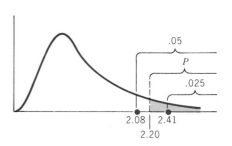

FIGURE 9.3b
.025 < P < .05

P Value: Since this is a right-tailed test, the P value is the area to the right of the observed value, 2.20. Information about the P value can be obtained by examining F values in Appendix B.6 when df = (24, 20). From this table we see that

$$F(.05) = 2.08 \qquad F(.025) = 2.41 \qquad F(.01) = 2.86$$

Thus

Area of right tail	.05	P ↓	.025	.01
F value	2.08	↑ 2.20	2.41	2.86

See Figure 9.3b. This means

$$.025 < P < .05$$

Even when we conduct a two-tailed F test, it will be unnecessary to find a left critical value. The reason for this is that since we are doing our labeling so that $s_1^2 > s_2^2$, then

$$F = \frac{s_1^2}{s_2^2} > 1$$

But left critical values are less than 1. Hence our value of F will not fall in the left end of the distribution. Therefore, when conducting a two-tailed test, we will not have to find the left critical value (even though there is a left tail to the critical region). The following example will illustrate this point.

Example 9.2

Cholesterol in the blood can lead to fatty deposits in the arteries, which in turn can lead to coronary heart disease. Recently, there has been a great deal of interest in a component of (total) cholesterol called "high-density lipoprotein" (HDL). HDL cholesterol is known as the "good cholesterol" because high levels of HDL are thought to lower the risk of coronary heart disease (Gordon et al., 1977, p. 707). Runners tend to have higher levels of HDL than nonrunners. We will examine this more carefully later in this chapter when we learn how to compare population means. For now, we investigate the question of whether the variability in HDL readings is the same for runners and nonrunners. Table 9.3 gives some data obtained concerning HDL (measured in milligrams per 100 millimeters) for elite runners* versus nonrunners:

TABLE 9.3
HDL Data

Samples	n	$\bar{x}$	s
Elite runners	20	56	12.1
Nonrunners	72	49	10.5

Source: Martin, R., W. Haskell, and P. Wood, "Blood Chemistry and Lipid Profiles of Elite Distance Runners," in *The Long Distance Runner*, ed. P. Milvy (New York: Urizen Books, 1977), p. 88.

Both groups were young males (30 and under), so we should restrict our attention to this category. (The nonrunners are referred to as the *control group*.) Test the claim that the (population) variance of HDL readings for young male elite runners is the same as that of young male nonrunners. Use the 5% level of significance.

Solution

Since s_1 will represent the larger sample standard deviation, $s_1 = 12.1$ and $s_2 = 10.5$. Thus, σ_1^2 and σ_2^2 will represent the population variances for young male elite runners and young male nonrunners, respectively.

1. Hypotheses: The question is whether the population variances are equal or not. Thus the hypotheses are

$$H_0: \sigma_1^2 = \sigma_2^2$$
$$H_a: \sigma_1^2 \neq \sigma_2^2$$

2. Level of significance: $\alpha = .05$.

---

*The term *elite runner* refers to one of the top 2.5% of runners in the world.

3. Test statistic and its observed value:

$$F = \frac{s_1^2}{s_2^2} = \frac{(12.1)^2}{(10.5)^2} = \frac{146.41}{110.25} \doteq 1.33$$

$$df = (19, 71)$$

4. Critical region: Since the alternate hypothesis is $\sigma_1^2 \neq \sigma_2^2$, we conduct a two-tailed test. (Very large or very small values of F favor H_a.) The right critical value is found in Appendix Table B.6. Note that this table does not contain the values 19 and 71 under degrees of freedom. In a case like this, it is customary to use the next *lower* values for degrees of freedom: (15, 60). The right critical value is $F(\alpha/2) = F(.025) = 2.06$. The left critical value, $F(1 - \alpha/2) = F(.975)$, is less than 1. But since our observed value of F is greater than 1, there is no question of its falling in the left part of the critical region. Therefore, we will not have to know the value of $F(.975)$. The critical region is shown in Figure 9.4.

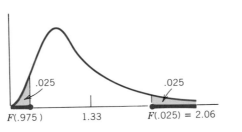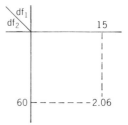

FIGURE 9.4

5. Decision: The observed value ($F = 1.33$) does not fall in the critical region, so we fail to reject H_0. This means that there is no evidence to conclude that σ_1^2 and σ_2^2 are different for the nonrunners and runners.

In the above two examples we performed a two-tailed test and a right-tailed test. What about a left-tailed test? We need not worry about left-tailed F tests. To see this, suppose the alternate hypothesis were $\sigma_A^2 < \sigma_B^2$. This looks like a left-tailed test, but observe that we can write this as $\sigma_B^2 > \sigma_A^2$. We would then use the statistic s_B^2/s_A^2 and do a right-tailed test. Of course we only perform the test if $s_B^2 > s_A^2$; otherwise there would be no evidence to support the alternate hypothesis ($\sigma_B^2 > \sigma_A^2$).

Some people don't follow the convention of putting the larger sample variance in the numerator. But if you don't, it may be necessary to find a left critical value. Although these are not in Appendix Table B.6, there is a way of finding them that is discussed on page 406.

We can summarize the essential features of an F test for variances as follows:

***F* test for Variances** To test hypotheses concerning two population variances, we use the test statistic

$$F = \frac{s_1^2}{s_2^2}$$

where labeling is done so that $s_1^2 > s_2^2$. This expression has degrees of freedom df $= (n_1 - 1, n_2 - 1)$ where n_1 and n_2 are the sample sizes associated with s_1^2 and s_2^2, respectively. Assume that α is the level of significance of the test.

(a) One-tailed test: Because of our labeling convention, the only one-tailed test we will conduct is a right-tailed test. To test

$$H_0: \sigma_1^2 = \sigma_2^2$$
$$H_a: \sigma_1^2 > \sigma_2^2$$

we compute the (observed) value of F from the sample data and find the critical value $F(\alpha)$ in Appendix Table B.6. If $F \geq F(\alpha)$, we reject H_0. The critical region is shown in Figure 9.5.

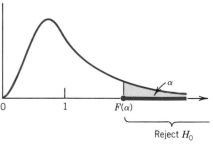

FIGURE 9.5
Right-Tailed Test

(b) Two-tailed test: To test the hypotheses

$$H_0: \sigma_1^2 = \sigma_2^2$$
$$H_a: \sigma_1^2 \neq \sigma_2^2$$

compute the value of F from the sample data and find the value of $F(\alpha/2)$ in Appendix Table B.6. If $F \geq F(\alpha/2)$, we reject H_0. (There is no need to find the left critical value, because F will not fall in the left tail as a result of our labeling convention. Nevertheless, this is a *bona fide* two-tailed test.) The critical region is shown in Figure 9.6.

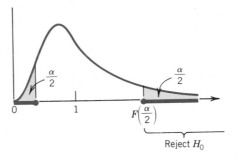

FIGURE 9.6
Two-Tailed Test

Assumptions for the test: The populations are normal and the samples are obtained independently.

Confidence Intervals for σ_1^2/σ_2^2 (Optional)

In order to estimate the relative size of σ_1^2 and σ_2^2, we can obtain a confidence interval for the ratio σ_1^2/σ_2^2. Using the F distribution, the following confidence interval can be established. (We omit the proof.)

A $1 - \alpha$ confidence interval for σ_1^2/σ_2^2 is

$$\frac{s_1^2}{s_2^2} \cdot \frac{1}{F(\alpha/2)} < \frac{\sigma_1^2}{\sigma_2^2} < \frac{s_1^2}{s_2^2} \cdot \frac{1}{F(1 - \alpha/2)}$$

For the F values, df $= (n_1 - 1, n_2 - 1)$. Assumptions: the samples are obtained independently and the populations are normal.

Note: When finding confidence intervals, it is not necessary that labeling be done so that s_1^2 is greater than s_2^2. The only time we follow this practice is when testing hypotheses concerning variances.

If our level of confidence is $1 - \alpha = .95$, then $\alpha/2 = .025$. So the critical values of F needed in the denominators above are $F(\alpha/2) = F(.025)$ and $F(1 - \alpha/2) = F(1 - .025) = F(.975)$, with df $= (n_1 - 1, n_2 - 1)$. Note that there is no F table corresponding to .975 in Appendix B.6. However, we can find $F(.975)$ indirectly as follows: Suppose that df $= (6, 8)$. To find $F(.975)$, that is, $F(1 - .025)$, we look up $F(.025)$ for the F distribution with degrees of freedom for the numerator and denominator interchanged, namely, df $= (8, 6)$. We then take the reciprocal of this. Using Appendix Table B.6, we see that when df = (8, 6), $F(.025)$ = 5.60. Therefore, when df = (6, 8)

$$F(.975) = \frac{1}{5.60} \doteq .18$$

In general, to find $F(1 - \alpha/2)$ when df = (a, b), we find $F(\alpha/2)$ for df = (b, a) and take the reciprocal of this.

Example 9.3

A college mathematics department was considering two mathematics screening examinations for the purpose of placing students. Students in two sections of an introductory mathematics course required of all students were given the test with the following results.

Examination	n	s^2
A	21	121
B	16	64

Let $s_1^2 = 121$ and $s_2^2 = 64$. Thus σ_1^2 is the population variance for exam A and σ_2^2 is the population variance for exam B. Find a 95% confidence interval for σ_1^2/σ_2^2.

Solution

The level of confidence is $1 - \alpha = .95$ so $\alpha = .05$. From Appendix Table B.6, $F(\alpha/2) = F(.025) = 2.76$ when df = (20, 15). To find $F(1 - \alpha/2) = F(.975)$, we look up $F(.025)$ for df = (15, 20) and take the reciprocal of this value. From Appendix Table B.6, when df = (15, 20), $F(.025) = 2.57$. Therefore, for df = (20, 15)

$$F(.975) = \frac{1}{2.57} \doteq .39$$

Our desired 95% confidence interval is

$$\left(\frac{121}{64}\right)\left(\frac{1}{2.76}\right) < \frac{\sigma_1^2}{\sigma_2^2} < \left(\frac{121}{64}\right)\left(\frac{1}{.39}\right)$$

$$.69 < \frac{\sigma_1^2}{\sigma_2^2} < 4.85$$

Thus we are 95% confident that σ_1^2/σ_2^2 is between .69 and 4.85. We can translate this into a 95% confidence interval for σ_1/σ_2 by taking the square root of each term, getting

$$.83 < \frac{\sigma_1}{\sigma_2} < 2.20$$

Note that the confidence interval for σ_1^2/σ_2^2 (and σ_1/σ_2) includes 1. When $\sigma_1^2/\sigma_2^2 = 1$, $\sigma_1^2 = \sigma_2^2$. Therefore, we cannot reject the possibility that $\sigma_1^2 = \sigma_2^2$.

Exercises

9.3   For the sample data given, (i) test the given claims and (ii) indicate the possible type of error committed (type I or type II).

(a)   Test the claim that the variances of populations A and B are different. Use a 5% level of significance.

	n	s^2
A	11	43
B	8	10

(b)   Test the claim that the variances of populations A and B are the same. Use a 5% level of significance.

	n	s^2
A	8	43
B	11	10

(c)   Test the claim that the variance of population A is smaller than the variance of population B. Use a 1% level of significance.

	n	s^2
A	21	5
B	31	20

(d)   Test the claim that the variance of population A is larger than the variance of population B. Use a 5% level of significance.

	n	s^2
A	31	20
B	21	8

(e)   Test the claim that the variances of populations A and B are different. Use a 10% level of significance.

	n	s^2
A	5	100
B	10	10

(f) Test the claim that the variance of population A is larger than the variance of population B. Use a 1% level of significance.

	n	s^2
A	10	100
B	5	10

9.4 Use the data in the corresponding parts of Exercise 9.3 to construct confidence intervals for σ_A^2/σ_B^2. The percent confidence is as follows:

(a) 95%.   **(b)** 95%.   **(c)** 98%.

(d) 90%.   **(e)** 90%.   **(f)** 98%.

9.5 Brands A and B of racquetballs were known to bounce to about the same height, on the average, under test conditions. The manufacturer of brand A claimed that the variability of bounce of brand A was smaller than that of brand B. Use the following sample data to test the manufacturer's claim at the 5% significance level:

	n	s^2
A	25	.50 inch
B	25	1.10 inches

9.6 An automobile executive was undecided whether to adopt brand A odometer or brand B odometer. The executive was convinced that the brands measured correct mileage, on the average, but that there may be a difference in their measurement variability. The executive had some of each brand tested on 100-mile automobile runs with the following results:

	n	s^2
A	21	.60
B	16	1.50

Using a 10% level of significance, can the executive conclude that there is a difference in population variability between brands A and B?

(a) Use the classical approach.   **(b)** Use the P value approach.

9.7 A manager of a supermarket wanted to compare the ability of clerks A and B to service customers at the check-out line. The manager felt that both clerks serviced the customers with approximately the same mean time. However, there were complaints that clerk A spent too much time with some customers and rushed with others. The manager timed the check-out times of nine customers for each clerk. Use the following data and a 1% level of significance to see if the data support the complaints.

	n	s^2
A	9	6 minutes
B	9	3 minutes

9.8   A biologist claimed that in the 20–24 age group, variability in males' systolic blood pressure was larger than females. Independent random samples of males and females in this age group were selected from the Framingham Heart Study. The data were as follows:

	n	$\bar{x}$	s
A (Males)	31	125	13.9
B (Females)	41	117	12.1

Test the biologist's claim using a 5% level of significance.

9.9   Consider the claim that the variance of population A is more than the variance of population B. With a 5% level of significance, one of the two following tables contains enough information to reject the null hypothesis; the other does not.

   (a)   Without testing, which table leads to rejection of H_0?

   (b)   Now complete the test for both cases and compare with part (a)

	n	s^2
A	41	100
B	41	50

	n	s^2
A	11	100
B	11	50

INFERENCE CONCERNING TWO POPULATION MEANS. DEPENDENT SAMPLES

The test statistic to be used when testing hypotheses concerning two population means can depend on a number of factors. One of these is the method of obtaining samples. We have already discussed independent samples in the previous section. Let us look at this idea more closely.

A high school developed a course in reading comprehension for its freshmen. The question is whether the course will be effective. Will the mean reading level for freshmen who are given the course be higher than that of those who are not given the course?

One way to study this question would be to select a group of freshmen, give them the course, and then compare their scores on a reading exam with the scores of a group of freshmen who have not taken the course. The two samples of scores in this case would be called **independent.** Another method would be to select one group of students and compare their scores on a reading test *before* they take the course with their scores on a test *after* they take the course. In this case the two samples of scores would be called **dependent.**

It is not always feasible to obtain dependent samples, but when we can obtain them, there are certain advantages. Dependent samples tend to reduce the effects of variability between the elements in a sample that can confound our results. For example, suppose we obtain independent samples of students, and one of the students who did not take the course has an extremely high level of reading comprehension. If we look at the sample means of the scores for each group, this student's score might distort our results, thus making the sample mean for the students who have not taken the course unrealistically high. However, suppose we have dependent samples. Here we would have a "before" and "after" score for each student. If a student in the study has a tendency to score high, this should appear in both samples and therefore will not distort just one of the samples. In other words, things will balance out; the effects of variability among the subjects in the sample will be reduced.

Definition Two samples are *independent* if the data values obtained from one are unrelated to the values from the other. The samples are *dependent* if each data value from one sample is paired in a natural way with a data value from the other sample.

When conducting tests for means in this chapter involving small sample sizes (less than 30), it will be necessary that the population be approximately normal. That is, it should have a roughly mound-shaped distribution. We will assume for the examples and exercises of this chapter that this requirement is met.

Inference from Dependent Samples

Suppose that we are interested in comparing the means μ_1 and μ_2 of populations 1 and 2, whose data values are represented by the symbols x_1 and x_2, respectively. Further, assume that the samples we obtain to investigate μ_1 and μ_2 are dependent. For example, an experimental automobile emission-control system was developed, and we wish to find out if it will increase gas mileage. Let μ_1 = mean gas mileage for cars using the experimental emission controls and μ_2 = mean gas mileage for cars with the standard emission controls. We are interested in whether $\mu_1 > \mu_2$. To compare the two emission-control systems, eight cars are selected. Each car is driven using one system and then the other. The gas mileage in each case is recorded in miles per gallon. The results are listed in Table 9.4.

We should look at the difference

$$d = x_1 - x_2$$

for each car. As one might expect, the mean of the difference is the difference of the means

$$\mu_d = \mu_1 - \mu_2$$

TABLE 9.4

	Cars							
	1	2	3	4	5	6	7	8
x_1 (mileage with experimental emission controls)	17	23	27	14	28	21	29	13
x_2 (mileage with standard emission controls)	9	17	21	16	22	17	25	13
$d = x_1 - x_2$ (difference)	8	6	6	−2	6	4	4	0

This is sometimes called the **mean difference.** We are asking if $\mu_1 > \mu_2$. This is the same as $\mu_d > 0$. Therefore the hypotheses we wish to investigate are

$$H_0: \mu_d = 0$$
$$H_a: \mu_d > 0$$

But this is a hypothesis test for a single mean, μ_d. We have investigated this subject in Chapters 7 and 8 (the only difference being that we used the symbol x for the random variable instead of d).

When investigating two population means using dependent samples, let

$$d = x_1 - x_2$$
$$\mu_d = \mu_1 - \mu_2$$

We can then conduct our investigation by studying μ_d using the methods developed in Chapters 7 and 8 to study a single mean.

For example, suppose that the sample size n is small (less than 30). If d is approximately normal, we can use a t test. The test statistic to be used would be

$$t = \frac{\bar{d} - \mu_d}{s_d/\sqrt{n}}$$

where $\bar{d}$ is the sample mean of the differences and s_d is the sample standard deviation of the differences. The statistic t will have Student's t distribution with degrees of freedom, df $= n - 1$.

Example 9.4

Do the data of Table 9.4 provide sufficient evidence to conclude that the experimental emission-control system will increase gas mileage? Use the 5% level of significance.

Solution

1. Hypotheses: We have seen that an increase in gas mileage will mean $\mu_1 > \mu_2$, which is the same as $\mu_d > 0$. Our hypotheses are

$$H_0: \mu_d = 0$$
$$H_a: \mu_d > 0$$

2. Level of significance: $\alpha = .05$.

3. Test statistic and its observed value: From Table 9.4, the sample values of d are 8, 6, 6, -2, 6, 4, 4, and 0. Since the sample size is small, we will use the test statistic

$$t = \frac{\bar{d} - \mu_d}{s_d/\sqrt{n}}$$

We need the values of $\bar{d}$ and s_d. Table 9.5 will help us compute these values.

$$\bar{d} = \frac{\Sigma d}{n} = \frac{32}{8} = 4$$

$$s_d = \sqrt{\frac{n(\Sigma d^2) - (\Sigma d)^2}{n(n-1)}}$$

$$= \sqrt{\frac{8(208) - (32)^2}{(8)(7)}} \doteq 3.38$$

TABLE 9.5

d	d^2	
8	64	
6	36	
6	36	
-2	4	
6	36	
4	16	
4	16	
0	0	
Sums: 32	208	$n = 8$

Therefore, the observed value of t is

$$t = \frac{4 - 0}{3.38/\sqrt{8}} \doteq 3.35 \qquad df = n - 1 = 7$$

4. Critical region: Since the alternate hypothesis is $\mu_d > 0$, we will perform a right-tailed test. Using Appendix Table B.4, we find $t(.05) = 1.895$. The critical region consists of values of $t \geq 1.895$. (See Figure 9.7.)

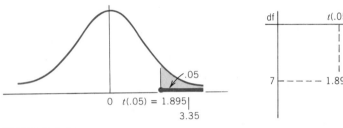

FIGURE 9.7

5. Decision: Since the observed value, $t = 3.35$, is in the critical region, we reject H_0. Therefore, it does appear that the experimental emission controls will increase gas mileage.

Confidence Intervals

We can find confidence intervals for the mean difference μ_d in the same way we found confidence intervals for a population mean in Chapters 7 and 8. For example, if we have a small sample, a $1 - \alpha$ confidence interval for μ_d is described by the endpoints:

$$\bar{d} \pm t\left(\frac{\alpha}{2}\right) \cdot \frac{s_d}{\sqrt{n}} \qquad df = n - 1$$

Example 9.5
Using the data in Table 9.4, find a 95% confidence interval for the mean difference in gas mileage (μ_d) between the two emission-control systems.

Solution
Since $1 - \alpha = .95$, $\alpha = .05$. Now $df = n - 1 = 8 - 1 = 7$. Using Appendix Table B.4, we see that $t(\alpha/2) = t(.025) = 2.365$. For the data in Table 9.4, $\bar{d} = 4$ and $s_d = 3.38$. Our $1 - \alpha$ confidence interval is described by the endpoints:

$$\bar{d} \pm t\left(\frac{\alpha}{2}\right) \cdot \frac{s_d}{\sqrt{n}}$$

Thus a 95% confidence interval is described by the endpoints:

$$4 \pm (2.365)\left(\frac{3.38}{\sqrt{8}}\right)$$

or

$$4 \pm 2.83$$

Therefore, we are 95% confident that

$$1.17 < \mu_d < 6.83$$

This means that we can be 95% confident that the new emission controls will increase gas mileage on the average between 1.17 and 6.83 miles per gallon, with a point estimate of 4 miles per gallon.

Exercises

9.10   For the following data, **(i)** test the given claims and **(ii)** indicate the possible type of error committed (I or II). Assume dependent samples (note that the mean difference is $\mu_d = \mu_1 - \mu_2$, where $d = x_1 - x_2$).

   (a)   Test the claim that the mean difference μ_d is not 0. Use a 5% level of significance.

x_1	x_2
37	34
42	36
34	38
37	43
40	37
43	38
42	35
41	36

   (b)   Test the claim that the mean difference μ_d is larger than 0. Use a 5% level of significance.

x_1	x_2
23	25
19	14
22	16
20	17
25	22
31	25

(c) Test the claim that the mean difference μ_d is less than 0. Use a 5% level of significance.

x_1	x_2
76	79
81	83
75	78
73	77
76	72
70	77
68	66

(d) Test the claim that the mean difference is 10. Use a 10% level of significance.

x_1	x_2
49	30
41	27
38	32
39	31
45	33
47	38
45	29
38	20
39	21

9.11 Twenty-four males aged 25–29 were selected from the Framingham Heart Study. Twelve were smokers and 12 were nonsmokers. The subjects were paired with one being a smoker and the other a nonsmoker. Otherwise, each pair was similar with regard to age and physical characteristics. Systolic blood pressure readings were as follows:

A Smokers	B Nonsmokers
122	114
146	134
120	114
114	116
124	138
126	110
118	112
128	116
130	132
134	126
116	108
130	116

Use a 5% level of significance to determine if the data indicate a difference in mean systolic blood pressure levels for the populations from which the two groups were selected.

9.12 The management of a large chain of stores wished to study whether advertising tends to increase sales of a product. Six pairs of stores, each pair of comparable size and comparable sales relative to the product, were selected. For each pair, one of the stores was randomly selected to advertise the product while the other store did not advertise the product. The following results represent the number of cases of the product sold over a week's period of time.

A Stores Advertising the Product	B Stores Not Advertising the Product
12	9
17	12
8	10
22	18
6	8
14	10

Is there sufficient evidence to suggest that the advertising program is effective? Use a 10% significance level.

9.13 A salesman for a shoe company claimed runners would record quicker times, on the average, with the company's brand of sneaker. A track coach decided to test the claim. The coach selected eight runners. Each runner ran two 100-yard dashes on different days. In one 100-yard dash, the runners wore the sneakers supplied by the school; in the other, the sneakers supplied by the salesman. Each runner was randomly assigned the sneakers to wear for the first run. Their times, measured in seconds, were as follows:

Race A With School's Sneakers	Race B With Shoe Company's Sneakers
11.4	10.8
12.5	12.3
10.8	10.7
11.7	12.0
10.9	10.6
11.8	11.5
12.2	12.1
11.7	11.2

(a) Find the P value. (You will not be able to find an exact P value, but indicate a possible range for the P value, such as $.05 < P < .10$ or $P > .25$.)

(b) For which of the following levels of significance would H_0 be rejected?
    (i) $\alpha = .10$?   (ii) $\alpha = .05$?   (iii) $\alpha = .01$?

[*Hint*: These can be answered using your answer in part (a).]

9.14   A teacher was interested in finding out whether a special study program would increase the scores of students on a national exam. Fourteen students were selected and paired according to IQ and scholastic performance. One student from each pair was randomly selected to participate in the special program, while the other student participated in the standard program. Both programs ended at the same time. Shortly thereafter, the students took the national exam. The results were

A Participated in the Special Program	B Participated in the Standard Program
65	60
83	79
94	92
72	73
78	75
82	80
67	68

Using a 5% significance level, is there sufficient evidence to indicate that the special study program is effective in raising the national exam scores, on the average?

9.15   Allied Foods, Inc. felt that a new recipe for chocolate cake mix would result in a thicker cake. Thirty-four cooks were asked to use the new recipe (A) and the old recipe (B). The difference (d) between thicknesses was recorded in each case ($d = x_A - x_B$). The results were: $\bar{d} = .15$ inches, $s_d = .09$ inches. At the 1% level of significance, does recipe A result in thicker cakes?

9.16   A random sample of 50 GMAT (Graduate Management Aptitude Test) scores gave a mean difference $\bar{d} = -1.12$ (d = verbal score − quantitative score) and standard deviation $s_d = 5.95$.

    (a)   With a 5% level of significance, is there sufficient evidence to indicate a difference between the mean verbal and quantitative scores?

    (b)   Construct a 95% confidence interval for $\mu_d = \mu_A - \mu_B$. Does the confidence interval support the conclusion reached in part (a)?

9.17   Use the data in the corresponding parts of Exercise 9.10 to construct confidence intervals for μ_d ($\mu_d = \mu_1 - \mu_2$, where $d = x_1 - x_2$). The percent confidence is as follows:

    (a)   95%.   **(b)**   90%.   **(c)**   90%.   **(d)**   90%.

9.18   Construct confidence intervals for $\mu_A - \mu_B$ in parts (a)–(c).

    (a)   Use the data in Exercise 9.12 to construct an 80% confidence interval.

    (b)   Use the data in Exercise 9.11 to construct a 95% confidence interval.

    (c)   Use the data in Exercise 9.14 to construct a 90% confidence interval.

9.19   Suppose we are testing the hypotheses: $H_0: \mu_A - \mu_B = 0$; $H_a: \mu_A - \mu_B > 0$ with a 5% level of significance. Two samples, each consisting of 6 pairs of data values, is given below. (*Note*: d is the difference between corresponding scores from A and B.)

A	32	29	29	30	31	29
B	28	25	24	27	28	24
d	4	4	5	3	3	5

A	32	29	29	30	31	29
B	34	20	21	38	20	23
d	−2	9	8	−8	11	6

Note that in both cases, $\bar{d} = \bar{x}_A - \bar{x}_B = 4$. One of the two data sets above leads to a rejection of the null hypothesis.

(a) Without testing, which of the two data sets leads to rejecting H_0? [*Hint*: A sketch of the data (perhaps a dot diagram) might be helpful.]

(b) Now complete the test using each data set and compare with part (a).

9.5 INFERENCE CONCERNING TWO POPULATION MEANS BASED ON INDEPENDENT SAMPLES— Large-Sample Case

Consider the following situation: A statistics instructor had two textbooks in mind (call them texts 1 and 2), one of which would be adopted by the mathematics department as the standard text for its introductory statistics course. The department wondered if there was any real difference in results between the two textbooks. In other words, if μ_1 is the mean numerical grade on a scale of 0 to 100 for the population of future students who may use textbook 1, and μ_2 is the mean grade for future students using textbook 2, will these means be equal or not? The instructor was teaching a large lecture class divided into two recitation sections, and decided to investigate this by using text 1 in section 1 and text 2 in section 2. Students were randomly selected for assignment to the two sections. These sections can be viewed as samples from the populations of interest. The results are summarized in Table 9.6.

We are interested in whether $\mu_1 = \mu_2$ or $\mu_1 \neq \mu_2$. This is equivalent to testing

$$H_0: \mu_1 - \mu_2 = 0$$
$$H_a: \mu_1 - \mu_2 \neq 0$$

TABLE 9.6
Mean Course Grades

Textbook	Number of Students	Sample Mean	Sample Standard Deviation
1	$n_1 = 35$	$\bar{x}_1 = 78$	$s_1 = 8$
2	$n_2 = 40$	$\bar{x}_2 = 75$	$s_2 = 6$

It would seem natural to look at the difference $\bar{x}_1 - \bar{x}_2$. We can think of $\bar{x}_1 - \bar{x}_2$ as a rough estimate for $\mu_1 - \mu_2$. If H_0 were true, $\bar{x}_1 - \bar{x}_2$ should not be too far away from 0. If the value of $\bar{x}_1 - \bar{x}_2$ is so far from 0 that such a value would be unlikely (if H_0 were true), we would reject H_0. To find out what values of $\bar{x}_1 - \bar{x}_2$ are likely (i.e., consistent with H_0) and what values are not likely, we need to know the probability distribution of $\bar{x}_1 - \bar{x}_2$.

Now $\bar{x}_1 - \bar{x}_2$ is a random variable, and it can be shown that when n_1 and n_2 are at least 30, it is approximately normal. Further, the mean and standard deviation of $\bar{x}_1 - \bar{x}_2$ are given by the formulas

$$\mu_{\bar{x}_1 - \bar{x}_2} = \mu_1 - \mu_2 \qquad \sigma_{\bar{x}_1 - \bar{x}_2} = \sqrt{\frac{\sigma_1^2}{n_1} + \frac{\sigma_2^2}{n_2}}$$

It follows that

$$z = \frac{(\bar{x}_1 - \bar{x}_2) - \mu_{\bar{x}_1 - \bar{x}_2}}{\sigma_{\bar{x}_1 - \bar{x}_2}} = \frac{(\bar{x}_1 - \bar{x}_2) - (\mu_1 - \mu_2)}{\sqrt{(\sigma_1^2/n_1) + (\sigma_2^2/n_2)}}$$

is approximately standard normal. We can use this to test the hypotheses given here in much the same way as we did for a single mean. We would calculate the value of z, substituting the value of $\mu_1 - \mu_2$ appearing in H_0 (0 in the above case). Values of z near 0 favor H_0. Values of z far to the right or left of 0 favor H_a.

Now we complete the test for the data in Table 9.6 using the 5% level of significance.

1. Hypotheses:

$$H_0: \mu_1 - \mu_2 = 0$$
$$H_a: \mu_1 - \mu_2 \neq 0$$

2. Level of significance: $\alpha = .05$.

3. Test statistic and its observed value: Note that σ_1 and σ_2 are unknown but we can use s_1 and s_2 as estimates, since n_1 and n_2 are large.

$$z = \frac{(\bar{x}_1 - \bar{x}_2) - (\mu_1 - \mu_2)}{\sqrt{(\sigma_1^2/n_1) + (\sigma_2^2/n_2)}} = \frac{(78 - 75) - 0}{\sqrt{\dfrac{8^2}{35} + \dfrac{6^2}{40}}} \doteq 1.82$$

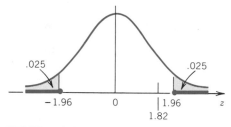

FIGURE 9.8

4. Critical region: Values of z far to the right or left of 0 favor H_a. The critical region consists of two tails, each of size $\alpha/2 = .025$. From Appendix Table B.3, $z(.025) = 1.96$. Therefore, the critical region consists of values of $z \geq 1.96$ or ≤ -1.96. (See Figure 9.8.)

5. Decision: The observed value of z is 1.82. This is not in the critical region. Hence we fail to reject H_0. This means that there is not enough evidence to conclude that the texts produce different results.

Statistic for Tests Concerning Two Population Means μ_1 and μ_2—Large-Sample Case   When both sample sizes n_1, n_2 are large (at least 30), use the test statistic

$$z = \frac{(\bar{x}_1 - \bar{x}_2) - (\mu_1 - \mu_2)}{\sqrt{(\sigma_1^2/n_1) + (\sigma_2^2/n_2)}}$$

This is approximately standard normal. When σ_1, σ_2 are unknown, use s_1 and s_2 as estimates.

The term $\mu_1 - \mu_2$ appears in the above statistic. As usual we will use the value of $\mu_1 - \mu_2$ given in the null hypothesis, since the null hypothesis is assumed true until there is strong evidence to the contrary. In general, if the null hypothesis were $H_0: \mu_1 - \mu_2 = k$, we substitute the value of k for $\mu_1 - \mu_2$. (The value of k need not always be 0. For example, if we were testing the claim that μ_1 was 10 more than μ_2, the null hypothesis would be $\mu_1 - \mu_2 = 10$.) To complete the test at a given level of significance, we evaluate the test statistic and conduct a z test. The critical region for the test looks the same as the z test for a single population mean. (See Section 7.4.)

In the example discussed in the beginning of this section, the alternate hypothesis was $\mu_1 \neq \mu_2$ and we did a two-tailed test. We can formulate a rule for determining critical regions in tests for two means similar to the one we used for single-mean tests. If we keep the order of μ_1, μ_2 appearing in H_a the same as the order of $\bar{x}_1$, $\bar{x}_2$ in the test statistic, we can use the following rule:

> H_a: $\mu_1 \neq \mu_2$ implies a two-tailed test
> H_a: $\mu_1 < \mu_2$ implies a left-tailed test
> H_a: $\mu_1 > \mu_2$ implies a right-tailed test

Confidence Intervals for $\mu_1 - \mu_2$

In Chapter 7 we discussed a statistic to be used for testing hypotheses about a single mean μ when the sample size is large ($n \geq 30$). This was the same statistic we used to develop a $1 - \alpha$ confidence interval for μ. (See Section 7.2.) The statistic and the confidence interval are as follows:

Test Statistic	**$1 - \alpha$ Confidence Interval for μ**
$z = \dfrac{\bar{x} - \mu}{\sigma/\sqrt{n}}$	$\bar{x} \pm z\left(\dfrac{\alpha}{2}\right) \cdot \dfrac{\sigma}{\sqrt{n}}$

By examining these relationships, we ought to be able to guess a confidence interval for $\mu_1 - \mu_2$, knowing the test statistic for testing hypotheses concerning $\mu_1 - \mu_2$. For example, we saw that the test statistic for investigating $\mu_1 - \mu_2$, when the sample sizes are large and the samples are independent, is

$$z = \frac{(\bar{x}_1 - \bar{x}_2) - (\mu_1 - \mu_2)}{\sqrt{(\sigma_1^2/n_1) + (\sigma_2^2/n_2)}}$$

Reasoning by analogy, a $1 - \alpha$ confidence interval for $\mu_1 - \mu_2$ ought to be (and indeed is)

$$(\bar{x}_1 - \bar{x}_2) \pm z\left(\frac{\alpha}{2}\right) \cdot \sqrt{\frac{\sigma_1^2}{n_1} + \frac{\sigma_2^2}{n_2}}$$

Example 9.6

A manufacturer of small engines concerned with the issue of noise pollution developed a new engine that would hopefully be quieter than the standard model. Forty-one of the new models were tested for noise level and compared with 65 of the standard model. The results are given in Table 9.7 (with noise measured in decibels).

(a) Find a 95% confidence interval for the difference in mean decibel levels $\mu_1 - \mu_2$, where μ_1 is the mean decibel level for the standard engine and μ_2 is the mean decibel level for the new engine.

(b) Do the population means appear to be different?

TABLE 9.7

Sample	Sample Size	Sample Mean (Decibels)	Sample Standard Deviation (Decibels)
Standard model	65	84	11.6
New model	41	72	9.2

Solution

(a) The sample sizes are large, so we use

$$(\bar{x}_1 - \bar{x}_2) \pm z\left(\frac{\alpha}{2}\right) \cdot \sqrt{\frac{\sigma_1^2}{n_1} + \frac{\sigma_2^2}{n_2}}$$

Now $1 - \alpha = .95$, so $\alpha = .05$. $z(\alpha/2) = z(.025) = 1.96$. We will use s_1 and s_2 for σ_1 and σ_2. We get

$$(84 - 72) \pm (1.96) \cdot \sqrt{\frac{(11.6)^2}{65} + \frac{(9.2)^2}{41}}$$

or

$$12 \pm 3.99$$

So we are 95% sure that the difference $\mu_1 - \mu_2$ is between 8.01 and 15.99 decibels.

(b) Since 0 is not contained in the confidence interval for $\mu_1 - \mu_2$, we are 95% sure that $\mu_1 - \mu_2 \neq 0$. Hence, $\mu_1 \neq \mu_2$. In fact we can say more. Since the confidence interval consists of positive numbers, $\mu_1 - \mu_2 > 0$, or $\mu_1 > \mu_2$.

Exercises

9.20 It was claimed that the mean of population A was not the same as the mean of population B. Assume independent samples and consider the following sample information:

	n	$\bar{x}$	s^2
A	40	175	360
B	50	165	350

(a) Complete the test at the 5% level of significance.

(b) Suppose that you had used the 1% level of significance in part (a). Without calculating, answer the following:

　(i) Would the value of the test statistic change?

　(ii) Would the critical values change?

　(iii) Might the decision you reached in part (a) change?

(c) Now complete the test at the 1% level of significance and compare with part (b).

9.21 It was claimed that the mean of population A was larger than the mean of population B. Assume independent samples and consider the following sample information:

	n	$\bar{x}$	s^2
A	35	400	210
B	35	396	105

(a) Complete the test at the 5% level of significance.

(b) Suppose that you had used the 10% level of significance in part (a). Without calculating, answer the following:

(i) Would the value of the test statistic change?

(ii) Would the critical value change?

(iii) Might the decision you reached in part (a) change?

(c) Now complete the test at the 10% level of significance and compare with part (b).

9.22 The personnel officer of a large corporation claimed that college graduates applying for jobs with the firm in the current year tended to have higher grade-point averages than those applying in the previous year. Samples from the groups of applicants gave the following results:

	n	$\bar{x}$	s
Preceding year (A)	52	2.80	.50
Current year (B)	60	2.98	.40

Using a 5% level of significance, is there sufficient evidence to justify the claim?

9.23 Business schools A and B reported the following summary of GMAT (Graduate Management Aptitude Test) verbal scores:

	n	$\bar{x}$	s^2
A	201	34.75	48.59
B	115	33.74	30.68

Using a 5% level of significance, do the data support the belief that the (population) mean score for school B is less than that of A?

(a) Use the classical approach. **(b)** Use the P value approach.

9.24 **(a)** A biologist suspected that in the 20–24 age group, males have a higher mean systolic blood pressure than females. Independent random samples of males and females were selected from the Framingham Heart Study. The data are:

	n	$\bar{x}$	s
A (Males)	31	125	13.9
B (Females)	41	117	12.1

Using a 1% level of significance, is there sufficient evidence to justify the biologist's suspicions? Use (i) the classical approach and (ii) the P value approach.

(b) The biologist also suspected that in the 20–24 age group, males have a higher mean diastolic blood pressure than females. The data are:

	n	$\bar{x}$	s
A (Males)	45	75	10.1
B (Females)	45	70	9.8

Using a 1% level of significance, is there sufficient evidence to justify the biologist's suspicions? Use (i) the classical approach and (ii) the P value approach.

9.25 Two types of sports cars were compared for acceleration rates. Forty test runs were done for each car, and elapsed time from 0 to 60 miles per hour was recorded for each run. The results in seconds were

	$\bar{x}$	s
Car A	7.4	1.5
Car B	7.1	1.8

At the 1% level of significance, does there appear to be a difference in mean elapsed time for the two types of vehicles?

9.26 A manager of a boat line that services a resort island wanted to know if more passengers, on a per trip average, traveled on the 8 A.M. or the 10 A.M. boat. Independent random samples were obtained. Note that n refers to the number of trips.

	n	$\bar{x}$	s^2
A (8 A.M. boat)	35	820	4900
B (10 A.M. boat)	40	850	6900

Using a 1% level of significance, what can the manager conclude?

9.27 Construct confidence intervals for $\mu_A - \mu_B$ in parts (a)–(c).

(a) Use the data in Exercise 9.22 to construct a 90% confidence interval.

(b) Use the data in Exercise 9.24 (a) to construct a 98% confidence interval.

(c) Use the data in Exercise 9.26 to construct a 95% confidence interval.

9.28 An administrator at a large university stated there was a difference in the mean grade-point averages of graduating males and females. Independent random samples of graduating males and females gave the following information:

	n	$\bar{x}$	s^2
A (Males)	45	2.10	.64
B (Females)	50	2.45	.70

(a)   Construct a 95% confidence interval for the difference $(\mu_A - \mu_B)$.

(b)   Using the confidence interval obtained in part (a), do the data support the administrator's belief? (*Hint*: Does the confidence interval contain 0?)

9.6 INFERENCE CONCERNING TWO POPULATION MEANS BASED ON INDEPENDENT SAMPLES— Small-Sample Case

Suppose that either or both of the sample sizes are small (less than 30). In this case we must require that the populations (of x_1 and x_2 values) be approximately normal for what follows to be correct. First, if σ_1 and σ_2 are known, we may use the very same statistic as given in the previous section. However, if σ_1 and/or σ_2 are unknown, we cannot reliably estimate them by s_1 and s_2 when the sample sizes are small. In practice we usually do not know the values of σ_1 and σ_2, so let us concentrate on this situation. The test statistic to be used depends on whether $\sigma_1 = \sigma_2$ or not. Sometimes a researcher will have a feel for whether this is true or not. However, an F test, discussed in Section 9.2, is often employed to investigate this situation.* (But keep in mind that if the F test is used, the populations should be normal.)

(1)   If $\sigma_1 = \sigma_2$, use the test statistic

$$t = \frac{(\bar{x}_1 - \bar{x}_2) - (\mu_1 - \mu_2)}{s_p\sqrt{(1/n_1) + (1/n_2)}}$$

where

$$s_p = \sqrt{\frac{(n_1 - 1)s_1^2 + (n_2 - 1)s_2^2}{n_1 + n_2 - 2}}$$

This is called the "pooled sample standard deviation." The statistic t has (approximately) a Student's t distribution with

$$df = n_1 + n_2 - 2$$

---

*When an F test is employed in this connection, we will follow the practice in this book of conducting a two-tailed test of the hypothesis $\sigma_1^2 = \sigma_2^2$ against $\sigma_1^2 \neq \sigma_2^2$ at the same level of significance that the test of μ_1, μ_2 uses.

(2) If $\sigma_1 \neq \sigma_2$, use the test statistic

$$t = \frac{(\bar{x}_1 - \bar{x}_2) - (\mu_1 - \mu_2)}{\sqrt{(s_1^2/n_1) + (s_2^2/n_2)}}$$

This has (approximately) a Student's t distribution with degrees of freedom, df, being the smaller of $n_1 - 1$ or $n_2 - 1$. This is called the *Behrens–Fisher statistic*.

Note: As usual, the value of $\mu_1 - \mu_2$ in these formulas is specified in the null hypothesis.

Example 9.7
In Example 9.2 we discussed some data concerning high-density lipoprotein (HDL) levels in the blood for young male elite runners versus young male nonrunners. Increased levels of HDL are thought to be associated with lower risk of coronary heart disease. It is felt that runners have increased HDL levels. Refer to the data given in Example 9.2 and test the claim that young male elite runners have a higher (population) mean HDL level than young male nonrunners. Use the 5% level of significance.

Solution
We summarize the sample HDL data from Example 9.2:

Elite Runners	Nonrunners
$n_1 = 20$	$n_2 = 72$
$\bar{x}_1 = 56$	$\bar{x}_2 = 49$
$s_1 = 12.1$	$s_2 = 10.5$

Since one of the sample sizes is less than 30, we must know whether $\sigma_1 = \sigma_2$ or not before picking a test statistic. But we have seen in Example 9.2 that we failed to reject the null hypothesis that $\sigma_1^2 = \sigma_2^2$. Hence we will assume that $\sigma_1 = \sigma_2$. This means that to conduct a test concerning population means, we will use the test statistic given in item 1 (p. 426).

1. Hypotheses: We will let μ_1 and μ_2 represent the population means of the runners and nonrunners, respectively. The claim is that $\mu_1 > \mu_2$. This is the alternate hypothesis. So the hypotheses are

$$H_0: \mu_1 = \mu_2 \quad (\text{or } \mu_1 - \mu_2 = 0)$$
$$H_a: \mu_1 > \mu_2 \quad (\text{or } \mu_1 - \mu_2 > 0)$$

2. Level of significance: $\alpha = .05$.

3. Test statistic and its observed value: We have said that since we concluded in Example 9.2 that $\sigma_1 = \sigma_2$, we will use the test statistic in item 1 mentioned above. We first calculate the pooled standard deviation.

$$s_p = \sqrt{\frac{(n_1 - 1)s_1^2 + (n_2 - 1)s_2^2}{n_1 + n_2 - 2}}$$

$$= \sqrt{\frac{(19)(12.1)^2 + (71)(10.5)^2}{20 + 72 - 2}} \doteq 10.857$$

$$t = \frac{(\bar{x}_1 - \bar{x}_2) - (\mu_1 - \mu_2)}{s_p\sqrt{(1/n_1) + (1/n_2)}} = \frac{(56 - 49) - 0}{10.857\sqrt{\dfrac{1}{20} + \dfrac{1}{72}}} = \frac{7}{2.744} \doteq 2.55$$

$$df = n_1 + n_2 - 2 = 20 + 72 - 2 = 90$$

4. Critical region: Since the alternate hypothesis is $\mu_1 > \mu_2$, we conduct a right-tailed test. (Values of $\bar{x}_1$ much larger than $\bar{x}_2$ favor H_a. These correspond to large values of z.) We use Appendix Table B.4 to find the critical value $t(.05)$ when $df = 90$. Recall that when $df > 29$ we use the last line of Appendix Table B.4. This gives $t(.05) = 1.645$. The critical region is shown in Figure 9.9.

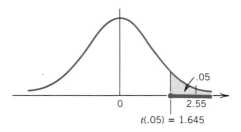

FIGURE 9.9

5. Decision: Since the observed value ($t = 2.55$) is in the critical region, we reject H_0. This means that there is sufficient evidence to conclude that $\mu_1 > \mu_2$.

Example 9.8
Researchers Morganroth and Maron have studied various heart abnormalities that occur in male athletes. One characteristic studied was left ventricular end diastolic volume (the volume of the left lower chamber of the heart when it is filled with blood). Do the data in Table 9.8 show a significant difference in volume between runners and nonrunners (at the 5% level of significance)? The measurements are in millimeters.

TABLE 9.8

Samples	n	$\bar{x}$	s
Runners	15	160.1	33.1
Nonrunners	16	100.8	17.6

The data are estimated from a graph in Morganroth, J., and B. Maron, "The Athletes Heart Syndrome: A New Perspective," in *The Long Distance Runner*, ed. P. Milvy (New York: Urizen Books, 1977), p. 218.

Solution

There will be a "significant difference" in volume between runners and non-runners, if the sample means are far enough apart to provide evidence that the (appropriate) populations from which the two groups were drawn have different population means.

Now the sample sizes are small, so we should know whether the population variances are equal or not. We will do an F test to investigate this. According to our labeling convention, $s_1 = 33.1$, and $s_2 = 17.6$. (Therefore, the population parameters μ_1, σ_1 will refer to runners and μ_2, σ_2 will refer to nonrunners.)

F Test

1. Hypotheses:

$$H_0: \sigma_1^2 = \sigma_2^2$$
$$H_a: \sigma_1^2 \neq \sigma_2^2$$

2. Level of significance: We will use the same level given to test the means, $\alpha = .05$.

3. Test statistic and its observed value:

$$F = \frac{s_1^2}{s_2^2} = \frac{(33.1)^2}{(17.6)^2} \doteq 3.54 \qquad df = (14, 15)$$

4. Critical region: Since H_a is $\sigma_1^2 \neq \sigma_2^2$, we do a two-tailed test. The critical region is shown in Figure 9.10. Since df $= (14, 15)$ is not in the table, we

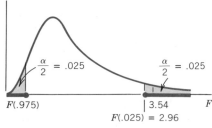
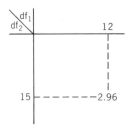

FIGURE 9.10

use df = (12, 15) when finding the critical value. When a value for degrees of freedom does not appear in the table, it is customary to use the next smaller value of degrees of freedom in the table. From Appendix Table B.6, $F(.025) = 2.96$.

5. Decision: The observed value ($F = 3.54$) is in the critical region; thus we reject H_0. Therefore, we conclude that $\sigma_1^2 \neq \sigma_2^2$.

Now we know what statistic to use to investigate the means μ_1, μ_2. We will use the statistic given in item 2 (p. 427).

t Test

1. Hypotheses: To say that the population means are different is to say $\mu_1 \neq \mu_2$; this is H_a. Thus

$$H_0: \mu_1 = \mu_2 \qquad (\text{or } \mu_1 - \mu_2 = 0)$$
$$H_a: \mu_1 \neq \mu_2 \qquad (\text{or } \mu_1 - \mu_2 \neq 0)$$

2. Level of significance: $\alpha = .05$.
3. Test statistic and its observed value:

$$t = \frac{(\bar{x}_1 - \bar{x}_2) - (\mu_1 - \mu_2)}{\sqrt{(s_1^2/n_1) + (s_2^2/n_2)}} = \frac{(160.1 - 100.8) - 0}{\sqrt{(33.1)^2/15 + (17.6)^2/16}} \doteq 6.17$$

df = the smaller of $n_1 - 1$, $n_2 - 1 = 14$

4. Critical region: Since the alternate hypothesis is H_a: $\mu_1 \neq \mu_2$, we do a two-tailed test. Critical values are obtained from Appendix Table B.4. See Figure 9.11.

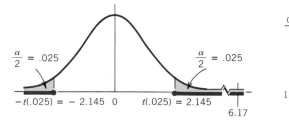

FIGURE 9.11

5. Decision: The observed value of t is 6.17. This is in the critical region; thus we reject H_0. Therefore, μ_1 and μ_2 do appear to be different. In fact, it would appear from the data that runners tend to have larger heart volume than nonrunners.

Note: At one time some insurance companies were reluctant to insure runners and other athletes because of their larger heart volume. This is no longer regarded as a pathological condition in athletes. But some questions still remain about the long-range effects of heart abnormalities in athletes.

In the above problem we used the minimum of $n_1 - 1$ and $n_2 - 1$ as the degrees of freedom, and will continue to do so in this book. Some statisticians, however, think this approach is too conservative. It gives a rather small value for the degrees of freedom. The smaller the degrees of freedom, the larger the critical value of t, meaning the observed value of t must be quite large for us to reject H_0. This may give too much of an advantage to H_0. An alternative value for degrees of freedom preferred by some statisticians is discussed in the Minitab section (9.9).

Confidence Intervals for $\mu_1 - \mu_2$

Suppose n_1 and/or n_2 are less than 30. In this case we require that the populations (of x_1 and x_2 values) be approximately normal. To find a $1 - \alpha$ confidence interval, we proceed as follows:

1. If $\sigma_1 = \sigma_2$, we use

$$(\bar{x}_1 - \bar{x}_2) \pm t\left(\frac{\alpha}{2}\right) \cdot s_p \cdot \sqrt{\frac{1}{n_1} + \frac{1}{n_2}}$$

where

$$df = n_1 + n_2 - 2$$

and

$$s_p = \sqrt{\frac{(n_1 - 1)s_1^2 + (n_2 - 1)s_2^2}{n_1 + n_2 - 2}}$$

2. If $\sigma_1 \neq \sigma_2$, we use

$$(\bar{x}_1 - \bar{x}_2) \pm t\left(\frac{\alpha}{2}\right) \cdot \sqrt{\frac{s_1^2}{n_1} + \frac{s_2^2}{n_2}}$$

where

$$df = \text{the smaller of } n_1 - 1, n_2 - 1$$

Note: It may be necessary to employ a two-tailed F test to investigate whether $\sigma_1 = \sigma_2$ or not, in which case the populations should be normal. When employing an F test, we will use a level of significance α if the level of confidence is $1 - \alpha$.

Example 9.9
Using the data in Example 9.7, find a 95% confidence interval for the difference in mean HDL level for elite runners and nonrunners.

Solution
Since we concluded in Example 9.2 that $\sigma_1 = \sigma_2$, the confidence interval for $\mu_1 - \mu_2$ can be described by

$$(\bar{x}_1 - \bar{x}_2) \pm t\left(\frac{\alpha}{2}\right) \cdot s_p \cdot \sqrt{\frac{1}{n_1} + \frac{1}{n_2}}$$

Now

$$\text{df} = n_1 + n_2 - 2 = 20 + 72 - 2 = 90$$
$$1 - \alpha = .95$$
$$\alpha = .05$$
$$t\left(\frac{\alpha}{2}\right) = t(.025) = 1.96$$

From Example 9.7, we found that $s_p = 10.857$. So the confidence interval is

$$(56 - 49) \pm (1.96) \cdot (10.857) \cdot \sqrt{\frac{1}{20} + \frac{1}{72}}$$
$$7 \pm 5.38$$

or

$$1.62 \text{ to } 12.38$$

Exercises
9.29 **(a)** Do the following data support the belief that the mean of population B is 10 more than the mean of population A?

	n	$\bar{x}$	s^2
A	6	120	100
B	10	135	81

Assume $\sigma_A^2 = \sigma_B^2$ and complete the test with a 5% level of significance.

(b) Do the following data support the belief that the mean of population A is less than the mean of population B?

	n	$\bar{x}$	s^2
A	10	180	70
B	7	200	340

Assume $\sigma_A^2 \neq \sigma_B^2$ and complete the test using the 10% level of significance.

9.30 The mayor of city A claimed that there was no difference in mean air quality between cities A and B, based on a measurement of air quality. Independent samples for each city gave the following data:

	n	$\bar{x}$	s^2
City A	11	3.8	.39
City B	11	3.5	.10

Using a 5% level of significance, test the claim.

9.31 A new actuarial test (A) was developed to possibly replace test B. An actuary claimed that the mean time to complete the test was more than the mean time required to complete test B. To test the claim, seven people were selected to take test A and nine people selected to take test B. The summary data for completion time (in minutes) are as follows:

	$\bar{x}$	s^2
Test A	166	240
Test B	150	205

Complete the test with a 10% level of significance.

9.32 A corporation, planning to expand its production facilities, did a study of executive salaries in two regions from which it might draw executives: Middle Atlantic and New England. The following sample data were obtained:

	n	$\bar{x}$	s
A: Middle Atlantic	27	$121,700	$8300
B: New England	22	$115,500	$7500

Assume $\sigma_A = \sigma_B$.

(a) Find a 95% confidence interval for $\mu_A - \mu_B$.

(b) Based on your answer in part (a), do you feel that there is a difference in mean salary level for the two regions?

9.33 The personnel office of an auto company felt that if workers were rotated to different positions on an assembly line, it would improve their psychological outlook. A sample of such workers was compared to a control group of workers who stayed at the same position. The following data relate to depression scores on a psychological inventory called the Depression Adjective Checklist (DACL):

	n	$\bar{x}$	s^2
A: Rotated Group	18	4.8	4.3
B: Control Group	16	6.9	12.7

Test the claim that rotated workers have lower mean depression level than non-rotated workers. Use the 5% level of significance and assume $\sigma_A^2 \neq \sigma_B^2$.

(a) Use the classical approach.   **(b)** Use the P value approach.

9.34 A real estate agent said the percentage of income needed to maintain two homes was, on average, less for residents of town A than for town B. Samples of residents from both towns gave the following data:

A	B
23.2	29.1
21.4	35.3
24.5	28.7
27.0	27.7
28.5	

Assuming $\sigma_A^2 = \sigma_B^2$, is there sufficient evidence to support the agent's belief? Use a 5% level of significance.

9.35 Assume the following data were obtained with independent random samples from populations A and B. Assume that $\sigma_A^2 = \sigma_B^2$.

A	B
27.2	28.5
19.7	27.7
14.6	30.9
20.4	23.1
23.1	24.8
	24.0
	22.3

Test the hypothesis that the means of populations A and B are different using a 5% level of significance.

9.36 **(a)** An automobile company wished to compare the mean performance of two brands of shock absorbers: brands A and B. Of interest was whether brand B lasted longer than brand A, on the average. A test was conducted to compare the mean life in miles of the two brands. Using independent samples, the following information was obtained.

	n	$\bar{x}$	s^2
A	13	25,200	80,000
B	13	25,375	40,000

Is there sufficient evidence to indicate that brand B lasts longer, on the average? Use a 5% level of significance.

(b) Suppose that the data had been

	n	$\bar{x}$	s^2
A	13	25,200	80,000
B	13	25,375	80,000

Using a 5% level of significance, what do you now conclude? Compare with part (a). Comment.

9.37 A physical education instructor believed the mean height of male runners is 8 centimeters more than the mean height of female runners. In a study, independent samples of 10 male and 10 female runners yielded the following data pertaining to the runners' heights in centimeters (*Source*: Daniels et al., 1977, p. 142):

	n	$\bar{x}$	s
A (Males)	10	179.3	6.1
B (Females)	10	166.2	4.7

Complete the test using a 2% level of significance.

9.38 Consider the following dialogue between two students in a statistics class:

Joe: I thought our textbook said that dependent sampling is often more desirable than independent sampling (when possible).

Mary: It did.

Joe: I take it that this implies that dependent sampling might make it easier to reject a false null hypothesis.

Mary: So do I.

Joe: Well, how do you account for this? Dependent samples of size n result in degrees of freedom $n - 1$, while independent samples of size n (each) result in degrees of freedom $n + n - 2 = 2n - 2 = 2(n - 1)$ if $\sigma_1 = \sigma_2$.

Mary: So what?

Joe: Critical values in the t table get smaller as n increases. Therefore, it will be easier to reject H_0.

Mary: Maybe. But didn't the authors say something about sample variability?

Now repeat Exercise 9.12 as if the samples were independent and comment on the above dialogue.

9.39 Use the data in the exercises named below to construct confidence intervals for the difference $(\mu_A - \mu_B)$. The percent confidence is as follows:

(a) 95% [Exercise 9.29 (b). Assume $\sigma_A \neq \sigma_B$.]

(b) 90% (Exercise 9.31).

(c) 98% (Exercise 9.37).

9.7

INFERENCE CONCERNING TWO POPULATION PROPORTIONS

Sometimes we are interested in comparing two (unknown) population proportions. For example, we may wish to know if in a voting district the proportion of men favoring an Equal Rights Amendment (E.R.A.) is less than the proportion of women favoring such an amendment. To study a topic such as this, we might obtain random samples from each population and record the number in each sample who favor E.R.A. Obtaining each sample is a binomial process, so the number in a sample who favor E.R.A. is the number of successes in a binomial experiment.

In general, when studying (unknown) population proportions p_1, p_2 for populations 1 and 2, respectively, we obtain independent samples of size n_1, n_2 and record the number of successes x_1, x_2 from populations 1 and 2, respectively. Now we can estimate p_1 and p_2 by

$$\hat{p}_1 = \frac{x_1}{n_1} \qquad \hat{p}_2 = \frac{x_2}{n_2}$$

For example, if 100 women and 100 men are interviewed and 80 women and 75 men favor an E.R.A., then

$$\hat{p}_1 = \frac{80}{100} = .80 \qquad \hat{p}_2 = \frac{75}{100} = .75$$

We can formulate hypotheses concerning p_1, p_2 as statements involving $p_1 - p_2$. For example, if H_0 is $p_1 = p_2$, we can rewrite this as $p_1 - p_2 = 0$. We can think of $\hat{p}_1 - \hat{p}_2$ as a rough estimate for $p_1 - p_2$. Hence if H_0 is true, $\hat{p}_1 - \hat{p}_2$ should not be too far from 0. In order to know what values of $\hat{p}_1 - \hat{p}_2$ would be reasonable or likely if H_0 is indeed true, we need to know something about the probability distribution of $\hat{p}_1 - \hat{p}_2$.

It can be shown that when n_1 and n_2 are sufficiently large*, the variable quantity $\hat{p}_1 - \hat{p}_2$ is approximately normal. Further, the mean and standard deviation are given by the formulas

$$\mu_{\hat{p}_1 - \hat{p}_2} = p_1 - p_2 \qquad \sigma_{\hat{p}_1 - \hat{p}_2} = \sqrt{\frac{p_1 q_1}{n_1} + \frac{p_2 q_2}{n_2}}$$

Thus

$$z = \frac{(\hat{p}_1 - \hat{p}_2) - (p_1 - p_2)}{\sqrt{(p_1 q_1 / n_1) + (p_2 q_2 / n_2)}}$$

will possess an approximately standard normal distribution.

When testing hypotheses concerning $p_1 - p_2$, we can use the above expression as a test statistic. The value of $p_1 - p_2$ is obtained from the null hypothesis H_0. The value of $\sigma_{\hat{p}_1 - \hat{p}_2}$ (the denominator) must be estimated from the sample

*As a rule of thumb, many statisticians will regard n_1 as sufficiently large if it is large enough so that the number of successes x_1 and the number of failures $n_1 - x_1$ are both at least 5. A similar remark applies to n_2.

data (since we do not know the values of the population parameters p_1, q_1, p_2, and q_2). The form of $\sigma_{\hat{p}_1 - \hat{p}_2}$ depends on the form of the null hypothesis.

1. Suppose that the null hypothesis is H_0: $p_1 = p_2$ (or $p_1 - p_2 = 0$). Assuming that H_0 is true, we will use the symbol p to represent the common value of p_1 and p_2; then $p = p_1 = p_2$. We pool the data in both samples to estimate p. It would seem reasonable to estimate p by

$$\hat{p} = \frac{x_1 + x_2}{n_1 + n_2} = \frac{\text{total number of successes in both samples}}{\text{total size of both samples taken together}}$$

We then estimate $\sigma_{\hat{p}_1 - \hat{p}_2}$ by

$$\sigma_{\hat{p}_1 - \hat{p}_2} \doteq \sqrt{\frac{\hat{p}\hat{q}}{n_1} + \frac{\hat{p}\hat{q}}{n_2}} = \sqrt{\hat{p}\hat{q}\left(\frac{1}{n_1} + \frac{1}{n_2}\right)}$$

The test statistic then becomes

$$z = \frac{(\hat{p}_1 - \hat{p}_2) - 0}{\sqrt{\hat{p}\hat{q}\left(\dfrac{1}{n_1} + \dfrac{1}{n_2}\right)}} = \frac{\hat{p}_1 - \hat{p}_2}{\sqrt{\hat{p}\hat{q}\left(\dfrac{1}{n_1} + \dfrac{1}{n_2}\right)}}$$

2. Suppose that the null hypothesis is H_0: $p_1 - p_2 = D$ (where $D \neq 0$). (This can occur, for example, if H_0 said that the proportion p_1 of women favoring E.R.A. is 10 percentage points higher than the proportion p_2 of men favoring E.R.A. In this case H_0 is $p_1 - p_2 = .10$.) In a situation like this we will replace p_1, p_2 in $\sigma_{\hat{p}_1 - \hat{p}_2}$ by the sample proportions $\hat{p}_1$, $\hat{p}_2$. The test statistic will then be

$$z = \frac{(\hat{p}_1 - \hat{p}_2) - D}{\sqrt{(\hat{p}_1\hat{q}_1/n_1) + (\hat{p}_2\hat{q}_2/n_2)}}$$

In the examples and exercises of this section the sample sizes will be large enough so that we may use the above statistics where they apply.

Example 9.10

An appliance dealer sells two brands of washing machines, each with a 1-year warranty. The dealer suspected that the proportion p_1 of brand 1 washing machines that would need servicing under the warranty was smaller than the proportion p_2 for brand 2. A random sample of sales of each brand was obtained and checked to see how many of each needed service under the warranty, with the results listed in Table 9.9.

TABLE 9.9

	Number Checked (n)	Number Serviced (x)
Brand 1	100	5
Brand 2	120	9

Is there sufficient evidence (at the 5% level of significance) to bear out the dealer's suspicion?

Solution

1. Hypotheses: The dealer feels that $p_1 < p_2$. This is H_a.

$$H_0: p_1 = p_2 \qquad (\text{or } p_1 - p_2 = 0)$$
$$H_a: p_1 < p_2 \qquad (\text{or } p_1 - p_2 < 0)$$

2. Level of significance: $\alpha = .05$.

3. Test statistic and its observed value: H_0 is $p_1 - p_2 = 0$; hence 0 is the value we will use for $p_1 - p_2$. This means that we use the test statistic (1) in the outline preceding this example. Now

$$x_1 = 5 \qquad\qquad x_2 = 9$$
$$n_1 = 100 \qquad\qquad n_2 = 120$$
$$\hat{p}_1 = \frac{x_1}{n_1} = \tfrac{5}{100} = .050 \qquad \hat{p}_2 = \frac{x_2}{n_2} = \tfrac{9}{120} = .075$$

$$\hat{p} = \frac{x_1 + x_2}{n_1 + n_2} = \frac{5 + 9}{220} \doteq .064$$

$$\hat{q} = 1 - \hat{p} = .936$$

$$z = \frac{\hat{p}_1 - \hat{p}_2}{\sqrt{\hat{p}\hat{q}[(1/n_1) + (1/n_2)]}} = \frac{.050 - .075}{\sqrt{(.064)(.936)(\tfrac{1}{100} + \tfrac{1}{120})}} \doteq -.75$$

4. Critical region: Since H_a is $p_1 - p_2 < 0$, we perform a left-tailed test. The critical region is shown in Figure 9.12.

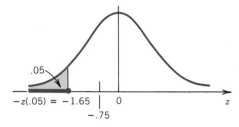

FIGURE 9.12

5. Decision: $z = -.75$ is not in the critical region; thus we fail to reject H_0. There is insufficient evidence to conclude that p_1 is less than p_2.

Confidence Intervals for $p_1 - p_2$

We can use the following procedure to find a $1 - \alpha$ confidence interval for $p_1 - p_2$:

> When the sample sizes n_1 and n_2 are sufficiently large, a $1 - \alpha$ confidence interval for the difference in two population proportions $p_1 - p_2$ is
>
> $$(\hat{p}_1 - \hat{p}_2) \pm z\left(\frac{\alpha}{2}\right) \cdot \sqrt{\frac{\hat{p}_1\hat{q}_1}{n_1} + \frac{\hat{p}_2\hat{q}_2}{n_2}}$$

Example 9.11
Democrats and Republicans in Boston were interviewed by a sociologist and asked if they favor mandatory minimum sentencing for violent crimes. The results were the following:

TABLE 9.10

Political Party	Number Interviewed (n)	Number Favoring Mandatory Sentences (x)
Republicans	400	280
Democrats	300	180

Find a 95% confidence interval for the difference $p_1 - p_2$ between the proportion p_1 of all Republicans in the city favoring mandatory sentences, and the proportion p_2 of all Democrats in the city favoring the same.

Solution

$$x_1 = 280 \qquad x_2 = 180$$

$$n_1 = 400 \qquad n_2 = 300$$

$$\hat{p}_1 = \frac{x_1}{n_1} = .70 \qquad \hat{p}_2 = \frac{x_2}{n_2} = .60$$

$$\hat{p}_1 - \hat{p}_2 = .10$$

$$1 - \alpha = .95, \ \alpha = .05$$

$$z\left(\frac{\alpha}{2}\right) = z(.025) = 1.96$$

Therefore, the 95% confidence interval is

$$(\hat{p}_1 - \hat{p}_2) \pm z\left(\frac{\alpha}{2}\right) \cdot \sqrt{\frac{\hat{p}_1 \hat{q}_1}{n_1} + \frac{\hat{p}_2 \hat{q}_2}{n_2}}$$

$$= .10 \pm (1.96) \cdot \sqrt{\frac{(.7)(.3)}{400} + \frac{(.6)(.4)}{300}}$$

$$= .10 \pm .071$$

Therefore, we are 95% sure that the difference is between .029 and .171 (i.e., from 2.9 to 17.1 percentage points higher for Republicans).

Remark   Keep in mind that the sample sizes n_1 and n_2 are regarded as large enough for the methods of this section to apply if x_1, x_2, $n_1 - x_1$, and $n_2 - x_2$ are all at least 5.

Exercises

9.40   In each of the following parts, assume independent samples. Note that n refers to sample size and x the number of successes.

(a)   It is claimed that proportions p_1 and p_2 for populations A and B, respectively, are the same. Consider the following sample information:

	n	x
A	250	175
B	175	135

Complete the test with a 5% level of significance.

(b)   It is believed that proportion p_1 for population A is larger than proportion p_2 for population B. Consider the following sample information:

	n	x
A	375	195
B	325	150

Complete the test with a 10% level of significance.

(c) It is suspected that the proportion p_1 for population A is .05 more than the proportion p_2 for population B. Consider the following sample information:

	n	x
A	400	205
B	425	175

Complete the test with a 5% level of significance.

9.41 Use the data in the corresponding parts of Exercise 9.40 to construct confidence intervals for the difference $(p_1 - p_2)$. The percent confidence is as follows:

(a) 95%.   (b) 80%.   (c) 90%.

9.42 At a large university A, a poll of 150 faculty members showed that 50 had a salary exceeding $32,000. At another large university, B, a study showed that 95 of 200 sampled faculty members had salaries exceeding $32,000.

(a) What is the estimated proportion of faculty members at universities A and B together earning more than $32,000?

(b) Does there appear to be a difference in the population proportions between faculty members at the two universities earning more than $32,000? Use a 1% level of significance.

9.43 A state politician wanted to determine if there was any difference in her support between the eastern and western parts of the state. In samples from each sector of size 150, 90 of the voters in the east and 80 in the west were satisfied with her performance.

(a) What is the estimated proportion of voters (combined) that are satisfied with her performance?

(b) Using a 5% level of significance, is there sufficient evidence to indicate a difference in her support between the eastern and western parts of the state?

9.44 A retailer received a large shipment of air conditioners. A sample of size 100 yielded 20 that did not conform to specification. The manufacturer believed that modifications in the production process would improve the quality of the product. Later the retailer received another shipment made with the modified process, and 10 of 100 sampled pieces did not conform to specification. At the 5% level, is there sufficient evidence supporting the manufacturer's belief?

(a) Use the classical approach.   (b) Use the P value approach.

9.45 In the case *Connecticut* v. *Teal*, 457 U.S. 440 (1982), the following data were given concerning job applicants:

	Selected	Rejected
Blacks	26	22
Whites	206	53

[*Source*: Data reported in DeGroot, M., S. Fienberg, and J. Kadane, *Statistics and the Law* (New York: Wiley, 1986), p. 38.]

(a) Test the claim that the population proportion of blacks selected is less than that of whites at the 2.5% level of significance (a level favored by some statisticians in a test of this type). Use the P value approach.

(b) Do the data meet the standards of the "80% rule"? This rule states that a substantial disparity between groups can ordinarily only be shown when the data are significant (i.e., lead to rejection of H_0) *and* the selection rate for the protected group ($\hat{p}_B$) is less than 80% of the selection rate for the highest group ($\hat{p}_W$).

9.46 A railroad company used two types of wheel mounts that differ in the way they handle track irregularities: type A (spring equalized) and type B (frame equalized). The following data give the repair record for the two types over a 1-year period:

	Number of Cars	Number Needing Service
Type A	150	20
Type B	180	18

(a) Find a 95% confidence interval for the difference in the population proportions $p_A - p_B$.

(b) Based on your results in part (a), do you feel there is a difference between p_A and p_B?

9.47 **(a)** Use the data in Exercise 9.42 to construct a 99% confidence interval for the difference in the population proportion p_A of faculty members earning more than \$32,000 at university A, and the population proportion p_B of faculty members earning more than \$32,000 at university B.

(b) Use the data in Exercise 9.43 to construct a 95% confidence interval for the difference in the population proportion p_1 of voters supporting the politician in the eastern part of the state, and the population proportion p_2 of voters supporting the politician in the western part of the state. Can you conclude that $p_1 \neq p_2$?

9.48 In the text (see Example 9.10) we considered the following data:

	Number Checked (n)	Number Serviced (x)
Brand 1	100	5
Brand 2	120	9

Also, p_1 and p_2 represented the proportion of washing machines of brands 1 and 2, respectively, that would need servicing under the warranty. Recall H_a was $p_1 - p_2 < 0$ and the value of the test statistic z was $-.75$.

Now consider the data:

	Number Checked (n)	Number Serviced (x)
Brand 1	600	30
Brand 2	720	54

(a)   Compute and compare $\hat{p}_1$ and $\hat{p}_2$ for both sets of data.

(b)   Using the second set of data, can we conclude that $p_1 - p_2 < 0$ at the 5% significance level? Note that in Example 9.10 we failed to reject H_0. Comment.

9.8 TOPICS IN THE DESIGN OF EXPERIMENTS (OPTIONAL)

Many colleges now offer short courses or seminars to improve study skills. In order to assess the effectiveness of such experiences, we might obtain a sample of students who took the course and compare them with a sample of students who didn't take the course. It would seem natural to compare the grade-point averages for the two groups. The skills course is referred to as the **treatment,** and the students who took the course as the **treatment group.** The other students are the **control group.**

> **Definition**
>
> - The *treatment* is the property being studied.
> - The *treatment group* is the group possessing the property.
> - The *control group* is the group not possessing the property.

There is a possible problem in the study discussed. The students who took the study skills course may have been more highly motivated than the other students, and this alone might cause them to study harder and therefore do better. We might conclude that the skills course was effective, when in fact it was the higher motivation that accounted for their better performance. Motivation is an example of a **confounding factor.**

To overcome this, we might have designed the study differently. We could obtain a group of students and ask half to take the course. This way motivation would not determine who took the course. When the researcher has control over which subjects obtain the treatment, it is a **controlled experiment.** When the researcher does not have this control, it is an **observational study** or **observational experiment.** Note that in both of the experiments described, there is a control group. *Just because there is a control group, it doesn't mean we have a controlled experiment.*

In the controlled experiment just described, there can still be a problem. The researcher involved may be partial to the skills course, and thus may assign the better students to take the course (perhaps even subconsciously). To overcome this, the students could be randomly assigned to the treatment and control groups, perhaps by the toss of a coin. This is called a **randomized controlled experiment.**

> **Definition**   When the researcher can control which subjects are assigned to the treatment, it is a *controlled experiment*. Otherwise, it is an *observational study*. When the subjects are assigned by chance to the treatment and control groups, it is a *randomized controlled experiment*. The purpose of a controlled experiment is to eliminate *confounding factors*, which are properties other than the treatment that can influence a study.

Even with randomized controlled experiments there can be problems. The students who are randomly assigned to take the skills course might feel compelled to do better, and thus study harder, thereby confounding the results. It's difficult to know how to overcome this in the study described, but in some situations, we can get around this difficulty. For example, if we wanted to study the effectiveness of vitamin C in combating colds, we might randomly assign some subjects to take regular doses of vitamin C (the treatment), and other subjects to take a placebo (the control). Further, we would not allow the subjects to know whether their pills were vitamin C or placebos. This way the subjects who took vitamin C would not know if they were "expected" to have fewer colds. At the end of a year's time, the subjects would be interviewed to determine how many colds each got. We should be sure that the interviewers do not know which subjects received the vitamin C and which received the placebos. This would be called a **double-blind randomized controlled experiment.** Whenever possible, this is the type of design we should use.

> **Definition**   In a *double-blind randomized controlled experiment*, neither the subjects, nor those who evaluate the effects, know which subjects are assigned to the treatment group and which are in the control group.

While controlled experiments are in general preferable to observational studies, we should keep in mind that a controlled experiment is not always possible or even ethical. If we were studying the effects of heavy drinking on health, it would hardly be ethical to tell some of the subjects to become alcoholics in the interest of science.

9.9

USING MINITAB (OPTIONAL)

Inference Concerning Means Based on Dependent Samples

In Table 9.4 we gave gas mileages for eight cars using experimental emission controls (x_1 values), and the same cars with standard emission controls (x_2

values). We looked at the differences $d = x_1 - x_2$ for each car. In Example 9.4 we carried out a t test to see if the (population) mean difference μ_d was greater than 0, which would indicate an improvement in mileage with the experimental system:

$$H_0: \mu_d = 0$$

$$H_a: \mu_d > 0$$

In the following Minitab printout we read the x_1 values into column 1 (C1) and the x_2 values into C2. We then place the differences (d) in C3 by the command

$$\text{LET C3} = \text{C1} - \text{C2}$$

We then perform a t test with 0 as the value of μ_d appearing in H_0. The 1 is a code for the symbol $>$ appearing in H_a. The data are in C3. Hence the command for the test is

TTEST OF MU = 0, DATA IN C3;

ALTERNATIVE = 1.

```
MTB > READ THE FOLLOWING DATA INTO C1 C2
DATA> 17 9
DATA> 23 17
DATA> 27 21
DATA> 14 16
DATA> 28 22
DATA> 21 17
DATA> 29 25
DATA> 13 13
DATA> END
        8 ROWS READ
MTB > LET C3 = C1 - C2
MTB > TTEST OF MU = 0, DATA IN C3;
SUBC>  ALTERNATIVE = 1.

TEST OF MU = 0.00 VS MU G.T. 0.00
```

	N	MEAN	STDEV	SE MEAN	T	P VALUE
C3	8	4.00	3.38	1.20	3.35	0.0062

The P value is .0062, so we would reject H_0 for any $\alpha \geq .0062$.

Inference Concerning Means Based on Independent Samples

A company planning a large purchase of computer terminals was interested in the temperature at which a computer terminal would malfunction for each of two brands under consideration (A and B). Ten brand A terminals and 11 brand B terminals were tested. In the printout below we place the breakdown temperatures for brand A and brand B into columns C1 and C2, respectively. We then conduct a hypothesis test to compare the (population) mean breakdown temperatures for the two brands:

$$H_0: \mu_1 = \mu_2$$
$$H_a: \mu_1 \neq \mu_2$$

The command used is the TWOSAMPLE command. This command uses the test statistic

$$\frac{(\bar{x}_1 - \bar{x}_2) - (\mu_1 - \mu_2)}{\sqrt{(s_1^2/n_1) + (s_2^2/n_2)}} = \frac{\bar{x}_1 - \bar{x}_2}{\sqrt{(s_1^2/n_1) + (s_2^2/n_2)}}$$

The degrees of freedom used is different from the one we used in this chapter. Whereas we used the smaller of $n_1 - 1$ or $n_2 - 1$ for degrees of freedom, Minitab uses the formula

$$df = \frac{(s_1^2/n_1 + s_2^2/n_2)^2}{\dfrac{(s_1^2/n_1)^2}{n_1 - 1} + \dfrac{(s_2^2/n_2)^2}{n_2 - 1}}$$

where df is rounded *down* to a whole number.

```
MTB > SET DATA FOR BRAND A IN C1
DATA> 100 103 97 112 99 94 109 95 99 105
DATA> END
MTB > SET DATA FOR BRAND B IN C2
DATA> 100 102 99 98 100 101 96 103 102 99 97
DATA> END
MTB > TWOSAMPLE, PERCENT CONFIDENCE 90, DATA IN C1 C2


TWOSAMPLE T FOR C1 VS C2
          N       MEAN      STDEV     SE MEAN
C1       10      101.30      5.91        1.9
C2       11       99.73      2.20        0.66

90 PCT CI FOR MU C1  -  MU C2: (-2.0, 5.13)

TTEST MU C1 = MU C2 (VS NE): T= 0.79 P=0.44 DF= 11
```

Notice that the command called for a 90% confidence interval for $\mu_1 - \mu_2$. If we don't specify a confidence interval, a 95% confidence interval is automatically given. A two-tailed test is performed unless we request otherwise with the usual ALTERNATIVE subcommand. The two-tailed test is reflected by the two-tailed P value of .44. If $\alpha < .44$ we fail to reject H_0. So if $\alpha = .05$, we fail to reject H_0, and conclude there is no difference in mean breakdown temperature.

The code

TWOSAMPLE, DATA IN C1 C2;

ALTERNATIVE = 1.

would produce a right-tailed test (a P value of .22) and a 95% confidence interval for $\mu_1 - \mu_2$.

In Section 9.5 we pointed out that if sample sizes n_1 and/or n_2 are small and $\sigma_1 = \sigma_2$, a test using the pooled standard deviation (s_p) is done. The test statistic given was

$$t = \frac{(\bar{x}_1 - \bar{x}_2) - (\mu_1 - \mu_2)}{s_p\sqrt{\dfrac{1}{n_1} + \dfrac{1}{n_2}}}, \qquad df = n_1 + n_2 - 2$$

To perform this test, we must use the POOLED subcommand. For example, to conduct a pooled right-tailed test, type

TWOSAMPLE, DATA IN C1 C2;

ALTERNATIVE = 1;

POOLED.

The developers of Minitab point out that this pooled test can be seriously in error if σ_1 is not the same as σ_2, hence the POOLED subcommand should not be used in most cases.* When the sample sizes are large, use the TWOSAMPLE command without the pooled subcommand.

Minitab provides a procedure for comparing two (or more) proportions, to be discussed in Chapter 12.

Finding P Values

In previous Minitab sections we have shown how to find exact P values for z tests, t tests, and chi-square tests based on summary data. We can also find P values for F tests. In Example 9.1 we examined data concerning voltages for two methods of producing 6-volt batteries. We tested

$$H_0: \sigma_1^2 = \sigma_2^2$$
$$H_a: \sigma_1^2 > \sigma_2^2$$

The observed value was

$$F = \frac{s_1^2}{s_2^2} = \frac{.33}{.15} = 2.20, \qquad df = (24, 20)$$

---

*Minitab does not provide an F test to compare σ_1 and σ_2. You could use the STDEV command on C1 and C2 to obtain s_1 and s_2 and then complete the F test with a hand calculator, or by finding the P value, as discussed later in this section. However, keep in mind that this F test is quite sensitive to departures from normality. So, in practice, unless you are fairly sure that the populations are normal, you are probably better off with the TWOSAMPLE command.

The P value will be the area to the right of 2.20. The following program computes the observed value, and then the P value:

```
MTB > LET K1 = .33/.15 # THE O.V.
MTB > CDF K1, STORE IN K2;
SUBC>   F, DF = 24, 20.
MTB > LET K3 = 1 - K2
MTB > PRINT K3 # THE P VALUE
K3          0.0388028
```

Minitab does not provide an F test command for variances. If you have raw data in, say, columns 1 and 2, you can use the STDEV command to find s_1 and s_2. Then you can compute the observed value and P value as in the above program.

Note that due to our convention of using the larger sample variance for the numerator, we will only be working the area of the right tail of an F distribution. In the case of a two-tailed test, we double this area.

Exercises

Suggested exercises for use with Minitab are 9.10 (a, b, c), 9.17 (a, b, c), 9.30, 9.31, and 9.72. Find P values for the following exercises: 9.3 (a, c, d), 9.23, and 9.29.

9.10 SUMMARY

In this chapter we developed methods for comparing two population means, two population variances, or two population proportions. In Table 9.11 (on pages 450 & 451) we give the parameter being investigated, the assumptions needed, the test statistic to be used in hypothesis testing about the parameter, and the form of a $1 - \alpha$ confidence interval for the parameter. For completeness we include the parameters studied in Chapters 7 and 8. It is understood that when the parameter being tested (such as μ or $\mu_1 - \mu_2$) appears in the formula for the test statistic, we substitute for this parameter the specific value appearing in the null hypothesis.

A **treatment** is a property being studied (such as smoking cigarettes). The **treatment group** is the group possessing the property. The **control group** does not possess the property. When the researcher can control which subjects can be assigned to the treatment, the study is called a **controlled experiment;** otherwise it is an **observational study.** When the subjects are randomly assigned to the treatment group, we have a **randomized controlled experiment.** A **confounding factor** is a property other than the treatment that can influence a study. In a **double-blind** study, neither the subjects, nor those evaluating the results, know which subjects are in the control group and which are in the treatment group.

Review Exercises

9.49 In each of the following parts find **(i)** $F(.01)$, **(ii)** $F(.025)$, and **(iii)** $F(.05)$.

    (a) Assume an F distribution with (4, 9) degrees of freedom.

    (b) Assume an F distribution with (9, 4) degrees of freedom.

    (c) Assume an F distribution with (30, 40) degrees of freedom.

9.50 A physical education instructor claimed there was no difference in weight variances between highly trained and talented male and female runners. A study of 10 male and 10 female runners gave the following result (Daniels et al., 1977, p. 142):

	s (Kilograms)
A (Males)	5.6
B (Females)	3.7

    Complete the test at the 2% significance level.

9.51 A nurse believed that in the 20–24 age group, the variability in males' diastolic blood pressure is larger than that of females. Independent samples of males' and females' diastolic blood pressure readings gave the following information:

	n	s
A (Males)	45	10.1
B (Females)	45	9.8

    Using a 5% level of significance, is there sufficient evidence to support the nurse's belief?

9.52 Construct confidence intervals for σ_A^2/σ_B^2 in parts (a) and (b).

    (a) Use a 90% level of confidence and the data in Exercise 9.50.

    (b) Use a 95% level of confidence and the data in Exercise 9.51.

9.53 Computer programmer A always drove to work while programmer B took the bus. Programmer A claimed that it was quicker, on the average, to go by car. The programmers recorded the following travel times in minutes for each of 10 days:

Day	A Time by Car	B Time by Bus
1	18.9	19.9
2	15.9	18.3
3	17.9	16.9
4	19.2	18.9
5	15.7	20.2
6	16.9	19.5
7	16.4	16.7
8	16.8	19.1
9	19.0	17.9
10	17.1	20.3

TABLE 9.11

Parameter	Assumptions	Test Statistic	$1 - \alpha$ Confidence Interval
(1) Single mean μ	Sample size $n \geq 30$	$z = \dfrac{\bar{x} - \mu}{\sigma/\sqrt{n}}$	$\bar{x} \pm z\left(\dfrac{\alpha}{2}\right) \cdot \dfrac{\sigma}{\sqrt{n}}$
	$n < 30$ x approximately normal	$t = \dfrac{\bar{x} - \mu}{s/\sqrt{n}}$ df $= n - 1$	If σ is unknown, use s $\bar{x} \pm t\left(\dfrac{\alpha}{2}\right) \cdot \dfrac{s}{\sqrt{n}}$
(2) Single proportion p	n sufficiently large[a]	$z = \dfrac{x - np}{\sqrt{npq}}$	$\hat{p} \pm z\left(\dfrac{\alpha}{2}\right) \cdot \sqrt{\dfrac{\hat{p}\hat{q}}{n}}$ where $\hat{p} = x/n$
(3) Single variance σ^2	x normal	$\chi^2 = \dfrac{(n - 1)s^2}{\sigma^2}$ df $= n - 1$	$\dfrac{(n - 1)s^2}{\chi^2(\alpha/2)} < \sigma^2 < \dfrac{(n - 1)s^2}{\chi^2[1 - (\alpha/2)]}$
(4) Two variances (equivalent to testing σ_1^2/σ_2^2)	x_1, x_2 normal. Independent samples	$F = \dfrac{s_1^2}{s_2^2}$ df $= (n_1 - 1, n_2 - 1)$ $(s_1^2 > s_2^2)$	$\dfrac{s_1^2}{s_2^2} \cdot \dfrac{1}{F(\alpha/2)} < \dfrac{\sigma_1^2}{\sigma_2^2} < \dfrac{s_1^2}{s_2^2} F[1 - (\alpha/2)]$

(5) Two means (equivalent to testing $\mu_1 - \mu_2$)

Condition	Test statistic	Confidence interval
Dependent samples	Let $d = x_1 - x_2$, then $\mu_d = \mu_1 - \mu_2$ Now use methods for single mean in (1) above	
Independent samples $n_1, n_2 \geq 30$	$z = \dfrac{(\bar{x}_1 - \bar{x}_2) - (\mu_1 - \mu_2)}{\sqrt{(\sigma_1^2/n_1) + (\sigma_2^2/n_2)}}$	$(\bar{x}_1 - \bar{x}_2) \pm z\left(\dfrac{\alpha}{2}\right) \cdot \sqrt{\dfrac{\sigma_1^2}{n_1} + \dfrac{\sigma_2^2}{n_2}}$ If σ_1, σ_2 unknown, use s_1, s_2.
n_1 and/or $n_2 < 30$, x_1, x_2 approx. normal, $\sigma_1 = \sigma_2$	$t = \dfrac{(\bar{x}_1 - \bar{x}_2) - (\mu_1 - \mu_2)}{s_p\sqrt{(1/n_1) + (1/n_2)}}$ $\mathrm{df} = n_1 + n_2 - 2$ $s_p = \sqrt{\dfrac{(n_1 - 1)s_1^2 + (n_2 - 1)s_2^2}{n_1 + n_2 - 2}}$	$(\bar{x}_1 - \bar{x}_2) \pm t\left(\dfrac{\alpha}{2}\right) \cdot s_p\sqrt{\dfrac{1}{n_1} + \dfrac{1}{n_2}}$
n_1 and/or $n_2 < 30$, x_1, x_2 approx. normal, $\sigma_1 \neq \sigma_2$	$t = \dfrac{(\bar{x}_1 - \bar{x}_2) - (\mu_1 - \mu_2)}{\sqrt{(s_1^2/n_1) + (s_2^2/n_2)}}$ $\mathrm{df} = $ smaller of $n_1 - 1, n_2 - 1$	$(\bar{x}_1 - \bar{x}_2) \pm t\left(\dfrac{\alpha}{2}\right) \cdot \sqrt{\dfrac{s_1^2}{n_1} + \dfrac{s_2^2}{n_2}}$

(6) Two proportions (equivalent to testing $p_1 - p_2$)

Condition	Test statistic	Confidence interval
n_1, n_2 sufficiently large[a] Independent samples	When $H_0: p_1 - p_2 = 0$ $z = \dfrac{\hat{p}_1 - \hat{p}_2}{\sqrt{\hat{p}\hat{q}[(1/n_1) + (1/n_2)]}}$ $\hat{p} = \dfrac{x_1 + x_2}{n_1 + n_2}, \hat{q} = 1 - \hat{p}$ When $H_0: p_1 - p_2 = D, D \neq 0$ $z = \dfrac{(\hat{p}_1 - \hat{p}_2) - D}{\sqrt{(\hat{p}_1\hat{q}_1/n_1) + (\hat{p}_2\hat{q}_2/n_2)}}$	$(\hat{p}_1 - \hat{p}_2) \pm z\left(\dfrac{\alpha}{2}\right) \cdot \sqrt{\dfrac{\hat{p}_1\hat{q}_1}{n_1} + \dfrac{\hat{p}_2\hat{q}_2}{n_2}}$

[a]The number of successes and the number of failures should in each case be at least 5.

With a 5% significance level, is there enough evidence to support A's claim?

(a)   Use the classical approach.   **(b)**   Use the P value approach.

9.54   A producer of corn wanted to know which of two fertilizers A or B was better for growing corn. To study this, seven plots of land were selected. Each plot was divided in half. For each plot, A was assigned to one half and B to the other using the toss of a coin. The yield of corn in bushels after a period of time gave the following data:

Plot	Fertilizer A	Fertilizer B
1	42	40
2	38	42
3	41	36
4	43	39
5	42	41
6	38	40
7	41	36

Using a 5% level of significance, is there sufficient evidence to conclude that one of the fertilizers is more effective?

9.55   **(a)**   Use the data in Exercise 9.53 to construct a 90% confidence interval for the difference in mean time between driving a car and taking the bus to work.

   (b)   Use the data in Exercise 9.54 to construct a 95% confidence interval for the difference in mean yield between the use of fertilizer A and fertilizer B.

9.56   In parts (a) and (b), assume the information was obtained with independent random samples from populations A and B.

   (a)   Test the claim that the means of populations A and B are the same. Use a 5% level of significance. The sample information is

	n	$\bar{x}$	s^2
A	40	315	400
B	60	324	360

   (b)   At the 1% significance level, is there sufficient evidence to believe the mean of B is larger than the mean of A? The sample information is

	n	$\bar{x}$	s^2
A	60	400	540
B	50	410	800

9.57 In a winter and spring study of men's diets on the island of Crete, the results of independent samples of percent calories from fats was as follows (*Source*: Keys, ed., 1970, p. I–166):

	n	$\bar{x}$	s
A (Winter)	30	37.4	4.9
B (Spring)	33	41.8	5.7

Is there sufficient evidence to indicate a difference in population means? Use a 1% level of significance.

9.58 The union president of factory A employees wanted to compare the wages of his union members with those of employees at factory B. Both factories manufactured the same product, and the union president thought workers at factory B had higher wages, on the average. Forty members were sampled from each factory and asked their hourly wage. The results were

	$\bar{x}$	s
A	13.10	1.90
B	13.95	1.65

Find the P value. Using this value, complete the test at the 1% level of significance.

9.59 Assume independent samples in parts (a) and (b).

(a) At the 5% significance level, is there sufficient evidence to conclude that the mean of A is smaller than the mean of B? The sample information is

	n	$\bar{x}$	s^2
A	10	74	60
B	13	81	40

(b) At the 10% significance level, is there sufficient evidence to conclude that the mean of A is larger than the mean of B? The sample information is

	n	$\bar{x}$	s^2
A	10	150	60
B	21	140	300

9.60 A study was done on six-site skin fold measurements [tricep, subscapular, suprailiac, umbilical, pectoral, and mid-thigh (anterior)], measured in millimeters (*Source*: Daniels et al., 1977, p. 142). The results of a study were

	n	$\bar{x}$	s
Males	10	39.6	6.8
Females	10	48.1	4.4

A track coach suspected that for highly trained and talented runners, the female six-site skin fold was 5 millimeters more than the males, on the average. Complete the test with a 2% level of significance.

9.61 A drug was tested on rats to see if the mean time in completing a maze was reduced. Twenty rats were randomly separated into groups A and B of 10 each. Each rat in group A was injected with the drug. The results of completing the maze, measured in seconds, were as follows:

	$\bar{x}$	s^2
A	5.1	1.25
B	5.9	1.40

Using a 5% level of significance, is there sufficient evidence to indicate that the drug reduces the mean time for rats to complete the maze? Assume $\sigma_A^2 = \sigma_B^2$.

(a) Use the classical approach. (b) Use the P value approach.

9.62 Repeat Exercise 9.14 as if the samples were independent. Comment on the dialogue in Exercise 9.38.

9.63 Construct confidence intervals for $\mu_A - \mu_B$ in parts (a)-(c).

(a) Use the data in Exercise 9.56 (a) to construct a 95% confidence interval.

(b) Use the data in Exercise 9.59(a) to construct a 98% confidence interval. Assume $\sigma_A = \sigma_B$.

(c) Use the data in Exercise 9.59 (b) to construct a 90% confidence interval.

9.64 At a large university, 200 students and 100 faculty members were selected and asked whether an additional science course should be required of all students. Seventy students and 45 members of the faculty supported the additional requirement. Do the populations of students and faculty members at the university appear to be in agreement? Use a 5% level of significance.

9.65 A pollster randomly sampled 250 men and 260 women in Massachusetts and asked whether they believed a capital punishment law should be reinstated. Of these people, 130 men and 104 women felt that the law should be reinstated. At the 1% significance level, is there sufficient evidence to indicate a difference in population proportion of Massachusetts men and women favoring capital punishment?

(a) Use the classical approach. (b) Use the P value approach.

9.66 A study showed that 70 of 125 workers sampled in a large factory A were satisfied with their job, while 130 of 200 workers sampled from a large factory B said they were satisfied. Using the 1% level of significance, does the evidence indicate a difference in population proportions p_A and p_B of workers satisfied in factories A and B, respectively?

(a) Use the classical approach. (b) Use the P value approach.

9.67   In the case of *Jackson* v. *Nassau County Civil Service Commission*,* 424 F. Supp. 1162, 1167 (E.D.N.Y. 1976), the following information was given concerning the results of an examination taken by job applicants:

	Passed	Failed
Blacks	40	15
Whites	99	14

[*Source*: Data reported in DeGroot, M., S. Fienberg, and J. Kadane, *Statistics and the Law* (New York: Wiley, 1986), p. 38.]

(a)   Test the claim that the population proportion of blacks who pass is less than that of whites at the 2.5% level of significance.

(b)   The "80% rule" says that a "substantial" disparity between groups can ordinarily only be shown when the data are significant (i.e., lead to rejection of the null hypothesis) *and* the pass rate for the protected group (blacks in this case) is less than 80% of the pass rate of the group with the highest pass rate. Do the data meet the requirements of the 80% rule?

9.68   A manufacturer wanted to compare the quality of work produced by two shifts. From each shift 300 items were selected. Six percent of the items manufactured by shift A were found to be defective, and 4% of the items manufactured by shift B were defective. Is there sufficient evidence to indicate a difference in the population proportion of defectives produced by the two shifts? Use a 5% level of significance.

9.69   Construct confidence intervals for $p_A - p_B$ in parts (a)-(c).

(a)   Use the data in Exercise 9.64 to construct a 95% confidence interval.

(b)   Use the data in Exercise 9.65 to construct a 99% confidence interval.

(c)   Use the data in Exercise 9.66 to construct a 95% confidence interval.

9.70   Suppose that you are testing H_0: $p_A - p_B = 0$, where p_A and p_B are proportions for populations A and B, respectively. Consider the following three sample results. Note that x is the number of successes.

(i)	n	x	(ii)	n	x	(iii)	n	x
A	100	40	A	200	80	A	500	200
B	100	50	B	200	100	B	500	250

(a)   Calculate the value of z for each sample and compare.

(b)   What do you notice about $\hat{p}_A - \hat{p}_B$ in each case?

(c)   Comment on your findings.

---

*As with most cases of this kind, *Jackson* drew upon concepts developed in the case of *Griggs* v. *Duke Power Company*, 401 U.S. 424 (1971), in which it was established that a qualification test that substantially disadvantages protected groups is illegal, unless the employer can show that use of the test is a "business necessity."

9.71 The purpose of this exercise is to demonstrate a relationship between the t and F distributions. To illustrate the relationship, complete the following table. Round off $[t(.025)]^2$ to two decimal places.

df for F	df for t	$F(.05)$	$t(.025)$	$[t(.025)]^2$
(1, 3)	3			
(1, 6)	6			
(1, 8)	8			
(1, 10)	10			
(1, 14)	14			
(1, 20)	20			

Note that $P[F > F(.05)] = .05$. Also note that $t^2 > [t(.025)]^2$ if $t > t(.025)$ or $t < -t(.025)$. So $P\{t^2 > [t(.025)]^2\} = .05$. Conjecture a relationship between the F distribution with $(1, k)$ degrees of freedom and the t distribution with k degrees of freedom.

9.72 A town engineer suspected that a power plant was substantially increasing the air pollution in the vicinity of the plant. A sample of 1-hour measurements of carbon monoxide in the vicinity of the plant was obtained for comparison with another sample obtained from other locations in the town. The results are given below. (Measurements are in milligrams per cubic meter of air.)

Near plant:	40	16	44	47	36	64	35	53
	45	52	31	38	44	29	45	47
Other locations:	34	27	27	28	35	32	35	
	25	32	26	25	31	28	30	

(a) At the 5% level of significance, is the engineer's suspicion justified?

(b) Find a 90% confidence interval for $\mu_1 - \mu_2$, where μ_1 = mean near the plant and μ_2 = mean elsewhere.

Notes

Daniels, J., G. Krakenbuhl, C. Foster, J. Gilbert, and S. Daniels, "Aerobic Responses of Female Distance Runners to Submaximal and Maximal Exercise," in P. Milvy, ed., *The Long Distance Runner* (New York: Urizen Books, 1977).

DeGroot, M. H., S. E. Fienberg, and J. B. Kadane, eds., *Statistics and the Law* (New York: Wiley, 1986).

Gordon, T., W. Castelli, M. Hjortland, W. Kannel, and T. Dawber, "High Density Lipoprotein as a Protective Factor Against Coronary Disease: The Framingham Study," *The American Journal of Medicine* 62 (1977).

Keys, A., ed., "Coronary Heart Disease in Seven Countries," *Circulation* 41(Suppl. 1):1 (1970).

Martin R., W. Haskell, and P. Wood, "Blood Chemistry and Lipid Profiles of Elite Distance Runners," in P. Milvy, ed., *The Long Distance Runner* (New York: Urizen Books, 1977).

Morganroth, J., and B. Maron, "The Athlete's Heart Syndrome: A New Perspective," in P. Milvy, ed., *The Long Distance Runner* (New York: Urizen Books, 1977).

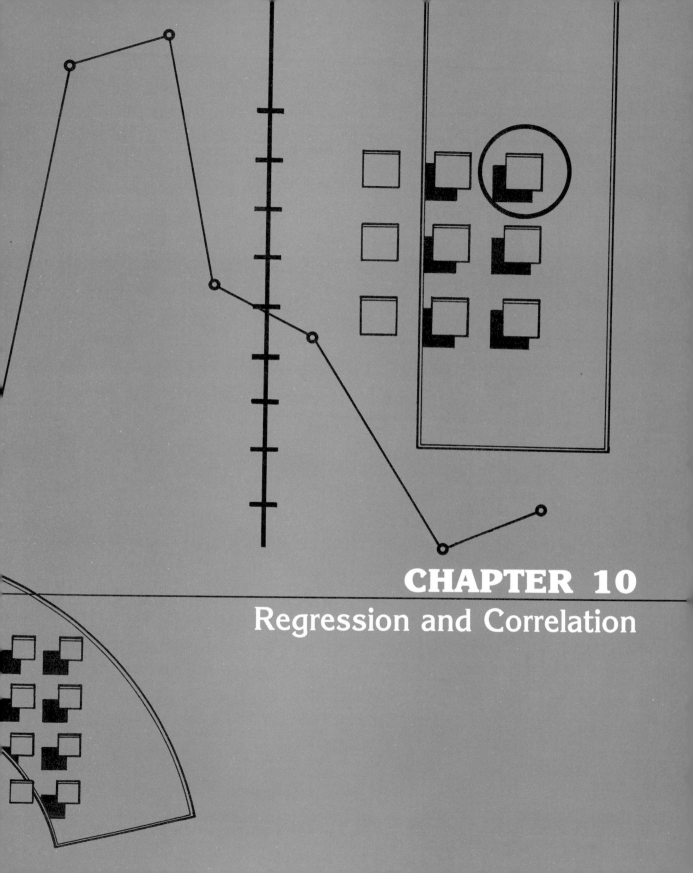

CHAPTER 10
Regression and Correlation

INTRODUCTION

Statistics is often used to investigate the relationship between two variables of interest. The following are examples of some of the kinds of issues that are often studied:

- Is there a relationship between years of schooling and level of income?
- What is the relationship between the inflation rate and the prime lending rate (the interest rate that banks charge their best customers)?

- Is there a relationship between high school grade-point average and college grade-point average? If so, what is the relationship?
- What is the relationship between a person's age and remaining years of life?

In the above examples we see that there are two basic kinds of questions of interest when studying a pair of variables:

1. Is there a relationship between the two variables?
2. What is the relationship (if any) between the two variables? This is the question asked in the last item in the list above. Actuarial scientists would be interested in the (approximate) relationship between age and the remaining years of life. To answer this, we might look for a formula that would enable us to predict the approximate expected remaining years of life for a person of any given age. Such information would be of use in determining the amount of money needed to provide a retirement annuity of a fixed amount for a person of a certain age.

In this chapter we will study these two questions. We will study **correlation analysis,** which is concerned with the question of whether there is a relationship between variables. We will also study **regression analysis,** where our objective is to find a relationship (or approximate relationship) between the variables. This relationship will take the form of an equation relating the two variables. Then for a given value of one variable, we can solve for the approximate value of the other variable.

10.2 THE LEAST SQUARES REGRESSION LINE

In many cases there is a linear or straight-line relationship between two variables of interest. For example, suppose a T.V. repair technician charges $15 for driving to your house, plus $20 for each hour of work on your T.V. If we let y = the total charge (excluding parts) and x = the number of hours of work, the relationship between x and y is

$$y = 15 + 20x$$

Then if $x = 2$ hours of work is done, the cost is

$$y = 15 + (20)(2) = \$55$$

Notice that if the problem is fixed immediately, then $x = 0$. But you still pay the amount $y = 15 + (20)(0) = \$15$. A few of the values of x and y are given in Table 10.1.

The graph of the equation $y = 15 + 20x$ is the collection of all points in the xy plane that satisfy the equation [such as the points (0, 15), (1, 35), (2, 55), etc.]. In Figure 10.1 is a graph of the equation showing the points in Table 10.1.

TABLE 10.1

Hours Worked (x)	Cost for Labor and Service Charge (y)
0	15
1	35
2	55
3	75

The graph in Figure 10.1 is, of course, a straight line. This is why we say that the relationship between x and y is **linear.** There are two terms that should be mentioned in connection with straight lines. Observe that an increase of 1 hour in the technician's labor corresponds to an increase of $20 in the fee. This value is called the **slope** of the line. Note that the slope is the coefficient of the x term in the equation of the line. This is true for any line. When the slope is a positive number, an increase in x is accompanied by an increase in y. In this case we say there is a *positive linear relationship* between x and y. If the slope were negative, it would mean that an increase in x was accompanied by a decrease in y. In this case we say there is a *negative linear relationship* between x and y. When the slope is 0, the line is horizontal. When the technician fixes the problem immediately, then $x = 0$ and $y = 15. This value is called the **y intercept,** because it is the value on the y axis where the line crosses the y axis. The y intercept is always the constant term in the equation of the line.

In general, the equation of a (non-vertical) straight line may be written in the form

$$y = b_0 + b_1 x$$

where b_0 and b_1 are constants. Conversely, any equation in this form has as its graph a straight line. The constant b_0 is the y intercept and b_1 is the slope.

If such a straight-line (or linear) relationship exists between two variables, and if we can find the values of b_0 and b_1, then it is a simple matter to find the

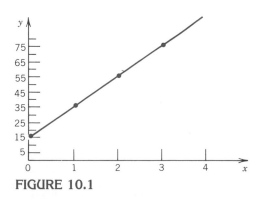

FIGURE 10.1

y value when we know the corresponding x value. We simply "plug" the x value into the equation. We sometimes call x the **independent variable** and y the **dependent variable**.

In the example discussed above, all the possible points [pairs (x, y)] lie exactly on the line. We say that there is a *perfect linear relationship* between the x and y values. However, in many situations the points are not exactly on a line, but rather tend to cluster around a line. In this case, the line and its equation may still be of use, because we might use the line to predict the approximate y value associated with a given x value.

To get an understanding of this, consider the following example: An electronics firm was planning to expand its product line and wanted to get an idea of the salary picture for the technicians it would hire in this field. A company official obtained the following information on annual salary in thousands of dollars (y) and years of experience (x) for 12 technicians employed at various firms involved in this field (see Table 10.2). In Figure 10.2a we have plotted the (x, y) values. This is called a **scatter diagram.**

Notice how the points in the scatter diagram tend to cluster about the straight line in Figure 10.2b. Using this line, the company could estimate the approximate salary it could expect to pay a technician with, say, 15 years of experience. Locate the value 15 on the x axis; then read up to the line and across to the value on the y axis as shown. This value appears to be about 30. This means that the company can expect to pay a salary of about \$30,000 per year to a technician with 15 years of experience in the field. (We also think of \$30,000 as an estimate of the mean salary of all technicians with 15 years experience in the field.) Instead of reading a y value from a graph, it would be more convenient to use the equation of the line and substitute the value $x = 15$ into the equation to obtain y. *When the points in the scatter diagram cluster about a straight line, it would be worthwhile to find the equation of the line.*

TABLE 10.2
Salary Data for 12 Technicians

Years of Experience (x)	Salary in Thousands of Dollars (y)
12	29
16	29
6	23
23	34
27	38
8	24
5	22
19	34
23	36
13	27
16	33
8	27

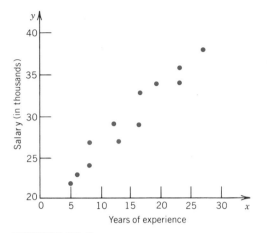

FIGURE 10.2*a*
Scatter Diagram for Salary Data

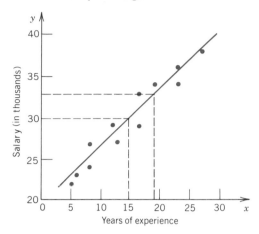

FIGURE 10.2*b*
Salary Data Cluster about a Line

The question is, how does one find the line in Figure 10.2*b*, and what is its equation? The line drawn there was not drawn by eye. There is a way of finding the equation of the line that "best fits" the data. Before showing how this is done, we should discuss what we mean when we say that a particular line best fits the data. When we use a straight line to predict or estimate a y value corresponding to a particular x value, we denote the estimate by $\hat{y}$. So we will denote the y values on the line by $\hat{y}$. The difference between the observed value y and the estimate $\hat{y}$ is called the **error** of the estimate. It is also called a **residual** and is denoted by e.

$$e = y - \hat{y}$$

(See Figure 10.3.)

For a concrete example, consider the salary data in Table 10.2. One of the technicians in the study has an annual salary of $y = 34$ (thousand dollars) with $x = 19$ years experience. If we examine Figure 10.2*b*, we see that for a technician

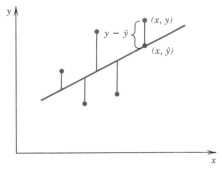

FIGURE 10.3
Residuals

with 19 years experience, the estimated salary (from the line) appears to be about $\hat{y} = 33$. Therefore, in this case the error or residual is

$$e = y - \hat{y} = 34 - 33 = 1 \quad \text{(thousand dollars)}$$

For a scatter diagram, we would want to choose the line that somehow minimizes the errors taken as a whole. But just what is to be minimized? Since some of the errors may be positive while others are negative, if we look at the sum of the errors for a particular line, we may get a very small value just due to the negative terms canceling (or reducing) the positive terms. Yet the errors might be quite large in magnitude. Therefore the sum of the errors would not be a good measure of how close the line fits the points. If we square the errors, we eliminate the problem of negative values. Thus the sum of the squares of the error terms will be a measure of how close the line fits the points: the smaller this sum, the better fit. We will agree that the line that best fits the data is the one for which the sum of the squares of the error terms is smallest. This is called the **least squares criterion**.

Least Squares Criterion   We define the sum of squares for error, SSE, to be the sum of the squares of the error terms

$$\text{SSE} = \sum e^2 = \sum (y - \hat{y})^2$$

We agree that the line that best fits the data is the line for which the sum of squares for error, SSE, is minimum. This line is called the *line of best fit* or the *regression line*.

There are may lines that can be drawn through the points in a scatter diagram. How do we find the line of best fit? Using techniques beyond the scope of this book, it can be shown that the line of best fit is obtained as follows:

Suppose that we have a scatter diagram with n points. *The line of best fit, also called the regression line,* has the equation

$$\hat{y} = b_0 + b_1 x$$

where

$$b_1 = \frac{n\left(\sum xy\right) - \left(\sum x\right)\left(\sum y\right)}{n\left(\sum x^2\right) - \left(\sum x\right)^2}$$

and

$$b_0 = \bar{y} - b_1 \bar{x}$$

The meaning of the terms in these formulas is explained in the following example.

Example 10.1
For the data in the accompanying table, plot the scatter diagram. Find the line of best fit and sketch the graph of the line.

x	y
1	2
2	4
3	4
4	6

Solution
The scatter diagram is given in Figure 10.4a. It suggests a clustering about a line. We now find the line of best fit. To find the slope b_1 and the y intercept b_0, it will be helpful to set up Table 10.3

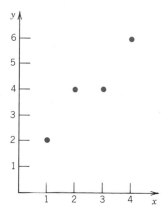

FIGURE 10.4a
Scatter Diagram

TABLE 10.3

x	y	xy	x^2
1	2	2	1
2	4	8	4
3	4	12	9
4	6	24	16
10	16	46	30
$\sum x$	$\sum y$	$\sum xy$	$\sum x^2$ $\quad n = 4$

From Table 10.3 we see that $\Sigma x = 10$, $\Sigma y = 16$, $\Sigma xy = 46$, and $\Sigma x^2 = 30$. Using these sums, we can find the line of best fit with the equation $\hat{y} = b_0 + b_1 x$, where

$$b_1 = \frac{n\left(\sum xy\right) - \left(\sum x\right)\left(\sum y\right)}{n\left(\sum x^2\right) - \left(\sum x\right)^2} = \frac{(4)(46) - (10)(16)}{4(30) - (10)^2}$$

$$= \frac{184 - 160}{120 - 100} = \frac{24}{20} = 1.2$$

$$b_0 = \bar{y} - b_1\bar{x} = \frac{16}{4} - (1.2)\left(\frac{10}{4}\right) = 1$$

Hence the line of best fit (regression line) has equation

$$\hat{y} = 1 + 1.2x$$

To graph the line, we need to find just two points that satisfy the equation. Then we draw the line through these two points. (Two points determine a line.) When $x = 0$, $\hat{y} = 1$ (the y intercept). When $x = 1$, $\hat{y} = 2.2$. So $(0, 1)$ and $(1, 2.2)$ are points on the line. The line is sketched in Figure 10.4b.

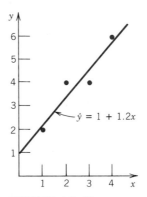

FIGURE 10.4b
Line of Best Fit

The line of best fit in Figure 10.4b is the line with the smallest possible sum of squares for error calculated for the points in Figure 10.4a. We will now calculate the SSE for the line of best fit, and compare it with the SSE for some other line through the scatter diagram. We will compare the following two lines (which are shown in Figure 10.5):

Line 1: $\hat{y} = 1 + 1.2x \leftarrow$ line of best fit

Line 2: $\hat{y} = 4 \leftarrow$ this is a horizontal line through the value 4 on the y axis

The SSE for line 1 should be smaller than the SSE for line 2.

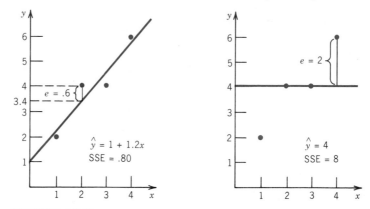

FIGURE 10.5
Line of Best Fit Has the Smallest Possible Value for SSE

The calculations are shown in Table 10.4. Note that in these calculations, the value of $\hat{y}$ corresponding to a particular x is obtained from the equation of the line. For example, for line 1 the value of $\hat{y}$ corresponding to $x = 1$ is obtained by substituting $x = 1$ into the equation of the line, so $\hat{y} = 1 + (1.2)(1) = 2.2$. Line 2 is a horizontal line; therefore each value of $\hat{y}$ is the same: $\hat{y} = 4$. This situation is displayed graphically in Figure 10.5.

TABLE 10.4

	Line 1: $\hat{y} = 1 + 1.2x$				Line 2: $\hat{y} = 4$				
x	y	$\hat{y}$	$e = y - \hat{y}$	e^2	x	y	$\hat{y}$	$e = y - \hat{y}$	e^2
1	2	2.2	−.2	.04	1	2	4	−2	4
2	4	3.4	.6	.36	2	4	4	0	0
3	4	4.6	−.6	.36	3	4	4	0	0
4	6	5.8	.2	.04	4	6	4	2	4
			SSE $= \sum e^2 = .80$					SSE $= 8$	

As expected, line 2 has a larger SSE than the line of best fit. This will always be the case, no matter what line is used in place of line 2.

We will now find the line of best fit for the salary data discussed at the beginning of this section. (See Table 10.2.)

Example 10.2
Find the line of best fit for the salary data in Table 10.2.

Solution
We have already seen that the points in the scatter diagram tend to cluster about a line. (See Figures 10.2*a* and *b*.) The calculations for the line of best fit follow.

TABLE 10.5

x	y	xy	x^2
12	29	348	144
16	29	464	256
6	23	138	36
23	34	782	529
27	38	1026	729
8	24	192	64
5	22	110	25
19	34	646	361
23	36	828	529
13	27	351	169
16	33	528	256
8	27	216	64
Sums: 176	356	5629	3162 $n = 12$

$$b_1 = \frac{n\left(\sum xy\right) - \left(\sum x\right)\left(\sum y\right)}{n\left(\sum x^2\right) - \left(\sum x\right)^2}$$

$$= \frac{(12)(5629) - (176)(356)}{(12)(3162) - (176)^2} = \frac{4892}{6968}$$

$$= .7020666 \doteq .70$$

$$b_0 = \bar{y} - b_1\bar{x} = \frac{356}{12} - (.7020666)\left(\frac{176}{12}\right) \doteq 19.37$$

(Note that we substituted the value .7020666 for b_1 in the formula for b_0, rather than the rounded value of .70. This is in keeping with the practice of rounding as little as possible during intermediate steps and only rounding the final answer.) The equation of the line of best fit is

$$\hat{y} = 19.37 + .70x$$

By using this equation, a salary schedule for technicians could be constructed. The equation suggests a base salary of 19.37 thousand dollars (or $19,370) plus .70 thousand dollars ($700) for each year of experience in the field. For a technician with $x = 15$ years, the company could expect to pay an annual salary (in thousands) of

$$\hat{y} = 19.37 + (.70)(15) = 29.87$$

or $29,870.

Example 10.3

During World War II there was a scarcity of foods rich in fat (meat, butter, eggs, etc.) in Norway (and other countries). Along with a decline in the consumption of fat, a decline in the death rate from atherosclerosis was also observed. (Atherosclerosis is the formation of fatty deposits on the walls of the arteries, which can lead to heart disease or stroke.) In Table 10.6, we give the data for Norway from 1938 to 1947.* In this table, x = consumption of fat in kilograms per year per person, and y = the number of deaths from atherosclerosis per 100,000 people in a year.

TABLE 10.6
Norway Data

Year	Fat Consumption (x)	Death Rate (y)
1938	14.4	29.1
1939	16.0	29.7
1940	11.6	29.2
1941	11.0	26.0
1942	10.0	24.0
1943	9.6	23.1
1944	9.2	23.0
1945	10.4	23.1
1946	11.4	25.2
1947	12.5	26.1

Source: The data are estimated from a graph in Williams, C. L., "Nutrition and Coronary Heart Disease," in *Bank of Epidemiology Exercises, Exercise 13*, 1st Ed. (New York: Dept. of Community and Preventive Medicine, New York Medical College, 1978).

(a) Plot the scatter diagram.

(b) Find the equation of the line of best fit for the data.

(c) If the consumption of fat were 13 kilograms per person per year, what would you estimate the annual death rate from atherosclerosis to be?

---

*Although atherosclerosis is a disease that takes years to develop, the decrease in mortality was immediate with the change in dietary habits. It has been suggested that this could be explained by the possibility that once fat in the arteries accumulates to a certain level, a triggering event (such as a blood clot) is associated with mortality. It is possible that the low-fat diet reduced the tendency of the blood to clot.

Solution

(a)   The scatter diagram shown in Figure 10.6a suggests a clustering of points about a line.

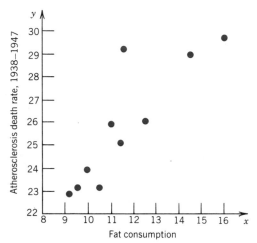

FIGURE 10.6a
Atherosclerosis Death Rate per 100,000
People per Year (y) versus Consumption
of Fat in Kg per Person per Year (x) for
10 Years

(b)

TABLE 10.7

	x	y	xy	x^2
	14.4	29.1	419.04	207.36
	16.0	29.7	475.20	256.00
	11.6	29.2	338.72	134.56
	11.0	26.0	286.00	121.00
	10.0	24.0	240.00	100.00
	9.6	23.1	221.76	92.16
	9.2	23.0	211.60	84.64
	10.4	23.1	240.24	108.16
	11.4	25.2	287.28	129.96
	12.5	26.1	326.25	156.25
Sums:	116.1	258.5	3046.09	1390.09	$n = 10$

$$b_1 = \frac{n\left(\sum xy\right) - \left(\sum x\right)\left(\sum y\right)}{n\left(\sum x^2\right) - \left(\sum x\right)^2}$$

$$= \frac{(10)(3046.09) - (116.1)(258.5)}{(10)(1390.09) - (116.1)^2} = \frac{449.05}{421.69}$$

$$= 1.0648818 \doteq 1.065$$

$$b_0 = \bar{y} - b_1\bar{x} = \frac{258.5}{10} - (1.0648818)\left(\frac{116.1}{10}\right) \doteq 13.487$$

The line of best fit is

$$\hat{y} = 13.487 + 1.065x$$

(c)   When $x = 13$, we estimate the death rate from atherosclerosis to be

$$\hat{y} = 13.487 + (1.065)(13) = 27.332$$

We display the scatter diagram and line of best fit in Figure 10.6b.

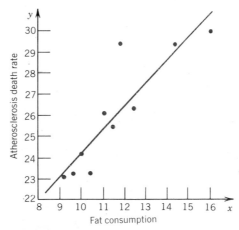

FIGURE 10.6b
Line of Best Fit for Data in Table 10.6

Example 10.4

An economist was interested in the production costs for companies supplying chemicals for use in fertilizers. The data in Table 10.8 represent the number of tons produced during a year in thousands (x) along with the production cost per ton (y) for seven companies. (Note that $x = 2.4$ means 2400 tons.)

TABLE 10.8

Number of Tons (in thousands) x	Cost per Ton (in dollars) y
3.0	40
4.0	40
2.4	50
5.0	35
2.6	55
4.0	35
5.5	30

Plot the points and find the line of best fit for the data. Sketch its graph on the scatter diagram.

Solution
The points of a scatter diagram are displayed in Figure 10.7 and suggest a clustering about a line.

TABLE 10.9

	x	y	xy	x^2	
	3.0	40	120	9.00	
	4.0	40	160	16.00	
	2.4	50	120	5.76	
	5.0	35	175	25.00	
	2.6	55	143	6.76	
	4.0	35	140	16.00	
	5.5	30	165	30.25	
Sums:	26.5	285	1023	108.77	$n = 7$

$$b_1 = \frac{n\left(\sum xy\right) - \left(\sum x\right)\left(\sum y\right)}{n\left(\sum x^2\right) - \left(\sum x\right)^2}$$

$$= \frac{(7)(1023) - (26.5)(285)}{(7)(108.77) - (26.5)^2} = \frac{7161 - 7552.5}{761.39 - 702.25}$$

$$= \frac{-391.5}{59.14} = -6.619885 \doteq -6.62$$

$$b_0 = \bar{y} - b_1\bar{x} = \frac{285}{7} - (-6.619885)\left(\frac{26.5}{7}\right) \doteq 65.78$$

The equation of the line of best fit is

$$\hat{y} = 65.78 - 6.62x$$

This line, along with the scatter diagram, is displayed in Figure 10.7. Note that the negative slope indicates that an increase in the number of tons is accompanied by a decrease in the cost per ton. (Of course, we would not expect this equation to apply to firms with a production capacity outside the range studied by the economist, which happened to be companies in the 2000- to 6000-ton range.)

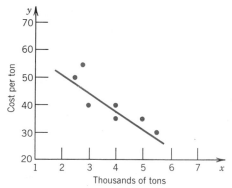

FIGURE 10.7
Production Costs

The regression line is useful only when the points in the scatter diagram cluster about a straight line. There are many situations, however, when this is not the case, as is shown in Figure 10.8.

In Figure 10.8 (*a*), the points appear to cluster about a curve; there appears to be some relationship between the *x* and *y* values, although not a linear

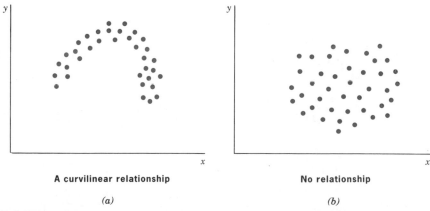

A curvilinear relationship

(*a*)

No relationship

(*b*)

FIGURE 10.8

relationship. In Figure 10.8 (*b*), the points do not appear to cluster about a curve or a straight line, indicating no relationship of any kind.

Our method of finding the line of best fit will work even for data in Figure 10.8. However, the line so obtained may be of little value in predicting a *y* value for a given *x* value (see Figure 10.9).

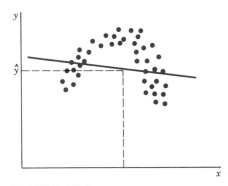

FIGURE 10.9

Up until now we have relied upon a scatter diagram to determine if the points are clustered about a straight line. This approach to the problem is rather imprecise. In the next section, we will study a more satisfactory approach. We will introduce the **linear correlation coefficient**. This is a number calculated from the data; it measures the degree to which the points in the scatter diagram cluster about a straight line. In other words, it measures the degree of linear relationship between the *x* and *y* values.

Exercises

10.1 Find the *y* intercept (b_0) and the slope (b_1) of each of the following lines given by the equations. Then graph the line.

(a) $y = 4 + 7x$.   (b) $y - 3 = 5(x + 2)$.   (c) $y = 8$.

(d) $4x + 2y - 9 = 0$.   (e) $3y = 9x + 12$.

In Exercises 10.2–10.6, (a) plot the scatter diagram, (b) find the line of best fit for the given data, and (c) sketch the line of best fit on the same paper as the scatter diagram.

10.2

x	y
2	9
2	7
4	5
4	5
6	3
6	1

10.3

x	y
0	2
0	4
0	6
1	6
2	7
2	8
2	9

10.4

x	y
0	5
0	4
1	3
2	2
2	1

10.5

x	y
1	2
2	2
3	1
4	0
5	0

10.6

x	y
1	2
2	6
4	14
5	18

10.7 For the following data one of these equations, $\hat{y} = 6$ or $\hat{y} = 4 + 2x$, represents the line of best fit:

x	y
0	2
0	4
0	6
1	6
2	7
2	8
2	9

Using the value of SSE for each line, determine which line is the line of best fit.

10.8 For the following data one of these equations, $\hat{y} = -1.2 + 1.2x$ or $\hat{y} = x$, represents the line of best fit:

x	y
2	2
4	2
6	6
8	10
10	10

Using the value of SSE for each line, determine which line is the line of best fit.

10.9 The data in the following table represent the inflation rate (x) and prime lending rate (y) over a 7-year period. Plot a scatter diagram. Find the line of best fit for predicting the prime lending rate from the inflation rate.

Inflation Rate (x)	Prime Lending Rate (y)
3.3	5.2
6.2	8.0
11.0	10.8
9.1	7.9
5.8	6.8
6.5	6.9
7.6	9.0

Source: *Statistical Abstract of the United States,* 100th Ed. (Washington, D.C.: U.S. Bureau of the Census), pp. 419 and 541.

10.10 The data in the following table represent trends in cigarette consumption (x) and lung cancer mortality (y) for Canadian males. The data have been rounded off for convenience. Find the line of best fit for predicting the mortality rate from the cigarette consumption.

Cigarette Consumption per Capita in Hundreds (x)	Mortality Rate per 100,000 (y)
11.8	10.4
12.5	16.5
15.7	22.9
19.2	26.6
21.9	33.8
23.3	42.8

Source: Phillips, A. J., "Smoking Control, Programs for Canadian Adults," in *Proceedings of the Third World Conference, Smoking and Health*, Vol. II, p. 273.

10.11 A state official was investigating the relationship between salary (x) and the number of absences (y) for state employees. The variable y in the following table represents the average number of absences per year for employees at that salary.

Salary in Thousands (x)	Number of Absences (y)
20.0	2.3
22.5	2.0
25.0	2.0
27.5	1.8
30.0	2.2
32.5	1.5
35.0	1.2
37.5	1.3
40.0	0.6

(a) Find and sketch the line of best fit.

(b) Estimate the average number of absences for employees earning $29,000.

10.12 The owner of a liquor store wanted to investigate the relationship between the intensity of advertising (x) and the number of cases (y) of Samuel Adams Boston Ale sold per day. The intensity of advertising was indexed from 1 for no advertising to 5 for the most intensive advertising. The data are given in the following table.

Intensity of Advertising (x)	Number of Cases Sold (y)
1	4
2	7
3	8
4	14
5	15

(a) Find and sketch the line of best fit for predicting the number of cases sold from the intensity of advertising.

(b) Estimate the number of cases sold for an intensity of advertising 3.5.

10.13 An academic dean wanted to study the relationship of a chairperson's evaluation (y) of a professor and number of A's issued (x) by the professor. The variable x in the following table represents the number of A's per 100 given by the professor, and the variable y is the evaluation of performance on a scale of 0 to 10.

Number of A's per 100 (x)	Evaluation (y)
4	9.4
6	9.2
6	8.0
7	8.5
14	7.4
15	6.0
19	6.1

(a) Find and sketch the line of best fit for predicting the evaluation of a faculty member for a particular value of x.

(b) Predict the approximate evaluation score for a professor with $x = 10$.

10.14 A town assessor was trying to determine a relationship between the size of a parcel of land (x) and the selling price (y). The assessor used the data in the following table.

Size in Acres (x)	Selling Price in Thousands of Dollars (y)
.5	25
1.0	40
1.5	55
2.0	65
2.5	75
3.0	80

(a) Find and sketch the line of best fit for predicting selling price from the size of the land.

(b) Estimate the selling price for a 1.7-acre piece of land.

10.15 Below are education and crime ratings for selected U.S. cities. (Education is a composite rating including pupil/teacher ratio, academic options in higher education, etc. The higher the rating, the better. Crime is the crime rate per 100 people.)

(a) Find and sketch the line of best fit for predicting crime rate from an education rating.

(b) Estimate the crime rate for an education rating of 34.

City	Education (x)	Crime (y)
Boston	35	12
Chicago	35	10
Detroit	31	16
Los Angeles	32	20
New York	30	25
Washington, D.C.	36	13

Source: Boyer, R., and D. Savageau, *Places Rated Almanac* (Rand McNally). As reported in *MacSpin Release 1.1.*

10.16 Let x and y be *Fortune* magazine's rankings (based on sales, assets, profits, etc.) of 30 randomly selected industrial firms in 1980 and 1985, respectively. Consider the following summary information:

$$\sum x = 4571 \qquad \sum x^2 = 941{,}239 \qquad \sum y = 3860$$
$$\sum y^2 = 648{,}562 \qquad \sum xy = 771{,}472$$

(a) Find the line of best fit for predicting a 1985 rank from a 1980 rank.

(b) Estimate the 1985 rank given a 1980 rank of 25. (*Source:* Data compiled by David Donoho and Liat Kulwarski from various issues of *Fortune* magazine. As reported in *MacSpin Release 1.1.*)

10.17 An automotive engineer, interested in the relationship between miles per gallon y (MPG) and horsepower x (HP), sampled 30 cars and obtained the following information:

$$\sum x = 2444 \qquad \sum x^2 = 204{,}232 \qquad \sum y = 960$$
$$\sum y^2 = 31{,}514 \qquad \sum xy = 76{,}641$$

(a) Find the line of best fit for predicting MPG from HP.

(b) Estimate MPG when HP is 88.

THE LINEAR CORRELATION COEFFICIENT

We said that we can always find the regression line (or line of best fit) for a scatter diagram. However, such a line will only be of use if the x and y values show some degree of linear relationship. By this we mean that the points in the scatter diagram tend to cluster about a straight line. The closer the points are to the line, the stronger the degree of linear relationship (also called *linear correlation*).

In this section, we will show how to calculate a number from the data, called the **linear correlation coefficient, r.** This number measures the degree to which the points in the scatter diagram tend to cluster about a straight line. In other words, r *measures the degree of linear relationship between the x and y values.* Using the value of r, we can determine if it is worthwhile to find the regression line. In addition, there are situations when we are not particularly interested in a regression line or predicting y values, but instead we may simply wish to know if some degree of linear relationship exists between the x and y values. The linear correlation coefficient can be used to study this question.

Definition   The (linear) *correlation coefficient r* for a collection of n pairs of data values is

$$r = \frac{\sum (x - \bar{x})(y - \bar{y})}{(n - 1)s_x s_y}$$

where s_x and s_y are the standard deviations of the x values and y values, respectively. Instead of using the above formula to find r, we will use an alternative formula that gives the same result. Although the alternate formula looks more complicated, it actually simplifies our computations.

$$r = \frac{n\left(\sum xy\right) - \left(\sum x\right)\left(\sum y\right)}{\sqrt{n\left(\sum x^2\right) - \left(\sum x\right)^2} \cdot \sqrt{n\left(\sum y^2\right) - \left(\sum y\right)^2}}$$

Before giving the properties of r, we illustrate how to use the formula on the data of Example 10.1 (see Figure 10.10 and Table 10.10).

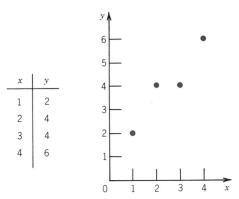

x	y
1	2
2	4
3	4
4	6

FIGURE 10.10
Scatter Diagram

TABLE 10.10

	x	y	xy	x^2	y^2	
	1	2	2	1	4	
	2	4	8	4	16	
	3	4	12	9	16	
	4	6	24	16	36	
Sums:	10	16	46	30	72	$n = 4$

$$r = \frac{n\left(\sum xy\right) - \left(\sum x\right)\left(\sum y\right)}{\sqrt{n\left(\sum x^2\right) - \left(\sum x\right)^2} \cdot \sqrt{n\left(\sum y^2\right) - \left(\sum y\right)^2}}$$

$$= \frac{(4)(46) - (10)(16)}{\sqrt{(4)(30) - (10)^2} \cdot \sqrt{(4)(72) - (16)^2}} = \frac{24}{\sqrt{20} \cdot \sqrt{32}}$$

$$= .9486833 \doteq .95$$

Properties of r

1. The value of r is always between -1 and 1: $-1 \leq r \leq 1$.
2. $r = 1$, provided that all of the points in the scatter diagram lie exactly on a line with a positive slope. (As would be expected, this is the regression line.) Also, $r = -1$, provided that all of the points lie exactly on a line with a negative slope. (See Figure 10.11.)

$r = 1$

Perfect positive correlation

$r = -1$

Perfect negative correlation

FIGURE 10.11

3. Suppose $r \neq 0$. Then the points in the scatter diagram are grouped around a nonhorizontal line (the regression line). If r is positive, the line has positive

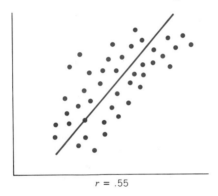

$r = .55$

Positive correlation

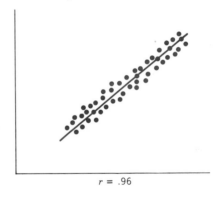

$r = .96$

Strong positive correlation

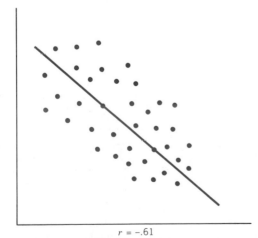

$r = -.61$

Negative correlation

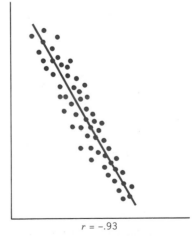

$r = -.93$

Strong negative correlation

FIGURE 10.12

slope (i.e., sloping upward to the right). If r is negative, the line has negative slope. The closer r is to 1 or -1, the closer the points tend to cluster about the line. Therefore, a value of r close to 1 or -1 indicates a strong degree of linear relationship. The closer r is to 0, the more dispersed the points will be from the line. (See Figure 10.12 for some values of r.)

4. If the value of $r = 0$, there is no linear relationship between the x and y values for the data. When $r = 0$, it can be shown that the line of best fit (the regression line) is a horizontal line through the points. (See Figure 10.13.) (If the value of r is near 0, there will only be a very weak linear relationship.)

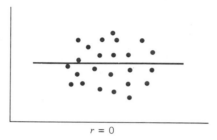

$r = 0$

FIGURE 10.13
No Linear Relationship

We should stress that when $r = 0$, we can say that there is no degree of *linear* relationship between the x and y values of the data. This does not mean that there can be no other kind of relationship for the data. The points in Figure 10.14 suggest a strong relationship between the x and y values, although not a linear relationship. In fact, this is a quadratic relationship. This is the type of scatter diagram that occurs if x represents a person's age, and y is the maximum weight the person can lift.

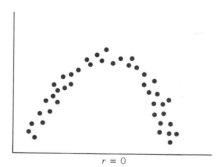

$r = 0$

FIGURE 10.14
A Nonlinear Relationship

When $r \neq 0$, the points are scattered about a nonhorizontal line, and there will be at least some degree of linear relationship between the x and y values for the data. But as we said, if r is close to 0, there will be only a very small degree of linear relationship, and the regression line will not be very helpful in predicting y

values. In this case, it may be advisable to look for some suitable nonlinear relationship. A scatter diagram might suggest such a relationship. We will not discuss nonlinear relationships in this book. If, on the other hand, the value of r is reasonably close to 1 or -1, a worthwhile linear relationship will exist. It would then make sense to find the regression line if we wanted to predict y values. We will examine what we mean by "close to ± 1" more carefully later in this section and in the next section. For now we rely on intuition.

Example 10.5
Find the linear correlation coefficient for the salary data in Table 10.2. Interpret the result.

Solution
Note that most of the calculations were worked out in Example 10.2. The only additional information we need is Σy^2; therefore, we add a column of y^2 values in Table 10.11.

TABLE 10.11

	x	y	xy	x^2	y^2	
	12	29	348	144	841	
	16	29	464	256	841	
	6	23	138	36	529	
	23	34	782	529	1156	
	27	38	1026	729	1444	
	8	24	192	64	576	
	5	22	110	25	484	
	19	34	646	361	1156	
	23	36	828	529	1296	
	13	27	351	169	729	
	16	33	528	256	1089	
	8	27	216	64	729	
Sums:	176	356	5629	3162	10,870	$n = 12$

$$r = \frac{n\left(\Sigma xy\right) - \left(\Sigma x\right)\left(\Sigma y\right)}{\sqrt{n\left(\Sigma x^2\right) - \left(\Sigma x\right)^2} \cdot \sqrt{n\left(\Sigma y^2\right) - \left(\Sigma y\right)^2}}$$

$$= \frac{(12)(5629) - (176)(356)}{\sqrt{(12)(3162) - (176)^2} \cdot \sqrt{(12)(10,870) - (356)^2}}$$

$$= \frac{4892}{\sqrt{6968} \cdot \sqrt{3704}} = .9629348 \doteq .96$$

This value is fairly close to $+1$, indicating a strong degree of positive linear relationship for the data. This is consistent with Figure 10.2b, which shows the points in the scatter diagram clustered closely about the line.

Note: The reader may have noticed that the numerator in the formula for r is the same as the numerator in the formula for b_1, and the term under the first square root in the denominator of r is the denominator of b_1. Thus, if you have already found b_1, these calculations can simplify the process for finding r even further. We have not made use of this fact in this section, because we have chosen to keep the calculations self-contained. But it is something to keep in mind.

Example 10.6
Find the linear correlation coefficient for the production cost data in Example 10.4.

Solution
We reproduce the calculations from Example 10.4 along with a column of y^2 values in Table 10.12.

TABLE 10.12

	x	y	xy	x^2	y^2	
	3.0	40	120	9.00	1,600	
	4.0	40	160	16.00	1,600	
	2.4	50	120	5.76	2,500	
	5.0	35	175	25.00	1,225	
	2.6	55	143	6.76	3,025	
	4.0	35	140	16.00	1,225	
	5.5	30	165	30.25	900	
Sums:	26.5	285	1023	108.77	12,075	$n = 7$

$$r = \frac{n\left(\sum xy\right) - \left(\sum x\right)\left(\sum y\right)}{\sqrt{n\left(\sum x^2\right) - \left(\sum x\right)^2} \cdot \sqrt{n\left(\sum y^2\right) - \left(\sum y\right)^2}}$$

$$= \frac{(7)(1023) - (26.5)(285)}{\sqrt{(7)(108.77) - (26.5)^2} \cdot \sqrt{(7)(12,075) - (285)^2}}$$

$$= \frac{-391.5}{\sqrt{59.14} \cdot \sqrt{3300}} = -.8862049 \doteq -.89$$

This is consistent with Figure 10.7, and indicates a negative linear relationship for the data.

The Coefficient of Determination

Before finding the regression line, we want to feel sure that there is a reasonable degree of linear relationship between x and y. The correlation coefficient r addresses this issue. A value of $r = .99$ would seem to indicate that a useful linear relationship exists. But what if $r = .70$? Is this value close enough to 1 to indicate a reasonable degree of linear relationship? The reason why it is difficult to answer this question is that, although r measures the degree of linear relationship, it does not do so in familiar terms. (In other words, what does $r = .70$ mean? How do we interpret it?)

We will discuss two approaches to this problem. One approach is to establish cutoff values for r so that if a value of r is within a certain distance of ± 1, we will conclude that there is a significant degree of linear relationship between the x and y values. This approach is discussed in the next section. The other approach is to look at the square of the correlation coefficient, r^2. It turns out that this quantity, called the **coefficient of determination**, enables us to interpret the degree of linear relationship in more familiar terms, so that we can make a judgment about whether a useful relationship between the x and y values exists (for the data).

Suppose that we consider the variables: age (x) and height (y) for males 18 years and older. We would expect little relationship between age and height for this group. After all, we would not expect a difference in height of 1 foot between two men to be "explained" by the fact that one of the men is 30 years old and the other is 40. However, if we considered age and height for males age 1 to 16, then we would expect some relation between these variables for this group. If we were told that one such male was 1 foot tall and the other 6 ft tall, it would be a pretty safe bet that the 6-footer was older. In other words, we believe that a good portion of the variability in heights for this group can be accounted for, or explained by, the relationship between height and age.

It turns out that for a collection of pairs of (x, y) values, the coefficient of determination r^2 gives the proportion of the variability in y values that is accounted for, or explained by, the degree of linear relationship with the x values. So if $r = .9$, then $r^2 = .81$. Thus 81% of the variability in y is explained by x.

We now clarify this idea using the data in Example 10.1. We reproduce these data along with the regression line in Figure 10.15.

The deviation of a y value from the mean $\bar{y}$, $y - \bar{y}$, can be thought of as consisting of two components:

$$y - \bar{y} = (y - \hat{y}) + (\hat{y} - \bar{y})$$

For example, consider the point $(x, y) = (4, 6)$. The mean of the y values is $\bar{y} = 4$. From the regression equation in Figure 10.15, we see that when $x = 4$, $\hat{y} = 5.8$. Observe that

$$y - \bar{y} = 2 = .2 + 1.8 = (y - \hat{y}) + (\hat{y} - \bar{y})$$

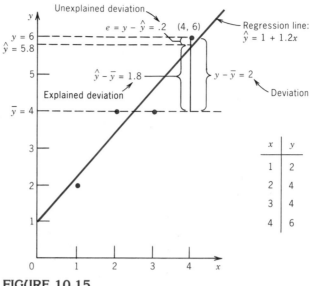

FIGURE 10.15

(See Figure 10.15.) The following list is a description of the terms of interest in Figure 10.15.

1. $y - \bar{y} = 2$ measures the deviation of $y = 6$ from the mean of 4.
2. $y - \hat{y} = .2$ measures the deviation of $y = 6$ from the regression line.
3. $\hat{y} - \bar{y} = 1.8$ measures the deviation of $\hat{y} = 5.8$ on the regression line from the mean.

If $\hat{y}$ were used as an estimate for y, then $\hat{y} - \bar{y}$ could be thought of as an estimate of $y - \bar{y}$. Hence $\hat{y} - \bar{y} = 1.8$ can be thought of as an estimate of $y - \bar{y} = 2$. In this sense, $\hat{y} - \bar{y}$ is that portion of the deviation that is "explained," or "accounted for," by the regression line (in this case, 1.8 out of 2 units). The error term ($e = y - \hat{y} = .2$ in this case) is that portion of the deviation that is "unexplained." When the points cluster close to the regression line, the error terms will tend to be small. When an error term is small, most of the deviation will be explained by the regression line. This corresponds to $\hat{y}$ being a pretty good estimator of y.

Instead of looking at individual points, we need to look at all the points taken together.

1. To measure the variation of the y values from the mean $\bar{y}$, we can calculate the sum of the squares of the deviations $y - \bar{y}$. This is called the **total variation**, or the **total sum of squares, TSS.**

$$\text{TSS} = \sum (y - \bar{y})^2$$

(This is similar to the variance, except that we do not divide by $n - 1$.)

2. The variation of the y values from the regression line is measured by the sum of the squares of the errors, $y - \hat{y}$. This is called the **unexplained**

variation, or the **sum of squares for error, SSE**.

$$\text{SSE} = \sum(y - \hat{y})^2$$

3. The variation of the values $\hat{y}$ from the mean $\bar{y}$ is measured by the sum of the squares of the $\hat{y} - \bar{y}$ terms. This is called the **explained variation**, or the **sum of squares for regression, SSR**.

$$\text{SSR} = \sum(\hat{y} - \bar{y})^2$$

Here again, if $\hat{y}$ were used as an estimate of y, then $\hat{y} - \bar{y}$ could be used as an estimate of $y - \bar{y}$. Therefore, SSR could be thought of as an estimate of TSS.

Now we calculate the values of TSS, SSE, and SSR for the points in Figure 10.15. (See Table 10.13.)

TABLE 10.13

	x	y	$\hat{y}$	$(y - \bar{y})^2$	$(y - \hat{y})^2$	$(\hat{y} - \bar{y})^2$
	1	2	2.2	$(2 - 4)^2 = 4$	$(2 - 2.2)^2 = .04$	$(2.2 - 4)^2 = 3.24$
	2	4	3.4	$(4 - 4)^2 = 0$	$(4 - 3.4)^2 = .36$	$(3.4 - 4)^2 = .36$
	3	4	4.6	$(4 - 4)^2 = 0$	$(4 - 4.6)^2 = .36$	$(4.6 - 4)^2 = .36$
	4	6	5.8	$(6 - 4)^2 = \underline{4}$	$(6 - 5.8)^2 = \underline{.04}$	$(5.8 - 4)^2 = \underline{3.24}$
Sums:				TSS $= 8$	SSE $= .80$	SSR $= 7.20$

Notice that

$$8 = .80 + 7.20$$

or

$$\text{TSS} = \text{SSE} + \text{SSR}$$

This result holds for *any* collection of data.

> The total sum of squares is equal to the sum of squares for error plus the sum of squares for regression:
>
> $$\text{TSS} = \text{SSE} + \text{SSR}$$
> $$\sum(y - \bar{y})^2 = \sum(y - \hat{y})^2 + \sum(\hat{y} - \bar{y})^2$$
> *Total Variation = Unexplained Variation + Explained Variation*

As we indicated before, SSR can be viewed as an estimate of TSS. In this sense, SSR is that portion of the total variation (TSS) that is explained by the regression line.

When the sum of squares for regression SSR accounts for a large portion of the total variation TSS, the sum of squares for error SSE will be small, meaning that the error terms, $y - \hat{y}$, will tend to be small. This in turn means that the points cluster fairly close to the regression line. In other words, there is a strong degree of linear relationship for the data, and the line will be useful in obtaining estimates of y values.

The ratio

$$\frac{\text{SSR}}{\text{TSS}} = \frac{\text{explained variation}}{\text{total variation}}$$

gives the proportion of the total variation of the y values that is accounted for, or explained, by the linear relationship with x (the regression line). In this sense, this ratio is a measure of the strength of the linear relationship. *The larger the proportion, the stronger the linear relationship.*

For example, for the data in Figure 10.15 we showed that TSS = 8 and SSR = 7.2. Therefore,

$$\frac{\text{SSR}}{\text{TSS}} = \frac{7.2}{8} = .90$$

This means that 90% of the variation in y values is explained by the linear relationship—a sizable portion.

It turns out that if we know the linear correlation coefficient r, we can calculate the value of the proportion SSR/TSS very quickly by squaring r. Let's check this out for the data in Figure 10.15. At the beginning of this section, we found that for this data, $r = .9486833$. Note that

$$r^2 = (.9486833)^2 = .90 = \frac{\text{SSR}}{\text{TSS}}$$

This result holds in general, that is,

$$r^2 = \frac{\text{SSR}}{\text{TSS}}$$

The square of the linear correlation coefficient is called the *coefficient of determination.*

r^2 = proportion of the variation in y values
that is explained by the linear relationship
with x (the regression line)

The coefficient of determination r^2 is a measure of the strength of the linear relationship between the x and y values. The greater the proportion of explained variation, the stronger the degree of linear relationship. Usually, the simplest way to find the proportion of explained variation is to calculate r and then square it.

The coefficient of determination is an intuitively appealing measure of linear relationship in that it measures the strength of the relationship in rather familiar terms.

Example 10.7

Calculate and interpret the coefficient of determination for the data in Example 10.5 (the salary data).

Solution

We have seen that for these data $r \doteq .96$. Therefore,

$$\text{coefficient of determination} = r^2 \doteq .92$$

This means that about 92% of the variation in the salaries is explained by years of experience (i.e., by the linear relationship). About 8% of the variability is unexplained. Perhaps this portion of the variation is due to chance or other variables not considered.

Example 10.8

Table 10.14 gives the size of the youth population, that is, the number of 14 to 24 year olds per 1000 of the population (x) versus the homicide rate measured in number of homicides per 100,000 of the population (y), in the six New England states in 1 year.

TABLE 10.14

State	Number of 14 to 24 Year Olds per 1000 (x)	Number of Homicides per 100,000 (y)
Maine	206	2.7
New Hampshire	204	1.4
Vermont	211	3.3
Massachusetts	211	3.7
Rhode Island	204	4.0
Connecticut	203	4.2

Source: Statistical Abstract of the United States, 100th Ed. (Washington, D.C.: U.S. Bureau of the Census), p. 178.

Find the linear correlation coefficient and the coefficient of determination. Interpret the results.

Solution

TABLE 10.15

	x	y	xy	x^2	y^2	
	206	2.7	556.2	42,436	7.29	
	204	1.4	285.6	41,616	1.96	
	211	3.3	696.3	44,521	10.89	
	211	3.7	780.7	44,521	13.69	
	204	4.0	816.0	41,616	16.00	
	203	4.2	852.6	41,209	17.64	
Sums:	1239	19.3	3987.4	255,919	67.47	$n = 6$

$$r = \frac{n\left(\sum xy\right) - \left(\sum x\right)\left(\sum y\right)}{\sqrt{n\left(\sum x^2\right) - \left(\sum x\right)^2} \cdot \sqrt{n\left(\sum y^2\right) - \left(\sum y\right)^2}}$$

$$= \frac{(6)(3987.4) - (1239)(19.3)}{\sqrt{(6)(255,919) - (1239)^2} \cdot \sqrt{(6)(67.47) - (19.3)^2}}$$

$$= \frac{11.7}{\sqrt{393} \cdot \sqrt{32.33}} = .1037975 \doteq .10$$

This value is quite small, indicating almost no linear relationship for the data.

coefficient of determination $= r^2 = (.1037975)^2 = .0107739 \doteq .01$

This confirms the fact that there is virtually no degree of linear relationship between the x and y values. It says that only about 1% of the variability in the homicide rate can be explained by (the linear relationship with) the number of 14 to 24 year olds per 1000. The accompanying scatter diagram (Figure 10.16) does not suggest any other kind of plausible relationship.

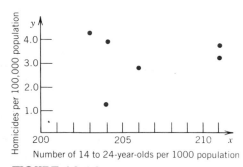

FIGURE 10.16

Exercises

In Exercises 10.18–10.22, (a) find the linear correlation coefficient, and (b) find the coefficient of determination and give its interpretation. The data, which were used in Exercises 10.2–10.6, follow.

10.18

x	y
2	9
2	7
4	5
4	5
6	3
6	1

10.19

x	y
0	2
0	4
0	6
1	6
2	7
2	8
2	9

10.20

x	y
0	5
0	4
1	3
2	2
2	1

10.21

x	y
1	2
2	2
3	1
4	0
5	0

10.22

x	y
1	2
2	6
4	14
5	18

10.23   A business school dean was interested in the relationship between quantitative GMAT scores (QUANT) and first-year grade-point average (GPA) at the school. Use the data below to find the porportion of variation in GPA explained by QUANT.

QUANT	GPA
38	3.2
40	3.7
25	3.3
32	3.1
33	2.9
36	3.7

10.24   In Exercise 10.15 we considered education and crime ratings from 6 U.S. cities. Find the proportion of variation in crime rating explained by education. The data are

City	Education (x)	Crime (y)
Boston	35	12
Chicago	35	10
Detroit	31	16
Los Angeles	32	20
New York	30	25
Washington, D.C.	36	13

10.25   Due to an overpopulation of deer in the Crane Reservation, located in the coastal community of Ipswich, Massachusetts, there was a local epidemic of Lyme disease. Lyme disease is transmitted by a tick (*Ixodes dammini*) that winters and mates on deer. In the table below, the zone is based on the distance from the reservation, with 1 being closest. Also given is the percentage of residents in each zone who have been infected with Lyme disease.

Lyme Disease Data

Zone (x)	% of Residents Infected (y)
1	66
2	42
3	32
4	13
5	21

Source: Lastavica, C. et al., "Rapid Emergence of a
Focal Epidemic of Lyme Disease in Coastal
Massachusetts," *New Engl. J. Med.*, 320:133–137, 1989.

Find the correlation coefficient and the coefficient of determination.

10.26   The information below is from Exercise 10.16, where x and y are *Fortune* magazine's rankings of 30 industrial firms in 1980 and 1985, respectively:

$$\sum x = 4571 \qquad \sum x^2 = 941{,}239 \qquad \sum y = 3860$$

$$\sum y^2 = 648{,}562 \qquad \sum xy = 771{,}472$$

Find the coefficient of determination and give its interpretation.

10.27   The following information, obtained from a sample of 30 cars, was used in Exercise 10.17 where x and y are horsepower and miles per gallon, respectively:

$$\sum x = 2444 \qquad \sum x^2 = 204{,}232 \qquad \sum y = 960$$

$$\sum y^2 = 31{,}514 \qquad \sum xy = 76{,}641$$

Find the coefficient of determination and give its interpretation.

10.28   Below is a scatter diagram with variables PRO being percentage approving the way President Reagan was doing his job (y), and UNEMP being the unemployment rate in percentage (x) (*Source:* Compiled by Paul Robertson from issues of *Public Opinion Quarterly and Business Conditions Digest*. As reported in *MacSpin Release 1.1*). The data are for the time period January 1981 through October 1985.

(a)   Do the data indicate a positive or negative relationship between the variables? Is this what you would expect?

(b)   Interpret the coefficient of determination $r^2 \doteq .814$.

(c)   What is the value of the linear correlation coefficient r?

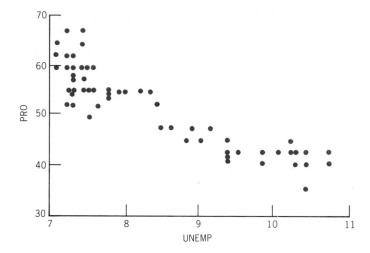

10.29 Below is a scatter diagram with variables PRO being percentage approving the way President Carter was doing his job (*y*), and UNEMP being the unemployment rate in percentage (*x*) (*Source:* Compiled by Paul Robertson from issues of *Public Opinion Quarterly and Business Conditions Digest.* As reported in *MacSpin Release 1.1*). The data are for the time period February 1977 through September 1980.

  (a) Do the data indicate a positive or negative relationship between the variables? Is this what you would expect?

  (b) Interpret the coefficient of determination $r^2 \doteq .08$.

  (c) What is the value of the linear correlation coefficient *r*?

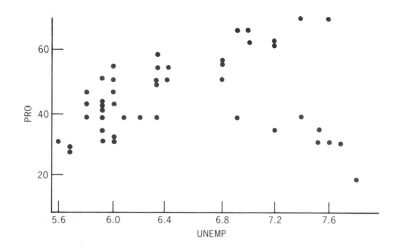

10.30   In Exercise 10.3 we considered the following data:

x	y
0	2
0	4
0	6
1	6
2	7
2	8
2	9

(a)   Complete the following table. Then find **(i)** TSS $= \Sigma(y - \bar{y})^2$ and **(ii)** SSR $= \Sigma(\hat{y} - \bar{y})^2$. Use TSS and SSR to find the coefficient of determination r^2. (*Note:* $\hat{y} = 4 + 2x$)

x	y	$\hat{y}$	$\bar{y}$	$(y - \bar{y})^2$	$(\hat{y} - \bar{y})^2$

(b)   Now use the formula given in the text to find r for the same data. Then compute r^2 and compare with your answer in part (a).

10.31   In Exercise 10.12 we looked at the variables Intensity of Advertising and Number of Cases Sold.

Intensity of Advertising (x)	Number of Cases Sold (y)
1	4
2	7
3	8
4	14
5	15

(a)   Complete the following table. Then find **(i)** TSS $= \Sigma(y - \bar{y})^2$ and **(ii)** SSR $= \Sigma(\hat{y} - \bar{y})^2$. Use TSS and SSR to find the coefficient of determination r^2. (*Note:* $\hat{y} = 9 + 2.9$)

x	y	$\hat{y}$	$\bar{y}$	$(y - \bar{y})^2$	$(\hat{y} - \bar{y})^2$

(b) Now use the formula given in the text to find r for the same data. Then compute r^2 and compare with your answer in part (a).

10.32 In parts (a)–(h), find the missing entries.

	SSR	SSE	TSS	r^2
(a)	40	60		
(b)	100		300	
(c)		20	50	
(d)	80			.4
(e)		27		.7
(f)			40	.2
(g)	25			1.0
(h)	30	0		

10.33 This exercise illustrates the relationship between the linear correlation coefficient r and the line of best fit. Consider

x	y
1	2
1	4
3	3
4	2
6	9

(a) Find the sample standard deviations s_x and s_y.

(b) Find the linear correlation coefficient r.

(c) Find the line of best fit.

(d) Using your results in parts (a), (b), and (c),

    (i) Verify that $r\left(\dfrac{s_y}{s_x}\right) = b_1$.

    (ii) Verify that $\bar{y} - r\left(\dfrac{s_y}{s_x}\right)\bar{x} = b_0$.

(e) Using part (d), verify that the equation of the line of best fit for these data may be written as

$$\hat{y} = \bar{y} + r\left(\frac{s_y}{s_x}\right)(x - \bar{x})$$

(*Note:* It can be shown that the line of best fit can always be written in this form.)

10.34 This exercise uses the results of Exercise 10.33. A company gave a course for its typists to improve typing speed. A pretest (x) and posttest (y) were given to each typist. The results were

$$\bar{x} = 60 \text{ words per minute} \qquad \bar{y} = 61 \text{ words per minute}$$

$$s_x = 5 \text{ words per minute} \qquad s_y = 5 \text{ words per minute}$$

$$r = .6$$

(a) Refer to Exercise 10.33 and write the equation of the regression line.

(b) What is the expected (average) posttest score for a typist who typed 80 words per minute on the pretest?

(c) What is the expected posttest score for a typist who typed 40 words per minute on the pretest?

Observe that the results of parts (b) and (c) suggest that those who did well on the pretest will not do as well on the posttest, and those who did not do well on the pretest show an improvement on the posttest. This is called the *regression effect*, and it is due to the structure of the regression equation rather than any effect due to the typing course. Specifically, this is due to the presence of r in the regression equation. If you replace .6 by the value 1 and repeat parts (b) and (c), you will get very different results.

10.4 INFERENCE CONCERNING CORRELATION AND REGRESSION

The coefficient of determination, discussed in the last section, gives us a measure of the degree of linear relationship between the x and y values in a particular scatter diagram.

Another way to study the issue of linearity is to consider the population from which the data were obtained. For example, consider the data in Examples 10.2 and 10.5. In these examples, x represented the years of experience and y the annual salary for 12 technicians in a certain field of electronics. We can view the 12 technicians in the study as a sample from a large population of technicians employed in this field. Experience (x) and salary (y) can assume a variety of values and may be regarded as random variables. We can also view the population as the collection of all the (x, y) values for all these technicians. As such, it is called a **bivariate population.** The linear correlation coefficient for the population is denoted by ρ (the Greek letter rho). If $\rho = 0$, there is no linear relationship between x and y. If $\rho \neq 0$, there is at least some degree of linear relationship between x and y for the population. In other words, the pairs of (x, y) values for the population are scattered about a nonhorizontal line—the population regression line.

When investigating linearity for a population, we wish to test

$$H_0: \rho = 0 \quad \text{(there is no linear relationship)}$$
$$H_a: \rho \neq 0 \quad \text{(there is a linear relationship)}$$

Now the sample coefficient r may be thought of as a rough estimate for ρ. If $\rho = 0$, the value of r should not be too far from 0. But if r is too far from 0 (i.e., too close to 1 or -1) to be consistent with H_0, we reject H_0 in favor of H_a.* The value of r will be considered too far from 0, if it is unlikely that we would observe a value so far from 0 (if H_0 were true).

To test the above hypotheses, we can find critical values of r from Appendix Table B.7. The table enables us to use either of two levels of significance: .01 or .05. Suppose that we decided to conduct a test using $\alpha = .01$ based on a sample of 12 pairs of (x, y) values. From Appendix Table B.7, we see that for $n = 12$, the critical value (call it c) is .708. This is the right critical value for a two-tailed test. See Figure 10.17.

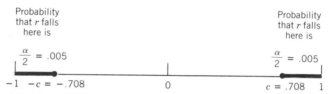

FIGURE 10.17
Critical Region for r When $\alpha = .01$ and $n = 12$

The critical region consists of the values of r from .708 to 1 and from $-.708$ to -1. If the value of r computed from the sample data falls in the critical region, we reject H_0 and conclude that the variables x and y tend to show a linear relationship ($\rho \neq 0$). In this case we say that the x and y values for the sample data showed a *significant degree of linear relationship*.

Recall, that for the salary data, we found that $r = .96$ for the 12 technicians in Example 10.5. We concluded that this value is quite close to 1 and does seem to indicate a strong linear relationship. Figure 10.17 seems to confirm this for the population, in that $r = .96$ falls in the critical region. That is, we reject H_0 (which asserts no linear relationship) in favor of H_a (which asserts a linear relationship). The value of .96 was quite close to 1, but sometimes it is not so obvious as to whether a significant linear relationship exists or not. The hypothesis testing procedure discussed in this section can be very useful in making a decision in these cases.

We now summarize the steps in a test for a linear relationship between two variables.

---

*Keep in mind that $-1 \leq r \leq 1$.

1. Hypotheses:

$$H_0: \rho = 0 \quad \text{(there is no linear relationship)}$$
$$H_a: \rho \neq 0 \quad \text{(there is a linear relationship)}$$

2. Determine the level of significance α. (We will use .01 or .05.)

3. Test statistic: We use the linear correlation coefficient r, obtained from the sample data.

4. We perform a two-tailed test. Using Appendix Table B.7, we find the right critical value c. The critical region consists of values of $r \geq c$ or $\leq -c$, as shown in Figure 10.18.

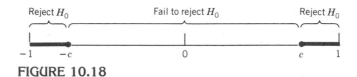

FIGURE 10.18

5. Decision: Using the value of r obtained in step 3, reject H_0 if $r \geq c$ or $\leq -c$. If H_0 is rejected, we conclude that there is a linear relationship between the variables x and y.

Example 10.9

A computer science professor recorded, for each student in a computer science seminar, the number of hours logged on the computer (x) while preparing for a final exam, along with the score on the final examination (y). The results are given in Table 10.16.

TABLE 10.16

Student	Hours on Computer (x)	Score on Final Examination (y)
1	5	60
2	8	55
3	10	70
4	12	65
5	14	85
6	15	70
7	18	80
8	20	90

Do these data indicate a significant degree of linear relationship between the variables x and y? Use the 1% level of significance.

Solution

The question is whether there would be a linear relationship between the x and y values for the population of all possible students who have taken, or will take, this course or a similar course. We are asking whether the population coefficient ρ is different from 0.

1. Hypotheses:

$$H_0: \rho = 0 \quad \text{(no linear relationship)}$$

$$H_a: \rho \neq 0 \quad \text{(linear relationship)}$$

2. Level of significance: $\alpha = .01$.

3. Test statistic and its observed value: We need to compute the value of r, the linear correlation coefficient for the sample.

TABLE 10.17

	x	y	xy	x^2	y^2
	5	60	300	25	3600
	8	55	440	64	3025
	10	70	700	100	4900
	12	65	780	144	4225
	14	85	1190	196	7225
	15	70	1050	225	4900
	18	80	1440	324	6400
	20	90	1800	400	8100
Sums:	102	575	7700	1478	42,375	$n = 8$

$$r = \frac{n\left(\sum xy\right) - \left(\sum x\right)\left(\sum y\right)}{\sqrt{n\left(\sum x^2\right) - \left(\sum x\right)^2} \cdot \sqrt{n\left(\sum y^2\right) - \left(\sum y\right)^2}}$$

$$= \frac{(8)(7700) - (102)(575)}{\sqrt{(8)(1478) - (102)^2} \cdot \sqrt{(8)(42,375) - (575)^2}}$$

$$= \frac{2950}{\sqrt{1420} \cdot \sqrt{8375}} = .8554319 \doteq .86$$

4. Critical region: From Appendix Table B.7 we see that for sample size $n = 8$ and $\alpha = .01$, the critical values are $\pm.834$, as shown in Figure 10.19.

FIGURE 10.19

5.   Decision: Since the observed value $r = .86$ is in the critical region, we reject H_0 in favor of H_a. This means that there is a linear relationship.

The Linear Regression Model

For the above test procedure to be valid, certain properties concerning the population must be satisfied. These are the properties that describe the so-called **linear regression model.**

Properties of the Linear Regression Model

1.   For each x value, the collection of all possible y values corresponding to that x value has a normal distribution.* For example, let $x = $ a person's high school grade-point average (GPA) and $y = $ college GPA. The collection of all the college GPA's for all students with a high school average of 2.50 has a normal distribution.

2.   There is a population regression line with equation

$$y = \beta_0 + \beta_1 x$$

where β_0 and β_1 are constants. β is the Greek letter beta. For a particular x value substituted into this equation, the resulting y value gives the mean for all the y values corresponding to that x value. This is sometimes denoted by $\mu_{y|x}$.

3.   The variance from the regression line of all the y values corresponding to a specific x value will be the same no matter what x value we choose. We denote this common variance by σ^2.

The linear regression model is shown in Figure 10.20a. The population regression line is usually unknown. However, if we obtain a sample, we can estimate the population regression line by the sample regression line: $\hat{y} = b_0 + b_1 x$.

The linear model can also be described by the equation

$$y = \beta_0 + \beta_1 x + \varepsilon$$

where ε is the Greek letter epsilon. Each point (x, y) in the population satisfies this equation for some value of ε. See Figure 10.20b. When $\varepsilon = 0$, the point is

---

*Moderate departures from normality are not very serious. In fact, departures from normality become less important as the sample size increases.

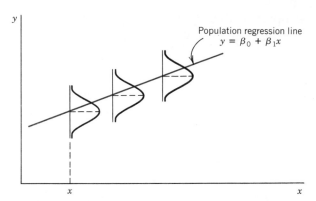

FIGURE 10.20a

Linear Model: for Each Value of x, the Population of Corresponding y Values Is Normal with the Mean on the Regression Line and with the Same Variance σ^2

FIGURE 10.20b

$y = \beta_0 + \beta_1 x + \epsilon$

exactly on the population regression line

$$y = \beta_0 + \beta_1 x$$

For a fixed x, ϵ can assume a variety of values, each corresponding to a different y value. For the linear model, we assume:

- ϵ (and hence y) is normal.
- The mean of ϵ is 0. (Hence the mean of y is $\beta_0 + \beta_1 x$.)
- The variance of ϵ (and hence y) is σ^2. This is the same for each value of x.

Sometimes a scatter diagram will reveal departures from the linear regression model. For two examples see Figure 10.21.

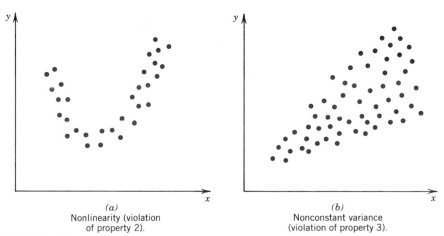

(a)
Nonlinearity (violation
of property 2).

(b)
Nonconstant variance
(violation of property 3).

FIGURE 10.21

Departures from the Linear Regression Model

For the exercises and examples in this section we will assume that the properties of the linear regression model are satisfied.

Exercises

In Exercises 10.35–10.39, do the sample data indicate a linear relationship between the x and y values for the population?

10.35

x	y
0	2
0	4
0	6
1	6
2	7
2	8
2	9

Use $\alpha = .01$. (You may have calculated r in Exercise 10.19.)

10.36

x	y
0	3
2	5
4	0
6	4

Use $\alpha = .05$.

10.37

x	y
1	2
2	2
3	1
4	0
5	0

Use $\alpha = .05$. (You may have calculated r in Exercise 10.21.)

10.38

x	y
0	5
0	4
1	3
2	2
2	1

Use $\alpha = .05$. (You may have calculated r in Exercise 10.20.)

10.39

x	y
1	2
2	6
4	14
5	18

Use $\alpha = .05$. (You may have calculated r in Exercise 10.22.)

10.40 The data in the following table represent domestic (x) and export (y) factory sales, in millions, of passenger cars for the years 1982–1987 (*Source: The 1989 Information Please Almanac, p. 76*).

Domestic (x)	Exports (y)
4.70	35
6.20	54
7.03	59
7.34	67
6.87	65
6.49	60

Is there sufficient evidence to indicate a linear relationship between the x and y variables? Use a 5% level of significance.

10.41 The data in the following table represent high school grade-point averages (x) and final introductory statistics grades (y).

Grade-Point Average (x)	Final Grade (y)
3.2	91
2.4	72
3.4	94
2.6	65
2.9	83
2.1	60
3.6	97

(a) Do the sample data indicate a linear relationship between the x and y values of the population? Use a 5% level of significance.

(b) Find the line of best fit.

(c) Estimate the final statistics grade for a student with a grade-point average of 2.5.

10.42 A developer intended to build a racquetball club in a city of 135,000 people but was undecided as to the number of courts to build. The data in the following table were obtained concerning a flourishing chain of racquetball clubs. The data represent the population of a city (x) and the number of courts (y) the chain has in that city.

Population in Thousands (x)	Number of Courts (y)
150	17
230	16
265	21
90	8
75	6
190	12
105	10
175	14

(a) Is there sufficient evidence to indicate a linear relationship between the x and y values for the population? Use a 1% level of significance.

(b) Find the line of best fit.

(c) What number of courts would you recommend the developer to build?

10.43 The data in the following table represent percent sugars (x) and cost (y) for eight hot cereals. (*Source: Consumer Reports, 1982, p. 71.*)

Percent Sugars (x)	Cost per Serving (in cents) (y)
5	4
31	11
3	9
4	6
11	6
29	11
8	8
5	4

Do the sample data indicate a linear relationship between the x and y variables? Use a 1% level of significance.

10.44 In Exercises 10.16 and 10.26, we discussed the ranks x and y of 30 randomly selected industrial firms in 1980 and 1985, respectively. Do the data below indicate a linear relationship between the x and y variables? Use a 5% level of significance.

$$\sum x = 4571 \qquad \sum x^2 = 941,239 \qquad \sum y = 3860$$
$$\sum y^2 = 648,562 \qquad \sum xy = 771,472$$

(You may have calculated r in Exercise 10.26.)

10.45   Consider the following data:

x	y
2	12
4	9
5	8
6	6
8	5

Is there sufficient evidence to indicate a linear relationship between the x and y values for the population? Use a 5% level of significance.

10.46   Consider the following data:

x	y
4	2
4	2
4	5
6	6
8	7
8	9
15	9

Is there sufficient evidence to indicate a linear relationship between the x and y values for the population? Use a 5% level of significance.

PREDICTION INTERVALS AND CONFIDENCE INTERVALS (OPTIONAL)

We have seen that the regression line for a sample of (x, y) values is often used to estimate or predict an unknown y value corresponding to a given x value. For example, at the beginning of this chapter we looked at salary data for 12 technicians (Table 10.2). The equation of the regression line was

$$\hat{y} = 19.37 + .70x$$

We saw that an estimate of the salary for a technician with 15 years experience was

$$\hat{y} = 19.37 + (.70)(15) = 29.87$$

(This is equivalent to $29,870.) If we wanted to predict the salary of a technician with 15 years experience, we would use this figure. The actual value of y will be hopefully not too far from $\hat{y}$. We can calculate a type of confidence interval for y, called a **prediction interval.** We select a level of confidence $1 - \alpha$. There is then a procedure for finding an interval such that the confidence is $1 - \alpha$ that the procedure will yield an interval containing y.

To estimate the mean salary of *all* technicians with 15 years experience in the field ($\mu_{y|15}$), we use the same figure, $29,870. We can also find an interval that (hopefully) contains the mean, a **1 − α confidence interval.** Since individual y values can be quite variable, we can expect prediction intervals to be wider than confidence intervals.

Prediction intervals and confidence intervals are both types of interval estimates, and we have seen that interval estimates often take the general form

$$\text{(estimate)} \pm t \cdot \text{(standard deviation)}$$

For example, we have seen that a confidence interval for a population mean based on a small sample is

$$\bar{x} \pm t\left(\frac{\alpha}{2}\right) \cdot \frac{s}{\sqrt{n}}$$

The estimate is $\bar{x}$ and the standard deviation of $\bar{x}$ is estimated by $s/\sqrt{n}$. Prediction intervals and confidence intervals take this same general form. In fact they differ from one another only in the standard deviation. We will have more to say about the structure of these intervals at the end of this section.

Let's first consider prediction intervals.

---

1 − α Prediction Interval for y when $x = x_0$ Calculate the value of

$$S = \sqrt{\frac{\text{SSE}}{n - 2}} = \sqrt{\frac{\sum(y - \hat{y})^2}{n - 2}}$$

This is called the *standard deviation from the regression line* and is an estimate of the standard deviation (σ) of y values from the population regression line. Using the regression line, calculate the predicted value of y when $x = x_0$,

$$\hat{y} = b_0 + b_1 x_0$$

Now use Appendix Table B.4 to find the t value, $t(\alpha/2)$, based on df = $n - 2$. Then a $1 - \alpha$ prediction interval for y when $x = x_0$ is defined by the endpoints:

$$\hat{y}_0 \pm t\left(\frac{\alpha}{2}\right) \cdot S \cdot \sqrt{1 + \frac{1}{n} + \frac{(x_0 - \bar{x})^2}{\left(\sum x^2\right) - \left(\sum x\right)^2 / n}}$$

Assumptions: The properties of the linear regression model are satisfied.

Example 10.10

Refer to the salary data of Example 10.2. Find a 95% prediction interval for the annual salary of a technician with 15 years experience.

Solution

We have seen that the regression equation is

$$\hat{y} = 19.37 + .70x$$

and the predicted salary for a technician with $x = x_0 = 15$ years experience is

$$\hat{y}_0 = 19.37 + (.70)(15) = 29.87 \text{ (thousand dollars)}$$

Table 10.18 will help in the computation of S. Each value of $\hat{y}$ in the table is obtained by substituting the corresponding value of x into the regression equation.

TABLE 10.18

x	y	$\hat{y}$	$y - \hat{y}$	$(y - \hat{y})^2$
12	29	27.77	1.23	1.5129
16	29	30.57	−1.57	2.4649
6	23	23.57	−.57	.3249
23	34	35.47	−1.47	2.1609
27	38	38.27	−.27	.0729
8	24	24.97	−.97	.9409
5	22	22.87	−.87	.7569
19	34	32.67	1.33	1.7689
23	36	35.47	.53	.2809
13	27	28.47	−1.47	2.1609
16	33	30.57	2.43	5.9049
8	27	24.97	2.03	4.1209
				22.4708 = SSE

$$S = \sqrt{\frac{SSE}{n - 2}} = \sqrt{\frac{22.4708}{12 - 2}} = \sqrt{2.24708} \doteq 1.50$$

From Example 10.2, $\Sigma x^2 = 3162$, $\Sigma x = 176$, and $\bar{x} = 176/12 \doteq 14.67$. Our level of confidence is 95%. So $1 - \alpha = .95$ and $\alpha = .05$. Using Appendix Table B.4, we find that when df $= n - 2 = 10$,

$$t\left(\frac{\alpha}{2}\right) = t(.025) = 2.228$$

The prediction interval has endpoints

$$\hat{y}_0 \pm t\left(\frac{\alpha}{2}\right) \cdot S \cdot \sqrt{1 + \frac{1}{n} + \frac{(x_0 - \bar{x})^2}{\left(\sum x^2\right) - \left(\sum x\right)^2 / n}}$$

or

$$29.87 \pm (2.228)(1.50)\sqrt{1 + \frac{1}{12} + \frac{(15 - 14.67)^2}{3162 - (176)^2/12}}$$

$$29.87 \pm (2.228)(1.50)\sqrt{1.0835209}$$

$$29.87 \pm (2.228)(1.50)(1.0409231)$$

$$29.87 \pm 3.48 \text{ or } 26.39 \text{ to } 33.35$$

Hence our confidence is 95% that the actual value of y will be between \$26,390 and \$33,350.

In the previous example, we found an interval estimate for the salary of an *individual* technician with 15 years of experience. But we may also be interested in estimating the mean salary for *all* technicians in the population with 15 years of experience, $\mu_{y|15}$. We said that the point estimate for the mean is the same as the point estimate for an individual y (\$29,870). We construct a **1 − α confidence interval** for the mean as follows:

---

1 − α Confidence Interval for $\mu_{y|x_0}$. (the mean of all y values corresponding to $x = x_0$) The point estimate for $\mu_{y|x_0}$ is obtained from our computed regression line

$$\hat{y}_0 = b_0 + b_1 x_0$$

The 1 − α confidence interval is defined by the endpoints

$$\hat{y}_0 \pm t\left(\frac{\alpha}{2}\right) \cdot S \cdot \sqrt{\frac{1}{n} + \frac{(x_0 - \bar{x})^2}{\left(\sum x^2\right) - \left(\sum x\right)^2 / n}}$$

where $t\left(\frac{\alpha}{2}\right)$ is obtained from Appendix Table B.4 based on df = $n - 2$, and S is the standard deviation from the regression line.
Assumptions: The properties of the linear regression model are satisfied.

Figure 10.22 shows the relationship between the estimate $\hat{y}_0$ and the quantities being estimated.

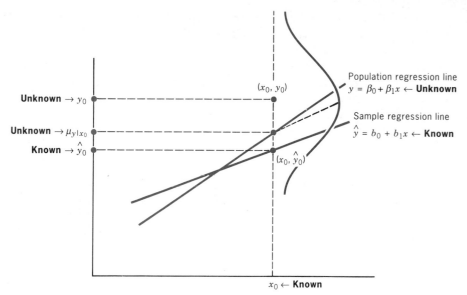

FIGURE 10.22
$\hat{y}_0$ **is a Point Estimate for Both an** *Individual y* **Value (** y_0 **) and the Mean of** *All y*
Values Corresponding to the Value $x_0(\mu_{y|x_0})$

Example 10.11
Referring to the salary data in the previous example, find a 95% confidence
interval for the mean salary of all technicians with 15 years experience ($\mu_{y|15}$).

Solution
As noted, the point estimate is obtained from the computed regression line
when $x = x_0 = 15$, getting

$$\hat{y}_0 = 19.37 + (.70)(15) = 29.87 \text{ (thousand dollars)}$$

From Example 10.10, we have seen that $S = 1.50$, $t\left(\dfrac{\alpha}{2}\right) = t(.025) = 2.228$,

$\Sigma x^2 = 3162$, $\Sigma x = 176$, and $\bar{x} = 14.67$. Thus the $1 - \alpha$ confidence interval for
$\mu_{y|15}$ is

$$\hat{y}_0 \pm t\left(\frac{\alpha}{2}\right) \cdot S \cdot \sqrt{\frac{1}{n} + \frac{(x_0 - \bar{x})^2}{\left(\Sigma x^2\right) - \left(\Sigma x\right)^2 / n}}$$

or

$$29.87 \pm (2.228)(1.50)\sqrt{\frac{1}{12} + \frac{(15 - 14.67)^2}{3162 - (176)^2/12}}$$

$$29.87 \pm (2.228)(1.50)\sqrt{.0835209}$$

$$29.87 \pm (2.228)(1.50)(.2889998)$$

$$29.87 \pm .97 \text{ or } 28.90 \text{ to } 30.84$$

Thus we are 95% confident that the mean salary for those with 15 years experience is between $28,900 and $30,840.

In Example 10.10 we found that a 95% prediction interval for the salary of a technician with 15 years experience went from 26.39 to 33.35 (thousand dollars). If you were to find a 95% prediction interval for a salary of a technician with $x_0 = 25$ years experience, you would obtain the endpoints:

$$36.87 \pm 3.76$$

$$33.11 \text{ to } 40.63$$

Note that the width of this interval is 7.52. In Example 10.10, where the number of years of experience was $x_0 = 15$, the width of the prediction interval was 6.96. The reason for this difference can be understood by examining the formula for the prediction interval. Note the term

$$\frac{(x_0 - \bar{x})^2}{\left(\sum x^2\right) - \left(\sum x\right)^2 / n}$$

As the value of x_0 gets farther from $\bar{x}$, this term gets larger, producing a wider prediction interval. In Example 10.10 we dealt with the value $x_0 = 15$, which is quite close to $\bar{x} = 14.67$. But the value 25 is farther from $\bar{x}$, accounting for the wider prediction interval. In general, the farther x_0 is from $\bar{x}$, the wider the prediction interval.

Similar remarks apply to the confidence interval: it gets wider the farther x_0 is from $\bar{x}$. See Figure 10.23. Another property we should note from Figure 10.23

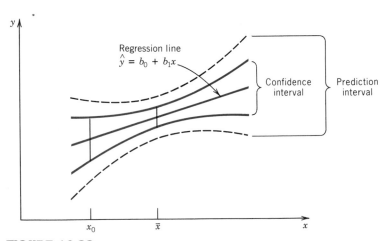

FIGURE 10.23
Prediction Intervals Are Wider than Confidence Intervals. Width of Both Intervals Increases as x_0 Gets Farther from $\bar{x}$

is that a prediction interval is wider than a confidence interval. In the previous examples, we found that a prediction interval for an individual salary for 15 years experience went from $26,390 to $33,350. But a confidence interval for the mean salary (for 15 years experience) went from $28,900 to $30,840. The reason for this can be seen by comparing the formulas. The formula for a prediction interval has an additional term (1) under the radical, making the prediction interval wider than the confidence interval.

Background Discussion

We noted earlier that our prediction and confidence intervals take the form

$$\text{(estimate)} \pm t \cdot \text{(standard deviation)}$$

Here we mean the standard deviation of the distance between the estimate and that which is being estimated. We used $\hat{y}_0$ to estimate $\mu_{y|x_0}$. See Figure 10.22. Now $\hat{y}_0$ can be thought of as a variable. (In repeated sampling its value could change because the regression line could change.) It can be shown that an estimate of the variance of $\hat{y}_0$ from the population regression line is

$$S_{\hat{y}_0}^2 = S^2 \left[\frac{1}{n} + \frac{(x_0 - \bar{x})^2}{\left(\sum x^2\right) - \left(\sum x\right)^2 / n} \right]$$

Therefore, the standard deviation is

$$S_{\hat{y}_0} = S \sqrt{\frac{1}{n} + \frac{(x_0 - \bar{x})^2}{\left(\sum x^2\right) - \left(\sum x\right)^2 / n}}$$

Thus our $1 - \alpha$ confidence interval for $\mu_{y|x_0}$ is

$$\hat{y}_0 \pm t\left(\frac{\alpha}{2}\right) S \sqrt{\frac{1}{n} + \frac{(x_0 - \bar{x})^2}{\left(\sum x^2\right) - \left(\sum x\right)^2 / n}}$$

What about the prediction interval? An examination of Figure 10.22 shows that the distance from y_0 to $\hat{y}_0$ consists of two components: the distance from y_0 to the population regression line, and the distance from the population regression line to $\hat{y}_0$. The variance of the first distance is estimated by S^2, the second by $S_{\hat{y}_0}^2$. The overall variance is the sum of these:

$$S^2 + S_{\hat{y}_0}^2 = S^2 + S^2 \left[\frac{1}{n} + \frac{(x_0 - \bar{x})^2}{\left(\sum x^2\right) - \left(\sum x\right)^2 / n} \right]$$

$$= S^2 \left[1 + \frac{1}{n} + \frac{(x_0 - \bar{x})^2}{\left(\sum x^2\right) - \left(\sum x\right)^2 / n} \right]$$

$$\text{standard deviation} = \sqrt{S^2 + S_{\hat{y}_0}^2} = S\sqrt{1 + \frac{1}{n} + \frac{(x_0 - \bar{x})^2}{\left(\sum x^2\right) - \left(\sum x\right)^2 / n}}$$

Thus our $1 - \alpha$ prediction interval is

$$\hat{y}_0 \pm t\left(\frac{\alpha}{2}\right) S \sqrt{1 + \frac{1}{n} + \frac{(x_0 - \bar{x})^2}{\left(\sum x^2\right) - \left(\sum x\right)^2 / n}}$$

Exercises

10.47   In Exercise 10.45 we used the data:

x	y
2	12
4	9
5	8
6	6
8	5

    (a)   Find the line of best fit.
    (b)   Estimate y when $x = 7$.
    (c)   Find a 95% prediction interval for y when $x = 7$.
    (d)   Find a 95% confidence interval for $\mu_{y|7}$.

10.48   In Exercise 10.46 we considered the data:

x	y
4	2
4	2
4	5
6	6
8	7
8	9
15	9

    (a)   Find the line of best fit.
    (b)   Estimate y when $x = 5$.
    (c)   Find a 95% prediction interval for y when $x = 5$.
    (d)   Find a 95% confidence interval for $\mu_{y|5}$.

10.49   In Exercises 10.17 and 10.27, we discussed horsepower (x) and miles per gallon (y) of 30 automobiles with

$$\sum x = 2444 \qquad \sum x^2 = 204{,}232$$

Assume $r \doteq -.777$, $\hat{y} = 56.90 - .31x$, and SSE $= 315.11$.

(a) Is there sufficient evidence to indicate a linear relationship between the x and y values for the population? Use a 5% level of significance.

(b) Find a 95% prediction interval for miles per gallon for a car with 90 horse-power.

(c) Find a 95% confidence interval for $\mu_{y|90}$.

10.50 Refer to Exercise 10.41 and find

(a) A 90% prediction interval for the final statistics grade, given a grade-point average of 2.5.

(b) A 90% confidence interval for $\mu_{y|2.5}$.

10.51 Refer to Exercise 10.42 and find

(a) A 90% prediction interval for the number of courts when the city size is 200,000.

(b) A 90% confidence interval for $\mu_{y|200,000}$.

SOME WORDS OF CAUTION CONCERNING CORRELATION AND REGRESSION

There are a number of misuses of the ideas we have discussed in this chapter. We now mention a few of them.

1. Equating correlation with causality: Suppose that we were to record for each month the number of snakebites (y) and the amount of ice cream consumed in the United States (x). It is a fact that a strong positive correlation exists between these two variables. Does it mean that eating ice cream causes snakebite? Hardly so. Instead, there is a third factor that explains the relationship. As the weather gets warmer, snakes come out of hibernation and become more active. At the same time our appetite for cool refreshments increases. Therefore, we cannot assert that correlation equals causality. This does not mean that correlation analysis may never be used in drawing conclusions about causal relationships, but some common sense should be used. For example, a correlation study contributed heavily to the Surgeon General's report linking cigarette smoking with lung cancer.* Furthermore, other studies have shown an association between smoking and lung cancer. Also, there does not appear to be a plausible third factor that might explain changes in both smoking and rate of lung cancer. There has been clinical evidence as well that shows the deleterious effects of smoking on the lungs, so a conclusion asserting a causal link would not be unreasonable.

2. Some assume that a significant value for r implies a strong linear relationship. For example, when testing the hypotheses

$$H_0: \rho = 0$$
$$H_a: \rho \neq 0$$

*See Doll, R., "Etiology of Lung Cancer," *Adv. Cancer Res.*, 3:1–50, 1955. Report of the U.S. Surgeon General, *Smoking and Health* (1964).

the value of r for the sample data is significant if we reject H_0. This implies that there is *some* degree of linear relationship between the x and y values for the population but not necessarily a *strong* relationship. For example, if we had $n = 100$ pairs of (x, y) values, and we were using the 5% level of significance, a value of $r = .2$ would lead to the rejection of H_0, indicating a linear relationship between the variables x and y. (See Appendix Table B.7. The critical values of r are $\pm.196$ when $n = 100$.) Therefore there is some degree of linear relationship. But notice that the coefficient of determination is $r^2 = .04$. This means that only 4% of the variability of the y values (in the sample) can be accounted for by the linear relationship. In this case the regression line may not be a reliable predictor of y values.

3. Unwarranted extrapolation: In Example 10.4 we found the regression line relating production cost per ton (y) to the number of tons of a chemical produced (x). The data used were obtained from companies in the range of 2000 to 6000 tons per year. It might be reasonable to use this line to predict the cost per ton for a company in this range. But to use this line to predict the cost per ton for a company producing, say, 1 million tons per year would be questionable.

10.7 USING MINITAB (OPTIONAL)

The Minitab printout below deals with the data in Table 10.2, where x represents years of experience and y represents annual salary (in thousands of dollars) for 12 technicians. We read the years (x) into column C1 and salaries (y) into C2. The PLOT command produces the scatter diagram that we saw in Figure 10.2*a*:

PLOT C2 VS C1

The REGRESS command produces the regression equation given in Example 10.2. Since the x values in C1 are used to predict y values in C2, we typed

REGRESS C2 ON 1 PREDICTOR IN C1

We remind the reader that Minitab requires only the bare essentials, and any other text is optional. We could have typed

REGR C2 1 C1

We also produce the correlation coefficient by use of the command

CORRELATION BETWEEN DATA IN C1 C2

```
MTB > READ THE FOLLOWING DATA INTO C1 C2
DATA> 12   29
DATA> 16   29
DATA> 6    23
```

```
DATA> 23  34
DATA> 27  38
DATA> 8  24
DATA> 5  22
DATA> 19  34
DATA> 23  36
DATA> 13  27
DATA> 16  33
DATA> 8  27
DATA> end
      12 ROWS READ
MTB > PLOT C2 VS Cl
```

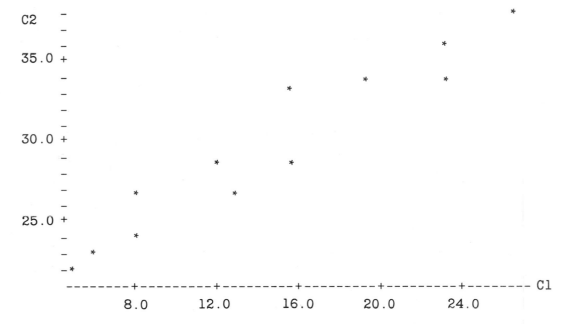

```
MTB > REGRESS C2 ON 1 PREDICTOR IN Cl

The regression equation is
C2 = 19.4 + 0.702 Cl


Predictor      Coef      Stdev     t-ratio        p
Constant     19.370      1.010       19.19    0.000
Cl          0.70207    0.06219       11.29    0.000

s = 1.499    R-sq = 92.7%    R-sq(adj) = 92.0%

Analysis of Variance
```

SOURCE	DF	SS	MS	F	p
Regression	1	286.21	286.21	127.44	0.000
Error	10	22.46	2.25		
Total	11	308.67			

```
MTB > CORRELATION BETWEEN DATA IN Cl C2

Correlation of Cl and C2 = 0.963
```

In addition to the regression equation, the REGRESS command produces a variety of other information, including b_0 and b_1 displayed under Coef. In this chapter we did not discuss t ratios for the coefficients b_0, b_1 given in the printout. But the important ratio is the t ratio of the coefficient of C1 (that is, the coefficient of x.) This t ratio is a type of standard score. It is the coefficient b_1 divided by an estimate of its standard deviation, and it has a t distribution with degrees of freedom

$$df = n - 2 = 12 - 2 = 10$$

In the above printout, the t ratio for the coefficient of x is

$$t = \frac{b_1}{\text{standard deviation of } b_1} = \frac{.70207}{.06219} \doteq 11.29$$

When b_1 (or equivalently, its t-ratio) is sufficiently far from 0, we conclude that the slope (β_1) of the *population* regression line is different from 0. This in turn implies a nonhorizontal population regression line, which means that there is a linear relationship between x and y (i.e. $p \neq 0$). So a two-tailed t test would be appropriate, and is equivalent to testing

$$H_0: \rho = 0$$

$$H_a: \rho \neq 0$$

Note that the P value for $t = 11.29$ is approximately 0, indicating we should reject the null hypothes and conclude that there is a significant linear relationship.

The value of S, the standard deviation from the regression line, is given as 1.499. We computed this in Example 10.10 from the formula

$$S = \sqrt{\frac{SSE}{n - 2}} = \sqrt{\frac{\Sigma(y - \hat{y})^2}{n - 2}}$$

We can think of S as an estimate of σ, the standard deviation from the *population* regression line. This is the standard deviation of all y values corresponding to any fixed x value.

The coefficient of determination R-SQUARED is given and (except for a difference in roundoff) agrees with the value we found in Example 10.7. The printout also gives R-SQUARED ADJUSTED FOR D.F. (degrees of freedom). (For a more complete discussion of this, consult a Minitab reference manual.)

It suffices to say that R-SQUARED ADJUSTED FOR D.F. is a certain type of estimate for the *population* coefficient of determination—an *unbiased estimator*.

Under ANALYSIS OF VARIANCE we see the breakdown of the total sum of squares (TSS) into sum of squares for regression (SSR) and sum of squares for residual or error (SSE). Observe that the proportion of variation explained by regression is

$$\frac{\text{SSR}}{\text{TSS}} = \frac{286.21}{308.67} \doteq .927 \text{ or } 92.7\%, \text{ which is } r^2$$

Under ANALYSIS OF VARIANCE we also see columns for DF (degrees of freedom) and MS (mean square). When discussed in connection with a sum of squares, we can think of degrees of freedom as a number we divide a sum of squares by to get a quantity called a "mean square." A mean square can be thought of as an estimate of some sort of variance, in this case the variance σ^2 from the population regression line. The **mean square for error** (or residual) is

$$\text{MSE} = \frac{\text{SSE}}{n-2} = \frac{22.46}{10} \doteq 2.25$$

This is an estimate for σ^2. The **mean square for regression** is

$$\text{MSR} = \frac{\text{SSR}}{1} = 286.21$$

This will only be a good estimate for σ^2 if there is no linear relationship between x and y, that is, if the population regression line is horizontal ($\beta_1 = 0$). If $\beta_1 \neq 0$, MSR will overestimate σ^2 and be much larger than MSE. This indeed appears to be the case for these data. Sometimes the ratio

$$F = \frac{\text{MSR}}{\text{MSE}} = \frac{286.21}{2.25} \doteq 127.2$$

is used to decide if there is a linear relationship or not. If there is no linear relationship, the numerator and denominator should be comparable in size, and F will have an F distribution with df $= (1, n - 2)$. If there is a linear relation, the numerator will be relatively large and hence F will be large. So a right-tailed F test can be done to test:

$$H_0: \text{there is no linear relationship}$$

$$H_a: \text{there is a linear relationship}$$

The value $F = 127.2$ seems quite large. In fact, from the printout, we see its P value is approximately 0. Hence, we reject H_0 and conclude there is a linear relationship.

This F test is equivalent to the t test described above. In fact, you will note that

$$F = 127.2 \doteq (11.29)^2 = t^2$$

The square of a t statistic with $n - 2$ degrees of freedom is an F statistic with $(1, n - 2)$ degrees of freedom. (See Exercise 9.69.)

Notice that the CORRELATION command produces the value of r (.963), which agrees with the value we found in Example 10.5.

The Subcommand PREDICT

In Section 10.4 we estimated the salary of a technician with 15 years experience to be 29.87 thousand dollars. We also obtained prediction and confidence intervals. If we typed

REGRESS C2 ON 1 PREDICTOR IN C1;

PREDICT FOR 15.

Minitab would have given the output displayed previously and in addition the following would be given:

```
   Fit   Stdev.Fit         95% C.I            95% P.I.
29.901      0.433   ( 28.935,  30.866)   ( 26.424,  33.377)
```

The "fit" is $\hat{y}_0$ when $x_0 = 15$. The standard deviation of the fit is calculated from

$$S_{\hat{y}_0} = S \sqrt{\frac{1}{n} + \frac{(x_0 - \bar{x})^2}{\left(\sum x^2\right) - \left(\sum x\right)^2 / n}} = .43$$

The prediction and confidence intervals are those we found in Examples 10.10 and 10.11 approximately. Minitab only gives 95% intervals.

Exercises

Suggested exercises for use with Minitab are 10.56 through 10.63, and 10.66.

SUMMARY

We use correlation analysis to study the question of whether or not there is a relationship between two variables x and y. In regression analysis we attempt to find the particular relationship. Such a relationship can be used to predict the approximate value for an unknown y associated with a known x value. In this chapter we restricted our attention to *linear* correlation and regression. In the linear case we investigate the possibility of a linear or straight-line relationship between the x and y values.

The **line of regression** (or line of best fit) for a collection of (x, y) values has the equation

$$\hat{y} = b_0 + b_1 x$$

where

$$b_1 = \frac{n\left(\sum xy\right) - \left(\sum x\right)\left(\sum y\right)}{n\left(\sum x^2\right) - \left(\sum x\right)^2}$$

$$b_0 = \bar{y} - b_1\bar{x}$$

To study the question of whether or not there is a linear relationship between x and y values, we use the **linear correlation coefficient, r**:

$$r = \frac{n\left(\sum xy\right) - \left(\sum x\right)\left(\sum y\right)}{\sqrt{n\left(\sum x^2\right) - \left(\sum x\right)^2} \cdot \sqrt{n\left(\sum y^2\right) - \left(\sum y\right)^2}}$$

Values of r near ± 1 indicate a strong linear relationship. Values of r near 0 indicate little or no degree of linear relationship.

The **coefficient of determination** is r^2.

$r^2 =$ the proportion of the variation in y values that is explained by the linear relationship (line of best fit)

A **$1 - \alpha$ prediction interval** for a y value corresponding to the x value x_0 is obtained as follows: Find the value $\hat{y}_0$ corresponding to the value x_0 from the regression line: $\hat{y}_0 = b_0 + b_1 x_0$. Compute the value of

$$S = \sqrt{\frac{\sum(y - \hat{y})^2}{n - 2}}$$

from the sample data. The $1 - \alpha$ prediction interval is

$$\hat{y}_0 \pm t\left(\frac{\alpha}{2}\right) \cdot S \cdot \sqrt{1 + \frac{1}{n} + \frac{(x_0 - \bar{x})^2}{\left(\sum x^2\right) - \left(\sum x\right)^2 / n}}$$

A **$1 - \alpha$ confidence interval** for the mean of *all* y values corresponding to the value x_0 is

$$\hat{y}_0 \pm t\left(\frac{\alpha}{2}\right) \cdot S \cdot \sqrt{\frac{1}{n} + \frac{(x_0 - \bar{x})^2}{\left(\sum x^2\right) - \left(\sum x\right)^2 / n}}$$

Review Exercises

10.52 Consider the data:

x	y
3	7
3	5
5	3
6	3
8	2

(a) Find the line of best fit.

(b) Find SSE $= \Sigma(y - \hat{y})^2$, with $\hat{y}$ corresponding to the line of best fit.

(c) Find SSE $= \Sigma(y - \hat{y})^2$, using the line with the equation $\hat{y} = \bar{y} = 4$.

(d) Compare the SSE in parts (b) and (c). Comment on the fact that SSE in part (b) is less than SSE in part (c).

10.53 Construct a set of four data points (x, y) so that

(a) $r = 1$.   (b) $r = -1$.   (c) $r = 0$.

10.54 Consider the data:

x	y
2	0
5	0
5	3
6	4
6	4
6	7

(a) Find the line of best fit.

(b) Complete the following table:

x	y	$\hat{y}$	$\bar{y}$	$(y - \bar{y})^2$	$(\hat{y} - \bar{y})^2$

(c) Find TSS and SSR.

(d) Use TSS and SSR to find the coefficient of determination and give the interpretation.

10.55 In parts (a)–(h), find the missing entries.

	SSR	SSE	TSS	r^2
(a)	50	50		
(b)	100		400	

(c)		40	50	
(d)	90			.6
(e)		28		.3
(f)			40	.8
(g)	50			1.0
(h)		0	75	

10.56 In Exercise 10.9, the line of best fit was found to be $\hat{y} = 3.175 + .654x$ for predicting the prime lending rate (y) from the inflation rate (x). The data are listed in the following table:

Inflation Rate (x)	Prime Lending Rate (y)
3.3	5.2
6.2	8.0
11.0	10.8
9.1	7.9
5.8	6.8
6.5	6.9
7.6	9.0

(a) Find the linear correlation coefficient.

(b) Is there sufficient evidence to indicate a linear relationship between the variables x and y? Use a 5% level of significance.

(c) Estimate the prime lending rate when the inflation rate is 7.0.

10.57 In Exercise 10.10, the line of best fit was found to be $\hat{y} = -15.474 + 2.355x$ for predicting lung cancer mortality rate (y) from cigarette consumption (x). The data are given in the following table:

Cigarette Consumption per Capita in Hundreds (x)	Mortality Rate per 100,000 (y)
11.8	10.4
12.5	16.5
15.7	22.9
19.2	26.6
21.9	33.8
23.3	42.8

(a) Find the linear correlation coefficient.

(b) Is there sufficient evidence to indicate a linear relationship between the variables x and y? Use a 1% level of significance.

(c) Estimate the lung cancer mortality rate when the cigarette consumption per capita is 2000.

10.58   The data in the following table represent the total number of Ph.D.'s in statistics (x), and the total number of Ph.D.'s in mathematics (y) awarded over a 12-year period (*Source:* Moore and Olkin, 1984, p. 2).

Number of Ph.D's in Statistics (in Hundreds) (x)	Number of Ph.D's in Mathematics (in Hundreds) (y)
1.5	12.2
1.6	12.4
2.1	12.8
2.5	12.2
2.2	12.0
2.3	11.5
2.5	10.0
2.5	9.6
2.6	8.4
2.3	7.7
2.3	7.5
2.5	7.3

(a)   Find the linear correlation coefficient.

(b)   Is there sufficient evidence to indicate a linear relationship between the variables x and y? Use a 5% level of significance.

10.59   A high school administrator wished to study the relationship between the number of years teaching (x) and the teacher's effectiveness rating (y). The effectiveness ratings are on a scale of 0–10. The administrator obtained the following data:

Number of Years Teaching (x)	Effectiveness Rating (y)
2	5.8
2	6.7
5	9.8
9	7.7
12	9.1
12	7.9
14	9.0

(a)   Find the linear correlation coefficient.

(b)   Do the sample data indicate a population linear relationship between x and y? Use a 5% level of significance.

10.60 The data in the following table represent average cholesterol intake (x) and male population death rate (y) from arteriosclerotic and degenerative heart disease. The males were aged 55–59 and were chosen from selected countries. (*Source:* Data estimated from a graph in Williams, 1978, p. 6.)

Country	Cholesterol Intake in Tens of mg/day (x)	Death Rate per 10,000 (y)
United States	59	72
Finland	31	65
Holland	30	30
Italy	19	21
Greece	15	9
Yugoslavia	12	9
Japan	8	12

(a) Find the linear correlation coefficient.

(b) Do the sample data indicate a linear relationship between the variables x and y? Use a 5% level of significance.

(c) Find the line of best fit.

(d) Estimate the death rate for a cholesterol intake of 40.

10.61 An owner of a health spa wished to study the relationship between the temperature (x) in degrees Fahrenheit at 11 A.M. and the number (y) of customers using the facilities at that time for randomly selected days during the summer.

Temperature (x)	Number of Customers (y)
65	27
67	25
75	20
81	22
85	16
87	10

(a) Find the linear correlation coefficient.

(b) Do the sample data indicate a linear relationship between the x and y values of the population? Use a 5% level of significance.

(c) Find and sketch the line of best fit.

(d) Estimate the number of customers using the facilities when the temperature is 70°F.

10.62 A farmer was interested in the relationship between the amount of fertilizer (x) and the number of bushels (y) of soybeans. The farmer conducted an experiment and obtained the following data:

Hundreds of Pounds per Acre (x)	Bushels per Acre (y)
1.0	25
2.5	32
3.0	35
3.0	32
3.4	35
4.0	39
4.0	41
4.5	40

(a) Find the linear correlation coefficient.

(b) Do the sample data indicate a linear relationship between the x and y values of the population? Use a 1% level of significance.

(c) Find and sketch the line of best fit.

(d) Estimate the number of bushels for $x = 3.5$.

10.63 The owner of a one-bedroom apartment was undecided what to charge per month. The apartment was located 1.5 miles from a rapid transit station. The owner obtained the following information pertaining to one-bedroom apartments in the city. The variables x and y represent the distance from a rapid transit station (x) and the rent (y).

Distance in Miles (x)	Rent in Hundreds of Dollars (y)
.3	5.5
.5	5.0
.7	5.2
1.1	4.1
1.2	4.6
2.3	3.8
2.9	4.0
3.0	3.8

(a) Find the linear correlation coefficient.

(b) Do the sample data indicate a linear relationship between the x and y values of the population? Use a 5% level of significance.

(c) Find the line of best fit.

(d) What should the owner charge for rent?

10.64 Let x and y be the content (mg/cigarette) of nicotine (N) and carbon monoxide (CO), respectively, from 203 brands of cigarettes (*Source:* Data compiled by David Donoho and Liat Kulwarski from *San Francisco Examiner*, April 4, 1983. As reported in *MacSpin Release 1.1.*). Consider the following summary information:

$$\sum x = 169.1 \qquad \sum x^2 = 173.6 \qquad \sum y = 2240.4$$
$$\sum y^2 = 30{,}390 \qquad \sum xy = 2243.7$$

(a) Find the coefficient of determination and give its interpretation.

(b) Find the line of best fit for predicting CO content from N content.

(c) Estimate the CO content for N = 1 mg/cigarette.

10.65 Refer to Exercise 10.61. Find

(a) A 99% prediction interval for the number of customers using the facilities at 11 A.M. when the temperature is 70°F.

(b) A 99% confidence interval for $\mu_{y|70}$.

10.66 Refer to Exercise 10.62. Find

(a) A 95% prediction interval for the number of bushels per acre when $x = 3.5$.

(b) A 95% confidence interval for $\mu_{y|3.5}$.

10.67 Refer to Exercise 10.63. Find

(a) A 90% prediction interval for the amount of rent when the apartment is located 1.5 miles from a rapid transit.

(b) A 90% confidence interval for $\mu_{y|1.5}$.

Notes

Boyer, R., and D. Savageau, *Places Rated Almanac*. Data in Donoho, A., D. Donoho, M. Gasko, A. Ledbetter, and C. Olson, *MacSpin Release 1.1* (Austin, TX: D² Software Inc.).

Consumer Reports, Buying Guide Issue, 1982.

Fortune magazine; Data compiled by Donoho, D. and L. Kulwarski, in Donoho, A., D. Donoho, M. Gasko, A. Ledbetter, and C. Olson, *MacSpin Release 1.1* (Austin, TX: D² Software Inc.).

Information Please Almanac (Boston: Houghton Mifflin Company, 1989).

Lastavica, C., M. Wilson, V. Berardi, A. Spielman, and R. Deblinger, "Rapid Emergence of a Focal Epidemic of Lyme Disease in Coastal Massachusetts," *New England Journal of Medicine*, 320: 1989.

Moore, D. S., and I. Olkin, "Academic Statistics: Growth, Change and Federal Support," *The American Statistician* 38, No. 1 (February 1984), pp. 1–7.

Phillips, A. J., "Smoking Control Programs for Canadian adults," Proceedings of the Third World Conference, *Smoking and Health*, Vol. II. Department of Health, Education and Welfare, Publication Number (NIH) 77-1413.

Public Opinion Quarterly and Business Conditions Digest; Data compiled by P. Robertson, in Donoho, A., D. Donoho, M. Gasko, A. Ledbetter and C. Olson, *MacSpin Release 1.1* (Austin, TX: D² Software Inc.).

San Francisco Examiner; Data compiled by Donoho, D. and L. Kulwarski, in Donoho, A., D. Donoho, M. Gasko, A. Ledbetter and C. Olson, *MacSpin Release 1.1* (Austin, TX: D² Software Inc.).

Statistical Abstract of the United States, 100th ed. (Washington, D.C.: U.S. Bureau of the Census, 1979).

Williams, C. L., "Nutrition and Coronary Heart Disease," *Bank of Epidemiology Exercises, Exercise 13*, edition 1 (1978), Dept. of Community and Preventive Medicine, New York Medical College (1978). Data obtained from G. Biorck.

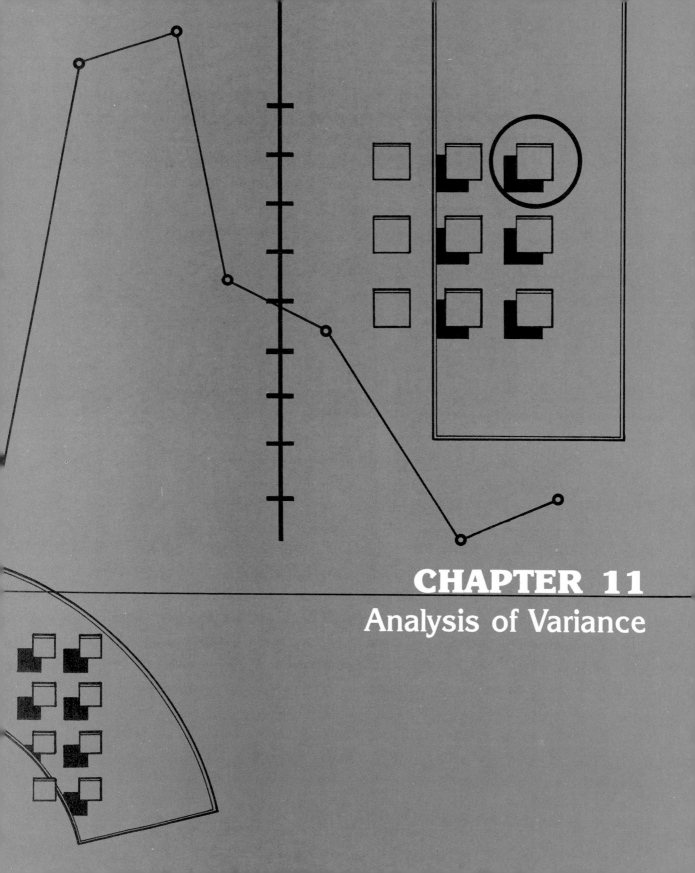

CHAPTER 11
Analysis of Variance

INTRODUCTION

In Chapters 7 and 8 we studied inference methods (decision-making and confidence intervals) concerning a single population mean. In Chapter 9 we showed how one may compare two population means. In this chapter we will develop a method for comparing several population means at the same time. This method is called **analysis of variance**, abbreviated ANOVA.

Suppose that there are k populations of interest. The method of *analysis of variance* enables us (under suitable conditions) to test the hypotheses

$$H_0: \mu_1 = \mu_2 = \cdots = \mu_k$$

H_a: not all the means are equal

A thought that occurs to many is to use the methods of Chapter 9 to compare each possible pair of means. For example, if we are interested in the three means μ_1, μ_2, and μ_3 we could test

$$\begin{Bmatrix} \mu_1 = \mu_2 \\ \mu_1 \neq \mu_2 \end{Bmatrix} \qquad \begin{Bmatrix} \mu_2 = \mu_3 \\ \mu_2 \neq \mu_3 \end{Bmatrix} \qquad \begin{Bmatrix} \mu_1 = \mu_3 \\ \mu_1 \neq \mu_3 \end{Bmatrix}$$

The trouble with this approach is that each time we conduct a test, there is a chance we will make an error. The chance of making at least one error when we conduct all these tests is usually unacceptably large. The usual approach is to use analysis of variance to conduct a single test concerning all the means

$$H_0: \mu_1 = \mu_2 = \mu_3$$

H_a: not all the means are equal

11.2 THE IDEA BEHIND ANALYSIS OF VARIANCE. THE CASE OF EQUAL SAMPLE SIZES

Analysis of variance is widely used in many different fields. In manufacturing, for example, quality control engineers can use it to compare the output from different assembly lines or different plants. To get an idea of how analysis of variance works, we will consider the following hypothetical example.

A large chemical company uses four manufacturing plants to produce the same fertilizer. The plants were designed to be equivalent, so (theoretically) they should each have the same mean output (and the same variability). The company wished to check to see if each of the four plants does have the same mean output. The output from a given plant is measured in terms of weekly production (tons of fertilizer produced during a week). This will, of course, vary somewhat from week to week. We are interested in the true mean weekly production for a plant. This would be the mean of the conceptual population consisting of weekly production figures for the plant for many, many weeks. Suppose that we let μ_1 represent the true (population) mean weekly production for plant 1. Similarly, μ_2, μ_3, μ_4 represent the true mean weekly production for plants 2, 3, 4. The company wishes to test the hypotheses

$$H_0: \mu_1 = \mu_2 = \mu_3 = \mu_4$$

H_a: not all the means are equal

TABLE 11.1
Weekly Production Figures for 5 Weeks along with
Their Mean and Variance for Four Fertilizer Plants

Plant	1	2	3	4
Weekly production in tons (x)	574	566	580	573
	578	576	570	570
	573	569	577	569
	568	571	575	577
	572	573	573	576
$\bar{x}$	573	571	575	573
s^2	13	14.5	14.5	12.5

To investigate this situation, the company obtains the weekly production figures for 5 weeks for each plant. The results are given in Table 11.1.

The data from Table 11.1 are displayed in Figure 11.1, where the data values are represented by dots and the sample means are represented by squares. We can think of the sample means as very rough estimates for the population means. Notice that in Figure 11.1 the sample means are clustered fairly close together, which would tend to support H_0.

If there were a great deal of variability between the sample means, this would suggest that not all the population means were equal, thus supporting H_a. To appreciate this, look at the data in Table 11.2 and its display in Figure 11.2. To obtain these data, we used the data in Table 11.1, except that we subtracted 20 from each figure for plant 2 and added 12 to the figures for plant 4. Note the greater variability in the sample means, suggesting that the true means are not all the same.

A key to testing for equality of several population means is to look at the variability between the sample means. If there is a large degree of variability

TABLE 11.2
Revised Weekly Production Figures for the Four
Plants of Table 11.1

Plant	1	2	3	4
Weekly production in tons (x)	574	546	580	585
	578	556	570	582
	573	549	577	581
	568	551	575	589
	572	553	573	588
$\bar{x}$	573	551	575	585
s^2	13	14.5	14.5	12.5

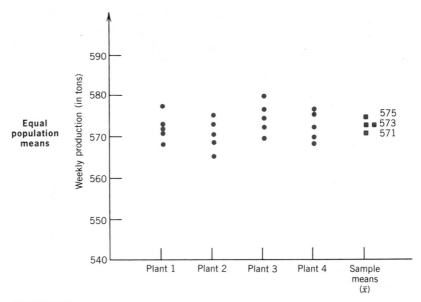

FIGURE 11.1
Weekly Production from Four Commercial Fertilizer Plants for 5 Weeks

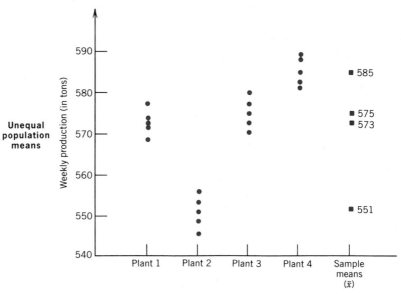

FIGURE 11.2
Weekly Production for the Four Plants in Figure 11.1 Obtained by Subtracting 20 Tons from the Figures for Plant 2 and Increasing the Figures for Plant 4 by 12 Tons/Week. Note How the Variability between the Sample Means Has Increased

between the sample means, this would suggest that not all the population means are equal. Hence we would reject H_0 in favor of H_a. If this variability is not large, we would not reject H_0. Of course, "large" is a relative term. Variability must be measured in relation to something. When we say that the variability between the sample means is large, we will mean that it is large in comparison with the variability of the data values within the samples. Therefore, when the variability *between* the sample means is large in relation to the variability *within* the samples, we will reject the null hypothesis and conclude that not all the population means are equal. In the example under discussion, this would mean that not all the plants have the same productive capacity.

Throughout this discussion we will assume that each plant has the same (true) variance σ^2 (whether the means are equal or not). The variance σ^2 is a measure of the variability we can expect within each sample. From the data in Table 11.1 we will obtain two estimates of σ^2. The first estimate is based on the variability within the samples of weekly production figures, and is called the **within samples estimate of σ^2**. The second estimate of σ^2 is based on the variability between the sample means, and is called the **between samples estimate of σ^2**. It turns out that these two estimates are all we need to test the hypothesis H_0, which asserts that each plant has the same true mean output ($\mu_1 = \mu_2 = \mu_3 = \mu_4$). The reason for this is that the first estimate (within samples) gives us a valid estimate of σ^2 whether H_0 is true or not. However, the second estimate (between samples) only gives a valid estimate of σ^2 if H_0 is true. If H_0 is false, the relatively large degree of variability between the sample means causes the second estimate (between samples) to be an inflated overestimate of σ^2, as we will see shortly. Thus by comparing the two estimates we can get an idea of whether H_0 is true or not: If H_0 is true, we expect the two estimates of σ to be comparable in size. But if the second estimate is much larger than the first, we would conclude that H_0 is false.

Estimate 1 (Within Samples Estimate)

All four sample variances in Table 11.1 can be thought of as estimates of the common variance σ^2. It would be natural to pool these estimates by averaging them. This gives us our first estimate of σ^2.

$$\text{estimate 1} = \frac{s_1^2 + s_2^2 + s_3^2 + s_4^2}{4}$$

$$= \frac{13 + 14.5 + 14.5 + 12.5}{4} = 13.625$$

Estimate 2 (Between Samples Estimate)

Let us assume (for the moment) that all four plants are equivalent and hence that H_0 is true. We may then view the samples of production figures as four samples of size 5 from the same population. The four sample means are four values of the random variable $\bar{x}$. We studied this random variable in Chapter 7.

Recall that the Central Limit Theorem (Section 6.7) told us that for random samples of size m, the mean and standard deviation of $\bar{x}$ are

$$\mu_{\bar{x}} = \mu \qquad \sigma_{\bar{x}} = \frac{\sigma}{\sqrt{m}}$$

so

$$\sigma_{\bar{x}}^2 = \frac{\sigma^2}{m}$$

This means that

$$\sigma^2 = m \cdot \sigma_{\bar{x}}^2$$

This gives us another way of estimating σ^2: We can obtain an estimate of $\sigma_{\bar{x}}^2$ from our four values of $\bar{x}$. Then multiply this by the sample size m of the production samples (which in this case is 5). We will use the sample variance of the four values of $\bar{x}$, which we call $s_{\bar{x}}^2$, as an estimate of $\sigma_{\bar{x}}^2$. The mean of the four values of $\bar{x}$ is called the *grand mean* and is denoted by $\bar{\bar{x}}$

$$\bar{\bar{x}} = \frac{573 + 571 + 575 + 573}{4} = 573$$

Now we find the sample variance $s_{\bar{x}}^2$.

$\bar{x}$	$\bar{x} - \bar{\bar{x}}$	$(\bar{x} - \bar{\bar{x}})^2$
573	0	0
571	-2	4
575	2	4
573	0	0
$\bar{\bar{x}} = 573$		$8 \leftarrow \sum(\bar{x} - \bar{\bar{x}})^2$

$$s_{\bar{x}}^2 = \frac{\sum(\bar{x} - \bar{\bar{x}})^2}{4 - 1} = \frac{8}{3}$$

We can now obtain our second estimate of σ^2 by multiplying the value of $s_{\bar{x}}^2$ by the sample size, $m = 5$. (Remember that we have four samples, each of size $m = 5$.)

$$\text{estimate 2} = m \cdot s_{\bar{x}}^2 = (5)(\tfrac{8}{3}) \doteq 13.333$$

Note that the two estimates of σ^2 (13.625 and 13.333) appear to be quite close together. Estimate 1 will be an estimate of σ^2 only if H_0 is true. However, estimate 2 will be a valid estimate of σ^2 whether H_0 is true or not. The fact that both estimates appear to be close, therefore, seems to support the truth of H_0. That is, the (true) means for the assembly lines appear equal.

To get an idea of what happens when H_0 is not true, look at the data in Table 11.2 and Figure 11.2. Recall that these data were obtained from the original data in Table 11.1 by subtracting 20 from the production figure for plant 2 and adding 12 to the figures for plant 4. For the new data it appears that the variability between the sample means is large in relation to the variability within the samples, suggesting that H_0 is not true. The variability within the samples for the new data has not changed. If you want to go to the trouble of calculating estimate 1 for the new data, you will get precisely the same answer we got for the original data, 13.625. What about estimate 2? Estimate 2 will give an estimate of σ^2 when the four samples come from essentially the same population, but this does not appear to be the case for the new data. It does not appear that each plant has the same productive capacity. Notice how far apart the sample means are. The variance for the new sample means will therefore be larger than for the original sample means. The new sample means are 573, 551, 575, and 585. You may wish to verify that $\bar{x} = 571$; the sample variance for these values is

$$s_{\bar{x}}^2 = \frac{\sum(\bar{x} - \bar{\bar{x}})^2}{4 - 1} = \frac{616}{3}$$

This means that estimate 2 (for the new data) is

$$\text{estimate } 2 = m \cdot s_{\bar{x}}^2 = (5)\left(\frac{616}{3}\right) \doteq 1026.667$$

Note how much larger this is than estimate 1. Thus, when H_0 is false, estimate 2 overestimates σ^2.

We can summarize our findings as follows: When H_0 is true, estimate 1 and estimate 2 should be roughly the same size. If estimate 2 is much greater than estimate 1, we would reject H_0. In practice, we usually look at the ratio of estimate 2 to estimate 1. Let

$$F = \frac{\text{estimate } 2}{\text{estimate } 1} = \frac{m \cdot s_{\bar{x}}^2}{(s_1^2 + s_2^2 + s_3^2 + s_4^2)/4}$$

If H_0 is true (i.e., the population means are equal), we expect the value of F to be not too far from 1. If F is too large, we reject H_0. We consider F "too large" if it would be unlikely that we would observe such a large value of F (if H_0 were true). To assess this, we must know something about the probability distribution of F. It can be shown that under appropriate conditions, F has an F distribution. (This was discussed in Section 9.2.) These conditions are given in the following general result.

Suppose that we obtain independent random samples of size m from each of k normal populations having the same variance. Then, if the null hypothesis of equal population means is true, the random variable

$$F = \frac{\text{estimate 2}}{\text{estimate 1}} = \frac{m \cdot s_{\bar{x}}^2}{(s_1^2 + s_2^2 + \cdots + s_k^2)/k}$$

has an F distribution. The degrees of freedom of the numerator is $k - 1$ and the degrees of freedom of the denominator is $n - k$, where n is the total number of data values in all the samples.

For degrees of freedom, we write df $= (k - 1, n - k)$.

We said that the null hypothesis H_0 (of equal population means) will be rejected if the value of F is so large that it is unlikely that we would observe such a large value (if H_0 were true). ("Unlikely" means having a low order of probability, which we call α, the level of significance. As before, the researcher chooses this value.) This means that we perform a right-tailed F test.

For example, we said that the new production data in Table 11.2 seemed to suggest unequal true population means. The F value for these data is

$$F = \frac{\text{estimate 2}}{\text{estimate 1}} = \frac{1026.667}{13.625} \doteq 75.35$$

For these data $k = 4$, and the total number of data values is $n = 20$. Hence df $= (k - 1, n - k) = (3, 16)$. Suppose that we decided on the 5% level of significance. From Appendix Table B.6 we find that $F(.05) = 3.24$. It would be unlikely (only a 5% chance) that we would observe a value of $F \geq 3.24$ if the null hypothesis were true. Since the observed value of 75.35 is greater than 3.24, we reject the null hypothesis of equal population means. This is precisely what we expected by examining Figure 11.2, which portrays these data.

We summarize the essential features of a test for equality of means as follows.

ANOVA Test   To test the hypotheses,

$$H_0: \mu = \mu_2 = \cdots = \mu_k$$

$$H_a: \text{not all the means are equal}$$

we perform a right-tailed F test. Suppose that the level of significance of the test is α. Calculate the value of F and find the critical value $F(\alpha)$ in Appendix Table B.6. If $F \geq F(\alpha)$, we reject H_0. Otherwise we fail to reject H_0. (See Figure 11.3.) Assumptions underlying the test: Populations are normal with equal variances and the samples are independent.

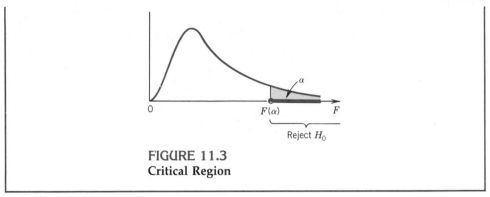

FIGURE 11.3
Critical Region

Thus the analysis of variance uses an F test for equality of population means. Note, however, that the following properties of the populations were assumed:

1. Normality.
2. Equality of the population variances.

Property 2 is called *homoscedasticity*. (The inequality of population variances is called *heteroscedasticity*.) Moderate departures from properties 1 and 2 do not seriously affect the test. In the next section we will discuss the analysis of variance when the sample sizes are unequal. In this case the test is more sensitive to departures from homoscedasticity, but moderate departures are still acceptable as long as the sample sizes do not differ by very large amounts.

Example 11.1
For the (original) production data of Table 11.1, test the hypothesis that the true mean weekly production is the same for all four plants. Use the 5% level of significance.

Solution
We use the usual five steps to complete the test.

1. Hypotheses:

$$H_0: \mu_1 = \mu_2 = \mu_3 = \mu_4$$

$$H_a: \text{not all the means are equal}$$

2. Level of significance: $\alpha = .05$.
3. Test statistic and observed value: We use the test statistic

$$F = \frac{m \cdot s_{\bar{x}}^2}{(s_1^2 + s_2^2 + s_3^2 + s_4^2)/4}$$

In the previous discussion we have already evaluated the numerator and denominator of F. The numerator is estimate 2 of σ^2 and has the value 13.333. Similarly, the denominator (estimate 1) was 13.625. Therefore,

$$F = \frac{13.333}{13.625} \doteq .98$$

Degrees of freedom $= (k - 1, n - k)$:

$$k = \text{number of means} = 4$$
$$n = \text{total number of data values} = 20$$

so

$$\text{degrees of freedom} = \text{df} = (3, 16)$$

4. Critical region: We perform a right-tailed test with $\alpha = .05$. From Appendix Table B.6 we find $F(.05) = 3.24$. (See Figure 11.4.)

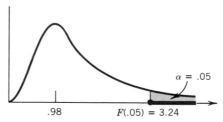

$\alpha = .05$

.98 $F(.05) = 3.24$

FIGURE 11.4

5. Decision: The value of the test statistic ($F = .98$) is not in the critical region; therefore, we fail to reject H_0. This means that there is no evidence to conclude that there is a difference in the true mean output levels of the four plants.

Example 11.2
A sociologist, studying the living habits of young adults, obtained the data in Table 11.3 on the percentage of budgets used for recreation for three groups of subjects grouped by social class. There were $m = 8$ subjects in each group.

TABLE 11.3

	Lower Working Class	Middle Class	Upper Class
$\bar{x}$	4.3	5.4	10.1
s	3.0	3.6	4.0

Do the data indicate that the mean percentages for the populations from which the three groups were obtained are different? Use the 1% level of significance.

Solution

1. Hypotheses: Let μ_1, μ_2, μ_3 represent the (population) mean percentages for the three groups.

$$H_0: \mu_1 = \mu_2 = \mu_3$$
$$H_a: \text{not all the means are equal}$$

2. Level of significance: $\alpha = .01$.

3. Test statistic and observed value: Before we evaluate the F statistic, we need the value of $s_{\bar{x}}^2$. Using the alternative formula for variance of Section 3.4, we get

$$s_{\bar{x}}^2 = \frac{3\left(\sum \bar{x}^2\right) - \left(\sum \bar{x}\right)^2}{(3)(2)} = \frac{3[(4.3)^2 + (5.4)^2 + (10.1)^2] - (4.3 + 5.4 + 10.1)^2}{6}$$

$$= \frac{448.98 - 392.04}{6} = \frac{56.94}{6} \doteq 9.49$$

$$F = \frac{m \cdot s_{\bar{x}}^2}{(s_1^2 + s_2^2 + s_3^2)/3} = \frac{(8)(9.49)}{[(3)^2 + (3.6)^2 + (4)^2]/3} = \frac{75.92}{12.653} \doteq 6$$

Degrees of freedom $= (k - 1, n - k) = (3 - 1, 24 - 3) = (2, 21)$.

4. Critical region: The analysis of variance always uses a right-tailed test. From Appendix Table B.6 we see that $F(.01) = 5.78$. (See Figure 11.5.)

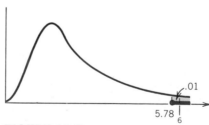

FIGURE 11.5

5. Decision: The observed value of F is in the critical region; thus we reject H_0. This means that there is sufficient evidence to conclude that the mean percentages are not the same for the three populations.

Before closing this section we should address a point that may have occurred to the reader. Suppose we conclude that not all the population means are equal. Then which is the largest, or the smallest? Which pairs of means are different? Such questions can be trickier than they may seem, and we will not treat them in this book. We refer the reader to Neter and Wasserman (1985, Chapter 17) for a discussion of these issues. This reference also has an excellent discussion of the questions of normality and homoscedasticity (Chapter 18).

The analysis of variance procedure used in this section requires that all the sample sizes be equal. However, the analysis of variance may also be used when the sample sizes are unequal, as we will see in the next section.

Exercises

11.1   Consider the hypothesis $H_0: \mu_1 = \mu_2 = \cdots = \mu_k$. Random samples of size m are selected from each of the k populations. Complete the following table. (*Note:* df = degrees of freedom.)

	m	k	α	$s_{\bar{x}}^2$	$(s_1^2 + s_2^2 + \cdots + s_k^2)$	Observed F	df	$F(\alpha)$	Decision
(a)	5	3	.05	50	100				
(b)	5	3	.05	20	100				
(c)	6	2	.01	25	50				
(d)	6	2	.01	25	25				
(e)	4	5	.05	40	200				
(f)	4	5	.05	40	400				

11.2   For each of the following parts, assume that random samples of size 4 were selected from each of six populations. Determine whether there is sufficient evidence to reject the hypothesis of equal population means using a 5% significance level.

	$s_{\bar{x}}^2$	$(s_A^2 + s_B^2 + s_C^2 + s_D^2 + s_E^2 + s_F^2)$
(a)	5	40
(b)	4	40
(c)	10	40
(d)	5	60
(e)	5	20

11.3   A psychologist was interested in the effects of three different kinds of drugs upon the mean time to complete a task. The psychologist used 15 subjects and randomly assigned 5 of them to each drug A, B, and C. The data represent the time in minutes to complete the task. Use a 5% level of significance to test that the population mean time to complete the task is the same with each drug. The data are

A	B	C
20	21	30
22	26	24
25	26	26
24	27	25
19	25	30

11.4 The owner of a large company wanted to compare the mean daily output of a particular item for five plants. For each plant, a random sample of 4 days gave the data listed in the following table. Do the sample data indicate a difference in the population means for the five plants? Use a 5% level of significance.

Plants	A	B	C	D	E
	29	22	24	23	15
	18	17	16	15	8
	18	12	14	14	9
	18	11	12	10	4
$\bar{x}$	20.75	15.50	16.50	15.50	9
s^2	30.25	25.67	27.67	29.67	20.67

11.5 A biology professor was interested in whether or not there was a difference in the mean heart rate maximum after exercise on a treadmill between highly trained male and female distance runners. Do the following data indicate a difference in the population means? Use the analysis of variance with a 1% level of significance.

	Sample Size	$\bar{x}$	s
Males	10	180.3	7.2
Females	10	190.8	10.7

(*Source:* Daniels et al., 1977, p. 142.)

11.6 Consider the sample data from populations A and B:

A	B
52	54
42	63
32	51
46	52
55	50
50	60

Suppose that you are to test H_0: $\mu_A = \mu_B$ with a 1% level of significance.

(a)   Use the analysis of variance.

   (i)   Find the degrees of freedom.

   (ii)   Find the critical value.

   (iii)   Find the observed value.

   (iv)   Is there sufficient evidence to reject H_0?

(b)   Use the two-sample t test that applies when the variances are equal.

   (i)   Find the degrees of freedom.

   (ii)   Find the critical values.

   (iii)   Find the observed value.

   (iv)   Is there sufficient evidence to reject H_0?

(c)   Compare each item of part (a) with the corresponding item of (b). Comment. [*Hint*: The square (or square root) is involved in the comparison of items (ii) and (iii).]

11.7   Consider the sample data from populations A, B, C, D, and E in the following table.

	A	B	C	D	E
	28	19	30	29	19
	27	28	23	26	19
	30	28	30	35	24
	22	27	16	24	16
	18	19	29	33	27
	19	29	22	33	27
$\bar{x}$	24	25	25	30	22
s	5.02	4.69	5.66	4.38	4.65

Suppose that you are to test H_0: $\mu_A = \mu_B = \mu_C = \mu_D = \mu_E$ with a 5% level of significance.

(a)   (i)   Find the degrees of freedom.

   (ii)   Find the critical value.

   (iii)   Find the observed value.

   (iv)   Is there sufficient evidence to reject H_0?

(b)   Decrease each observed value in population E by 4.

   (i)   Without calculating, how does the variability within the samples compare with that obtained in part (a)? (Does it increase, decrease, or remain the same?)

   (ii)   Without calculating $s_{\bar{x}}^2$, how does the variability between sample means compare with that obtained in part (a)?

   (iii)   Without calculating, and making use of your answers in parts (b)(i) and (b)(ii), how does the observed F value compare with that obtained in part (a)?

    (iv)   Now calculate the new value of F. Is there sufficient evidence to reject H_0 using a 5% significance level?

11.8   Consider the sample data from populations A, B, C, and D.

A	B	C	D
27	27	27	23
19	32	19	27
18	31	21	26
20	34	25	20

Suppose that you are to test H_0: $\mu_A = \mu_B = \mu_C = \mu_D$ with a 1% level of significance.

(a) **(i)**   Find the degrees of freedom.

      (ii)   Find the critical value.

      (iii)   Find the observed value.

      (iv)   Is there sufficient evidence to reject H_0?

(b)   Increase each observed value in population A by 5.

      (i)   Without calculating, how does the variability within the samples compare with that obtained in part (a)? (Does it increase, decrease, or remain the same?)

      (ii)   Without calculating $s_{\bar{x}}^2$, how does the variability between sample means compare with that obtained in part (a)?

      (iii)   Without calculating, and making use of your answers in parts (b)(i) and (b)(ii), how does the observed F value compare with that obtained in part (a)?

      (iv)   Now calculate the new value of F. Is there sufficient evidence to reject H_0 using a 1% significance level?

11.3

ANALYSIS OF VARIANCE WHEN SAMPLE SIZES ARE NOT NECESSARILY EQUAL. CONVENTIONAL TERMINOLOGY

Until now our analysis of variance technique required that the size of the sample from each population be the same. But analysis of variance may still be carried out when the sample sizes are unequal. We now give the appropriate formula for F that will work whether the sample sizes are equal or not.

The F statistic used to test for equality of population means is still based on the two estimates of the common population variance σ^2.

$$F = \frac{\text{estimate 2}}{\text{estimate 1}} = \frac{\text{between samples estimate of } \sigma^2}{\text{within samples estimate of } \sigma^2}$$

Now assume that we have samples from k populations. We represent the data values from the first population by x_1, those from the second population by x_2, and so on. We represent the size of the sample from the first population by n_1, the sample size from the second population by n_2, and so on. We let n denote the total number of data values ($n = n_1 + n_2 + \cdots + n_k$).

We will give general formulas for estimate 1 and estimate 2, at the same time showing how these formulas apply to the production data in Table 11.1.

Estimate 1

First calculate the *sum of squares* for each sample. For example, the sum of squares for the first sample is

$$\sum (x_1 - \bar{x}_1)^2$$

In Table 11.1 we were given the sample variances. It would be a simple matter to obtain the sum of squares for each sample from these. But since we usually are not given the sample variance, we will calculate the sum of squares for each sample from scratch, as shown in Table 11.4.

TABLE 11.4

| | Plant 1 | | | Plant 2 | |
x_1	$x_1 - \bar{x}_1$	$(x_1 - \bar{x}_1)^2$	x_2	$x_2 - \bar{x}_2$	$(x_2 - \bar{x}_2)^2$
574	1	1	566	−5	25
578	5	25	576	5	25
573	0	0	569	−2	4
568	−5	25	571	0	0
572	−1	1	573	2	4
$\bar{x}_1 = 573$		$\sum (x_1 - \bar{x}_1)^2 = 52$	$\bar{x}_2 = 571$		$\sum (x_2 - \bar{x}_2)^2 = 58$

| | Plant 3 | | | Plant 4 | |
x_3	$x_3 - \bar{x}_3$	$(x_3 - \bar{x}_3)^2$	x_4	$x_4 - \bar{x}_4$	$(x_4 - \bar{x}_4)^2$
580	5	25	573	0	0
570	−5	25	570	−3	9
577	2	4	569	−4	16
575	0	0	577	4	16
573	−2	4	576	3	9
$\bar{x}_3 = 575$		$\sum (x_3 - \bar{x}_3)^2 = 58$	$\bar{x}_4 = 573$		$\sum (x_4 - \bar{x}_4)^2 = 50$

We now add up the sum of squares for each sample. The result is called the **sum of squares within,** denoted by SSW.

$$SSW = \sum(x_1 - \bar{x}_1)^2 + \sum(x_2 - \bar{x}_2)^2 + \cdots + \sum(x_k - \bar{x}_k)^2$$

We associate a number with SSW, called its *degrees of freedom*. The number of degrees of freedom for SSW is $n - k$. Think of degrees of freedom as a number we divide a sum of squares by to get an estimate for a population variance, in this case the common population variance, σ^2. For the production data (in Table 11.4)

$$SSW = 52 + 58 + 58 + 50 = 218$$

The number of degrees of freedom is $n - k = 20 - 4 = 16$. To find estimate 1, we divide SSW by its degrees of freedom, $n - k$. The result is sometimes called the **mean square within,** denoted by MSW.

$$\text{estimate 1} = MSW = \frac{SSW}{n - k}$$

For the production data

$$\text{estimate 1} = MSW = \frac{218}{16} = 13.625$$

This is the same value we obtained in Section 11.2.

Estimate 2

First we find the mean of *all* the data values in the study. We saw that this is called the *grand mean* and is denoted by $\bar{\bar{x}}$ (For the data in Table 11.1, $\bar{\bar{x}} = 573$.) Now we calculate the **sum of squares between,** denoted by SSB. This is defined to be

$$SSB = n_1(\bar{x}_1 - \bar{\bar{x}})^2 + n_2(\bar{x}_2 - \bar{\bar{x}})^2 + \cdots + n_k(\bar{x}_k - \bar{\bar{x}})^2$$

The number of degrees of freedom for SSB is defined to be $k - 1$. For the production data $n_1 = 5$, $n_2 = 5$, and so on. Therefore,

$$SSB = 5(573 - 573)^2 + 5(571 - 573)^2 + 5(575 - 573)^2 + 5(573 - 573)^2 = 40$$

(Note that the sample sizes happen to be equal for these data sets, but this need not be the case.) The degrees of freedom $= k - 1 = 4 - 1 = 3$.

Now to obtain estimate 2 divide SSB by its degrees of freedom, $k - 1$. The result is sometimes called the **mean square between,** denoted by MSB.

$$\text{estimate 2} = \text{MSB} = \frac{\text{SSB}}{k - 1}$$

This will be an estimate of σ^2 only when the population means are equal. For the production data

$$\text{estimate 2} = \text{MSB} = \frac{40}{3} \doteq 13.333$$

This is the same value we obtained in Section 11.2.

The test statistic used to test for the equality of several population means (for normal populations with equal variances) is

$$F = \frac{\text{MSB}}{\text{MSW}}$$

If the means are equal, this has the F distribution with $k - 1$ numerator degrees of freedom and $n - k$ denominator degrees of freedom. If the level of significance of the test is α, we reject the null hypothesis of equality of population means if $F \geq F(\alpha)$.

As we pointed out in Section 11.2, moderate departures from the requirement of normality and equal variances for the populations do not seriously affect the test.

Notes:

1. When calculating SSW, we use the sum of squares for each sample. Sometimes the sample variances are known, and when this is the case, we can calculate the sum of squares for each sample quickly. For example,

$$s_1^2 = \frac{\sum(x_1 - \bar{x}_1)^2}{n_1 - 1}$$

Multiplying both sides by $n_1 - 1$, we get

$$(n_1 - 1)s_1^2 = (n_1 - 1) \cdot \frac{\sum(x_1 - \bar{x}_1)^2}{(n_1 - 1)} = \sum(x_1 - \bar{x}_1)^2$$

If we multiply the sample variance by one less than the sample size, we get the sum of squares for the sample. For instance, the production data for plant 1 discussed above gave a sum of squares of 52. From Table 11.1, we saw that $s_1^2 = 13$ and $n_1 = 5$; thus

$$(4)(13) = 52 = \text{sum of squares}$$

2. The grand mean $\bar{\bar{x}}$ is the mean of all the data values in the samples. If we know each sample mean, we can make use of these to find $\bar{\bar{x}}$. But we must be careful not to average the sample means unless the sample sizes are equal. What we can do is multiply each sample mean by the sample size. For example,

$$n_1 \bar{x}_1 = n_1 \cdot \frac{\sum x_1}{n_1} = \sum x_1$$

This gives the sum of the data values in the first sample. Now we just add these sums and divide by the total number of data values to get $\bar{\bar{x}}$.

Example 11.3
In a study of various heart characteristics of athletes, the data in Tables 11.5, 11.6, and 11.7 were obtained concerning heart size for samples of swimmers, runners, and wrestlers.* The characteristic measured was left ventricular end diastolic volume, in milliliters. (This is the volume of the left lower chamber when the heart is filled with blood.) Test the hypothesis that the mean size is the same for the three populations of swimmers, runners, and wrestlers. Use the 1% level of significance.

Solution

TABLE 11.5
Swimmers ($n_1 = 15$)

x_1	$x_1 - \bar{x}_1$	$(x_1 - \bar{x}_1)^2$
140	−41.600	1730.560
140	−41.600	1730.560
140	−41.600	1730.560
147	−34.600	1197.160
147	−34.600	1197.160
175	−6.600	43.560
185	3.400	11.560
194	12.400	153.760
194	12.400	153.760

Source: Data estimated from a graph in Morganroth, J., and B. Maron, 1977, p. 218.

TABLE 11.5 (Continued)
Swimmers ($n_1 = 15$)

x_1	$x_1 - \bar{x}_1$	$(x_1 - \bar{x}_1)^2$
203	21.400	457.960
203	21.400	457.960
214	32.400	1049.760
214	32.400	1049.760
214	32.400	1049.760
214	32.400	1049.760
$\sum x_1 = 2724$		$\sum (x_1 - \bar{x}_1)^2 = 13{,}063.600$
$\bar{x}_1 = 181.600$		

TABLE 11.6
Runners ($n_2 = 15$)

x_2	$x_2 - \bar{x}_2$	$(x_2 - \bar{x}_2)^2$
125	−34.267	1174.227
125	−34.267	1174.227
133	−26.267	689.955
133	−26.267	689.955
133	−26.267	689.955
140	−19.267	371.217
140	−19.267	371.217
147	−12.267	150.479
156	−3.267	10.673
165	5.733	32.867
165	5.733	32.867
194	34.733	1206.381
194	34.733	1206.381
214	54.733	2995.701
225	65.733	4320.827
$\sum x_2 = 2389$		$\sum (x_2 - \bar{x}_2)^2 = 15{,}116.929$
$\bar{x}_2 = 159.267$		

TABLE 11.7
Wrestlers ($n_3 = 12$)

x_3	$x_3 - \bar{x}_3$	$(x_3 - \bar{x}_3)^2$
83	−27.167	738.046
91	−19.167	367.374
97	−13.167	173.370
97	−13.167	173.370
108	−2.167	4.696
111	0.833	0.694

TABLE 11.7 (Continued)
Wrestlers ($n_3 = 12$)

x_3	$x_3 - \bar{x}_3$	$(x_3 - \bar{x}_3)^2$
111	0.833	0.694
117	6.833	46.690
117	6.833	46.690
125	14.833	220.018
125	14.833	220.018
140	29.833	890.008

$\sum x_3 = 1322$ $\sum(x_3 - \bar{x}_3)^2 = 2881.668$

$\bar{x}_3 = 110.167$

1. Hypotheses:

$$H_0: \mu_1 = \mu_2 = \mu_3$$

H_a: not all the population means are equal

2. Level of significance: $\alpha = .01$.

3. Test statistic and value

$$F = \frac{MSB}{MSW}$$

degrees of freedom $= (k - 1, n - k)$

To find MSW and MSB, we first find SSW and SSB. The sum of squares within is

$$SSW = \sum(x_1 - \bar{x}_1)^2 + \sum(x_2 - \bar{x}_2)^2 + \sum(x_3 - \bar{x}_3)^2$$
$$= 13063.600 + 15116.929 + 2881.668 = 31062.197$$

degrees of freedom $= n - k = 42 - 3 = 39$

To find the sum of squares between, we first find $\bar{\bar{x}}$, the mean of all 42 data values. We have already added the data values in each group. Hence, we add these three sums and divide by 42.

$$\bar{\bar{x}} = \frac{2724 + 2389 + 1322}{42} = \frac{6435}{42} \doteq 153.214$$

$$SSB = n_1(\bar{x}_1 - \bar{\bar{x}})^2 + n_2(\bar{x}_2 - \bar{\bar{x}})^2 + n_3(\bar{x}_3 - \bar{\bar{x}})^2$$

$$= 15(181.6 - 153.214)^2 + 15(159.267 - 153.214)^2$$
$$+ 12(110.167 - 153.214)^2$$

$$= 12086.475 + 549.582 + 22236.531 = 34872.588$$

degrees of freedom $= k - 1 = 3 - 1 = 2$

$$MSW = \frac{SSW}{n-k} = \frac{31062.197}{39} \doteq 796.467$$

$$MSB = \frac{SSB}{k-1} = \frac{34872.588}{2} = 17436.294$$

$$F = \frac{MSB}{MSW} = \frac{17436.294}{796.467} \doteq 21.89$$

degrees of freedom $= (2, 39)$

4. Critical region: The degrees of freedom of the denominator, 39, does not appear in Appendix Table B.6, so we follow the custom of using the next lower value in the table, 30. For degrees of freedom (2, 30), the critical value is $F(.01) = 5.39$. (See Figure 11.6.)

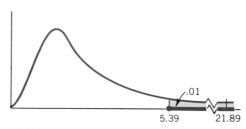

FIGURE 11.6

5. Decision: The observed value ($F = 21.89$) is in the critical region; thus we reject H_0. Therefore, we conclude that not all three populations of athletes have the same (mean) heart size.

ANOVA Tables

The various quantities used in the analysis of variance can be summarized in an *ANOVA table*. Computer statistical packages often print out such tables. Before explaining the table, we need to introduce one more term, the **total sum of squares**, denoted by TSS. To compute this, we calculate the mean of *all* the data values in our samples, the grand mean $\bar{\bar{x}}$. We then compute the difference between each data value x and $\bar{\bar{x}}$, and add the squares of these differences.

$$TSS = \sum(x - \bar{\bar{x}})^2$$

The number of degrees of freedom for TSS is $n - 1$. It can be shown that

$$TSS = SSB + SSW$$

TABLE 11.8
ANOVA Table

Source	Sum of Squares (SS)	Degrees of Freedom (df)	Mean Square (MS) $MS\left(=\dfrac{SS}{df}\right)$	F statistic
Between samples	SSB	$k - 1$	$MSB\left(=\dfrac{SSB}{k-1}\right)$	$F\left(=\dfrac{MSB}{MSW}\right)$
Within samples	SSW	$n - k$	$MSW\left(=\dfrac{SSW}{n-k}\right)$	
Total	TSS	$n - 1$		

For the data in Example 11.3,

$$TSS = 34{,}872.588 + 31{,}062.197 = 65{,}934.785$$

This relationship shows us that the total variation among all the data values is broken down into two components: one representing the variability between samples, the other the variability within samples.

An ANOVA table takes the above general form (Table 11.8). The ANOVA table for Example 11.3 is given in Table 11.9 below.

TABLE 11.9
ANOVA Table for Example 11.3

Source	Sum of Squares (SS)	Degrees of Freedom (df)	Mean Square (MS)	F statistic
Between samples	34,872.588	2	17,436.294	21.89
Within samples	31,062.197	39	796.467	
Total	65,934.785	41		

Terminology

Some statisticians refer to the populations of interest in the analysis of variance as *treatments*. They refer to the sum of squares between (SSB) as the *sum of squares for treatments*, denoted by SSTR. The deviations of the data values from their sample means (such as $x_1 - \bar{x}_1$) are referred to as *errors*, and SSW is referred to as the *sum of squares for error*, denoted by SSE. The displayed table (Table 11.10) shows the correspondence between the two types of terminology.

TABLE 11.10
Terminology

SSB (Sum of Squares Between)	=	SSTR (Sum of Squares for Treatments)
SSW (Sum of Squares Within)	=	SSE (Sum of Squares for Error)
MSB (Mean Square Between)	=	MSTR (Mean Square for Treatments)
MSW (Mean Square Within)	=	MSE (Mean Square for Error)

Exercises

In Exercises 11.9–11.12 compute the sums of squares to find SSW and SSB. Then complete the ANOVA table and the test for equality of population means using a 5% significance level.

11.9

A	B
30	35
27	37
24	33
	30
	40

11.10

A	B
29	26
38	31
25	25
36	18

11.11

A	B	C
10	9	8
16	3	10
10	6	4
		6

11.12

A	B	C	D
12	10	6	22
2	2	6	16
4		14	16
		6	13
			13

11.13 Consider $H_0: \mu_A = \mu_B = \mu_C = \mu_D = \mu_E = \mu_F$

(a) Complete the following ANOVA table.

Source	Sum of Squares (SS)	Degrees of Freedom (df)	Mean Square (MS)	F
Between samples	250			
Within samples				
Total	400	30		

(b) Using a 5% significance level, would you reject H_0?

11.14 Consider H_0: $\mu_A = \mu_B = \mu_C = \mu_D$

(a) Complete the following ANOVA table.

Source	Sum of Squares (SS)	Degrees of Freedom (df)	Mean Square (MS)	F
Between samples	180			
Within samples	120	12		
Total				

(b) Using a 1% significance level, would you reject H_0?

11.15 Consider H_0: $\mu_A = \mu_B = \mu_C = \mu_D = \mu_E$

(a) Complete the following ANOVA table.

Source	Sum of Squares (SS)	Degrees of Freedom (df)	Mean Square (MS)	F
Between samples	128			
Within samples	160		16	
Total				

(b) Using a 5% significance level, would you reject H_0?

11.16 Consider H_0: $\mu_A = \mu_B = \mu_C$. Sample sizes: $n_A = 26$, $n_B = 11$, $n_C = 6$.

(a) Complete the following ANOVA table.

Source	Sum of Squares (SS)	Degrees of Freedom (df)	Mean Square (MS)	F
Between samples	100			
Within samples				
Total	260			

(b) Using a 1% significance level, would you reject H_0?

11.17 The data in the following table represent final grades given by three professors in an advanced statistics course.

	A	B	C
	63	67	97
	45	45	97
	73	76	87
	77	80	87
	72	70	84
		70	74
			74
			64
$\bar{x}$	66	68	83
s	12.806	12.215	11.637

Does the difference in sample means appear to be due to chance variation, or is there sufficient evidence to indicate that not all population means are the same? Use a 5% level of significance.

11.18 Refer to Exercise 11.17. Prior to obtaining the data, the chairperson suspected that there might be a difference in grading between A and C. Test H_0: $\mu_A = \mu_C$ using a 5% significance level by

(a) The analysis of variance procedure.

(b) Using the two-sample t test that applies when the variances are equal. Compare the observed values in parts (a) and (b). [*Hint*: The square (or square root) is involved.]

11.19 The data in the following table represent the starting weekly wages (in hundreds of dollars) for skilled workers in selected companies in the regions. Do the data indicate a difference in population means, or do the data appear to be due to random fluctuation? Use a 5% level of significance.

	Pacific	East North Central	West North Central	South Atlantic
	7.6	7.3	6.4	6.6
	7.2	7.2	7.0	6.6
	6.4	6.8	6.2	6.7
	5.9	8.1	5.5	5.1
		6.7	5.2	6.7
			5.8	4.5
				4.7
				4.9
				5.0
$\bar{x}$	6.775	7.220	6.017	5.644
s	.768	.554	.652	.970

11.20 A physical fitness expert claimed that there was a difference in the mean HDL cholesterol (milligrams per 100 milliliters) between elite runners, good runners, and nonrunners. Do the following data support the claim? Use a 1% level of significance.

	Sample Size	$\bar{x}$	s
Elite runners	20	56	12.1
Good runners	8	52	10.9
Nonrunners	72	49	10.5

(*Source*: Martin, 1977, p. 93.)

11.21 Test the claim that there is no difference in the mean percent of calories from fats between men in the Crevalcore, Montegiorgio, and Corfu areas of Greece. Use a 1% level of significance with the following data:

	Sample Size	$\bar{x}$	s
Crevalcore	28	26.5	4.5
Montegiorgio	34	25.5	4.6
Corfu	34	31.2	5.2

(*Source*: Keys, 1970, p. I-166.)

11.22 Consider the following data from populations A, B, C:

A	B	C
0	7	6
4	6	7
5	2	8

Now SSB = 24, SSW = 30

(a) Complete the ANOVA table and test $H_0: \mu_A = \mu_B = \mu_C$ with a 5% level of significance.

Source	Sum of Squares (SS)	Degrees of Freedom (df)	Mean Square (MS)	F
Between samples				
Within samples				
Total				

(b)   Replace the sample data {6, 7, 8} from population C with {12, 13, 14}. Note that 6 has been added to each of the sample values in population C.

   (i)   Without calculating, which (if any) of the nine entries in the new ANOVA table would be larger than the corresponding entry from part (a)? Smaller? The same?

   (ii)   Compute and complete the new ANOVA table.

(c)   Replace the sample data {0, 4, 5} from population A with {3, 7, 8}. Note that 3 has been added to each of the sample values in population A. (Use the original values from population C.)

   (i)   Without calculating, which (if any) of the nine entries in the new ANOVA table would be larger than the corresponding entry from part (a)? Smaller? The same?

   (ii)   Compute and complete the new ANOVA table.

11.4 ALTERNATE FORMULAS (OPTIONAL*)

In Example 11.3 you may have noticed that there were a large number of calculations, many of which involved rounding off. There are some shortcut formulas that can simplify the computations considerably. Also, since these formulas involve less rounding off, this can improve the accuracy. We will introduce these formulas in an example.

A cereal producer wanted to find out if the market shelf position of the cereal boxes had any effect on sales. The company made an arrangement with the owner of a chain of 14 supermarkets (similar in sales and clientele) to conduct an experiment to investigate this problem. Each supermarket placed the cereal boxes on one of four shelves. We number the shelves 1 for the bottom up to 4 for the top shelf. Table 11.11 contains 14 numbers representing the number of cases of cereal sold by each store during the experimental period. Each figure is recorded under the shelf position assigned to the store. Notice that 3 stores were assigned shelf 1, four stores were assigned shelf 2, and so on. In Table 11.11 we also include the total sales for each shelf position: T_1 = total sales for shelf 1, T_2 = total sales for shelf 2, and so on.

Table 11.11 contains a sample of sales for each of the four shelf positions. (Not all samples sizes are equal.) The company is interested in whether the (true population) mean sales for each shelf are equal or not.† The hypotheses to be tested are

$$H_0: \mu_1 = \mu_2 = \mu_3 = \mu_4$$

$$H_a: \text{not all the means are equal}$$

---

*The formulas in this section can simplify the calculations in analysis of variance. The increasing use of computers has made it less important to cover these formulas, however.

†The population for position 1, for example, would be the collection of data values representing the cases sold for each store that might in the future use that shelf position for the same experimental time period.

TABLE 11.11
Number of Cases of Cereal Sold at Each Shelf Position during an Experimental Time Period

	Shelf Position				
	1	2	3	4	
Cases sold	20	24	39	33	
	23	25	44	39	
	17	29	47	27	
		22	38		
Total	$T_1 = 60$	$T_2 = 100$	$T_3 = 168$	$T_4 = 99$	$k = 4$
Number of stores	$n_1 = 3$	$n_2 = 4$	$n_3 = 4$	$n_4 = 3$	$n = 14$

Our shortcut formulas will be stated in terms of the following symbols:

n = total number of data values in all the samples

k = number of samples or groups of data

T_i = sum of data values in sample i (i may be 1, 2, 3, etc.)

n_i = number of data values in sample i

The alternate formulas for SSW and SSB are

$$\text{SSB} = \sum \frac{T_i^2}{n_i} - \frac{\left(\sum T_i\right)^2}{n}$$

$$\text{SSW} = \sum x^2 - \sum \frac{T_i^2}{n_i}$$

The symbol $\sum x^2$ represents the sum of the squares of all the data values in the study.

Let us see how these formulas work for the data in Table 11.11.

$$\sum \frac{T_i^2}{n_i} = \frac{T_1^2}{n_1} + \frac{T_2^2}{n_2} + \frac{T_3^2}{n_3} + \frac{T_4^2}{n_4}$$

$$= \frac{60^2}{3} + \frac{100^2}{4} + \frac{168^2}{4} + \frac{99^2}{3}$$

$$= 1200 + 2500 + 7056 + 3267 = 14{,}023$$

$$\frac{\left(\sum T_i\right)^2}{n} = \frac{(T_1 + T_2 + T_3 + T_4)^2}{n}$$

$$= \frac{(60 + 100 + 168 + 99)^2}{14} = \frac{(427)^2}{14}$$

$$= \frac{182{,}329}{14} = 13{,}023.5$$

$$\sum x^2 = 20^2 + 23^2 + 17^2 + 24^2 + 25^2 + 29^2 + 22^2 + 39^2 + 44^2$$
$$+ 47^2 + 38^2 + 33^2 + 39^2 + 27^2$$
$$= 14{,}193$$

$$\text{SSB} = \sum \frac{T_i^2}{n_i} - \frac{\left(\sum T_i\right)^2}{n}$$

$$= 14{,}023 - 13{,}023.5 = 999.5$$

The degrees of freedom for SSB $= k - 1 = 4 - 1 = 3$.

$$\text{SSW} = \sum x^2 - \sum \frac{T_i^2}{n_i}$$

$$= 14{,}193 - 14{,}023 = 170$$

The degrees of freedom for SSW $= n - k = 14 - 4 = 10$.

We can now complete the test for the equality of means. We use the 5% level of significance.

1. Hypotheses:

$$H_0: \mu_1 = \mu_2 = \mu_3 = \mu_4$$

$$H_a: \text{not all the means are equal}$$

2. Level of significance: $\alpha = .05$.

3. Test statistic and value:

$$F = \frac{\text{MSB}}{\text{MSW}} = \frac{\text{SSB}/(k - 1)}{\text{SSW}/(n - k)} = \frac{999.5/3}{170/10} \doteq 19.60$$

Degrees of freedom $= (3, 10)$.

4. Critical region: From Appendix Table B.6 we see that $F(.05) = 3.71$. (See Figure 11.7.)

5. Decision: The observed value of $F = 19.60$, which is in the critical region; thus we reject H_0. Therefore, it appears that the (population) means for the four shelf positions are not all equal.

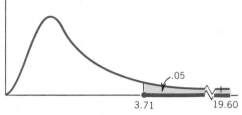

FIGURE 11.7

Exercise

11.23   In each of the following parts, use the alternate formulas of Section 11.4 to find SSB and SSW.

  (a)   Use the data in Exercise 11.9.

  (b)   Use the data in Exercise 11.10.

  (c)   Use the data in Exercise 11.11.

  (d)   Use the data in Exercise 11.12.

11.5 USING MINITAB (OPTIONAL)

The Minitab printout below is a computerization of the analysis of variance we carried out on the data in Table 11.11. This table reported the sales of cases of cereal for four shelf positions over an experimental time period. The question was whether shelf position has an effect on sales. More precisely, we are interested in whether the mean sales for each shelf position is the same.

   Fourteen similar stores were involved in the study; some placed the cereal in shelf position 1, some in position 2, and so on. The number of cases sold was recorded for each store (in Table 11.11). The type of analysis of variance treated in this chapter is sometimes called *one-way analysis of variance*, and the command that performs this test is the AOVONEWAY command.

```
MTB > SET DATA FOR FIRST POSITION IN C1
DATA> 20 23 17
DATA> END
MTB > SET DATA FOR SECOND POSITION IN C2
DATA> 24 25 29 22
DATA> END
MTB > SET DATA FOR THIRD POSITION IN C3
DATA> 39 44 47 38
DATA> END
MTB > SET DATA FOR FOURTH POSITION IN C4
DATA> 33 39 27
DATA> END
MTB > AOVONEWAY ON DATA IN C1 C2 C3 C4
```

```
ANALYSIS OF VARIANCE
SOURCE     DF        SS       MS        F        p
FACTOR      3     999.5    333.2    19.60    0.000
ERROR      10     170.0     17.0
TOTAL      13    1169.5
                                INDIVIDUAL 95 PCT CI'S FOR
                                MEAN BASED ON POOLED STDEV
LEVEL      N      MEAN     STDEV ------+---------+---------+---------+
C1         3    20.000    3.000      (----*----)
C2         4    25.000    2.944        (----*----)
C3         4    42.000    4.243                           (----*----)
C4         3    33.000    6.000               (----*----)
                                      ------+---------+---------+---------+
POOLED STDEV = 4.123                    20        30        40        5
```

The ANOVA table printed out looks slightly different from the ones displayed previously in this chapter—the printout uses the term "factor" where we used "between," and the term "error" where we used "within." Note that the P value is approximately 0, meaning we should reject the null hypothesis of equal sales (on the average). This means that shelf position does appear to affect sales.

The printout also gives some statistics for the data in columns C1, C2, C3, and C4 (the data for the four shelf positions). Also, 95% confidence intervals are displayed graphically for the (population) mean sales for the four shelf positions (for the length of time studied). These confidence intervals give a rough idea of the location of the population means. The intervals are of the form:

$$\bar{x}_i \pm t(.025) \cdot S_p/\sqrt{n_i} \qquad \text{df for } t = n - k = 10$$

where the $\bar{x}_i$ is the sample mean for sample i ($i = 1, 2, 3,$ or 4), n_i is the size of sample i, and S_p is the pooled standard deviation given in the printout. This is an estimate of the common standard deviation σ. We saw that MSW is an estimate of σ^2 (estimate 1). Minitab computes S_p as follows:

$$S_p = \sqrt{MSW} = \sqrt{MSError} = \sqrt{17} \doteq 4.123$$

Great care must be taken in interpreting these confidence intervals *together*. Whereas each one taken *individually* has a 95% level of confidence, we can't be 95% sure that *all* of them will *simultaneously* be correct. (Each time you toss a coin, you're 50% sure of getting heads. But if you toss 4 coins you can't be 50% sure of getting 4 heads. In fact, your level of confidence in 4 heads should only be 6.25%.)

Exercises

Suggested exercises for use with Minitab are 11.29, 11.31 through 11.34, and 11.36.

11.6

SUMMARY

In this chapter we discussed how one investigates the question of whether several population means are equal or not. That is, we learned how one tests the hypotheses:

$$H_0: \mu_1 = \mu_2 = \cdots = \mu_k$$

H_a: not all the means are equal

The method used to test these hypotheses is called **analysis of variance** (ANOVA).

The statistic used to test these hypotheses is computed as follows. Let

k = number of means to be tested

n_i = number of data values in sample i from population i (i = 1, 2, 3, etc.)

x_i = any data value from sample i

T_i = sum of the data values in sample i

x = any data value in the study

n = total number of data values in the study

$\bar{\bar{x}}$ = grand mean (mean of all the data values in the study)

1. Find the **sum of squares within,** SSW.

$$SSW = \sum(x_1 - \bar{x}_1)^2 + \sum(x_2 - \bar{x}_2)^2 + \cdots + \sum(x_k - \bar{x}_k)^2$$

A formula that is sometimes more convenient is

$$SSW = \sum x^2 - \sum \frac{T_i^2}{n_i}$$

The number of degrees of freedom for SSW is defined to be $n - k$.

2. Find the **sum of squares between,** SSB.

$$SSB = n_1(\bar{x}_1 - \bar{\bar{x}})^2 + n_2(\bar{x}_2 - \bar{\bar{x}})^2 + \cdots + n_k(\bar{x}_k - \bar{\bar{x}})^2$$

A formula that is sometimes more convenient is

$$SSB = \sum \frac{T_i^2}{n_i} - \frac{\left(\sum T_i\right)^2}{n}$$

The number of degrees of freedom for SSB is defined as $k - 1$.

3. Find the **mean square within,** MSW.

$$MSW = \frac{SSW}{n - k}$$

4. Find the **mean square between**, MSB.

$$MSB = \frac{SSB}{k - 1}$$

5. Find the value of F.

$$F = \frac{MSB}{MSW}$$

This is our test statistic. Large values of F favor H_a.

The statistic F has the F distribution with df $= (k - 1, n - k)$. Appendix Table B.6 contains critical values of F. If α is the level of significance of the test, find $F(\alpha)$ from this table. If the calculated value of F is $\geq F(\alpha)$, we reject H_0. Otherwise, we fail to reject H_0.

Review Exercises

11.24 For each of the following parts, assume that random samples of size 6 were selected from each of 4 populations. Determine whether there is sufficient evidence to reject the hypothesis of equal population means using a 1% significance level.

	$s_{\bar{x}}^2$	$s_A^2 + s_B^2 + s_C^2 + s_D^2$
(a)	4	24
(b)	2.5	24
(c)	6	24
(d)	4	48
(e)	4	12

In Exercises 11.25–11.26, compute the sums of squares to find SSW and SSB. Then complete the ANOVA table and test the hypothesis of equal population means with a 5% significance level.

11.25

A	B	C
5	15	20
15	20	20
	25	25
		30
		30

11.26

A	B	C	D
0	5	12	13
0	10	12	21
9	15	6	
		2	

11.27   Consider H_0: $\mu_A = \mu_B = \mu_C = \mu_D = \mu_E$

(a)   Complete the following ANOVA table.

Source	Sum of Squares (SS)	Degrees of Freedom (df)	Mean Square (MS)	F
Between samples	132			
Within samples				
Total	532	44		

(b)   Using a 5% level of significance, would you reject H_0?

11.28   Consider H_0: $\mu_A - \mu_B = \mu_C = \mu_D$. Sample sizes: $n_A = n_B = 5$, $n_C = 6$, $n_D = 8$.

(a)   Complete the following ANOVA table.

Source	Sum of Squares (SS)	Degrees of Freedom (df)	Mean Square (MS)	F
Between samples	54			
Within samples	240			
Total				

(b)   Using a 1% level of significance, would you reject H_0?

11.29   Use the following data to test H_0: $\mu_A = \mu_B = \mu_C$ with a 5% significance level.

A	B	C
72	47	50
48	55	40
60	60	30
60		40

11.30   Consider the following data to test H_0: $\mu_A = \mu_B$ with a 5% significance level. Use the analysis of variance method. The data are

A	B
25	15
45	15
50	20
	30
	40

11.31 A college administrator claimed that there was no difference in (population) mean college grade-point averages for students coming from three high schools A, B, and C. Use the following data to test the administrator's claim with a 5% significance level. The data are

A	B	C
1.9	2.3	2.8
2.3	2.7	2.8
2.8	3.2	2.9
2.4	2.8	3.5
2.5	2.9	3.0
2.5	2.9	

11.32 A production plant manager claimed that there was no time difference to complete an assembly line job between plants A, B, C, and D. Samples from each of the plants yielded the following data, where a data value represents the time in minutes to complete the job.

A	B	C	D
18	20	23	12
11	14	16	18
14	16	21	17
12	18		13
15			

Test the claim using a 5% level of significance.

11.33 A corporation owned three large department stores A, B, and C located in three cities in a state. The manager of store A claimed that the mean daily percentage of sales over $50 was the same for the three stores. A sample yielded the following data, where each data value represents the percentage of sales over $50 for a particular day. The data are

A	B	C
32	38	64
40	42	46
24	26	42
48	44	48
36	40	50

Test the manager's claim using a 5% level of significance.

11.34 Refer to Exercise 11.33. Prior to the data being obtained, the manager of store C claimed that there was a difference in population means between stores B and C. Test the claim using the analysis of variance procedure. Use a 5% level of significance.

11.35 Test the claim that there is a difference in mean LDL cholesterol (mg/100 ml)* between elite runners, good runners, and nonrunners using the following data:

	Sample Size	$\bar{x}$	s
Elite runners	20	108	24.5
Good runners	8	121	29.5
Nonrunners	72	124	35.6

(*Source*: Martin, 1977, p. 93).

Use a 5% level of significance.

11.36 Test the claim that there is no difference in the mean percent of calories from protein between men in the Crevalcore and the Montegiorgio areas of Greece and Crete using a 5% level of significance with the following data:

	Sample Size	$\bar{x}$	s
Crevalcore	28	13	1.7
Montegiorgio	34	11.7	1.7
Crete	30	11.2	2.2

(*Source*: Keys, 1970, p. I–166.)

11.37 Suppose that you are to test the hypothesis that the means of populations A, B, C, D, E, and F are the same with a 5% level of significance. Consider the following sample data:

A	B	C	D	E	F
0	4	8	4	5	5
7	10	15	11	12	12
20	22	28	24	25	22

(a) Sketch the data (as in Figure 11.1). Based on the sketch of the data, do you believe that the hypothesis of equal population means should be rejected?

(b) Carry out the test. Does this substantiate your conclusion in part (a)?

11.38 Suppose that you are to test the hypothesis that the means of populations A, B, and C are the same with a 5% level of significance. Consider the following sample data:

A	B	C
0	5	17
2	8	20
8	13	25
2	9	21
3	9	23
3	10	20

*LDL stands for low-density lipoprotein.

(a)   Sketch the data (as in Figure 11.1). Based on the sketch of the data, do you believe that the hypothesis of equal population means should be rejected?

(b)   Carry out the test. Does this substantiate your conclusion in part (a)?

11.39   Using the data in Exercise 11.38, the following ANOVA table is obtained:

Source	Sum of Squares (SS)	Degrees of Freedom (df)	Mean Square (MS)	F
Between samples	1008	2	504	70
Within samples	108	15	7.2	
Total	1116	17		

Suppose that 2 is added to each data value. The following data are obtained:

A	B	C
2	7	19
4	10	22
10	15	27
4	11	23
5	11	25
5	12	22

(a)   Construct an ANOVA table for this data and compare with the ANOVA table above.

(b)   Explain why such a relationship holds between the two ANOVA tables.

Notes

Daniels, J., G. Krakenbuhl, C. Foster, J. Gilbert, and S. Daniels, "Aerobic Responses of Female Distance Runners to Submaximal and Maximal Exercise," in P. Milvy, ed., *The Long Distance Runner* (New York: Urizen Books, 1977).

Keys, A., ed., "Coronary Heart Disease in Seven Countries," *Circulation* 41(Suppl. 1): 1 (1970).

Martin, R., W. Haskell, and P. Wood, "Blood Chemistry and Lipid Profiles of Elite Distance Runners," in P. Milvy, ed., *The Long Distance Runner* (New York: Urizen Books, 1977).

Morganroth, J., and B. Maron, "The Athlete's Heart Syndrome: A New Perspective," in P. Milvy, ed., *The Long Distance Runner* (New York: Urizen Books, 1977).

Neter, J., W. Wasserman, and M. H. Kutner, *Applied Linear Statistical Models*, 2nd ed. (Homewood, IL: Richard D. Irwin Inc., 1985).

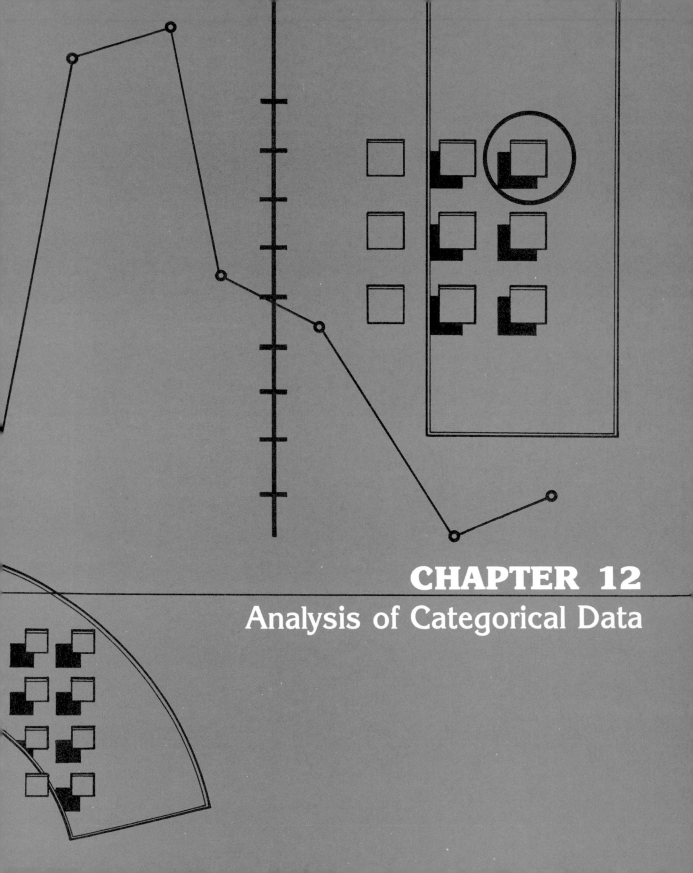

CHAPTER 12
Analysis of Categorical Data

12.1 INTRODUCTION

In Chapter 7 we discussed inference problems concerning a population proportion. For example, we learned how to test a claim that the proportion of all voters favoring a certain candidate is more than .5. This was done by sampling a number of voters and placing them into two categories (also called cells), namely, those in favor of the candidate and those not in favor of the candidate. We then calculated the appropriate test statistic and carried out the test. The important data in this study are the numbers of voters placed in each category. Frequency data of this type are sometimes called **categorical data.**

In some situations we may wish to study several proportions. For example, we might be interested in questions concerning the proportion of television viewers watching each of the three major networks: ABC, CBS, and NBC. In this chapter we will discuss such questions. The technique we will introduce to study these questions is called the **chi-square test for goodness-of-fit.** A variation of this test will also be used to conduct **tests of independence.** Such tests can be used to determine whether two characteristics (such as political preference and income) are related or independent. Still another variation of the chi-square test for goodness-of-fit that we will consider is a **test of homogeneity.** Such a test is used to study whether different populations are similar (or homogeneous) with respect to some characteristic. For example, we may wish to decide if several different countries have the same incidence of coronary heart disease.

12.2 THE CHI-SQUARE TEST FOR GOODNESS-OF-FIT

To get an understanding of how the chi-square test for goodness-of-fit works, consider the following example: A department store decided to market a new toy that comes in three different color schemes. The manufacturer claimed that each color scheme should sell in the same proportion. The department store wanted to check this claim and decided to randomly observe 300 sales to see how many of each color scheme was sold. (The color schemes are the categories in this study.)

We are actually interested in the proportion of each color scheme that will be sold in the population consisting of all these toys that may be sold by the store. The 300 toys sold may be thought of as a sample from this population. There are three proportions of interest associated with this population: p_1 is the proportion of all the toys sold in color scheme 1, p_2 is the proportion sold in color scheme 2, and p_3 is the proportion sold in color scheme 3. If each color scheme is equally popular, then each proportion has the value $\frac{1}{3}$. The department store wanted to test the hypotheses

H_0: $p_1 = p_2 = p_3 = \frac{1}{3}$

H_a: H_0 is not true (not all the color schemes will sell in the same proportion)

(Note that H_0 states that the proportions are equal, but the technique we will develop will enable us to test for any values of the proportions as long as they add up to 1. For example, if the situation warranted, we could check to see if $p_1 = \frac{1}{2}$ and $p_2 = p_3 = \frac{1}{4}$.)

Now if H_0 is true, then we can expect about one-third of the sample of 300 sales to be in color scheme 1, one-third in color scheme 2, and one-third in color scheme 3. Therefore, we expect about $(300)(\frac{1}{3}) = 100$ sales for each color scheme. The idea is to compare these expected frequencies with the actual observed frequencies (number of sales) for each color scheme. In the second and third columns of Table 12.1 we display the observed frequencies of sales (denoted by the symbol O) together with the expected frequencies (denoted by E) for each color scheme. The formula for E is

$$E = np$$

where n = the sample size and p is the proportion specified in the null hypothesis for a particular category.

If the observed frequencies deviate too much from the expected frequencies, we would reject the null hypothesis. Hence our first impulse would be to calculate the differences $O - E$. If H_0 is true, then these differences taken together should not be too large. We want to calculate one number (a test statistic) that will measure this. Note, however, that the sum of these differences (column 4 of Table 12.1) is 0. This is always the case. Therefore, the sum of the differences will not help us.

If we square the differences and look at $(O - E)^2$, we eliminate negative signs so that we will not have some terms canceling or diminishing other terms when we add up these terms. (See the fifth column in Table 12.1.)

However, one thing to keep in mind when looking at differences or squares of differences is that these should be assessed in relation to some standard. For example, a difference in weight of 5 pounds between two men is not much, but a difference of 5 pounds between two mice is a great deal. If we measure such differences in relation to the average weight of men and the average weight of mice, we get a more accurate measure of these differences. We could accomplish this by dividing the difference by the average or expected weight.

A similar situation occurs with the terms $(O - E)^2$. We should divide by the expected frequency E. That is, we should look at the terms $(O - E)^2/E$. (See column 6 of Table 12.1.) We will use the sum of these terms (which we denote by χ^2) as our test statistic.

$$\chi^2 = \sum \frac{(O - E)^2}{E}$$

From Table 12.1, we see that $\chi^2 = 1.86$. If H_0 is true, we expect the observed frequencies to be not too far from the expected frequencies. This means that the terms $(O - E)^2/E$, and therefore χ^2, should not be too large. Note that the strongest possible evidence in favor of H_0 would occur if each observed frequency were the same as the expected frequency ($O = E$) for each case. When this occurs, $\chi^2 = 0$.

Very large values of χ^2 would constitute evidence against H_0. If the value of χ^2 is "too large," we reject H_0. This would mean that not all three color schemes

TABLE 12.1

Color Scheme	O	$E = np$	$O - E$	$(O - E)^2$	$\frac{(O - E)^2}{E}$
1	89	100	−11	121	1.21
2	107	100	7	49	.49
3	104	100	4	16	.16
Totals	300	300	0	186	1.86

$$\chi^2 = \sum \frac{(O - E)^2}{E}$$

are equally popular. The value of χ^2 will be considered "too large" if it would be unlikely that we would observe such a large value of χ^2 (if H_0 were true). To assess this, we must know something about the probability distribution of χ^2. It can be shown that under appropriate conditions, χ^2 has an approximate chi-square distribution. These conditions are given in the following general result.

If the null hypothesis (specifying values for certain proportions) is true, the expression

$$\chi^2 = \sum \frac{(O - E)^2}{E}$$

has an approximate chi-square distribution provided the sample size n is sufficiently large. We will consider n sufficiently large, if it is large enough so that the expected frequency (E) of each category is at least 5. The degrees of freedom is

$$df = k - 1$$

where k = number of categories. The expected frequency of a category is

$$E = np$$

where p is the proportion for the category specified in the null hypothesis.

We said that the null hypothesis H_0 will be rejected when the value of χ^2 is so large that it would be unlikely that we would observe such a large value (if H_0 were true). "Unlikely" means having a low order of probability. We denote this probability by the symbol α; this is the level of significance. (As we stated before, the researcher usually chooses this value.) This means that we perform a right-tailed chi-square test. For example, suppose that we had decided on the value $\alpha = .05$ for the problem at hand. Now $df = k - 1 = 3 - 1 = 2$. We find the critical value from Appendix Table B.5 to be

$$\chi^2(\alpha) = \chi^2(.05) = 5.991$$

Therefore, it would be unlikely (only a 5% chance) that we would observe a value of $\chi^2 \geq 5.991$ if H_0 were true. This means that the critical region consists of values of $\chi^2 \geq 5.991$. We have seen in Table 12.1 that the observed value of $\chi^2 = 1.86$. This is not in the critical region; thus we fail to reject H_0. Hence there is no evidence that the three color schemes will not sell in the same proportion.

We can now summarize the essential features of this chi-square test for goodness-of-fit.

Chi-Square Test for Goodness-of-Fit To test the null hypothesis, which specifies certain (population) proportions associated with each category, we perform a right-tailed chi-square test. Suppose that the level of significance of the test is α. We calculate the value of χ^2 and find the critical value $\chi^2(\alpha)$ in Appendix Table B.5. If $\chi^2 \geq \chi^2(\alpha)$, we reject H_0. Otherwise, we fail to reject H_0.

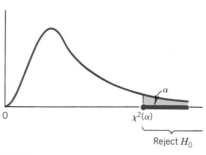

FIGURE 12.1
Critical Region

Assumption for the test: The expected frequency of each category is at least 5.

We should keep in mind the following nomenclature: We said that the categories into which the elements of the sample are classified are sometimes called cells. Therefore, the observed frequencies are also called the **observed cell frequencies**. The expected frequencies are also called the **expected cell frequencies**.

Example 12.1

A manufacturer of electronic instruments uses four assembly lines to produce the same instrument. Each assembly line is theoretically equivalent; hence all should have the same rate of production of instruments needing service under the warranty. The company wished to check this; therefore, a decision was made to look at the next 100 instruments returned to the factory as defective and determine how many came from each line. Now assembly line 1 is used for two work shifts per day, while assembly lines 2, 3, and 4 are each used for one shift per day. The defectives attributed to lines 1, 2, 3, and 4, respectively, are 53, 18, 14, and 15. Carry out the test for equivalence of the assembly lines using the 10% level of significance.

Solution

The population of interest is all defective instruments produced by the four assembly lines. Suppose the assembly lines are equivalent; that is, they produce defectives at the same rate. Since assembly lines 2, 3, and 4 are used for one shift and assembly line 1 is used for two shifts, we would expect the proportion of all defectives produced to be the same for assembly lines 2, 3, and 4 but twice this proportion for assembly line 1. Another way to look at it is this: There are a total of five work shifts per day using the four assembly lines. Assembly line 1 is used for two of these shifts; lines 2, 3, and 4 are each used for one shift. Therefore, the proportion p_1 of all defectives from line 1 should be $\frac{2}{5}$ and the proportions, p_2, p_3, and p_4 from lines 2, 3, and 4 should be $\frac{1}{5}$ each. This is the null hypothesis.

1. Hypotheses:

$$H_0: p_1 = \tfrac{2}{5} = .4, \, p_2 = p_3 = p_4 = \tfrac{1}{5} = .2$$
$$H_a: H_0 \text{ is not true}$$

2. Level of significance: $\alpha = .10$.

3. Test statistic and observed value: We have a sample of $n = 100$ defectives. To calculate the expected cell frequencies, we multiply this total by the cell probabilities (proportions) asserted in the null hypothesis H_0. The expected frequencies are given in Table 12.2.

TABLE 12.2

Assembly Line	p	$np = E$
1	.4	$(100)(.4) = 40$
2	.2	$(100)(.2) = 20$
3	.2	$(100)(.2) = 20$
4	.2	$(100)(.2) = 20$

Table 12.3, which is used to calculate χ^2, follows.

TABLE 12.3

Assembly Line	O	E	$O - E$	$(O - E)^2$	$\dfrac{(O - E)^2}{E}$
1	53	40	13	169	4.225
2	18	20	-2	4	.200
3	14	20	-6	36	1.800
4	15	20	-5	25	1.250

$$\chi^2 = \sum \frac{(O - E)^2}{E} = 7.475$$

degrees of freedom $= k - 1 = 4 - 1 = 3$

4. Critical region: We perform a right-tailed test. From Appendix Table B.5 we find that for df $= 3$, $\chi^2(.10) = 6.251$. The critical region is sketched in Figure 12.2.

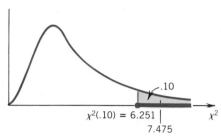

$$\chi^2(.10) = 6.251$$
7.475

FIGURE 12.2

5. Decision: The observed value of χ^2 is in the critical region; thus we reject H_0. This means that the evidence suggests that the assembly lines are not equivalent.

Sometimes the expected frequencies will not be whole numbers, which is perfectly all right, as the following illustrates.

Example 12.2

The manager of a theater complex with four theaters wanted to see if there was a difference in popularity of the four movies currently showing for Saturday afternoon matinees. The number of customers for each movie was recorded for one Saturday afternoon with the following results: 63, 55, 75, and 77 customers viewed movies 1, 2, 3, and 4, respectively. Complete the test to see if there is a difference at the 5% level of significance.

Solution

1. Hypotheses: If the four movies are equally popular, the proportion of the population of moviegoers in town who favor movie 1 is .25. Hence $p_1 = .25$. Similarly, $p_2 = p_3 = p_4 = .25$. Thus the hypotheses are

$$H_0: p_1 = p_2 = p_3 = p_4 = .25$$
$$H_a: H_0 \text{ is not true}$$

2. Level of significance: $\alpha = .05$.

3. Test statistic and its observed value: Our sample size is the sum of the observed frequencies:

$$n = 63 + 55 + 75 + 77 = 270$$

The proportion specified for each movie in the null hypothesis happens to be the same, namely, .25. The expected frequency of each movie is

$$E = np = (270)(.25) = 67.5$$

Table 12.4, which is used to calculate χ^2, follows.

TABLE 12.4

Movie	O	E	$O - E$	$(O - E)^2$	$\dfrac{(O - E)^2}{E}$
1	63	67.5	−4.5	20.25	.30
2	55	67.5	−12.5	156.25	2.31
3	75	67.5	7.5	56.25	.83
4	77	67.5	9.5	90.25	1.34

$$\chi^2 = \sum \frac{(O - E)^2}{E} = 4.78$$

degrees of freedom, df = 3

4. Critical region: We perform a right-tailed test. From Appendix Table B.5, we find that for df = 3, $\chi^2(.05) = 7.815$. (See Figure 12.3.)

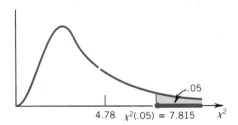

FIGURE 12.3

5. Decision: The observed value ($\chi^2 = 4.78$) is not in the critical region; thus we fail to reject H_0. This means that there is insufficient evidence to conclude that the four movies will not be equally popular.

We said that to apply the chi-square test for goodness-of-fit, the sample size should be large enough so that each expected cell frequency is at least 5. When this condition is not met, it may be possible to combine two or more cells into one cell for which the expected frequency is of the required size.

The requirement that the expected cell frequencies be at least 5 is thought by some statisticians to be overly conservative. It has been observed that when the sample size is about four or five times the number of cells, the chi-square test is still applicable even when some of the expected cell frequencies are much smaller than 5 (Lindgren, 1976, p. 424).

Exercises

In Exercises 12.1–12.3, you are given the null hypothesis H_0, the level of significance α, and the observed frequency for each of k categories. Complete the test in each exercise.

12.1 $H_0: p_1 = p_2 = p_3 = p_4 = \frac{1}{4}; \alpha = .05$

Category	O
1	68
2	65
3	77
4	90

12.2 $H_0: p_1 = p_2 = p_3 = \frac{1}{5}; p_4 = \frac{2}{5}; \alpha = .01$

Category	O
1	68
2	65
3	77
4	90

12.3 $H_0: p_1 = p_2 = \frac{1}{10}; p_3 = p_4 = \frac{1}{5}; p_5 = \frac{2}{5}; \alpha = .10$

Category	O
1	12
2	20
3	20
4	30
5	38

12.4 Two friends were playing a board game in which a die played a big role. One of the players believed the die was not fair. Sixty tosses of the die produced the results below. Test the hypothesis that the die is fair. Use a 5% level of significance.

Number of Dots	O
1	7
2	6
3	7
4	18
5	15
6	7

12.5 An official of a plastics industry claimed that the industry employed 30% white women, 5% minority women, 50% white men, and 15% minority men. To test the claim, an affirmative action committee randomly sampled 150 employees and obtained the following information:

Category	O
White females	40
Minority females	15
White males	80
Minority males	15

Test the official's claim with a 5% level of significance.

12.6 The accompanying table gives the percent distribution of U.S. exports:

Bought by	Percent
Canada	22
European Community	23.6
Japan	11.7
Newly industrialized countries	10.8
Other	31.9

(*Source*: U.S. News & World Report, February 5, 1990, p. 54.)

A computer executive wondered if exports of the company's newest computer model would differ from the above distribution. Exports of 200 units were as follows: Canada, 62; European Community, 58; Japan, 15; newly industrialized countries, 14; other, 51. At the 5% level of significance, will the company's sales differ from U.S. export distribution?

12.7 The owner of a large company claimed that $\frac{1}{10}$ of the personnel earned less than $25,000; $\frac{3}{10}$ earned at least $25,000 but less than $30,000; $\frac{3}{10}$ earned at least $30,000 but less than $35,000; and $\frac{3}{10}$ earned at least $35,000. Using a 10% level of significance, do the following sample data give sufficient evidence to refute the owner's claim?

Category	O
Less than $25,000	19
At least $25,000 but less than $30,000	56
At least $30,000 but less than $35,000	51
At least $35,000	40

12.8 A sports enthusiast theorized that 50% of college basketball games are decided by less than 10 points; 25% are decided by at least 10 and not more than 19 points; 20% by at least 20 and not more than 29 points; and 5% by more than 29 points. A survey of 112 games gave the following information:

Category	O
Less than 10 points	70
At least 10 but not more than 19 points	20
At least 20 but not more than 29 points	13
More than 29 points	9

Using a 5% significance level, is there sufficient evidence to refute the claim?

12.9 A computer science major claimed to have written a program that would randomly generate integers from 1 to 100. The program generated the following data. Use a 5% level of significance to test the claim.

Integers	O
1–10	6
11–20	6
21–30	13
31–40	9
41–50	13
51–60	11
61–70	8
71–80	12
81–90	10
91–100	12

12.10 Test the hypothesis that the data below were sampled from the standard normal distribution. Use a 10% level of significance. The data, which have been rounded to two decimal places, are as follows:

Integers	O
Less than or equal to -1.16	7
More than -1.16 but not more than $-.68$	7
More than $-.68$ but not more than $-.32$	6
More than $-.32$ but not more than 0	7
More than 0 but not more than .32	13
More than .32 but not more than .68	5
More than .68 but not more than 1.16	9
More than 1.16	6

(*Hint*: For a standard normal variable, what proportion of the data would you expect to be less than -1.16, at least -1.16 but not more than $-.68$, etc.?)

12.3 TESTS OF INDEPENDENCE. CONTINGENCY TABLES

At the beginning of this chapter we said that the chi-square test for goodness-of-fit can be used to study the question of whether two characteristics are related (dependent) or independent.

Now we examine what is meant by the *independence of two characteristics.* Suppose that two candidates, call them A and B, are running for political office and that in fact 75% of all the voters favor A and 25% favor B. Consider the two characteristics: the choice of candidates and the sex of a voter. These characteristics are independent if the percentages favoring candidates A and B are the same for male voters as for female voters. (In other words, 75% of male voters favor A and 75% of female voters favor A, etc.) If the percentage of men favoring a particular candidate is greater than the percentage of women favoring that candidate, the characteristics are related—men are more attracted to that candidate than are women. Note that the two characteristics enable us to place voters into four classes or cells: those who are female and favor candidate A, female and favor candidate B, male and favor candidate A, male and favor candidate B. See Table 12.5.

TABLE 12.5

	Favor A	Favor B
Females	Female and favoring A	Female and favoring B
Males	Male and favoring A	Male and favoring B

Suppose that 60% of the voters in this election are female. What proportion of voters would we expect in each cell if the two characteristics are independent? If 100 voters are randomly selected, we expect about 60 to be women. If candidate preference is independent of sex, and 75% of all the voters favor candidate A, then we expect about 75% of the 60 women, or 45 of them, to be in favor of candidate A. This reasoning leads us to the conclusion that if the two characteristics are independent, 45% of the voters will be both female and in favor of candidate A. Suppose that we let

p_F = proportion of voters who are female

p_A = proportion of voters who favor candidate A

p_{FA} = proportion of voters who are female and favor A

We said that $p_F = .60$, $p_A = .75$, and if the two characteristics of sex and choice of candidate are independent, $p_{FA} = .45$. Note that $.45 = (.60)(.75)$, so when we have independence

$$p_{FA} = p_F \cdot p_A$$

We call this a **multiplication rule for independent characteristics.***

Now let us calculate the proportions for the other three cells. We will call p_M the proportion of male voters, p_B the proportion of voters favoring candidate B, p_{MB} the proportion of voters who are male and favor B, and so on. Then, if the two characteristics are independent, the multiplication rule says that

$$p_{MA} = p_M \cdot p_A = (.40)(.75) = .30$$
$$p_{FB} = p_F \cdot p_B = (.60)(.25) = .15$$
$$p_{MB} = p_M \cdot p_B = (.40)(.25) = .10$$

The various proportions are displayed in Table 12.6.

TABLE 12.6
Independent Characteristics

	Favor A	Favor B	
Females	$p_{FA} = p_F \cdot p_A = .45$	$p_{FB} = p_F \cdot p_B = .15$	$p_F = .60$
Males	$p_{MA} = p_M \cdot p_A = .30$	$p_{MB} = p_M \cdot p_B = .10$	$p_M = .40$
	$p_A = .75$	$p_B = .25$	

If the multiplication rule does not hold for each cell, the two characteristics (sex and choice of candidate in this case) are not independent—they are related. As an extreme example of a situation where the characteristics would be related, consider Table 12.7. In this table we observe that although the overall proportions favoring A and B remain the same as in Table 12.6, all of the men, who constitute 40% of the voting population, favor candidate A. Note that the multiplication rule does not hold here. For example,

$$p_{FA} = .35 \neq (.60)(.75) = p_F \cdot p_A$$

---

*Those who have read Section 4.5 may recognize this as equivalent to the multiplication rule for independent events. Keep in mind that proportions can be viewed as probabilities. Now suppose that we let F be the event that a voter selected is a female and A be the event that the voter favors A. In Section 4.5 we saw that when the two events are independent, the multiplication rule for independent events tells us that

$$p(F \text{ and } A) = p(F) \cdot p(A)$$

TABLE 12.7
Dependent Characteristics

	Favor A	Favor B	
Females	$p_{FA} = .35$	$p_{FB} = .25$	$p_F = .60$
Males	$p_{MA} = .40$	$p_{MB} = 0$	$p_M = .40$
	$p_A = .75$	$p_B = .25$	

When testing for independence of two characteristics, our null hypothesis will be that the characteristics are independent. (This is equivalent to asserting that the multiplication rule holds for each category.) The alternate hypothesis will be that the characteristics are related.

A Test of Independence

To see how a chi-square test may be used to determine if two characteristics are related, we will examine data obtained in a survey of student opinions conducted at an eastern college. We will try to determine if there is a relationship between students' political views and their opinion about nuclear power for production of consumer energy. The population would be all students at the college (and at similar colleges). The questions asked of the students were as follows:

Question 1.   What label most closely describes your political views?
Democratic   Republican   Independent

Question 2.   What is your opinion of nuclear power for production of consumer energy?
Approve   Disapprove   Undecided

Table 12.8, called a *contingency table*, summarizes the result of the survey. In the square in the upper left-hand corner of Table 12.8 we see the value 10, meaning that 10 students in the survey see themselves as Democrats and favor nuclear power. The number in the lower left-hand corner means that a total of $10 + 9 + 8 = 27$ students see themselves as Democrats. The row total of 45 in the upper right-hand corner means that a total of 45 students approve of nuclear power. The total number of students in the survey (grand total) was 100, and this appears in the lower right-hand corner. This number can be obtained by adding either the row or column totals.

The hypotheses we wish to test for the student population are

H_0: political view and opinion about nuclear power are independent

H_a: political view and opinion of nuclear power are related

We will use the 5% level of significance and apply a chi-square test. We need to calculate the expected frequencies for the nine cells in Table 12.8. The expected frequencies are the frequencies we would expect if H_0 were true. Recall that

TABLE 12.8
Survey of Students' Political Views versus Their Opinions on Nuclear Power

Opinion	Political View			Row Total
	Democrat	Republican	Independent	
Approve	10	15	20	45
Disapprove	9	2	16	27
Undecided	8	2	18	28
Column Total	27	19	54	100 ← Grand Total

independence is equivalent to saying that the multiplication rule for independent characteristics holds. For example, this means that the proportion of students (in the population) who approve of nuclear power and are also Democrats is equal to the product of the proportion of students who approve of nuclear power (p_A) times the proportion of students who are Democrats (p_D). Therefore, if the characteristics are independent,

$$\text{proportion who approve and are Democrats} = p_A \cdot p_D$$

Using this proportion, we can find the expected frequency of the cell (assuming independence) by multiplying the proportion by the sample size of 100. The problem is that the proportions discussed (p_A and p_D) are the proportions for the entire population (of all students at similar colleges). These proportions are unknown.

We will estimate the proportions that we will need from the sample data. Now a total of 45 out of a grand total of 100 students in the survey said they approved of nuclear power. (See Table 12.8.) Hence we will estimate the proportion of the population who approve of nuclear power (p_A) as $\frac{45}{100}$. Note that

$$p_A \doteq \frac{45}{100} = \frac{\text{row total}}{\text{grand total}}$$

Similarly, a total of 27 out of 100 students interviewed were Democrats. Thus we estimate p_D by

$$p_D \doteq \frac{27}{100} = \frac{\text{column total}}{\text{grand total}}$$

Therefore, the expected proportion for the cell, assuming independence, is

proportion who approve of nuclear power and are Democrats

$$= p_A \cdot p_D \doteq \left(\frac{45}{100}\right)\left(\frac{27}{100}\right) = .1215$$

The expected cell frequency is obtained by multiplying the sample size by this proportion:

expected number who approve of nuclear power and are Democrats

$$= E = (100)\left(\frac{45}{100}\right)\left(\frac{27}{100}\right) = \frac{(45)(27)}{100} = 12.15$$

Note that for this cell the expected frequency is

$$\frac{(45)(27)}{100} = \frac{(\text{row total}) \cdot (\text{column total})}{\text{grand total}}$$

Therefore, the expected frequency E is

$$E = \frac{(\text{row total}) \cdot (\text{column total})}{\text{grand total}}$$

This same formula is used to calculate the expected frequency for each cell. For example, the expected frequency for the cell consisting of those who approve of nuclear power and are Republican is (from Table 12.8)

$$E = \frac{(45)(19)}{100} = 8.55$$

In Table 12.9 we display the observed frequencies O, along with the expected frequencies E (in parentheses).

TABLE 12.9
Table of Observed (and Expected) Frequencies

Opinion	Democrat	Political View Republican	Independent	Row Total
Approve	10 (12.15)	15 (8.55)	20 (24.30)	45
Disapprove	9 (7.29)	2 (5.13)	16 (14.58)	27
Undecided	8 (7.56)	2 (5.32)	18 (15.12)	28
Column Total	27	19	54	100

We now calculate the observed value of the test statistic

$$\chi^2 = \sum \frac{(O - E)^2}{E}$$

$$= \frac{(10 - 12.15)^2}{12.15} + \frac{(15 - 8.55)^2}{8.55} + \frac{(20 - 24.30)^2}{24.30}$$

$$+ \frac{(9 - 7.29)^2}{7.29} + \frac{(2 - 5.13)^2}{5.13} + \frac{(16 - 14.58)^2}{14.58}$$

$$+ \frac{(8 - 7.56)^2}{7.56} + \frac{(2 - 5.32)^2}{5.32} + \frac{(18 - 15.12)^2}{15.12}$$

$$\doteq .3805 + 4.8658 + .7609 + .4011 + 1.9097$$

$$+ .1383 + .0256 + 2.0719 + .5486$$

$$\doteq 11.10$$

Normally, the number of degrees of freedom associated with χ^2 is one less than the number of cells $(k - 1)$. However, whenever a population proportion is estimated, as we have done, the degrees of freedom must be reduced by 1 for each proportion estimated. We estimated the row and column proportions. For example, the first row proportion, the proportion who approve was estimated as $p_A \doteq \frac{45}{100} = .45$. There are three row proportions. However, since they must add up to 1, when we know two, the third is known. So, in effect, we estimated two row proportions and two column proportions. This means that we must reduce the usual degrees of freedom by 4.

$$\text{df} = (k - 1) - (\text{number of proportions estimated})$$
$$= (9 - 1) - (4) = 4$$

Note that if we let

$$r = \text{number of rows} = 3$$
$$c = \text{number of columns} = 3$$

then we can write

$$\text{df} = 4 = (2)(2) = (3 - 1)(3 - 1) = (r - 1)(c - 1)$$

This formula holds in general for contingency tables.* If r is the number of rows and c is the number of columns

$$\text{df} = (r - 1)(c - 1)$$

As usual, large departures of the observed frequencies from the expected frequencies favor the alternate hypothesis. These large departures also cause the value of χ^2 to be large; hence we reject H_0 when χ^2 is very large. This means that we perform a right-tailed test. From Appendix Table B.5 we see that the critical value of χ^2 when $\alpha = .05$ and df $= 4$ is $\chi^2(.05) = 9.488$. The critical region is shown in Figure 12.4. The observed value $\chi^2 = 11.10$ is in the critical region, so we reject the null hypothesis of independence. Thus the characteristics of political viewpoint and opinion on nuclear power seem to be related.

We can now summarize the essential features of a test for independence of two characteristics.

---

*When counting the number of rows and columns, we do not include the last row and column that contain the column and row totals.

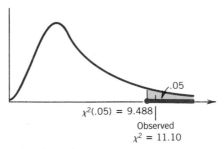

$\chi^2(.05) = 9.488$

Observed
$\chi^2 = 11.10$

FIGURE 12.4

Chi-square Test for Independence of Two Characteristics   To test the null hypothesis H_0, which states that the two characteristics are independent, against an alternative hypothesis that states that the characteristics are dependent, we conduct a right-tailed chi-square test. The number of observations (O) in each category is obtained and placed in a contingency table (such as Table 12.8). The expected frequency of a cell is

$$E = \frac{(\text{row total}) \cdot (\text{column total})}{\text{grand total}}$$

The test statistic is

$$\chi^2 = \sum \frac{(O - E)^2}{E}$$

This has an approximate chi-square distribution. If the contingency table has r rows and c columns, the degrees of freedom is

$$df = (r - 1)(c - 1)$$

If α is the level of significance, find the critical value $\chi^2(\alpha)$ in Appendix Table B.5. If $\chi^2 \geq \chi^2(\alpha)$, we reject H_0. (See Figure 12.5.)

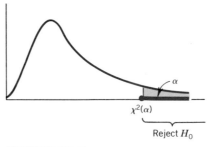

$\chi^2(\alpha)$

Reject H_0

FIGURE 12.5
Critical Region

Assumption for the test: Each expected frequency E is ≥ 5.

Example 12.3

Prior to the time that the germ theory of disease was established (around 1870) the mortality rate from surgery was very high due to infection. Louis Pasteur and Joseph Lister were largely responsible for the germ theory. Lister felt that if carbolic acid were used as a disinfectant, the patient's chance of survival might be improved. He used it to scrub everything in the operating room that came in contact with the patient. He even sprayed the air with carbolic acid and used it on the patient's dressing. Lister compared 40 operations (amputations) in which this procedure was used, with 35 amputations in which it was not used. The results are summarized in the contingency table (Table 12.10).

TABLE 12.10
Results of Carbolic Acid Used in 40 out of 75 Amputations

	Patient Lived	Patient Died	Row Total
Carbolic Acid Used	34	6	40
Carbolic Acid Not Used	19	16	35
Column Total	53	22	75

Source: Winslow, C., *The Conquest of Epidemic Disease* (Princeton, N.J.: Princeton University Press, 1943), p. 303.

At the 1% level of significance, test whether the outcome of the surgery (patient lived or died) is independent of the use of carbolic acid or not.

Solution

1. Hypotheses:

    H_0: A patient's survival (or death) does not depend on whether carbolic acid is used in surgery.

    H_a: A patient's survival is related to whether carbolic acid is used or not.

2. Level of significance: $\alpha = 01$.

3. Test statistic and observed value: We will use the chi-square statistic. We must first calculate the expected cell frequencies; these are the frequencies that we would expect if the null hypothesis of independence were true. Using the multiplication rule, we calculate these frequencies from the formula

$$E = \frac{(\text{row total}) \cdot (\text{column total})}{\text{grand total}}$$

For example, the expected number of patients who lived and for whom carbolic acid was used (assuming independence) is

$$E = \frac{(40)(53)}{75} \doteq 28.27$$

In Table 12.11 we record the expected frequencies along with the observed frequencies.

TABLE 12.11
Observed (and Expected) Frequencies for Lister's Data

	Patient Lived	Patient Died	Row Total
Carbolic Acid Used	34 (28.27)	6 (11.73)	40
Carbolic Acid Not Used	19 (24.73)	16 (10.27)	35
Column Total	53	22	75

$$\chi^2 = \sum \frac{(O - E)^2}{E} = \frac{(34 - 28.27)^2}{28.27} + \frac{(6 - 11.73)^2}{11.73}$$
$$+ \frac{(19 - 24.73)^2}{24.73} + \frac{(16 - 10.27)^2}{10.27}$$
$$\doteq 1.1614 + 2.7991 + 1.3277 + 3.1970$$
$$\doteq 8.49$$
$$df = (r - 1)(c - 1) = (2 - 1)(2 - 1) = 1$$

4. Critical region: Large values of χ^2 indicate large deviations of the observed values from the expected values and therefore favor the alternate hypothesis; thus we perform a right-tailed test. From Appendix Table B.5 we find that with df = 1, $\chi^2(.01) = 6.635$. The critical region with observed value is shown in Figure 12.6. (Note that chi-square distributions with df = 1 or 2 have somewhat different shapes from other chi-square distributions.)

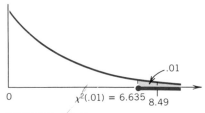

FIGURE 12.6

5. Decision: The observed value of χ^2 is in the critical region. Therefore, we reject H_0. This means that there does appear to be a relation between the use of carbolic acid and the survival of a patient.

Exercises

12.11 A sample of 50 moviegoers was taken to see if the rating of a controversial movie and age of viewer were related. The following data were obtained:

	Excellent	Fair	Poor
Less than 30	6	12	8
30 or more	10	6	8

Test whether the characteristics of rating and age are independent using a 5% level of significance.

12.12 A random sample of students at a small liberal arts college was obtained to test the claim that sex and political affiliation were independent at this college. Test the claim with a 10% level of significance. The data are

	Democrat	Republican	Independent
Females	20	10	25
Males	10	9	38

12.13 A study of CHD (coronary heart disease) among diabetics aged 45–74 with congestive heart failure yielded the following information (*Source*: Kannel, 1974, p. 31). Is there sufficient evidence to indicate a relationship between the absence or presence of CHD and gender? Use a 5% level of significance. The data are

	With CHD	Without CHD
Males	5	6
Females	10	7

12.14 A random sample of 200 voters living in a precinct of a large city was undertaken to see if a dependent relationship existed between union affiliation and political preference. Using a 5% level of significance, is there sufficient evidence to indicate a relationship? The data are

	Democrat	Republican	Independent
Union Member	80	15	35
Not a Union Member	25	20	25

12.15 A group of employees in a large firm claimed that sex and yearly salary were dependent. For the following data, test the claim using a 10% level of significance.

	Less than 30	Salaries (in thousands of dollars) At Least 30 but Less than 35	At Least 35 but Less than 40	At Least 40
Females	12	30	20	13
Males	7	26	31	27

12.16 A political science professor claimed that the political affiliation of students and their belief concerning public funding of abortions were independent. Test the claim using a 5% level of significance. The sample data are

	Democrat	Republican	Independent
Approve	9	6	21
Disapprove	16	11	23
Undecided	6	5	15

12.17 (a) Calculate the observed χ^2 value for each of the following contingency tables:

(i)

	B_1	B_2	B_3
A_1	10	20	30
A_2	20	40	60

(ii)

	B_1	B_2
A_1	120	60
A_2	40	20

(iii)

	B_1	B_2
A_1	20	30
A_2	40	60
A_3	200	300

(b) Compare the observed χ^2 values in part (a). Note the relationship between the rows (and columns) in each of the contingency tables in part (a).

(c) (i) Complete the following contingency table so that the observed χ^2 value is zero. (There is more than one correct way to complete the table.)

	B_1	B_2	B_3
A_1	20	10	30
A_2	60		
A_3			

(ii) In the table in item (c)(i), fill in the A_3B_1 cell with the number 30. Now complete the table so that the observed χ^2 value is zero. (There is only one correct way to complete the table.)

(d) What conclusion can you draw about contingency tables and the strongest evidence to support H_0?

12.4 TESTS OF HOMOGENEITY

In the tests of independence considered in Section 12.3, we were interested in whether two characteristics were related for individuals of the *same* population. For example, suppose we wanted to see if there were a relationship between political affiliation (Democrat, Republican, or Independent) and work status (self-employed or employed by another). We could interview a certain number of people from the population of employed individuals and classify them as to work status and political affiliation. The number of self-employed people in the survey might be so small that we would be unable to complete the test of independence.

A test of homogeneity provides another approach to the problem. Such a test considers one characteristic for *different* populations. We could consider two populations: those who are self-employed and those who are employed by another. The issue is whether these populations are similar (homogeneous) with respect to the characteristic of political affiliation. The sampling procedure would be different. We would obtain a sample (of desired size) of self-employed people and a sample of workers who are not self-employed, and would classify each person according to political affiliation. Similarity would exist if each population had the same political make-up (i.e., the same percentage of Democrats, the same percentage of Republicans, and the same percentage of Independents). Hence a test of homogeneity tests a null hypothesis that asserts that different populations are homogeneous with respect to some characteristic of interest, against an alternative hypothesis that asserts that they are not.

Tests of homogeneity involve tables that are similar to contingency tables and, in fact, the procedure for carrying out such tests is exactly the same as the chi-square test used in connection with contingency tables. We illustrate the procedure in the following example.

Example 12.4

In a study of voting characteristics of adult whites, blacks, and Hispanics in Massachusetts, samples of 100 individuals were obtained from each of these populations. Each individual was asked if he or she voted in the midterm congressional elections of 1982. The results of this survey are summarized in Table 12.12. Test to see if the populations of white, black, and Hispanic adults in Massachusetts are similar (homogeneous) with respect to the characteristic of voting or not voting in the 1982 midterm congressional election. [To put it another way, did (about) the same proportion from each population vote?] Use the 5% level of significance.

TABLE 12.12
Survey of Voting-Age Individuals from Three Racial Groups in the 1982 Midterm Congressional Elections

	Voted	Did Not Vote	Row Total
Whites	47	53	100
Blacks	40	60	100
Hispanics	27	73	100
Column Total	114	186	300

Solution

We will use the chi-square test for goodness-of-fit. It turns out that this test is used exactly the same way as it was used in analyzing contingency tables in Section 12.3.

1. Hypotheses:

   H_0: The three populations are homogeneous with respect to the proportion voting.

   H_a: The three populations are not homogeneous with respect to this characteristic.

2. Level of significance: $\alpha = .05$.

3. Test statistic and observed value: We will conduct a chi-square test. We first need the expected frequencies. To see that these can be calculated in the same manner as we did for contingency tables in the last section, we consider the expected number of whites who voted. This is the expected frequency of the cell in the upper left-hand corner. Note that a total of 114 individuals out of a total of 300 individuals from all three populations voted. That is,

$$\frac{114}{300} = .38 \text{ or } 38\% \text{ voted}$$

Now if all three populations are homogeneous (i.e., if H_0 is true), we would expect the same proportion (38%) to have voted in each population. Since there are 100 whites in the survey, we expect 38% of those, or 38 whites, to have voted. Observe that we can express this expected frequency as follows:

$$E = 38 = (100)(.38) = (100)\left(\frac{114}{300}\right) = \frac{(100)(114)}{300}$$

$$= \frac{(\text{row total}) \cdot (\text{column total})}{\text{grand total}}$$

Thus we can use the same formula to calculate expected frequencies as we did in Section 12.3:

$$E = \frac{(\text{row total}) \cdot (\text{column total})}{\text{grand total}}$$

In Table 12.13 we have recorded the observed and expected frequencies.

TABLE 12.13
Observed (and Expected) Frequencies

	Voted	Did Not Vote	Row Total
Whites	47 (38)	53 (62)	100
Blacks	40 (38)	60 (62)	100
Hispanics	27 (38)	73 (62)	100
Column Total	114	186	300

The test statistic is the usual chi-square statistic:

$$\chi^2 = \sum \frac{(O - E)^2}{E} = \frac{(47 - 38)^2}{38} + \frac{(53 - 62)^2}{62}$$

$$+ \frac{(40 - 38)^2}{38} + \frac{(60 - 62)^2}{62}$$

$$+ \frac{(27 - 38)^2}{38} + \frac{(73 - 62)^2}{62}$$

$$\doteq 2.1316 + 1.3065 + .1053 + .0645 + 3.1842 + 1.9516$$

$$\doteq 8.74$$

The degrees of freedom is the same as for contingency tables.

$$\text{df} = (r - 1)(c - 1) = (3 - 1)(2 - 1) = 2$$

4. Critical region: As usual, large departures of the observed frequencies from the expected frequencies favor the alternate hypothesis. These large departures also cause the value of χ^2 to be large; therefore, we conduct a right-tailed test. The critical value of χ^2 when df = 2 and α = .05 is from Appendix Table B.5: $\chi^2(.05)$ = 5.991. The critical region is shown in Figure 12.7 with the observed value.

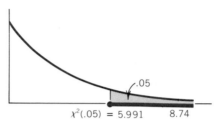

$$\chi^2(.05) = 5.991 \qquad 8.74$$

FIGURE 12.7

5. Decision: The observed value is in the critical region; thus we reject H_0. The three populations do not appear to be homogeneous with respect to the characteristic under investigation.

Note that in the previous example the sample size from each of the three populations happened to be the same. However, in tests of homogeneity this is not a requirement.

Exercises

12.18 In 1970, 1980, and 1990, women aged 20–24 living in a large city were sampled and asked whether they had ever been married. Use a 5% level of significance and test whether the three populations are homogeneous with respect to the proportion of women aged 20–24 who had ever married.

	Had Been Married Yes	No
1970	54	21
1980	58	32
1990	42	38

12.19 Two dice were tossed to see if they gave the "same results." Recorded were the frequency of the number of dots.

	1	2	3	4	5	6
First Die	4	6	7	5	5	8
Second Die	7	4	8	6	5	5

Using a 5% level of significance, test whether the two dice are homogeneous with respect to the number of dots showing.

12.20 A survey of medical doctors and student nurses was undertaken with the participants being asked whether they were smokers. Test whether the appropriate populations are homogeneous using a 10% significance level.

	Yes	No
Medical Doctors	18	32
Student Nurses	10	30

12.21 The results of a sample of male and female workers in a large factory are given below. Those sampled were asked whether they felt that there was discrimination in salaries between males and females. Test the hypothesis that the appropriate populations are homogeneous using a 1% level of significance.

	Yes	No
Males	55	70
Females	65	40

12.22 Some former patients from four hospitals were sampled and asked whether they were satisfied with the care they received while in the hospital. The results were as follows:

	Satisfied	Not Satisfied
Hospital 1	85	15
Hospital 2	82	18
Hospital 3	71	29
Hospital 4	68	32

Test the hypothesis that the appropriate populations are homogeneous with respect to satisfaction with hospital care. Use a 5% level of significance.

USING MINITAB (OPTIONAL)

In Table 12.8 we saw the results of a survey comparing political view versus opinion on nuclear power (for the production of consumer energy). The data were as follows:

	Political View		
Opinion	Democrat	Republican	Independent
Approve	10	15	20
Disapprove	9	2	16
Undecided	8	2	18

In the Minitab printout below, we read the data in the table into columns C1, C2, and C3. The CHISQUARE command then produces the table of observed and expected frequencies (which we have seen in Table 12.9). The observed value of χ^2 is also given (11.10) along with the degrees of freedom (4). The user may then look up a critical value corresponding to a desired value for α. From Appendix Table B.5, we found that when $\alpha = .05$, the critical value is $\chi^2(.05) = 9.488$. So the null hypothesis of independence (of political view versus opinion on nuclear power) is rejected.

```
MTB > READ THE FOLLOWING DATA INTO C1 C2 C3
DATA> 10 15 20
DATA> 9 2 16
DATA> 8 2 18
DATA> END
      3 ROWS READ
MTB > CHISQUARE ANALYSIS OF TABLE IN C1 C2 C3

Expected counts are printed below observed counts

              C1       C2       C3     Total
      1       10       15       20        45
            12.15     8.55    24.30

      2        9        2       16        27
             7.29     5.13    14.58

      3        8        2       18        28
             7.56     5.32    15.12

  Total       27       19       54       100
```

```
ChiSq = 0.380 + 4.866 + 0.761 +
        0.401 + 1.910 + 0.138 +
        0.026 + 2.072 + 0.549 = 11.102
df = 4
```

You may wish to use the P value approach. However, Release 7 of Minitab does not print a P value for the chi-square test. Nevertheless, we have seen that the CDF command can be useful in finding P values. The following program computes the P value. Keep in mind that the chi-square test here is a right-tailed test.

```
MTB > CDF 11.10 STORE IN K1;
SUBC>    CHISQUARE DF=4.
MTB > #K1 IS THE AREA TO LEFT OF 11.10
MTB > LET K2 = 1-K1 # THE P VALUE
MTB > PRINT K2
K2       0.0254628
```

Thus the P value is .0254628.

Exercises

Suggested exercises for use with Minitab are 12.28 through 12.32. Give the P value for these exercises.

12.6 SUMMARY

In this chapter we considered three types of statistical tests: the **chi-square test for goodness-of-fit,** the **test of independence,** and the **homogeneity test.** The test statistic used in all these tests is

$$\chi^2 = \sum \frac{(O - E)^2}{E}$$

where O represents the observed frequencies and E represents the expected frequencies. This statistic has approximately a chi-square distribution if the different values of E are at least 5. Large departures of the observed frequencies from the expected frequencies favor the alternate hypothesis in all three tests. These large departures cause the value of χ^2 to be large; therefore, a large value of χ^2 favors the alternate hypothesis. This means that we always conduct a right-tailed test.

The expected values, E, and degrees of freedom, df, of the test statistic for each of the three tests are calculated as follows.

1. The chi-square test for goodness-of-fit:
$$E = np \qquad \text{df} = k - 1$$
where n = sample size, p = proportion for a particular category or cell stated in the null hypothesis, and k = number of cells.

2. Test of independence and homogeneity test:
$$E = \frac{(\text{row total}) \cdot (\text{column total})}{\text{grand total}} \qquad \text{df} = (r - 1)(c - 1)$$
where r = number of rows and c = number of columns in the contingency table.

Review Exercises

12.23   Discuss the differences between tests for goodness-of-fit, independence, and homogeneity.

12.24   (a)   What is the smallest value χ^2 can be?

   (b)   Construct a 2 × 2 contingency table so that $\chi^2 = 0$.

   (c)   Construct a 2 × 4 contingency table so that $\chi^2 = 0$.

12.25   The distribution of final grades given by a mathematics department in the past was 10% A's, 20% B's, 30% C's, 25% D's, and 15% F's. A new teacher gave the following grades for the first semester:

Category	O
A	12
B	20
C	26
D	14
F	8

Is there sufficient evidence to suggest that the new teacher's grading policy is different from that of the department? Use a 5% level of significance.

12.26   Test the hypothesis that the following data were sampled from a normal distribution with mean 0 and standard deviation 34.92. Use a 10% level of significance. The data, rounded to two decimal places, are

Interval	O
Less than or equal to -23.40	9
More than -23.40 but not more than 0	18
More than 0 but not more than 23.40	19
More than 23.40	14

(*Hint*: If the population is normal with $\mu = 0$ and $\sigma = 34.92$, what proportion of data values would you expect to be less than or equal to -23.40, more than -23.40 but not more than 0, etc.?)

12.27   In an industrial complex, a committee was formed to study traffic conditions. The committee wanted to see if the modes of transportation to work by workers had changed over the past 5 years. At that time, 70% of the workers had driven alone; 20% had been in a carpool; 8% used public transportation; and the rest used other modes. The committee obtained the following information from a sample of 500 workers.

Mode of Transportation	O
Drive alone	320
Carpool	130
Public Transportation	35
Other means	15

Is there sufficient evidence to indicate that the modes of transportation to work have changed? Use a 5% level of significance.

12.28   A random sample of students at a public college was obtained to see if a relationship existed between political affiliation and attitude concerning racial quotas in hiring. Using a 5% level of significance, is there sufficient evidence to indicate a relationship?

	Democrat	Republican	Independent
Approve	7	7	9
Disapprove	19	13	38
Undecided	6	6	12

12.29   A study of CHD (coronary heart disease) among nondiabetics aged 45–74 with congestive heart failure yielded the following information (*Source*: Kannel, 1974, p. 31). Is there sufficient evidence to indicate a relationship between the absence or presence of CHD and gender? Use a 10% level of significance.

	With CHD	Without CHD
Males	54	32
Females	39	30

12.30   The transportation committee (discussed in Exercise 12.27) wanted to know if there was a relationship between the mode of transportation to work and the starting time for work. The following sample data were obtained:

	Drive Alone	Carpool	Public Transportation	Other Means
Start at 8:00	65	39	10	4
Start at 8:30	95	45	11	6
Start at 9:00	160	46	14	5

Using a 1% level of significance, determine if the characteristics of starting time and mode of transportation to work are related.

12.31   Seventy-five male and 100 female students at a large university were randomly sampled from the populations of enrolled males and females, respectively. Use the following data to test whether the male and female populations are homogeneous with respect to age. Use a 5% level of significance.

	Less than 22	Between 22 and 30	More than 30
Males	45	20	10
Females	40	40	20

12.32   The transportation committee (discussed in Exercise 12.27) wished to study the mode of transportation used by male and female employees. Two hundred and fifty men and 250 women were sampled and the following data were obtained:

	Drive Alone	Carpool	Public Transportation	Other Means
Men	173	58	15	4
Women	147	72	20	11

Using a 1% level of significance, determine if male and female employees are homogeneous with respect to their choice of mode of transportation.

12.33   In Exercise 12.28, the sampling scheme consisted of selecting 117 students. It turned out that there were 32 Democrats, 26 Republicans, and 59 Independents. Suppose instead that 32 Democrats were selected from the population of all Democrats on campus. Similarly, 26 Republicans and 59 Independents were chosen from appropriate populations.

(a)   What would be the appropriate hypotheses to test in this situation?

(b)   Would the value of the test statistic be the same as or different from the one obtained in Exercise 12.28?

Notes

Kannel, W. B., et al., "Role of Diabetes in Congestive Heart Failure, Framingham Heart Study," in *The American Journal of Cardiology* 34: (1974), pp. 29–34.

Lindgren, B., *Statistical Theory*, 3rd ed. (New York: Macmillan Publishing Company, 1976).

U.S. News & World Report, February 5, 1990.

Winslow, C., *The Conquest of Epidemic Disease* (Princeton, N.J.: Princeton University Press, 1943).

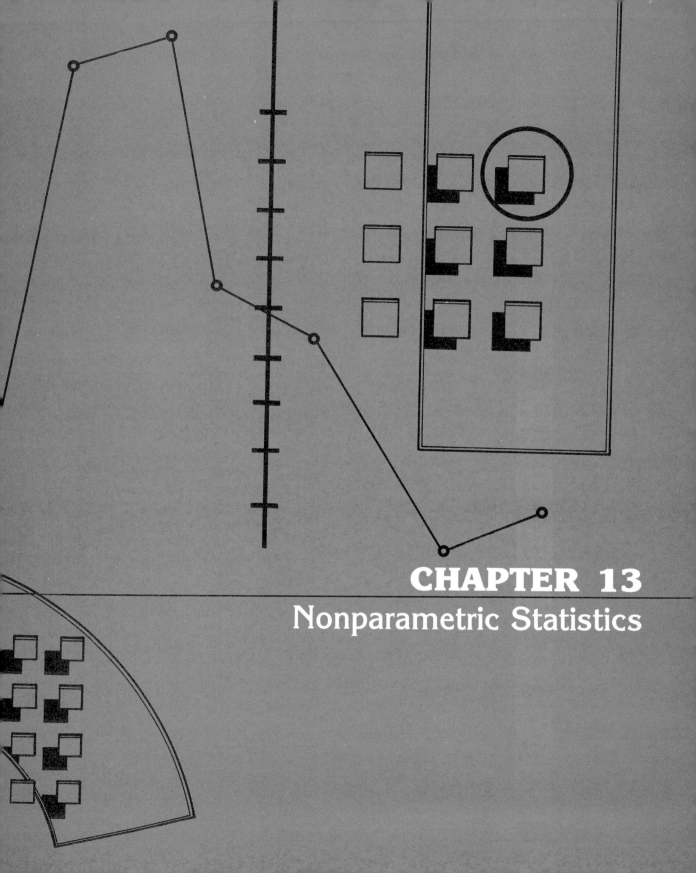

CHAPTER 13
Nonparametric Statistics

13.1

INTRODUCTION

The statistical tests we have developed thus far in this book often require that the population distributions be normal or approximately so. This is the case, for instance, with the small-sample t tests discussed in Chapters 8 and 9. Yet in some cases it is not known whether a distribution is approximately normal. In other cases it is known that the distribution departs substantially from normality. For example, systolic blood pressure readings are (approximately) normally distributed with a mean of 120 (millimeters of mercury). But suppose that we consider only blood pressure readings over 140. This population has a distribution that resembles the shape of the right tail of a normal distribution, as indicated in Figure 13.1.

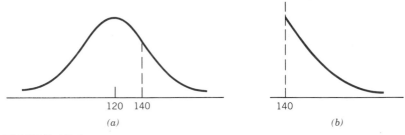

FIGURE 13.1
(a) Systolic Blood Pressures. (b) Systolic Blood Pressures above 140

Nonparametric tests were developed to deal with situations where the population distributions are nonnormal or unknown, especially when the sample size is small (< 30). They make no assumptions or few assumptions about the populations under consideration.

> **Definition**   Statistical methods that do not depend on a knowledge of the population distribution or its parameters are called *nonparametric* or distribution-free methods. All other methods are called *parametric* methods.

Actually, many nonparametric tests are not completely distribution free, since they place some (although minimal) requirements on the distribution (such as requiring that the distribution be continuous). For this reason some describe these methods as "distribution freer."

Among statisticians there is less than unanimous agreement on the definition of nonparametric methods, and the line between parametric and nonparametric tests is not a sharp one. For example, the large-sample z tests for means discussed in Chapters 7 and 9 are actually distribution free, yet they are usually classified as parametric tests.

Even though nonparametric tests are easy to understand, simple to apply, and place minimal requirements on the form of the population distributions involved, they are not always superior to parametric tests. When the conditions on the population are met, the parametric tests are better than their nonparametric counterparts (e.g., they are more likely to reject a false null hypothesis). We will have more to say about this in Section 13.6.

There are many nonparametric tests. In this chapter we will present just a few of the more commonly encountered tests.

THE SIGN TEST

We will illustrate the logic behind the sign test with an example. A consulting geographer was studying various aspects of a large city. It was claimed that the median value for all homes in the city was $70,000. Recall that the median for a collection of data has the property that half the data values are larger than the median and half are smaller. The geographer wanted to check this claim. The population under consideration is the collection of the values of all the homes in the city. The question is whether the median (Md) for this population is $70,000 or not. The hypotheses to be tested are

$$H_0: \text{Md} = 70,000$$

$$H_a: \text{Md} \neq 70,000$$

The geographer conducted a small random survey of real estate agencies from which he obtained the selling prices of 14 recently sold homes in the city. They were arranged in an increasing order of magnitude, as shown in Table 13.1. Notice that for the sample of 14 homes given below, only 2 prices are below $70,000, while 12 are above. If the median for the population were really $70,000, we would expect closer to one-half of the values in the sample (or 7) to be above $70,000 and one-half below. Intuitively, it seems unlikely that 12 prices would be above $70,000 and only 2 would be below $70,000 if the median were really $70,000.

TABLE 13.1
Selling Prices
of 14 Homes
(in Dollars)

61,300	87,000
66,200	95,000
71,000	96,100
76,500	99,000
77,100	105,000
77,300	120,300
81,000	140,000

In such problems we need not rely on intuition alone. In the problem at hand we could use x, the number of homes in the sample above $70,000 in value, as a test statistic. We may view x as a binomial random variable associated with a binomial experiment, where

$$\text{trial} = \text{randomly select a home}$$

$$\text{success} = \text{value of home is above } \$70,000$$

If the null hypothesis is true, then one-half the homes in the city should be valued above $70,000. Therefore,

$$P(\text{success}) = p = \tfrac{1}{2}$$

In the problem at hand, the number of trials $n = 14$ and $x = $ number of successes. If the null hypothesis is true, the expected value of x is

$$E(x) = \mu = np = (14)(\tfrac{1}{2}) = 7$$

If we observe a value of x so far from 7 (above or below) that such a value would be unlikely (if H_0 were true), we reject H_0. "Unlikely" means having a low order of probability. This probability is called the "level of significance" of the test (α), and we choose its value. Suppose that *prior* to obtaining our data we had decided to choose α to be as close to .01 as possible. The critical region will consist of values of x so far from 7 that there would only be about a 1% chance of observing a value of x in this region if H_0 were true.

We can use Appendix Table B.2 to find the critical region. We find the column corresponding to $n = 14$, $p = .5$. This gives probabilities corresponding to the various possible values of x. Certainly, the extreme values of 0 and 14 should be in the critical region. We keep on adding values of x near both extremes until the probabilities of these x's add up to approximately .01. If we choose for our critical region the x values:

0   1   2   12   13   14

their probabilities add up to

$$0 + .001 + .006 + .006 + .001 + 0 = .014$$

This is reasonably close to .01. Hence we will use these values of x as our critical region. (If we took 2 and 12 away from the critical region, the probability would only be .002. If we added 3 and 11 to the critical region, the probability would be .058.)

Since the observed value, $x = 12$, is in the critical region, we reject H_0. This means that the median home value does not appear to be $70,000. (It, in fact, appears to be higher.) The above test is called a **sign test** because sometimes we label the values in the sample above the (conjectured) value of the median with a plus sign, and those below with a minus sign. We then count the number of pluses.

Now we review the steps in the above test:

1. Hypotheses:

H_0: Md = \$70,000

H_a: Md $\neq$ \$70,000

2. Level of significance: Choose α as close to .01 as possible.

3. Test statistic and observed value: Refer to Table 13.1.

x = number of values in sample above \$70,000 = 12

4. Critical region: From Appendix Table B.2 we choose a two-tailed region that has a probability as close to .01 as possible. This is 0, 1, 2, 12, 13, 14. With this critical region, $\alpha = .014$. (See Figure 13.2.)

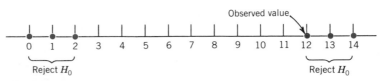

FIGURE 13.2

n the samp
airs). The
-tailed tes

The observed value ($x = 12$) is in the critical region; thus we reject

that may have occurred to the reader is why a t test was not
ve problem. A t test was not used because the geographer was
population of selling prices was approximately normal, which
if the t test is to be used. If we could be sure that the population
normal (roughly mound shaped) a t test should be used, be-
appropriate, a t test is a more sensitive test than the sign test.
vith a t test, there is a better chance of detecting (and rejecting)
nesis. If a t test could have been used in the previous problem,
be tested would have been

Sign of D
+
+
−
−
+
+
−
+
+
−
+
+
−

$$H_0: \mu = \$70{,}000$$

$$H_a: \mu \neq \$70{,}000$$

opulation is approximately normal, its distribution is sym-
dian. (See Figure 3.3(c).) It is also worth pointing out here
involving central tendency are used to study means while
e often used to study medians.]
orise that, when appropriate, a t test is more sensitive
test throws away a great deal of information—it only
value is above the conjectured value for the median,
the data value. The t test uses the magnitudes of all

findings as follows:

t seed that it felt
n farmers agreed
on another acre.

e two varieties of
Complete the test,

is test concerning the value of the median of a
t statistic:

lues in the sample above the value of the
e null hypothesis H_0

omial distribution with n = the sample size
x Table B.2 to find the critical region.

H_a involves the symbol $\neq$, we find a two-

ls)

s)

, we find a right-tailed critical region.
we find a left-tailed critical region.

(population) median

icance, we choose critical values so that
ritical region is as close to α as possible.
are continuous.

The sign test may be used to compare two populations whe[re]
are dependent (i.e., the values from the two samples occur in p[...]
lowing example illustrates this. It also illustrates the use of a on[e ...]

Example 13.1

TABLE 13.2
Bushels of Wheat from Two Types of Seed

Farm	New Variety (y_1)	Old Variety (y_2)	Difference $D = y_1 - y_2$
1	34	27	7
2	45	25	20
3	30	38	−8
4	30	42	−12
5	48	21	27
6	35	22	13
7	32	37	−5
8	46	30	16
9	41	32	9
10	23	38	−15
11	42	26	16
12	43	33	10
13	27	30	−3

A seed company considered marketing a new variety of whe[at ...]
would produce a greater yield than its current variety. Thirte[en ...]
to use the new type of seed on one acre and the old variety [...]
The resulting yields in bushels are given in Table 13.2.

The company was unsure about the distributions for th[e ...]
wheat, and so decided to use the sign test instead of a t test. [...]
using a level of significance as close to 5% as possible.

Solution
We call

$$y_1 = \text{yield from the new variety (in bush[els ...])}$$

$$y_2 = \text{yield from the old variety (in bushe[ls ...])}$$

$$D = y_1 - y_2$$

The new variety will be considered an improvement if the [...]
of D is positive: $Md_D > 0$. This is the alternate hypothesi[s ...]

1.   Hypotheses:

$$H_0: \text{Md}_D = 0$$

$$H_a: \text{Md}_D > 0$$

2.   Level of significance: $\alpha = .05$.

3.   Test statistic and observed value: We use the number of values of D from the sample that are above 0 (the number of pluses next to the D values in Table 13.2). Note that it was not necessary to calculate the values of D. We only need to know the signs of the D values. From Table 13.2,

$$x = \text{number of pluses} = 8$$

4.   Critical region: We perform a right-tailed test. Looking at Appendix Table B.2 when $n = 13$ and $p = .5$, we see that if we use the values 10, 11, 12, and 13 for the critical region, then

$$\alpha = P(10) + P(11) + P(12) + P(13)$$

$$= .035 + .010 + .002 + 0 = .047$$

which is close to .05. Hence the critical region will consist of the following x values: 10, 11, 12, and 13. (See Figure 13.3.)

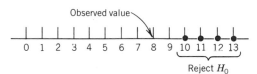

FIGURE 13.3

5.   Decision: The observed value ($x = 8$) is not in the critical region; thus we fail to reject H_0. This means that there is insufficient evidence to conclude that the new variety of seed is more effective than the old one.

The following question may have occurred to the reader: What happens if one (or more) of the data values in the sample is exactly equal to the value of the median in the null hypothesis? How is it classified? If we encounter a data value equal to the median, the usual procedure is to remove it from the sample and reduce the sample size by 1.

Normal Approximation

Appendix Table B.2 can be used in the sign test to find the critical region for values of x, when the sample size n is up to 25. For values of n greater than 25, we may use the normal approximation to the binomial variable x. We have seen in Chapter 6 that x may be viewed as approximately normal with

$$\mu = np = n \cdot \tfrac{1}{2} \text{ since } p = \tfrac{1}{2} \text{ if } H_0 \text{ is true}$$

$$\sigma = \sqrt{npq} = \sqrt{n \cdot \tfrac{1}{2} \cdot \tfrac{1}{2}} = \tfrac{1}{2}\sqrt{n}$$

Therefore,

$$z = \frac{x - \mu}{\sigma} = \frac{x - n/2}{\sqrt{n}/2}$$

is approximately standard normal. The following example illustrates the use of the normal approximation.

Example 13.2

A sports enthusiast claimed that the median weight of college football players was less than 250 pounds. A sample of 28 such players showed 16 weighing less than 250 pounds, 11 weighing more than 250 pounds, and one weighing 250 pounds. Test the claim at the 10% level of significance.

Solution

1. Hypotheses: It is claimed that Md $<$ 250. The hypotheses are

$$H_0: \text{Md} = 250$$
$$H_a: \text{Md} < 250$$

2. Level of significance: $\alpha = .10$.

3. Test statistic and observed value: One observation is 250 pounds. Since this equals the value in question, we discard it. Of the remaining $n = 27$ values, $x = 11$ are above 250. We use the standardized form of x:

$$z = \frac{x - n/2}{\sqrt{n}/2} = \frac{11 - 27/2}{\sqrt{27}/2} \doteq -.96$$

4. Critical region: Since H_a contains the symbol $<$, we do a left-tailed test. From Appendix Table B.3 we find that $z(.10) = 1.28$. Therefore, the critical region, shown in Figure 13.4, consists of values of $z \leq -1.28$.

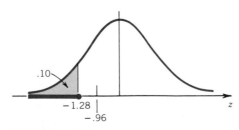

FIGURE 13.4

5. Decision: The observed value ($z = -.96$) is not in the critical region; thus we fail to reject H_0. This means that there is insufficient evidence to support the claim that Md < 250.

Exercises

13.1 In parts (a)–(l) assume a sign test is used. Complete the following table by finding either an appropriate critical region or the level of significance α. Note that n refers to the sample size.

	Hypothesis	n	Critical Region	α
(a)	Md $= 100$	12		.038
(b)	Md > 100	7	{6, 7}	
(c)	Md < 100	9		.090
(d)	Md $\neq 100$	8	{0, 1, 7, 8}	
(e)	Md > 100	15		.059
(f)	Md < 100	13	{0, 1, 2, 3}	
(g)	Md < 100	17		.071
(h)	Md > 100	11	{9, 10, 11}	
(i)	Md $= 100$	4	{0, 4}	
(j)	Md < 100	5	{0}	
(k)	Md $\neq 100$	16		.078
(l)	Md > 100	19		.031

13.2 The median time for runners in a specific age group to complete a 3-mile run is 30 minutes. A runner in this age group claimed to have a faster median time. A sample of this runner's past 3-mile runs gave the following times:

28.5   28.0   29.5   28.5   29.0   29.5   32.0   30.5   29.0   29.5   30.5

Is there sufficient evidence to justify the runner's claim? Use the sign test with a 5% significance level.

13.3 A shift in a factory was being evaluated on the basis of the number of hours to complete a job. The median time, set by management, to complete the job was 2.8 hours, but the plant manager believed that the median time of the shift was longer. A sample of 10 completed jobs gave the following times:

2.0   3.2   3.7   3.4   3.3   2.5   4.1   3.1   3.6   3.5

Use the sign test with a 5% significance level to test the manager's belief.

13.4 (a) A newspaper reported that the median age for drivers issued tickets for speeding in a large town was less than 20 years. A sample of 13 tickets gave the following ages of the drivers:

27   17   19   18   43   18   19   17   26   36   19   34   19

Use the sign test and a 5% significance level to test the newspaper's report.

(b) Suppose that a sample of size $n = 42$ had been obtained and 15 tickets were issued to drivers over 20, 24 to drivers under 20, and 3 to drivers recorded as 20 years old. Now use the normal approximation to test the hypothesis discussed in part (a) at the 5% level of significance.

13.5 To test the claim of a nutritionist that a new diet was effective, eight people were selected and their weights before and after the diet were recorded. The data are

Weight before Diet	171	183	162	196	151	209	198	215
Weight after Diet	166	187	155	198	140	206	192	210

Test the claim using the sign test and a 5% level of significance.

13.3 THE WILCOXON SIGNED-RANK TEST

Although the sign test discussed in Section 13.2 is a very simple test to use, it is not a very sensitive test. Sometimes it will fail to reject a false null hypothesis when another test would be successful in detecting the falsity of the null hypothesis. This is not very surprising, since the sign test throws away a good deal of information about the data—the only information it uses about a data value is whether it is above the conjectured value of the median or not, ignoring the magnitude of the data value.

In this section we will study another nonparametric test that is a better test than the sign test because it uses more information: the **Wilcoxon signed-rank test.** We illustrate this test with an example.

A public school official felt that high school seniors in a large school system would tend to score higher on a standard reading examination than the national median of 50. She randomly selected 13 seniors and gave them the examination. The results were as follows:

$$57 \quad 70 \quad 42 \quad 48 \quad 72 \quad 63 \quad 45 \quad 66 \quad 59 \quad 39 \quad 73 \quad 78 \quad 47$$

Since the official wanted to see if the median for the school system (the population median) was higher than the national median, she wished to test the hypotheses:

$$H_0: \text{Md} = 50$$

$$H_a: \text{Md} > 50$$

If we were to conduct a right-tailed sign test at an (approximate) level of significance of 5%, we would fail to reject H_0. (This is not hard to see. Notice that the number of scores above 50 is 8 out of 13, so the value of the test statistic would be 8. But recall that Example 13.1 was also a right-tailed sign test at the 5% level with a sample of size 13. The value of the test statistic was also 8. In Example 13.1 we failed to reject the null hypothesis; thus we would come to the same conclusion in the present case.)

The school official still felt that the null hypothesis should be rejected. There is a more sensitive test that does not throw away as much information as the sign test. This test can be applied because, based on scores on similar tests, it seemed that the seniors' test scores would have a distribution that was symmetric (although the distribution itself was unknown). When the distribution is symmetric, the Wilcoxon signed-rank test (for a single population) may be applied. (Symmetry means that a vertical line through the median will divide the distribution curve into halves that are mirror images of one another.) See Figure 13.5.

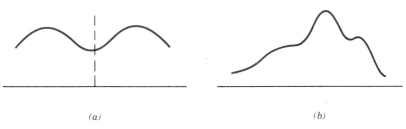

(a)   *(b)*

FIGURE 13.5
(a) Symmetric (b) Not Symmetric

When applying the Wilcoxon signed-rank test, we also require that the data be continuous. (Even though the test scores are whole numbers, we regard these scores as continuous data that have been rounded off.)

In Table 13.3 we record the data values in the sample, and the difference between each data value (x) and the value of the median appearing in the null hypothesis, namely, 50. We then rank the absolute value of these differences. We put a minus sign ($-$) in front of those ranks that correspond to data values below the median of 50. See column 4 of Table 13.3.

TABLE 13.3

Score x	Difference $D = x - 50$	Magnitude $|D|$	Signed Rank
57	7	7	4
70	20	20	10
42	-8	8	-5
48	-2	2	-1
77	27	27	12
63	13	13	8
45	-5	5	-3
64	14	14	9
59	9	9	6
39	-11	11	-7
73	23	23	11
78	28	28	13
47	-3	3	-2

Note: The minus sign ($-$) means that the rank corresponds to a difference for a data value to the left of 50.

Suppose for the moment that the population median is actually 50. Since the sample comes from a population of scores that are symmetric with respect to the population median, we expect the sample itself to be more or less symmetrically distributed about this median. This means that if we look at the ranks of the magnitudes of the differences (D), the ranks corresponding to the data values on one side of 50 should be comparable in magnitude to those corresponding to data values on the other side. Thus the sum of the ranks for one side should be comparable in magnitude to the sum of the ranks for the other side. Looking at the fourth column of Table 13.3, we let

$$W^+ = \text{sum of positive ranks} = 73$$

$$W^- = \text{absolute value of sum of negative ranks} = 18$$

As we said, if the true median is 50, we expect these two values to be of comparable size. If W^+ were much smaller than W^-, this would suggest that the data values were spread farther below 50 than above 50. This would imply that Md $<$ 50. If, on the other hand, W^- were much smaller than W^+, this would suggest that the data values were spread farther above 50 than below 50. This would imply that Md $>$ 50.

Small values of W^+ suggest Md $<$ 50.

Small values of W^- suggest Md $>$ 50.

For the data under consideration the value $W^- = 18$ seems small (in relation to $W^+ = 73$) and Figure 13.6 does seem to suggest that Md $>$ 50.

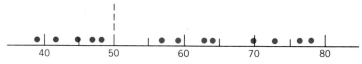

FIGURE 13.6
Spread of the Data Suggests Md $>$ 50

If we were to test

$$H_0: \text{Md} = 50$$

$$H_a: \text{Md} > 50$$

we could use W^- as a test statistic. If this is "too small," we reject H_0 in favor of H_a. We consider W^- to be too small if it would be unlikely that we would observe such a small value (if H_0 were true). "Unlikely" means having a low order of probability. As usual, this probability is denoted by α and is specified by the researcher. It is the level of significance of the test. Appendix Table B.8

enables us to determine the values of W^- that are too small. This table give us critical values for various values of α and the sample size. When W^- is less than or equal to the critical value, it is too small and therefore leads to rejection of the null hypothesis. A portion of Appendix Table B.8 is reproduced in Table 13.4. (This table is also used to find critical values of W^+.)

TABLE 13.4
Some Critical Values for the Wilcoxon Signed-Rank Test. The Boxed Value is the Critical Value c for a One-Sided Test with $\alpha = .05$ and Sample Size $n = 13$.

One-Sided α	Two-Sided α	$n = 11$	$n = 12$	$n = 13$	$n = 14$	$n = 15$	$n = 16$
.05	.10	14	17	21	26	30	36
.025	.05	11	14	17	21	25	30
.01	.02	7	10	13	16	20	24
.005	.01	5	7	10	13	16	19

We will use this table while completing the test in question at the 5% level of significance.

1. Hypotheses:

$$H_0: \text{Md} = 50$$
$$H_a: \text{Md} > 50$$

2. Level of significance: $\alpha = .05$.

3. Test statistic and observed value:

$$W^- = |(-5) + (-1) + (-3) + (-7) + (-2)| = |-18| = 18$$

4. Critical region: Values of W^- that are too small lead to rejection of H_0. From Table 13.4 we see that when $n = 13$ and $\alpha = .05$ with a one-sided test, the critical value is $c = 21$. Therefore, the critical region consists of (integral) values of W^- less or equal to 21: $W^- \leq 21$. (See Figure 13.7.)

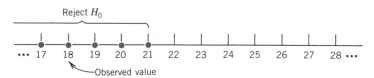

FIGURE 13.7

5. Decision: The observed value of W^- is 18 and is in the critical region; thus we reject H_0. This means that the median does appear to be greater than 50.

We can now summarize the procedure for the Wilcoxon signed-rank test. Assume that the null hypothesis involves some specific value M for the median. Keep in mind that small values of W^- favor Md $> M$, while small values of W^+ favor Md $< M$.

Wilcoxon Signed-Rank Test for a Population Median Suppose that M is the value of the median in question that appears in the null hypothesis. Calculate $D = x - M$ for each data value x. We then rank the values of $|D|$ and place a minus sign in front of each rank corresponding to a negative difference D (as in Table 13.3). Let

$$W^+ = \text{sum of the positive ranks}$$
$$W^- = \text{absolute value of sum of negative ranks}$$

(a) To test

$$H_0: \text{Md} = M$$
$$H_a: \text{Md} > M$$

use W^- as a test statistic. Find the critical value c for a one-sided test with desired significance level in Appendix Table B.8. If $W^- \leq c$, reject H_0.

(b) To test

$$H_0: \text{Md} = M$$
$$H_a: \text{Md} < M$$

use W^+ as a test statistic. Find the critical value c for a one-sided test with desired significance level in Appendix Table B.8. If $W^+ \leq c$, reject H_0.

(c) To test

$$H_0: \text{Md} = M$$
$$H_a: \text{Md} \neq M$$

find the critical value c for a two-sided test with desired significance level in Appendix Table B.8. If either $W^+ \leq c$ or $W^- \leq c$, reject H_0. This means that we can use the minimum of W^+ or W^- as a test statistic. We denote this value by W. If $W \leq c$, reject H_0.

Assumptions for the test:

1.   The data are continuous.
2.   The data come from a population with an approximately symmetric distribution

Tied Ranks

There is a problem that can arise in the above test that we have not considered. When we calculate the magnitude of the differences $|D|$ and rank them, what happens if some values of $|D|$ are the same? For example, if we list the values of $|D|$ in increasing order of magnitude, what happens if, say, the 5th and 6th values are the same? In this case we assign the rank 5.5 to both of them (the average of ranks 5 and 6). The next higher value of $|D|$ is assigned the rank 7. If the 5th, 6th, and 7th values of $|D|$ are the same, we assign the rank 6 to each (the average of the ranks 5, 6, and 7), and so on.

Zero Differences

If any of the values of D are 0, we will use the following procedure: If there are an even number of zeros, each zero is assigned the average rank for the set and then half of them are assigned a plus sign and half a minus sign. For example, if there are four zeros, we would average the ranks 1, 2, 3, 4, getting 2.5. Then we would end up with the signed ranks: -2.5, -2.5, 2.5, 2.5. If there are an odd number of zero values for D, we discard one of them and reduce the sample size by 1. We then have an even number of zeros and we can proceed as described above.

Comparing Two Populations Using a Paired Experiment

The Wilcoxon signed-rank test can be used to compare two populations when the data values from the two populations are obtained in pairs. In this case the samples are dependent. In addition to requiring that the data be continuous, it is required that the two populations have distributions with similar shapes. We illustrate the procedure with an example.

    Two experimental drugs were developed for the treatment of hypertension; call them drug 1 and drug 2. The drugs were administered to seven pairs of patients with hypertension. Each pair of subjects was matched for medical history, age, and level of blood pressure. One subject in each pair was given drug 1, the other drug 2. For each patient the drop in diastolic blood pressure was recorded. The two drugs were chemically similar, and so it was thought that their distributions (representing drop in diastolic pressure) would have similar shapes, but it was not known whether their medians would be the same. Two distributions may have the same *shape* but a different *location* (Figure 13.8). In Table 13.5 the drop in diastolic pressure for each patient (in millimeters of mercury) is recorded.

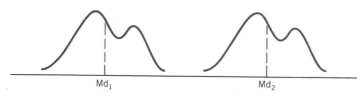

FIGURE 13.8

TABLE 13.5
Drop in Diastolic
Blood Pressure
for Seven Pairs
of Patients

Drug 1	Drug 2
10	6
16	20
10	8
4	12
2	8
14	4
4	15

Suppose that we use the symbol x_1 to represent the drop in diastolic pressure for a patient given drug 1 and x_2 to represent the drop for a patient given drug 2. For each pair in the sample we look at the difference

$$D = x_1 - x_2$$

This gives rise to a new variable D associated with a conceptual population of differences.

If the researchers studying the drugs wanted to see if there was any difference in the effectiveness of the drugs, they would test the hypotheses

$$H_0: \text{Md}_D = 0$$

$$H_a: \text{Md}_D \neq 0$$

This test concerns a single median (where the value in question is $M = 0$). We have seen that the Wilcoxon signed-rank test may be applied if the population of D values is symmetric. It turns out that as long as the x_1 values and x_2 values have similarly shaped distributions, it can be shown that the distribution of D values is symmetric (even if the populations of x_1 and x_2 values are not symmetric).* Therefore, the Wilcoxon signed-rank test can be applied.

In the following example we complete the test of the hypotheses discussed above. Note that D plays the same role as it did before.

Example 13.3
We will now apply this test to the data in Table 13.5. Use the 10% level of significance.

---

*In fact, it can also be shown that the median of the difference D is the difference of the medians, $\text{Md}_D = \text{Md}_1 - \text{Md}_2$ (assuming that the x_1 values and x_2 values have similarly shaped distributions).

Solution

1. Hypotheses: The question is whether or not drug 1 and drug 2 are equally effective. The hypotheses concern the difference D between the drop in blood pressure for drug 1 versus drug 2.

$$H_0: \text{Md}_D = 0$$
$$H_a: \text{Md}_D \neq 0$$

2. Level of significance: $\alpha = .10$.

3. Test statistic and observed value: If *either* W^+ or W^- are very small, we reject H_0. Therefore, we can use the minimum of these values, called W, as a test statistic. Table 13.6 is used to compute W.

TABLE 13.6

x_1	x_2	$D = x_1 - x_2$	$\|D\|$	Signed Rank	
10	6	4	4	2.5	tied for 2nd
16	20	-4	4	-2.5	and 3rd ranks
10	8	2	2	1	
4	12	-8	8	-5	
2	8	-6	6	-4	
14	4	10	10	6	
4	15	-11	11	-7	

(Note that if we rank the values of $|D|$, we get 2, 4, 4, 6, 8, 10, 11. Since the 4's occupy the 2nd and 3rd position, we assign them the rank 2.5.)

$$W^+ = 2.5 + 1 + 6 = 9.5$$
$$W^- = |(-2.5) + (-5) + (-4) + (-7)| = |-18.5| = 18.5$$
$$W = \text{minimum of } W^+ \text{ and } W^- = 9.5$$

4. Critical region: Using Appendix Table B.8, we find that when $n = 7$ and $\alpha = .10$, the critical value c for a two-sided test is 4. So the critical region consists of values of $W \leq 4$.

5. Decision: The observed value 9.5 is not in the critical region, so we fail to reject H_0. This means that there is no evidence that the drugs have different effects.

Example 13.4

A college professor developed an English course for college students in need of remediation. All students entering the college are given a screening examination. Of those receiving scores below 50, 25 students were obtained who agreed to take the course. Their scores on the pretest and on a posttest given after completing the course are given in Table 13.7 along with the differences D and the signed ranks of $|D|$.

TABLE 13.7
Before and After Scores for 25 Students

Pretest (x_1)	Posttest (x_2)	Difference $D = x_1 - x_2$	$|D|$	Signed Rank
46	76	-30	30	-25
27	36	-9	9	-7
37	53	-16	16	-12.5
34	55	-21	21	-18
20	12	8	8	6
38	50	-12	12	-10
10	36	-26	26	-22
24	18	6	6	4
20	21	-1	1	-1
39	57	-18	18	-15
16	27	-11	11	-9
20	48	-28	28	-23
47	70	-23	23	-19
45	25	20	20	17
40	50	-10	10	-8
46	39	7	7	5
32	51	-19	19	-16
49	33	16	16	12.5
45	69	-24	24	-20
49	52	-3	3	-2
44	60	-16	16	-12.5
45	20	25	25	21
16	12	4	4	3
41	70	-29	29	-24
48	64	-16	16	-12.5

Do these data justify the professor's claim that the course improves English skills? Use the 5% level of significance.

Solution

1. Hypotheses: The professor's claim is that scores on the pretest (x_1) tend to be less than scores on the posttest (x_2). This means that values of the differences $D = x_1 - x_2$ tend to be less than 0 $(\text{Md}_D < 0)$. Therefore, the hypotheses to be tested are

$$H_0: \text{Md}_D = 0$$

$$H_a: \text{Md}_D < 0$$

2. Level of significance: $\alpha = .05$.

3. Test statistic and observed value: We use W^+ as a test statistic. From the above table

$$W^+ = 68.5$$

4. Critical region: From Appendix Table B.8 we find the critical value for a one-tailed test with $\alpha = .05$ and $n = 25$ is $c = 101$. Hence the critical region would consist of values of $W^+ \le 101$.

5. Decision: The observed value of 68.5 is in the critical region, so we reject H_0. This means that the professor's claim ($Md_D < 0$) appears to be correct. In other words, scores on the pretest tend to be lower than on the posttest, suggesting that the course is effective.

The following is a summary of the procedure for a Wilcoxon signed-rank test for a paired experiment.

Wilcoxon Signed-Rank Test for a Paired Experiment
Let $D = x_1 - x_2$ for each pair of data values. To test a null hypothesis $Md_D = 0$, we proceed by conducting the Wilcoxon test for a single median (where $M = 0$). Here D plays the same role as before.

Assumptions for the test:

- The data are continuous.

- The two sets of data (the x_1 values and the x_2 values) come from distributions with similar shapes.

Normal Approximation

Appendix Table B.8 is applicable for sample sizes up to $n = 50$. We may wish to apply the Wilcoxon signed-rank test for larger sample sizes. It turns out that the test statistic used in this test is approximately normal when n is sufficiently large (greater than or equal to 15). Suppose that we let W represent the test statistic. This will either be W^+ or W^-, depending on the situation. It can be shown that

(a) $\quad \mu_W = \dfrac{n(n + 1)}{4}$

(b) $\quad \sigma_W = \sqrt{\dfrac{n(n + 1)(2n + 1)}{24}}$

(c) Since W is approximately normal when $n \geq 15$,

$$z = \frac{W - \mu_W}{\sigma_W} = \frac{W - n(n+1)/4}{\sqrt{n(n+1)(2n+1)/24}}$$

is approximately standard normal. Therefore, when the sample size is large enough, we can find the critical region in terms of z.

We will show how this works by repeating Example 13.4, this time using the normal approximation. Steps 1 and 2 will be the same as in Example 13.4. The remaining steps are as follows.

- We calculate the test statistic in step 3 using the standardized form of W^+, namely,

$$z = \frac{W^+ - n(n+1)/4}{\sqrt{n(n+1)(2n+1)/24}} = \frac{68.5 - (25)(26)/4}{\sqrt{(25)(26)(51)/24}} \doteq -2.53$$

- Critical region (step 4): Small values of W^+ favor H_a. These values correspond to values of z far to the left of 0, so we perform a left-tailed test. The critical region consists of values of $z \leq -z(.05) = -1.65$. (See Figure 13.9.)

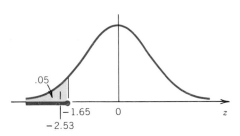

FIGURE 13.9

- Decision: The observed value -2.53 is in the critical region. Therefore, we reject the null hypothesis. (This is the same conclusion we reached in Example 13.4.)

If we had been testing

$$H_0: \mathrm{Md}_D = 0$$
$$H_a: \mathrm{Md}_D > 0$$

we would have used W^- when calculating z. Since small values of W^- favor H_a, we would still conduct a left-tailed test. The rejection region would consist of values of $z \leq -z(\alpha)$. If we had been interested in testing

$$H_0: \mathrm{Md}_D = 0$$
$$H_a: \mathrm{Md}_D \neq 0$$

we would reject H_0 if the observed value z is $\geq z(\alpha/2)$ or $\leq -z(\alpha/2)$. But since we have agreed to deal with the minimum of W^+ or W^- when conducting a two-tailed test, it can be shown that the value of z will always be ≤ 0; therefore, we will reject H_0 if $z \leq -z(\alpha/2)$.

Exercises

13.6   In parts (a)–(j) assume a Wilcoxon signed-rank test is used. Complete the following table. Note that n and α refer to the sample size and level of significance, respectively.

	Hypothesis	n	α	Observed Value	Critical Value	Decision
(a)	Md $>$ 50	6	.05	$W^- = 4$		
(b)	Md $<$ 50	6	.025	$W^+ = 4$		
(c)	Md $=$ 50	7	.05	$W = 2$		
(d)	Md $<$ 50	8	.01	$W^+ = 3$		
(e)	Md $=$ 50	8	.10	$W = 5$		
(f)	Md $>$ 50	10	.01	$W^- = 5$		
(g)	Md $=$ 50	11	.05	$W = 13$		
(h)	Md $<$ 50	15	.05	$W^+ = 24$		
(i)	Md $>$ 50	16	.01	$W^- = 22$		
(j)	Md $=$ 50	20	.10	$W = 68$		

In Exercises 13.7–13.9, you are given the alternate hypothesis, the level of significance α, and the sample data.

(a)   Complete the table.

(b)   Use the Wilcoxon signed-rank test and complete the test.

13.7 H_a: Md $>$ 10, $\alpha = .05$.

Sample Data x	Difference $D = x - 10$	Magnitude $|D|$	Signed Rank
18			
12			
4			
1			
15			
17			
26			
0			
14			

13.8 H_a: Md $\neq$ 60, α = .10.

Sample Data x	Difference $D = x - 60$	Magnitude $|D|$	Signed Rank
49			
54			
68			
72			
65			
51			
45			
58			

13.9 H_a: Md $<$ 50, α = .05.

Sample Data x	Difference $D = x - 50$	Magnitude $|D|$	Signed Rank
42			
39			
37			
55			
46			
52			
33			

13.10   A real estate salesman claimed that the median price of house lots in a remote resort area was $30,000. A sample of 10 lots for sale gave the following prices (in thousands of dollars):

28.9   31.5   42.5   28.0   34.9   32.5   36.9   38.5   37.9   32.9

Test the salesman's claim using the Wilcoxon signed-rank test with a 5% level of significance.

13.11   A lotion was advertised as relieving muscular pain in 10 minutes, on average. Believing that it might take longer, a "truth in advertising" group tested 15 people and obtained the data below. A data value represents the time in minutes that the symptom was relieved after application of the lotion. The data are

10.3   9.2   10.6   11.1   9.5   12.0   12.2   9.1
10.7   13.1   12.1   10.1   10.2   11.3   11.0

Use the Wilcoxon signed-rank test with a 1% significance level to complete the test.

13.12 An appliance dealer was concerned that the median time before an appliance needs to be serviced might be less than 3 years. A survey of 12 such appliances gave the following times (in years) to the first repair:

$$2.7 \quad 3.4 \quad 3.2 \quad 1.8 \quad 2.5 \quad 2.4$$
$$4.0 \quad 2.3 \quad 2.2 \quad 3.1 \quad 1.1 \quad 1.5$$

Use the Wilcoxon signed-rank test with a 5% level of significance to investigate the dealer's concern.

13.13 Past records at an eastern college indicated that the median MSAT score was 520 for incoming mathematics majors. The chairperson believed that a strong advertising campaign would help recruit majors with a higher median score. Following the advertising campaign, the MSAT scores of incoming mathematics majors were noted. Do the following data provide sufficient evidence to conclude that such advertising campaigns are effective? Use the Wilcoxon signed-rank test with a 1% level of significance.

$$535 \quad 560 \quad 500 \quad 550 \quad 560 \quad 540 \quad 610 \quad 490 \quad 545 \quad 550$$
$$510 \quad 620 \quad 460 \quad 575 \quad 485 \quad 630 \quad 600 \quad 515 \quad 640 \quad 650$$

13.14 A nutritionist claimed that a liquid diet was effective in reducing weight. A paired experiment gave the following data:

Weight Before Diet	Weight After Diet
171	166
183	187
162	155
196	198
151	140
209	206
198	192
215	210

(a) Use the Wilcoxon signed-rank test for a paired experiment with a 5% level of significance to test the claim.

(b) Test the claim using the sign test with a 5% significance level. (*Note*: You may have already done this in Exercise 13.5.)

(c) Compare your answers in parts (a) and (b). Comment on the different conclusions reached using the two tests.

13.15 An official of an athletic competition felt that judge A tended to give lower performance ratings than judge B. A sample of 26 pairs of ratings was obtained. Let x_1 and x_2 be the evaluation of judges A and B, respectively.

x_1	5.9	7.0	9.1	4.3	3.8	8.0	3.3	6.6	9.3	7.6	8.4	7.3	7.4
x_2	5.8	7.2	9.8	3.9	3.3	8.3	4.7	5.6	8.0	6.5	8.6	6.5	9.0

x_1	5.1	4.0	3.3	4.5	3.9	7.0	6.5	8.1	3.8	6.5	4.5	3.0	7.7
x_2	7.0	4.3	2.3	6.8	5.4	6.8	8.6	8.9	4.7	8.3	6.2	3.1	8.3

Using the Wilcoxon signed-rank test with a 5% level of significance, test the official's feeling by using:

(a) Appendix Table B.8 to obtain a critical value.

(b) The normal approximation.

13.4 THE MANN–WHITNEY U TEST

In the last section we showed how the Wilcoxon signed-rank test can be used to compare two populations when the sample data from the populations are paired—one value in the pair coming from each population. In this case the samples are *dependent*. Such samples were discussed in Chapter 9. We also discussed *independent* samples in Chapter 9. When the samples are independent, we simply have a collection of data values from each population, and there is no natural way to pair up the data values. F. Wilcoxon also proposed a test for dealing with independent samples in 1945. This test is called the *Wilcoxon rank-sum test*. Instead of treating this test, in this section we will discuss a test that is equivalent to the Wilcoxon rank-sum test, namely, the **Mann–Whitney U test.** This test enables us to compare two population medians. We illustrate this test with the following example.

A feed and grain company wanted to compare the effectiveness of two types of fertilizer it stocked. Four farmers who were using fertilizer x and six farmers who were using fertilizer y agreed to cooperate in the study. Each farmer reported the yield in corn obtained (in bushels per acre). The results are given in Table 13.8. We will assume at the outset that the yields for both fertilizers have similar distributions, although the distributions may differ in location. That is, they may have different medians.

TABLE 13.8
Yields in Bushels per Acre
for Two Types of Fertilizer

Fertilizer x	Fertilizer y
103	89
85	101
92	108
110	93
	90
	82

The Mann–Whitney test is based on a ranking of all the sample data taken together. The following is a list of all the data values ranked in ascending order of magnitude (with an x or y placed beneath the data value depending on whether it was a yield from fertilizer x or y):

82	85	89	90	92	93	101	103	108	110
y	x	y	y	x	y	y	x	y	x

The fertilizer yields and ranks are given in Table 13.9.

TABLE 13.9
Fertilizer Yields with Overall Rankings

Fertilizer x	Rank	Fertilizer y	Rank
103	8	89	3
85	2	101	7
92	5	108	9
110	10	93	6
		90	4
		82	1

If the x values have low ranks, this would suggest that fertilizer x was less effective than fertilizer y. That is, the median for fertilizer x is less than that of fertilizer y, $Md_x < Md_y$. Low ranks for the y values would suggest that fertilizer y was less effective than fertilizer x ($Md_y < Md_x$). We ought to look at the sum of the ranks for each fertilizer. These are given in Table 13.10.

TABLE 13.10
Rank Sums

Ranks for Fertilizer x	Ranks for Fertilizer y
8	3
2	7
5	9
10	6
$S_x = 25$	4
	1
	$S_y = 30$

Small values of S_x suggest that fertilizer x is less effective than fertilizer y. Small values of S_y suggest that fertilizer y is less effective than fertilizer x. But we should be careful in using the word "small." This should be measured in relation to something. The rank sum S_x can be smaller than S_y just because there are fewer x values and therefore fewer ranks for x values. We should look at how close S_x comes to the smallest possible value it could conceivably assume. The worst-case scenario for fertilizer x would occur if all the x values were less than all the y values:

$$
\begin{array}{ccccccccccc}
& x & x & x & x & y & y & y & y & y & y \\
\textit{Ranks:} & 1 & 2 & 3 & 4 & 5 & 6 & 7 & 8 & 9 & 10
\end{array}
$$

For this arrangement, $S_x = 1 + 2 + 3 + 4 = 10$. This is the smallest possible value for S_x. You may verify for yourself that the smallest possible value for S_y is 21. If we let $n_1 =$ the number of x values and $n_2 =$ number of y values, the following expressions give the smallest possible values for S_x and S_y, respectively.

$$
\frac{n_1(n_1 + 1)}{2} = \frac{(4)(5)}{2} = 10 \qquad \frac{n_2(n_2 + 1)}{2} = \frac{(6)(7)}{2} = 21
$$

It would seem reasonable to look at the amount by which S_x (and S_y) actually exceed their smallest possible values. To this end, we define

$$
U_x = S_x - \frac{n_1(n_1 + 1)}{2} = 25 - 10 = 15 \qquad \left\{ \begin{array}{l} \text{Amount by which } S_x \text{ exceeds} \\ \text{its smallest possible value} \end{array} \right.
$$

$$
U_y = S_y - \frac{n_2(n_2 + 1)}{2} = 30 - 21 = 9 \qquad \left\{ \begin{array}{l} \text{Amount by which } S_y \text{ exceeds} \\ \text{its smallest possible value} \end{array} \right.
$$

In general, the Mann–Whitney test makes use of U_x or U_y when comparing two population medians Md_x and Md_y according to the following principle:

> Small values of U_x suggest $\text{Md}_x < \text{Md}_y$.
> Small values of U_y suggest $\text{Md}_x > \text{My}_y$.

We still need to further clarify the meaning of the word "small" in this context. Therefore, we continue the fertilizer discussion. Suppose that before gathering the data, the company felt that perhaps fertilizer x was more effective than fertilizer y ($\text{Md}_x > \text{Md}_y$). The company decided to test

$$
H_0: \text{Md}_x = \text{Md}_y
$$

$$
H_a: \text{Md}_x > \text{Md}_y
$$

We would use U_y as a test statistic. Small values of U_y favor H_a. But how small should U_y be for us to reject H_0 in favor of H_a? We answer in the usual way—so small that it would be unlikely that we would observe such a small value if H_0 were true. Appendix Table B.9 contains critical values for U_x or U_y. Using this table, we can determine how small U_y must be for us to reject H_0, because this table enables us to find a critical value c for a given level of significance. We reject H_0 if $U_y \leq c$.

A portion of Appendix Table B.9 is shown in Table 13.11. Note that Table 13.11 gives critical values for:

1. A one-tailed test with: $\alpha = .05$ in roman type.
$\alpha = .025$ in boldface type.

2. A two-tailed test with: $\alpha = .10$ in roman type.
$\alpha = .05$ in boldface type.

(*Note*: Appendix Table B.9 includes additional values of α as well.)

TABLE 13.11
Some Critical Values for the Mann–Whitney Test. The Boxed Value is for the One-tailed Test with $\alpha = .05$

n_2 \ n_1	1	2	3	4	5	6	7	8	⋯	19	20
.	.	.	.	.	.	.	.	.			
.	.	.	.	.	.	.	.	.			
.	.	.	.	.	.	.	.	.			
4	—[a]	—	0	1	2	3					
	—	—	—	0	1	2					
5	—	0	1	2	4	5					
	—	—	0	1	2	3					
6	—	0	2	3	5	7					
	—	—	1	2	3	5					
.			.	.							
.			.	.							
.			.	.							

[a]Blanks indicate that the table is not applicable for these values of n_1 and n_2.

To use this table, we determine the values of n_1 and n_2, the two sample sizes. For the fertilizer problem $n_1 = 4$ and $n_2 = 6$. If we are interested in the one-tailed test discussed above with, say, $\alpha = .05$, we read the value in roman type in the fourth column and sixth row, namely, 3. Hence we reject H_0 if $U_y \leq 3$. But the observed value of U_y is 9, so we fail to reject H_0. This means that there is insufficient evidence that fertilizer x is more effective than fertilizer y.

Example 13.5
An educator wanted to see if college physics majors tended to have different quality-point averages (QPAs) from chemistry majors. He obtained 10 students majoring in physics and 10 majoring in chemistry. Although the QPAs of *all* students tended to be normally distributed, the educator felt that the QPAs for students in a given discipline might not be normal. Therefore, he decided to use a nonparametric test: the Mann–Whitney test. The sample data values along with their ranks are given in Table 13.12.

TABLE 13.12
Quality-Point Averages for Physics and
Chemistry Majors

Physics (x)	Ranks	Chemistry (y)	Ranks
2.72	11	3.10	13.5
3.10	13.5	2.10	5
3.75	19	2.62	9
1.92	3	1.62	2
2.41	8	2.85	12
3.72	18	1.95	4
2.35	6	3.28	16
3.80	20	1.50	1
3.25	15	3.71	17
2.71	10	2.37	7
	$S_x = 123.5$		$S_y = 86.5$

Note that the value 3.10 occurs twice, occupying the 13th and 14th positions in a ranked list. The rank assigned to each QPA is therefore the average of these two positions, namely, 13.5. The next rank assigned is 15, and so on. We will complete the test at the 5% level of significance.

Solution

1. Hypotheses:

$$H_0: \text{Md}_x = \text{Md}_y$$
$$H_a: \text{Md}_x \neq \text{Md}_y$$

2. Level of significance: $\alpha = .05$.

3. Test statistic and observed value: Small values of *either* U_x or U_y favor H_a. Therefore, we can use the minimum of these values as our test statistic. From Table 13.12

$$U_x = S_x - \frac{n_1(n_1 + 1)}{2} = 123.5 - \frac{(10)(11)}{2} = 123.5 - 55 = 68.5$$

$$U_y = S_y - \frac{n_2(n_2 + 1)}{2} = 86.5 - 55 = 31.5$$

Our test statistic is

$$U = \text{minimum of } U_x \text{ and } U_y = 31.5$$

4. Critical region: Find the critical value c for a two-tailed test when $n_1 = 10$, $n_2 = 10$, and $\alpha = .05$. From Appendix Table B.9 this value is $c = 23$. The critical region consists of values of $U \leq 23$.

5. Decision: Since the observed value ($U = 31.5$) is not in the critical region, we fail to reject H_0. This means that there is no evidence that physics majors' QPAs have a different median from that of chemistry majors.

We can now summarize the essential features of the Mann–Whitney test. Keep in mind that small values of U_x favor the relation $\text{Md}_x < \text{Md}_y$, while small values of U_y favor $\text{Md}_x > \text{Md}_y$.

Mann–Whitney Test This test is used for comparing two population medians based on independent samples.

- Decide on the hypotheses to be tested and the level of significance α.
- Rank all the data values in the two samples taken *together*. Call S_x the sum of the ranks of the x values, and S_y the sum of the ranks of the y values.

(a) To test the hypotheses

$$H_0: \text{Md}_x = \text{Md}_y$$
$$H_a: \text{Md}_x > \text{Md}_y$$

the test statistic U is

$$U = U_y = S_y - \frac{n_2(n_2 + 1)}{2} \qquad n_2 = \text{number of } y \text{ values}$$

Find the critical value c for a one-tailed test in Appendix Table B.9. If $U \le c$, reject H_0.

(b) To test

$$H_0: \text{Md}_x = \text{Md}_y$$
$$H_a: \text{Md}_x < \text{Md}_y$$

the test statistic U is

$$U = U_x = S_x - \frac{n_1(n_1 + 1)}{2} \qquad n_1 = \text{number of } x \text{ values}$$

Find the critical value c for a one-tailed test in Appendix Table B.9. If $U \le c$, reject H_0.

(c) To test

$$H_0: \text{Md}_x = \text{Md}_y$$
$$H_a: \text{Md}_x \ne \text{Md}_y$$

the test statistic is

$$U = \text{minimum of } U_x \text{ and } U_y$$

Find the critical value c for a two-tailed test in Appendix Table B.9. If $U \le c$, reject H_0.

Assumptions for the test:

- The data are continuous.
- The x and y values come from distributions with similar shapes.

Normal Approximation

Appendix Table B.9 allows sample sizes up to 20. We can still perform the Mann–Whitney test for larger sample sizes because of the following result: Let U represent either U_x or U_y. Then

(a) The mean is $\mu_U = \dfrac{n_1 n_2}{2}$

(b) The standard deviation is $\sigma_U = \sqrt{\dfrac{n_1 n_2 (n_1 + n_2 + 1)}{12}}$

(c) If $n_1 \geq 10$ and $n_2 \geq 10$, U is approximately normal. Therefore,

$$z = \frac{U - \mu_U}{\sigma_U} = \frac{U - \dfrac{n_1 n_2}{2}}{\sqrt{\dfrac{n_1 n_2 (n_1 + n_2 + 1)}{12}}}$$

is approximately standard normal. This means that when the sample sizes are large enough, we can find the critical region in terms of z.

Now we will see how the normal approximation would work if it were used in the previous example. Here the sample sizes are $n_1 = n_2 = 10$; therefore we may use the normal approximation. Steps (1) and (2) will be the same. The remaining steps are as follows:

- We calculate the test statistic in step (3) as follows: First find the value of U. We saw in Example 13.5 that $U = 31.5$. Now calculate the value of the test statistic z.

$$z = \frac{U - \dfrac{n_1 n_2}{2}}{\sqrt{\dfrac{n_1 n_2 (n_1 + n_2 + 1)}{12}}} = \frac{31.5 - \dfrac{(10)(10)}{2}}{\sqrt{\dfrac{(10)(10)(10 + 10 + 1)}{12}}}$$

$$= \frac{-18.5}{\sqrt{175}} \doteq -1.40$$

- Now find the critical region (step 4): We are conducting a two-tailed test with $\alpha = .05$. Usually, the critical values for a two-tailed z test are (from Appendix Table B.3): $\pm z(\alpha/2) = \pm z(.025) = \pm 1.96$. Normally the critical region would consist of values of $z \geq 1.96$ or ≤ -1.96. But since we have used the minimum of U_x and U_y to compute z, there will not be any question of z falling in the right tail (≥ 1.96). Hence, in effect the critical region consists of values of z such that $z \leq -z(\alpha/2) = -1.96$.

- Our decision is the same as that in Example 13.5: we fail to reject the null hypothesis because the observed value ($z = -1.40$) is not in the critical region.

Note: When using z to conduct a two-tailed Mann-Whitney test, we reject H_0 if $z \leq -z(\alpha/2)$. For a one-tailed test we reject H_0 if $z \leq -z(\alpha)$.

Exercises

13.16   In parts (a)–(i) assume a Mann-Whitney test is used. Complete the following table. Note that n_1 and n_2 refer to the x and y sample sizes, respectively, and α refers to the level of significance.

	Hypothesis	n_1	n_2	α	Observed Value U	Critical Value	Decision
(a)	$\text{Md}_x = \text{Md}_y$	6	10	.05	14		
(b)	$\text{Md}_x < \text{Md}_y$	9	12	.01	18		
(c)	$\text{Md}_x > \text{Md}_y$	15	20	.025	79		
(d)	$\text{Md}_x < \text{Md}_y$	14	8	.05	36		
(e)	$\text{Md}_x \neq \text{Md}_y$	13	13	.05	41		
(f)	$\text{Md}_x > \text{Md}_y$	17	11	.01	49		
(g)	$\text{Md}_x \neq \text{Md}_y$	5	7	.01	4		
(h)	$\text{Md}_x < \text{Md}_y$	16	12	.05	55		
(i)	$\text{Md}_x > \text{Md}_y$	9	5	.025	5		

In Exercises 13.17 and 13.18, you are given the alternate hypothesis, the level of significance α, and the sample data.

(a)   Complete the table.
(b)   Use the Mann–Whitney U test and complete the test.

13.17 $H_a: \text{Md}_x \neq \text{Md}_y$, $\alpha = .10$.   **13.18** $H_a: \text{Md}_x > \text{Md}_y$, $\alpha = .05$.

x	Rank (x)	y	Rank (y)
43		55	
47		62	
51		68	
59		75	
65			
71			

x	Rank (x)	y	Rank (y)
33		20	
36		23	
42		28	
45		34	
49		37	

13.19   A high school soccer coach believed that the median endurance level of soccer players was larger than that for football players. The data in the following table represent the results (in minutes) of an endurance test taken by seven football players (x) and eight soccer players (y). Test the belief using the Mann–Whitney U test with a 5% level of significance.

x	y
14.7	17.3
15.3	17.9
17.2	20.5
13.6	14.5
13.9	18.4
15.2	19.2
18.1	16.1
	17.5

13.20   A consumer protection agency wanted to test the hypothesis that the median cost of an automobile repair was the same in two nearby cities x and y. The agency obtained estimates from garages in the two cities for a particular job. The estimates (in dollars) were

x	y
77	68
80	71
85	73
90	97
92	101
	106
	108

Do the data provide sufficient evidence to indicate a difference in the median cost? Use the Mann–Whitney U test with a 10% level of significance.

13.21   A spokesman for automobile manufacturer x claimed that a model made by x tends to get better gas mileage than a comparable model made by y. Ten cars from each manufacturer were tested. The following data represent miles per gallon.

x	y
36.2	34.7
36.5	34.9
36.9	35.8
37.3	36.3
37.6	36.6
37.7	36.8
37.9	37.1
38.3	37.5
38.5	38.0
38.8	38.2

Test the claim using the Mann–Whitney U test with a 5% level of significance.

13.22 A teacher was assigned to teach two sections of a calculus course. Section x was to meet twice a week with 2-hour classes, and section y was to meet four times a week with 1-hour classes. Twenty-seven students were taking the course; 13 were randomly assigned to section x. The chairperson claimed that there would be no statistically significant difference in median performance. The final grades were

x	y
40	34
54	52
55	58
59	63
62	68
64	70
67	71
77	73
78	76
88	79
90	84
91	86
92	89
	95

With a 5% level of significance, test the claim using

(a) Appendix Table B.9 to select a critical value.

(b) The normal approximation.

13.5 THE RUNS TEST

The **runs test** is a test for randomness. We use the following example to illustrate the test: It was decided to conduct a survey of workers in a town, the purpose of which was to estimate the proportion of union members and to compare views of union members versus nonunion members on various issues. Each employee of the survey firm was instructed to randomly interview 12 workers and record whether the worker was a member of a union (M) or not (N), along with various other items of information. In addition, the interview forms were to be numbered in the order interviewed. Suppose an interviewer had the following results as to the order in which union members (M) and nonunion members (N) were interviewed:

$$M \quad M \quad M \quad M \quad M \quad M \quad N \quad N \quad N \quad N \quad N \quad N$$

This means that 6 union members were interviewed followed by 6 nonunion members. Such a process does not appear to be random. On the other hand, suppose that the interviewer had obtained the following sequence:

$$M \quad N \quad M \quad N \quad M \quad N \quad M \quad N \quad M \quad N \quad M \quad N$$

This process is not random either. The interviewer carefully alternated interviews, interviewing a union member followed by a nonunion member. Both interviewing procedures discussed above involve a systematic procedure and are nonrandom.

One can investigate such nonrandomness by examining the number of **runs** in a sequence.

Definition A *run* is a sequence of the same letter written one or more times. There is a different letter (or no letter) before and after the sequence.

For the sequences discussed above, the number of runs are

$$M \; M \; M \; M \; M \; M \qquad N \; N \; N \; N \; N \; N \qquad R = \text{total number of runs} = 2$$

$$M \; N \; M \; N \; M \; N \; M \; N \; M \; N \; M \; N \qquad R = 12$$

A very large or very small number of runs would imply nonrandomness. As usual, we would conclude nonrandomness if the number of runs is so large or small that it would be unlikely that we would observe such a large or small number, if the process were truly random. "Unlikely" means having a low order of probability, which we denote by α.

Appendix Table B.10 contains critical values for R, the number of runs, when $\alpha = .05$. For example, in the above interviewing process, $n_1 =$ number of M's $= 6$ and $n_2 =$ the number of N's $= 6$. Then, if $\alpha = .05$, the left critical value for R is 3 and the right critical value is 11. See Appendix Table B.10. In general, we denote the left critical value by c_1 and the right critical value by c_2.

This means that if we are considering the hypotheses:

$$H_0: \text{the process is random}$$

$$H_a: \text{the process is not random}$$

then we would reject H_0 if $R \le 3$ or $R \ge 11$. For example, for the sequence

$$M \quad M \quad M \quad M \quad M \quad M \quad N \quad N \quad N \quad N \quad N \quad N$$

the number of runs is $R = 2$. This value is in the critical region; therefore, we would reject H_0.

The procedure for testing for randomness is called the **runs test** and is summarized as follows:

The Runs Test   Suppose that a process results in a sequence of observations, and that each observation can be placed in either one of two categories (such as M for union member or N for nonunion member).

1. The hypotheses to be tested are

$$H_0: \text{the process is random}$$

$$H_a: \text{the process is not random}$$

2. For this test we will use .05 as the level of significance.
3. The test statistic to be used is

$$R = \text{number of runs in the sequence}$$

4. Let n_1 = number of occurrences of one of the symbols and n_2 = number of occurrences of the other. Using these values of n_1 and n_2, find the lower and upper critical values c_1 and c_2 from Appendix Table B.10.
5. If $R \leq c_1$ or $R \geq c_2$, reject H_0 and conclude nonrandomness.

One of the applications of the runs test in the area of industrial quality control involves **time series**.

Definition   A *time series* is a sequence of numerical measurements obtained over a period of time.

If we were studying some measurable characteristic of an industrial product obtained over time (such as the thickness of automobile rocker gaskets selected from a production line), deviations of these measurements above or below their mean can be examined to detect a possible departure from randomness. For example, we could label a gasket A if its measurement is above the mean, and B if it is below the mean. The sequence of A's and B's can then be examined for nonrandomness. Departures from randomness involving occasional surges of A's, for example, might be caused by a machine that periodically malfunctions.

Example 13.6

Every 5 minutes a gasket is selected from a production line and its thickness measured. Twenty-three such measurements are obtained. An A is recorded if the gasket measures above the (sample) mean, and a B is recorded if it measures below, with the following results:

$$A \quad B \quad A \quad B \quad B \quad B \quad B \quad A \quad A \quad B \quad B \quad B$$
$$A \quad A \quad B \quad B \quad A \quad A \quad A \quad B \quad A \quad A \quad B$$

Do these values indicate a lack of randomness? Use the 5% level of significance.

Solution

1. Hypotheses:

    H_0: the process (occurrence of A's and B's) is random

    H_a: it is not random

2. Level of significance: $\alpha = .05$.

3. Test statistic and observed value: R = number of runs calculated below:

$$\underbrace{A} \quad \underbrace{B} \quad \underbrace{A} \quad \underbrace{B\,B\,B\,B} \quad \underbrace{A\,A} \quad \underbrace{B\,B\,B} \quad \underbrace{A\,A} \quad \underbrace{B\,B} \quad \underbrace{A\,A\,A} \quad \underbrace{B} \quad \underbrace{A\,A} \quad \underbrace{B}$$
$$R = 12$$

4. Critical region: Let n_1 = number of A's = 11; n_2 = number of B's = 12. Using Appendix Table B.10, we find that the left and right critical values are $c_1 = 7$ and $c_2 = 18$, respectively, so the critical region consists of values of $R \leq 7$ or ≥ 18.

5. Decision: The observed value ($R = 12$) is not in the critical region. Therefore, we fail to reject H_0. This means that the results do not indicate a departure from randomness.

Normal Approximation

Appendix Table B.10 only allows for values of n_1 and n_2 up to 15. We can still apply the runs test for larger values of n_1 and n_2 because of the following result:

(a) $\quad \mu_R = \dfrac{2n_1 n_2}{n_1 + n_2} + 1$

(b) $\quad \sigma_R = \sqrt{\dfrac{2n_1 n_2 (2n_1 n_2 - n_1 - n_2)}{(n_1 + n_2)^2 (n_1 + n_2 - 1)}}$

(c) If n_1 and n_2 are both greater than 10, then R is approximately normal, so

$$z = \frac{R - \mu_R}{\sigma_R}$$

is approximately standard normal.

This means that for sufficiently large values of n_1 and n_2, we can accomplish the runs test by means of a z test.

We will show how the normal approximation can be used in the previous example. Steps 1 and 2 will be the same. The remaining steps are as follows:

- Test statistic and observed value: First find μ_R and σ_R.

$$\mu_R = \frac{2n_1n_2}{n_1 + n_2} + 1 = \frac{(2)(11)(12)}{11 + 12} + 1 = 12.48$$

$$\sigma_R = \sqrt{\frac{2n_1n_2(2n_1n_2 - n_1 - n_2)}{(n_1 + n_2)^2(n_1 + n_2 - 1)}} = \sqrt{\frac{(2)(11)(12)[(2)(11)(12) - 11 - 12]}{(11 + 12)^2(11 + 12 - 1)}}$$

$$= \sqrt{\frac{63,624}{11,638}} \doteq 2.34$$

so

$$z = \frac{R - \mu_R}{\sigma_R} = \frac{12 - 12.48}{2.34} \doteq -.21$$

- Critical region: Large or small values of R favor H_a. These correspond to values of z far to the right or left of 0. Therefore, we conduct a two-tailed test. Since $\alpha = .05$, the critical region will consist of values of

$$z \le -z\left(\frac{\alpha}{2}\right) = -1.96 \quad \text{or} \quad z \ge z\left(\frac{\alpha}{2}\right) = 1.96$$

(See Figure 13.10.)

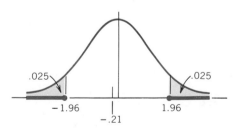

FIGURE 13.10

- Decision: The observed value, $z = -.21$, is not in the critical region, thus we fail to reject H_0. (This is the same conclusion that was reached in Example 13.6.)

Exercises

13.23   A plant manager is to test a hypothesis that a process is random. In a sequence of observations, each observation is recorded as A (acceptable) or B (not acceptable). For each sequence of observations given, complete the table. Use a 5% level of significance. Note that R is the number of runs; c_1 and c_2 are the lower and upper critical values, respectively.

													c_1	c_2	R	Decision
(a)	A	A	A	A	A	B	B	B	B							
(b)	A	B	A	B	A	B	A	B	A	B	A	B				
(c)	A	A	A	B	B	B	A	A	A	B	B	B	B			
(d)	B	A	A	B	B	B	A	B	B	A	B	A				
(e)	B	B	A	B	B	A	B	A	A	A	A	B	B	B		
(f)	A	A	B	B	A	A	B	B	A	A	B	B	A	B	A	

13.24   The owner of a small business computed the average revenue per day over a period of 12 days. For each day an L was recorded if the revenue was less than the average. Otherwise, an M was recorded. Do the data below indicate a lack of randomness at the 5% level of significance?

$$L \quad L \quad L \quad L \quad M \quad M \quad L \quad L \quad L \quad L \quad M \quad M$$

13.25   The following data represent a sample of 15 electrical components inspected at 5-minute intervals. A D was recorded if the component failed to meet the required specifications. Otherwise, an N was recorded. Do the data indicate that the process is not random? Use a 5% level of significance.

$$N \quad N \quad N \quad N \quad N \quad D \quad D \quad D \quad N \quad N \quad N \quad N \quad N$$

13.26   Students evaluated a college professor over a period of 25 classes. An S was recorded if more than 50% of the class indicated the teacher's performance was satisfactory. Otherwise, a U was recorded. Using a 5% level of significance, test the hypothesis that the following data indicate a lack of randomness by

(a)   Obtaining critical values from Appendix Table B.10.

(b)   The normal approximation.

$$U \quad S \quad S \quad S \quad S \quad U \quad U \quad U \quad S \quad S \quad S \quad S$$
$$S \quad S \quad S \quad U \quad U \quad U \quad S \quad U \quad S \quad U \quad U \quad U$$

13.6   A DISCUSSION OF PARAMETRIC VERSUS NONPARAMETRIC TESTS

Two statistical tests are often assessed by comparing the **power** of each test. In order to understand this concept, we should recall the types of errors that can occur in hypothesis testing. A type I error occurs if we reject the null hypothesis, H_0, when it is true; a type II error consists of failing to reject H_0 when it is false. The probabilities of committing these errors are denoted by

$$\alpha = P(\text{type I error}) \leftarrow \text{level of significance of the test}$$
$$\beta = P(\text{type II error})$$

The researcher selects the level of significance α. If two statistical tests have the same level of significance α, then the one with the smaller value for β is a better test. Suppose that for some test the probability of failing to reject H_0 when it is false is $\beta = .10$. Then the probability of rejecting H_0 when it is false is .90 ($= 1 - .10$). We call this the **power of the test**.

> **Definition** For a statistical test applied in a given situation, the *power of the test* is the probability of rejecting a false null hypothesis H_0
>
> $$\text{power} = 1 - \beta$$

If two statistical tests have the same level of significance α, then the one with the greater power is the better test.

In Chapters 7, 8, and 9 we considered various parametric tests concerning population means: t tests for small samples and z tests for large samples. The t tests required that the populations involved be approximately normal. (The z tests involving large samples did not require normality, but, still, the closer the population distributions are to normality the better these tests are in the sense that they have greater power.) When we are considering any tests involving measures of central tendency, we should certainly consider the parametric tests because when conditions are appropriate, these tests are more powerful than nonparametric tests. It is not surprising that these tests (when appropriate) would be better tests than the sign test, or the Wilcoxon signed-rank test, or the Mann-Whitney test. After all, they use more information. For example, a t test uses the magnitude of the data values, while the Mann–Whitney test uses only the ranks of the data values. However, when the population(s) departs substantially from normality, nonparametric tests should be considered, especially when the sample sizes are small. Even when the populations are normal, the Wilcoxon signed-rank test and the Mann–Whitney tests are almost as good as their parametric counterparts, and when the populations depart considerably from normality, the Wilcoxon signed-rank test and the Mann–Whitney test can be better tests.

USING MINITAB (OPTIONAL)

We will illustrate Minitab commands for nonparametric tests in the context of some of the examples discussed in this chapter. One-sided tests are done by means of the usual ALTERNATIVE subcommand.

1. **Sign Test** With a sample of new home prices in column 1, we can test the following hypotheses concerning median price for a new home:

$$H_0\text{: Md} = 70000$$
$$H_a\text{: Md} \neq 70000$$

We use the command

$$\text{STEST MEDIAN} = 70000, \text{DATA IN C1}$$

If the value of the median is not specified, Minitab assumes the value 0. A two-sided test is automatically done if no ALTERNATIVE subcommand is given. Among other things, Minitab will print the P value.

In Example 13.1 we used a sign test to compare the yield for two fertilizers based on dependent samples. Thirteen farms used both fertilizers. We could place the yields for two fertilizers in columns C1, C2. We then conduct the following test on the median of the differences (D):

$$H_0: \text{Md}_D = 0$$
$$H_a: \text{Md}_D > 0$$

Type

$$\text{LET C3} = \text{C1} - \text{C2}$$
$$\text{STEST C3};$$
$$\text{ALTERNATIVE} = 1.$$

2. **Wilcoxon Signed-Rank Test** With the test scores in Table 13.3 placed in column C1, we can conduct the following test:

$$H_0: \text{Md} = 50$$
$$H_a: \text{Md} > 50$$

Type

$$\text{WTEST OF MEDIAN 50, DATA IN C1};$$
$$\text{ALTERNATIVE} = 1.$$

Minitab will print an estimate of the median and a P value.

As in the case of the sign test, we can do a test involving dependent samples with data values in columns C1, C2. For example, to conduct the following test on the median of the differences (D):

$$H_0: \text{Md}_D = 0$$
$$H_a: \text{Md}_D \neq 0$$

type

$$\text{LET C3} = \text{C1} - \text{C2}$$
$$\text{WTEST C3}$$

(If the value of the median is not specified, Minitab uses 0.)

3. **Mann–Whitney U Test** This test compares two population medians based on independent samples. In Section 13.4 we compared median yield for two fertilizers in Table 13.8. We tested

$$H_0: \text{Md}_1 = \text{Md}_2$$
$$H_a: \text{Md}_1 > \text{Md}_2$$

The Minitab code is

$$\text{MANN-WHITNEY, ALTERNATIVE} = 1, \text{DATA IN C1, C2}$$

Note: Unlike previous tests, ALTERNATIVE is not a subcommand with MANN-WHITNEY. But it will be a subcommand in Release 8 of Minitab. The printout is given below.

```
MTB > SET FERTILIZER X IN Cl
DATA> 102 85 92 110
DATA> END
MTB > SET FERTILIZER Y IN C2
DATA> 89 101 108 93 90 82
DATA> END
MTB > MANN-WHITNEY, ALTERNATIVE = 1, DATA IN Cl, C2

Mann-Whitney Confidence Interval and Test
C1          N =    4    MEDIAN =     97.00
C2          N =    6    MEDIAN =     91.50
POINT ESTIMATE FOR ETA1-ETA2 IS           2.50
95.7 PCT C.I. FOR ETA1-ETA2 IS (     -16.00,     20.00)
W =     25.0
TEST OF ETA1 = ETA2 VS. ETA1 G.T. ETA2 IS SIGNIFICANT
AT 0.2970

CANNOT REJECT AT ALPHA = 0.05
```

Note that an approximate 95% confidence interval is given for eta 1 − eta 2. This is $Md_1 - Md_2$.

Notice that the printout states that the test is significant at .2970. This means that .2970 is the smallest level of significance for which the null hypothesis can be rejected. This is another way of describing the P value, because we reject H_0 for any $\alpha \geq P$.

In this chapter we used the test statistic U for the Mann–Whitney test (U is either U_x or U_y, depending on the situation). Minitab uses W. This is the sum of the ranks in the first sample (the x values, in this case). We called this S_x.

4. **The Runs Test** The Minitab Runs test assumes that the observations are numbers. So in a situation like the one discussed at the beginning of Section 13.5, where the observations were M (union member) and N (not a union member), we would convert these symbols to numbers. For example we could use 1 for M and 0 for N. Choose a value in the middle of the data, say, .5. A **run** can then be defined as one or more consecutive observations above .5, or one or more consecutive observations below .5. For example, the sequence

$$1,1,0,1,1,1,0,0,1,1,0$$

has 6 runs. To do a Runs test we set the observations in a column, say C1, then type

RUNS .5, DATA IN C1

If the .5 is omitted, Minitab automatically uses the average of the data values in C1. Minitab prints out the observed and expected number of runs, the number of runs above and below K (we used $K = .5$), and the P value.

A word of caution on using Minitab for nonparametric tests: For most of these tests Minitab uses the normal approximation to find P values. (An exception is the sign test, which only uses the normal approximation if $n > 50$.) We have seen that there are some minimum requirements on sample sizes in order to use the normal approximations.

Exercises

Suggested exercises for use with Minitab are 13.2, 13.5, 13.15 (b), 13.22 (b), 13.26 (b). (*Note*: Minitab will automatically use the normal approximation for the last three.

SUMMARY

Nonparametric tests are useful when little is known about the distributions of the populations under investigation, or when these distributions do not meet the requirements necessary for the use of parametric tests (such as approximate normality).

In this chapter we studied only a few nonparametric tests. These tests, along with their test statistics, were as follows:

1. **The Sign Test.** We used this test to study the median of a population and to compare two populations when the samples are dependent.

    Test statistic:

    x = number of data values in the sample that are greater than
            the value of the median appearing in the null hypothesis

    We treat x as a binomial variable with $p = .5$ and n = sample size.

2. **The Wilcoxon Signed-Rank Test.** We used this test to investigate a single population median and to compare two populations using a paired experiment.

    Test statistic:

    For a left-tailed test (with $<$ in the alternate hypothesis) use

    $$W^+ = \text{sum of the positive ranks}$$

    For a right-tailed test (with $>$ in alternate hypothesis) use

    $$W^- = \text{absolute value of sum of negative ranks}$$

For a two-tailed test (with $\neq$ in alternate hypothesis) use

$$W = \text{minimum of } W^+, W^-$$

Look up the critical value c in Appendix Table B.8. Reject H_0 if the test statistic $\leq c$.

3. **Mann–Whitney Test.** We used this to compare two population medians when the samples are independent. Let $n_1 =$ sample size of x values and $n_2 =$ sample size of y values.

Test statistic:

When the alternate hypothesis is $\text{Md}_x > \text{Md}_y$, use

$$U = U_y = S_y - \frac{n_2(n_2 + 1)}{2}$$

where $S_y =$ sum of the ranks of the y values.
When the alternate hypothesis is $\text{Md}_x < \text{Md}_y$, use

$$U = U_x = S_x - \frac{n_1(n_1 + 1)}{2}$$

where $S_x =$ sum of ranks of x values.
When the alternate hypothesis is $\text{Md}_x \neq \text{Md}_y$, use

$$U = \text{minimum of } U_x, U_y$$

Look up the critical value c in Appendix Table B.9. Reject H_0 if $U \leq c$.

4. **The Runs Test.** This is a test for randomness. The test statistic is

$$R = \text{number of runs}$$

Look up critical values c_1, c_2 in Appendix Table B.10. Reject H_0 (which asserts randomness) if $R \leq c_1$ or $R \geq c_2$.

Review Exercises

13.27   A real estate agent in a large city claimed that the median rent for two-bedroom apartments was more than \$625. To test this claim, a tenants' organization sampled 20 such apartments and obtained the following rents (in dollars):

| 635 | 660 | 670 | 615 | 620 | 660 | 600 | 610 | 700 | 675 |
| 615 | 605 | 650 | 600 | 665 | 670 | 645 | 640 | 655 | 680 |

Using a 5% level of significance, test the claim.

(a)   Use the sign test.   (b)   Use the Wilcoxon signed-rank test.
(c)   Compare parts (a) and (b) and comment.

13.28 A government official believed that the median prime interest rate for banks in Massachusetts was 9.3%. A sample of 10 banks gave the following rates (in percent):

$$9.6 \quad 9.8 \quad 9.1 \quad 10.2 \quad 10.1 \quad 9.7 \quad 8.5 \quad 10.3 \quad 9.9 \quad 10.0$$

Use a 5% level of significance to test the official's belief.

(a) Use the sign test. (b) Use the Wilcoxon signed-rank test.

13.29 A sports reporter commented that the median heart rate (beats/minute) of marathoners is less than 200, where the rate is measured after exercise on a treadmill. Eight marathoners were tested, and the following data were obtained:

$$193 \quad 203 \quad 196 \quad 192 \quad 187 \quad 206 \quad 190 \quad 195$$

Using a 5% level of significance, test the instructor's comment with

(a) The sign test. (b) The Wilcoxon signed-rank test.
(c) Compare parts (a) and (b) and comment.

13.30 Refer to Exercise 13.27. Test the claim that the median rent is more than $625. Use the Wilcoxon signed-rank test with the normal approximation and a 5% level of significance.

13.31 A fourth-grade teacher claimed that girls (y) tend to read with more comprehension than boys (x) upon entering the fourth grade. The results of a test designed to measure comprehension were as follows:

x	y
84	82
57	55
74	88
70	76
72	78
71	77
65	92
40	
45	

Test the teacher's claim using a nonparametric test and a 5% level of significance.

13.32 A type of electrical component was made by competing companies x and y. Four components were sampled from x and five sampled from y. The time to failure (in months) was recorded for each component. Use a nonparametric test with a 10% level of significance to test the hypothesis that the population medians are the same. The data are

x	y
7.4	8.9
8.7	7.8
7.5	8.2
8.0	8.8
	8.5

13.33 The Chamber of Commerce in a resort area A claimed that the median price of building lots was less in A than building lots in another resort area B. A survey of the cost of comparable lots from the two resorts gave the following data (in thousands):

Resort A	Resort B
39.5	42.5
48.9	47.9
35.0	40.9
32.5	61.0
30.9	53.9
44.3	36.0

Test the claim using a nonparametric test with a 1% level of significance.

13.34 A city bus was scheduled to reach a particular stop at noon each day. The bus was considered on time if it reached the stop within 2 minutes of noon. Over a 15-day period, an A was recorded if the bus was on time. Otherwise a B was recorded. The results were as follows:

A A A B B A B A A B B B A B A

Do the data indicate a lack of randomness? Use a 5% level of significance.

13.35 A college basketball team played 25 games. The following data represent the sequence of wins (W) and losses (L) for the season:

L L W W W L L L L L W L L

L W W W W W W L L W W W

Do the data indicate a lack of randomness at the 5% level?

(a) Use the critical values from Appendix Table B.10.

(b) Use the normal approximation.

13.36 Complete the test in Exercise 13.21 using the normal approximation.

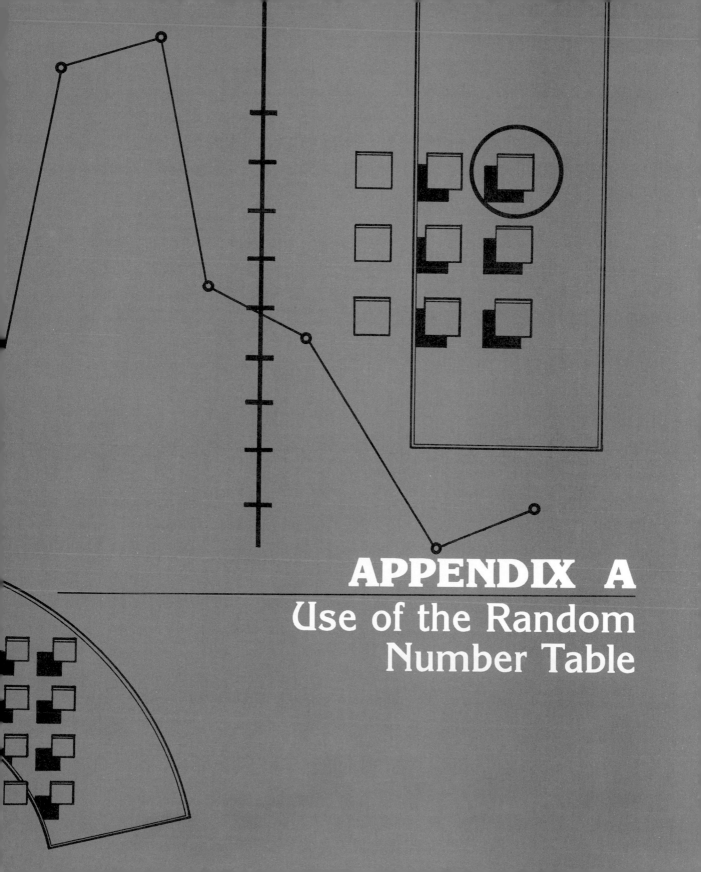

APPENDIX A
Use of the Random Number Table

Random numbers can be obtained from Appendix Table B.1. Suppose that we wish to choose a random sequence of 5 four-digit numbers. Randomly choose some starting point in the table. You could toss your pencil on the page and choose the nearest four digits. Then read down the table until you have 5 four-digit numbers. If you come to the bottom of the page, go to the top of the next column of four digits. The process is displayed in Figure A.1.

34	0 7	27	68	50
43	57	18	24	06
02	05	16	56	92
05	32	54	70	48
03	52	96	47	78
14	90	56	86	07
39	80	82	77	32

FIGURE A.1

A portion of Appendix Table B.1 showing a Random Sample of Five 4-Digit ID numbers

Often the elements of a population are assigned identification numbers. When this is the case, we can obtain a random sample from the population by choosing the desired ID numbers from the Random Number Table. Suppose, for example, that we wanted a random sample of 5 students from some university with 9000 students with ID numbers ranging from 0001 to 9000. We could select the students with ID numbers that we obtained above

2406
5692
7048
4778
8607

A random sample of 5 quality point averages (QPAs) could be obtained by looking up the QPAs of these students. If we had obtained a four-digit number in the Random Number Table that was out of range (like 9500), we would just ignore it and continue until the desired number of ID's is obtained. When we do not want any repetitions in our sample, we ignore any IDs that repeat.

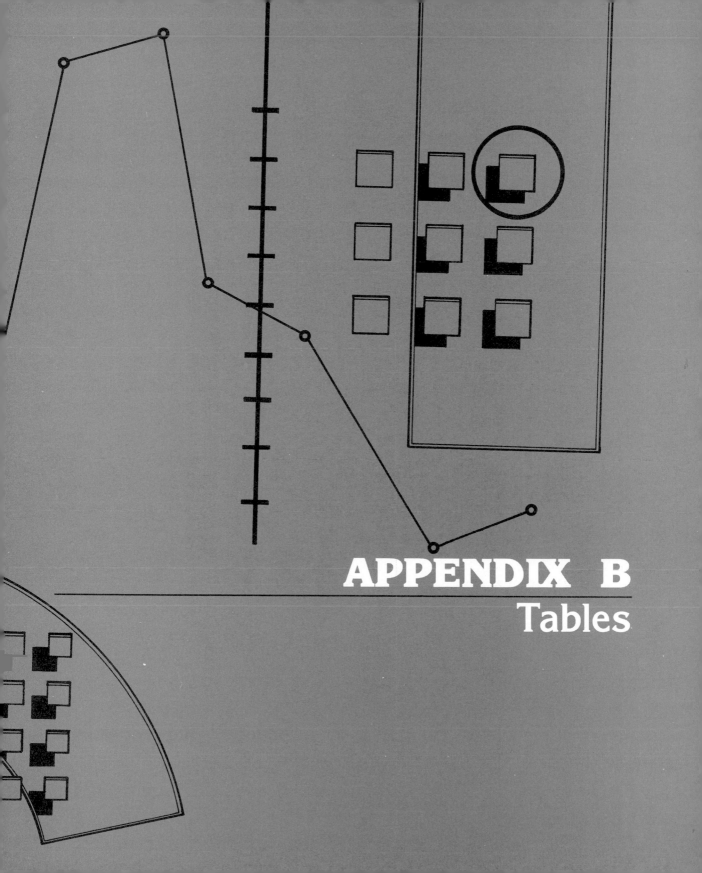

APPENDIX B
Tables

TABLE B.1
Random Numbers

10 09 73 25 33	76 52 01 35 86	34 67 35 48 76	80 95 90 91 17	39 29 27 49 45
37 54 20 48 05	64 89 47 42 96	24 80 52 40 37	20 63 61 04 02	00 82 29 16 65
08 42 26 89 53	19 64 50 93 03	23 20 90 25 60	15 95 33 47 64	35 08 03 36 06
99 01 90 25 29	09 37 67 07 15	38 31 13 11 65	88 67 67 43 97	04 43 62 76 59
12 80 79 99 70	80 15 73 61 47	64 03 23 66 53	98 95 11 68 77	12 17 17 68 33
66 06 57 47 17	34 07 27 68 50	36 69 73 61 70	65 81 33 98 85	11 19 92 91 70
31 06 01 08 05	45 57 18 24 06	35 30 34 26 14	86 79 90 74 39	23 40 30 97 32
85 26 97 76 02	02 05 16 56 92	68 66 57 48 18	73 05 38 52 47	18 62 38 85 79
63 57 33 21 35	05 32 54 70 48	90 55 35 75 48	28 46 82 87 09	83 49 12 56 24
73 79 64 57 53	03 52 96 47 78	35 80 83 42 82	60 93 52 03 44	35 27 38 84 35
98 52 01 77 67	14 90 56 86 07	22 10 94 05 58	60 97 09 34 33	50 50 07 39 98
11 80 50 54 31	39 80 82 77 32	50 72 56 82 48	29 40 52 42 01	52 77 56 78 51
83 45 29 96 34	06 28 89 80 83	13 74 67 00 78	18 47 54 06 10	68 71 17 78 17
88 68 54 02 00	86 50 75 84 01	36 76 66 79 51	90 36 47 64 93	29 60 91 10 62
99 59 46 73 48	87 51 76 49 69	91 82 60 89 28	93 78 56 13 68	23 47 83 41 13
65 48 11 76 74	17 46 85 09 50	58 04 77 69 74	73 03 95 71 86	40 21 81 65 44
80 12 43 56 35	17 72 70 80 15	45 31 82 23 74	21 11 57 82 53	14 38 55 37 63
74 35 09 98 17	77 40 27 72 14	43 23 60 02 10	45 52 16 42 37	96 28 60 26 55
69 91 62 68 03	66 25 22 91 48	36 93 68 72 03	76 62 11 39 90	94 40 05 64 18
09 89 32 05 05	14 22 56 85 14	46 42 75 67 88	96 29 77 88 22	54 38 21 45 98
91 49 91 45 23	68 47 92 76 86	46 16 28 35 54	94 75 08 99 23	37 08 92 00 48
80 33 69 45 98	26 94 03 68 58	70 29 73 41 35	53 14 03 33 40	42 05 08 23 41
44 10 48 19 49	85 15 74 79 54	32 97 92 65 75	57 60 04 08 81	22 22 20 64 13
12 55 07 37 42	11 10 00 20 40	12 86 07 46 97	96 64 48 94 39	28 70 72 58 15
63 60 64 93 29	16 50 53 44 84	40 21 95 25 63	43 65 17 70 82	07 20 73 17 90
61 19 69 04 46	26 45 74 77 74	51 92 43 37 29	65 39 45 95 93	42 58 26 05 27
15 47 44 52 66	95 27 07 99 53	59 36 78 38 48	82 39 61 01 18	33 21 15 94 66
94 55 72 85 73	67 89 75 43 87	54 62 24 44 31	91 19 04 25 92	92 92 74 59 73
42 48 11 62 13	97 34 40 87 21	16 86 84 87 67	03 07 11 20 59	25 70 14 66 70
23 52 37 83 17	73 20 88 98 37	68 93 59 14 16	26 25 22 96 63	05 52 28 25 62
04 49 35 24 94	75 24 63 38 24	45 86 25 10 25	61 96 27 93 35	65 33 71 24 72
00 54 99 76 54	64 05 18 81 59	96 11 96 38 96	54 69 28 23 91	23 28 72 95 29
35 96 31 53 07	26 89 80 93 54	33 35 13 54 62	77 97 45 00 24	90 10 33 93 33
59 80 80 83 91	45 42 72 68 42	83 60 94 97 00	13 02 12 48 92	78 56 52 01 06
46 05 88 52 36	01 39 09 22 86	77 28 14 40 77	93 91 08 36 47	70 61 74 29 41
32 17 90 05 97	87 37 92 52 41	05 56 70 70 07	86 74 31 71 57	85 39 41 18 38
69 23 46 14 06	20 11 74 52 04	15 95 66 00 00	18 74 39 24 23	97 11 89 63 38
19 56 54 14 30	01 75 87 53 79	40 41 92 15 85	66 67 43 68 06	84 96 28 52 07
45 15 51 49 38	19 47 60 72 46	43 66 79 45 43	59 04 79 00 33	20 82 66 95 41
94 86 43 19 94	36 16 81 08 51	34 88 88 15 53	01 54 03 54 56	05 01 45 11 76
98 08 62 48 26	45 24 02 84 04	44 99 90 88 96	39 09 47 34 07	35 44 13 18 80
33 18 51 62 32	41 94 15 09 49	89 43 54 85 81	88 69 54 19 94	37 54 87 30 43
80 95 10 04 06	96 38 27 07 74	20 15 12 33 87	25 01 62 52 98	94 62 46 11 71
79 75 24 91 40	71 96 12 82 96	69 86 10 25 91	74 85 22 05 39	00 38 75 95 79
18 63 33 25 37	98 14 50 65 71	31 01 02 46 74	05 45 56 14 27	77 93 89 19 36
74 02 94 39 02	77 55 73 22 70	97 79 01 71 19	52 52 75 80 21	80 81 45 17 48
54 17 84 56 11	80 99 33 71 43	05 33 51 29 69	56 12 71 92 55	36 04 09 03 24
11 66 44 98 83	52 07 98 48 27	59 38 17 15 39	09 97 33 34 40	88 46 12 33 56
48 32 47 79 28	31 24 96 47 10	02 29 53 68 70	32 30 75 75 46	15 02 00 99 94
69 07 49 41 38	87 63 79 19 76	35 58 40 44 01	10 51 82 16 15	01 84 87 69 38

TABLE B.1 (Continued)

Random Numbers

09 18 82 00 97	32 82 53 95 27	04 22 08 63 04	83 38 98 73 74	64 27 85 80 44
90 04 58 54 97	51 98 15 06 54	94 93 88 19 97	91 87 07 61 50	68 47 66 46 59
73 18 95 02 07	47 67 72 62 69	62 29 06 44 64	27 12 46 70 18	41 36 18 27 60
75 76 87 64 90	20 97 18 17 49	90 42 91 22 72	95 37 50 58 71	93 82 34 31 78
54 01 64 40 56	66 28 13 10 03	00 68 22 73 98	20 71 45 32 95	07 70 61 78 13
08 35 86 99 10	78 54 24 27 85	13 66 15 88 73	04 61 89 75 53	31 22 30 84 20
28 30 60 32 64	81 33 31 05 91	40 51 00 78 93	32 60 46 04 75	94 11 90 18 40
53 84 08 62 33	81 59 41 36 28	51 21 59 02 90	28 46 66 87 95	77 76 22 07 91
91 75 75 37 41	61 61 36 22 69	50 26 39 02 12	55 78 17 65 14	83 48 34 70 55
89 41 59 26 94	00 39 75 83 91	12 60 71 76 46	48 94 97 23 06	94 54 13 74 08
77 51 30 38 20	86 83 42 99 01	68 41 48 27 74	51 90 81 39 80	72 89 35 55 07
19 50 23 71 74	69 97 92 02 88	55 21 02 97 73	74 28 77 52 51	65 34 46 74 15
21 81 85 93 13	93 27 88 17 57	05 68 67 31 56	07 08 28 50 46	31 85 33 84 52
51 47 46 64 99	68 10 72 36 21	94 04 99 13 45	42 83 60 91 91	08 00 74 54 49
99 55 96 83 31	62 53 52 41 70	69 77 71 28 30	74 81 97 81 42	43 86 07 28 34
33 71 34 80 07	93 58 47 28 69	51 92 66 47 21	58 30 32 98 22	93 17 49 39 72
85 27 48 68 93	11 30 32 92 70	28 83 43 41 37	73 51 59 04 00	71 14 84 36 43
84 13 38 96 40	44 03 55 21 66	73 85 27 00 91	61 22 26 05 61	62 32 71 84 23
56 73 21 62 34	17 39 59 61 31	10 12 39 16 22	85 49 65 75 60	81 60 41 88 80
65 13 85 68 06	87 64 88 52 61	34 31 36 58 61	45 87 52 10 69	85 64 44 72 77
38 00 10 21 76	81 71 91 17 11	71 60 29 29 37	74 21 96 40 49	65 58 44 96 98
37 40 29 63 97	01 30 47 75 86	56 27 11 00 86	47 32 46 26 05	40 03 03 74 38
97 12 54 03 48	87 08 33 14 17	21 81 53 92 50	75 23 76 20 47	15 50 12 95 78
21 82 64 11 34	47 14 33 40 72	64 63 88 59 02	49 13 90 64 41	03 85 65 45 52
73 13 54 27 42	95 71 90 90 35	85 79 47 42 96	08 78 98 81 56	64 69 11 92 02
07 63 87 79 29	03 06 11 80 72	96 20 74 41 56	23 82 19 95 38	04 71 36 69 94
60 52 88 34 41	07 95 41 98 14	59 17 52 06 95	05 53 35 21 39	61 21 20 64 55
83 59 63 56 55	06 95 89 29 83	05 12 80 97 19	77 43 35 37 83	92 30 15 04 98
10 85 06 27 46	99 59 91 05 07	13 49 90 63 19	53 07 57 18 39	06 41 01 93 62
39 82 09 89 52	43 62 26 31 47	64 42 18 08 14	43 80 00 93 51	31 02 47 31 67
59 58 00 64 78	75 56 97 88 00	88 83 55 44 86	23 76 80 61 56	04 11 10 84 08
38 50 80 73 41	23 79 34 87 63	90 82 29 70 22	17 71 90 42 07	95 95 44 99 53
30 69 27 06 68	94 68 81 61 27	56 19 68 00 91	82 06 76 34 00	05 46 26 92 00
65 44 39 56 59	18 28 82 74 37	49 63 22 40 41	08 33 76 56 76	96 29 99 08 36
27 26 75 02 64	13 19 27 22 94	07 47 74 46 06	17 98 54 89 11	97 34 13 03 58
91 30 70 69 91	19 07 22 42 10	36 69 95 37 28	28 82 53 57 93	28 97 66 62 52
68 43 49 46 88	34 47 31 36 22	62 12 69 84 08	12 84 38 25 90	09 81 59 31 46
48 90 81 58 77	54 74 52 45 91	35 70 00 47 54	83 82 45 26 92	54 13 05 51 60
06 91 34 51 97	42 67 27 86 01	11 88 30 95 28	63 01 19 89 01	14 97 44 03 44
10 45 51 60 19	14 21 03 37 12	91 34 23 78 21	88 32 58 08 51	43 66 77 08 83
12 88 39 73 43	65 02 76 11 84	04 28 50 13 92	17 97 41 50 77	90 71 22 67 69
21 77 83 09 76	38 80 73 69 61	31 64 94 20 96	63 28 10 20 23	08 81 64 74 49
19 52 35 95 15	65 12 25 96 59	86 28 36 82 58	69 57 21 37 98	16 43 59 15 29
67 24 55 26 70	35 58 31 65 63	79 24 68 66 86	76 46 33 42 22	26 65 59 08 02
60 58 44 73 77	07 50 03 79 92	45 13 42 65 29	26 76 08 36 37	41 32 64 43 44
53 85 34 13 77	36 06 69 48 50	58 83 87 38 59	49 36 47 33 31	96 24 04 36 42
24 63 73 87 36	74 38 48 93 42	52 62 30 79 92	12 36 91 86 01	03 74 28 38 73
83 08 01 24 51	38 99 22 28 15	07 75 95 17 77	97 37 72 75 85	51 97 23 78 67
16 44 42 43 34	36 15 19 90 73	27 49 37 09 39	85 13 03 25 52	54 84 65 47 59
60 79 01 81 57	57 17 86 57 62	11 16 17 85 76	45 81 95 29 79	65 13 00 48 60

From tables of the RAND Corporation. Reprinted from Dixon, W. J. and J. Massey, Jr., *Introduction to Statistical Analysis*, 3rd ed. (New York: McGraw-Hill, 1969), pp. 446–447. With permission of the Rand Corporation.

TABLE B.2

Binomial Probabilities

n	x	.01	.05	.10	.20	.30	.40	p .50	.60	.70	.80	.90	.95	.99	x
2	0	980	902	810	640	490	360	250	160	090	040	010	002	0+	0
	1	020	095	180	320	420	480	500	480	420	320	180	095	020	1
	2	0+	002	010	040	090	160	250	360	490	640	810	902	980	2
3	0	970	857	729	512	343	216	125	064	027	008	001	0+	0+	0
	1	029	135	243	384	441	432	375	288	189	096	027	007	0+	1
	2	0+	007	027	096	189	288	375	432	441	384	243	135	029	2
	3	0+	0+	001	008	027	064	125	216	343	512	729	857	970	3
4	0	961	815	656	410	240	130	062	026	008	002	0+	0+	0+	0
	1	039	171	292	410	412	346	250	154	076	026	004	0+	0+	1
	2	001	014	049	154	265	346	375	346	265	154	049	014	001	2
	3	0+	0+	004	026	076	154	250	346	412	410	292	171	039	3
	4	0+	0+	0+	002	008	026	062	130	240	410	656	815	961	4
5	0	951	774	590	328	168	078	031	010	002	0+	0+	0+	0+	0
	1	048	204	328	410	360	259	156	077	028	006	0+	0+	0+	1
	2	001	021	073	205	309	346	312	230	132	051	008	001	0+	2
	3	0+	001	008	051	132	230	312	346	309	205	073	021	001	3
	4	0+	0+	0+	006	028	077	156	259	360	410	328	204	048	4
	5	0+	0+	0+	0+	002	010	031	078	168	328	590	774	951	5
6	0	941	735	531	262	118	047	016	004	001	0+	0+	0+	0+	0
	1	057	232	354	393	303	187	094	037	010	002	0+	0+	0+	1
	2	001	031	098	246	324	311	234	138	060	015	001	0+	0+	2
	3	0+	002	015	082	185	276	312	276	185	082	015	002	0+	3
	4	0+	0+	001	015	060	138	234	311	324	246	098	031	001	4
	5	0+	0+	0+	002	010	037	094	187	303	393	354	232	057	5
	6	0+	0+	0+	0+	001	004	016	047	118	262	531	735	941	6
7	0	932	698	478	210	082	028	008	002	0+	0+	0+	0+	0+	0
	1	066	257	372	367	247	131	055	017	004	0+	0+	0+	0+	1
	2	002	041	124	275	318	261	164	077	025	004	0+	0+	0+	2
	3	0+	004	023	115	227	290	273	194	097	029	003	0+	0+	3
	4	0+	0+	003	029	097	194	273	290	227	115	023	004	0+	4
	5	0+	0+	0+	004	025	077	164	261	318	275	124	041	002	5
	6	0+	0+	0+	0+	004	017	055	131	247	367	372	257	066	6
	7	0+	0+	0+	0+	0+	002	008	028	082	210	478	698	932	7
8	0	923	663	430	168	058	017	004	001	0+	0+	0+	0+	0+	0
	1	075	279	383	336	198	090	031	008	001	0+	0+	0+	0+	1
	2	003	051	149	294	296	209	109	041	010	001	0+	0+	0+	2
	3	0+	005	033	147	254	279	219	124	047	009	0+	0+	0+	3
	4	0+	0+	005	046	136	232	273	232	136	046	005	0+	0+	4
	5	0+	0+	0+	009	047	124	219	279	254	147	033	005	0+	5
	6	0+	0+	0+	001	010	041	109	209	296	294	149	051	003	6
	7	0+	0+	0+	0+	001	008	031	090	198	336	383	279	075	7
	8	0+	0+	0+	0+	0+	001	004	017	058	168	430	663	923	8

TABLE B.2 (Continued)

Binomial Probabilities

n	x	.01	.05	.10	.20	.30	.40	p .50	.60	.70	.80	.90	.95	.99	x
9	0	914	630	387	134	040	010	002	0+	0+	0+	0+	0+	0+	0
	1	083	299	387	302	156	060	018	004	0+	0+	0+	0+	0+	1
	2	003	063	172	302	267	161	070	021	004	0+	0+	0+	0+	2
	3	0+	008	045	176	267	251	164	074	021	003	0+	0+	0+	3
	4	0+	001	007	066	172	251	246	167	074	017	001	0+	0+	4
	5	0+	0+	001	017	074	167	246	251	172	066	007	001	0+	5
	6	0+	0+	0+	003	021	074	164	251	267	176	045	008	0+	6
	7	0+	0+	0+	0+	004	021	070	161	267	302	172	063	003	7
	8	0+	0+	0+	0+	0+	004	018	060	156	302	387	299	083	8
	9	0+	0+	0+	0+	0+	0+	002	010	040	134	387	630	914	9
10	0	904	599	349	107	028	006	001	0+	0+	0+	0+	0+	0+	0
	1	091	315	387	268	121	040	010	002	0+	0+	0+	0+	0+	1
	2	004	075	194	302	233	121	044	011	001	0+	0+	0+	0+	2
	3	0+	010	057	201	267	215	117	042	009	001	0+	0+	0+	3
	4	0+	001	011	088	200	251	205	111	037	006	0+	0+	0+	4
	5	0+	0+	001	026	103	201	246	201	103	026	001	0+	0+	5
	6	0+	0+	0+	006	037	111	205	251	200	088	011	001	0+	6
	7	0+	0+	0+	001	009	042	117	215	267	201	057	010	0+	7
	8	0+	0+	0+	0+	001	011	044	121	233	302	194	075	004	8
	9	0+	0+	0+	0+	0+	002	010	040	121	268	387	315	091	9
	10	0+	0+	0+	0+	0+	0+	001	006	028	107	349	599	904	10
11	0	895	569	314	086	020	004	0+	0+	0+	0+	0+	0+	0+	0
	1	099	329	384	236	093	027	005	001	0+	0+	0+	0+	0+	1
	2	005	087	213	295	200	089	027	005	001	0+	0+	0+	0+	2
	3	0+	014	071	221	257	177	081	023	004	0+	0+	0+	0+	3
	4	0+	001	016	111	220	236	161	070	017	002	0+	0+	0+	4
	5	0+	0+	002	039	132	221	226	147	057	010	0+	0+	0+	5
	6	0+	0+	0+	010	057	147	226	221	132	039	002	0+	0+	6
	7	0+	0+	0+	002	017	070	161	236	220	111	016	001	0+	7
	8	0+	0+	0+	0+	004	023	081	177	257	221	071	014	0+	8
	9	0+	0+	0+	0+	001	005	027	089	200	295	213	087	005	9
	10	0+	0+	0+	0+	0+	001	005	027	093	236	384	329	099	10
	11	0+	0+	0+	0+	0+	0+	0+	004	020	086	314	569	895	11
12	0	886	540	282	069	014	002	0+	0+	0+	0+	0+	0+	0+	0
	1	107	341	377	206	071	017	003	0+	0+	0+	0+	0+	0+	1
	2	006	099	230	283	168	064	016	002	0+	0+	0+	0+	0+	2
	3	0+	017	085	236	240	142	054	012	001	0+	0+	0+	0+	3
	4	0+	002	021	133	231	213	121	042	008	001	0+	0+	0+	4
	5	0+	0+	004	053	158	227	193	101	029	003	0+	0+	0+	5
	6	0+	0+	0+	016	079	177	226	177	079	016	0+	0+	0+	6
	7	0+	0+	0+	003	029	101	193	227	158	053	004	0+	0+	7
	8	0+	0+	0+	001	008	042	121	213	231	133	021	002	0+	8
	9	0+	0+	0+	0+	001	012	054	142	240	236	085	017	0+	9

TABLE B.2 (Continued)

Binomial Probabilities

n	x	.01	.05	.10	.20	.30	.40	p .50	.60	.70	.80	.90	.95	.99	x
12	10	0+	0+	0+	0+	0+	002	016	064	168	283	230	099	006	10
	11	0+	0+	0+	0+	0+	0+	003	017	071	206	377	341	107	11
	12	0+	0+	0+	0+	0+	0+	0+	002	014	069	282	540	886	12
13	0	878	513	254	055	010	001	0+	0+	0+	0+	0+	0+	0+	0
	1	115	351	367	179	054	011	002	0+	0+	0+	0+	0+	0+	1
	2	007	111	245	268	139	045	010	001	0+	0+	0+	0+	0+	2
	3	0+	021	100	246	218	111	035	006	001	0+	0+	0+	0+	3
	4	0+	003	028	154	234	184	087	024	003	0+	0+	0+	0+	4
	5	0+	0+	006	069	180	221	157	066	014	001	0+	0+	0+	5
	6	0+	0+	001	023	103	197	209	131	044	006	0+	0+	0+	6
	7	0+	0+	0+	006	044	131	209	197	103	023	001	0+	0+	7
	8	0+	0+	0+	001	014	066	157	221	180	069	006	0+	0+	8
	9	0+	0+	0+	0+	003	024	087	184	234	154	028	003	0+	9
	10	0+	0+	0+	0+	001	006	035	111	218	246	100	021	0+	10
	11	0+	0+	0+	0+	0+	001	010	045	139	268	245	111	007	11
	12	0+	0+	0+	0+	0+	0+	002	011	054	179	367	351	115	12
	13	0+	0+	0+	0+	0+	0+	0+	001	010	055	254	513	878	13
14	0	869	488	229	044	007	001	0+	0+	0+	0+	0+	0+	0+	0
	1	123	359	356	154	041	007	001	0+	0+	0+	0+	0+	0+	1
	2	008	123	257	250	113	032	006	001	0+	0+	0+	0+	0+	2
	3	0+	026	114	250	194	085	022	003	0+	0+	0+	0+	0+	3
	4	0+	004	035	172	229	155	061	014	001	0+	0+	0+	0+	4
	5	0+	0+	008	086	196	207	122	041	007	0+	0+	0+	0+	5
	6	0+	0+	001	032	126	207	183	092	023	002	0+	0+	0+	6
	7	0+	0+	0+	009	062	157	209	157	062	009	0+	0+	0+	7
	8	0+	0+	0+	002	023	092	183	207	126	032	001	0+	0+	8
	9	0+	0+	0+	0+	007	041	122	207	196	086	008	0+	0+	9
	10	0+	0+	0+	0+	001	014	061	155	229	172	035	004	0+	10
	11	0+	0+	0+	0+	0+	003	022	085	194	250	114	026	0+	11
	12	0+	0+	0+	0+	0+	001	006	032	113	250	257	123	008	12
	13	0+	0+	0+	0+	0+	0+	001	007	041	154	356	359	123	13
	14	0+	0+	0+	0+	0+	0+	0+	001	007	044	229	488	869	14
15	0	860	463	206	035	005	0+	0+	0+	0+	0+	0+	0+	0+	0
	1	130	366	343	132	031	005	0+	0+	0+	0+	0+	0+	0+	1
	2	009	135	267	231	092	022	003	0+	0+	0+	0+	0+	0+	2
	3	0+	031	129	250	170	063	014	002	0+	0+	0+	0+	0+	3
	4	0+	005	043	188	219	127	042	007	001	0+	0+	0+	0+	4
	5	0+	001	010	103	206	186	092	024	003	0+	0+	0+	0+	5
	6	0+	0+	002	043	147	207	153	061	012	001	0+	0+	0+	6
	7	0+	0+	0+	014	081	177	196	118	035	003	0+	0+	0+	7
	8	0+	0+	0+	003	035	118	196	177	081	014	0+	0+	0+	8
	9	0+	0+	0+	001	012	061	153	207	147	043	002	0+	0+	9

TABLE B.2 (Continued)

Binomial Probabilities

n	x	.01	.05	.10	.20	.30	.40	p .50	.60	.70	.80	.90	.95	.99	x
15	10	0+	0+	0+	0+	003	024	092	186	206	103	010	001	0+	10
	11	0+	0+	0+	0+	001	007	042	127	219	188	043	005	0+	11
	12	0+	0+	0+	0+	0+	002	014	063	170	250	129	031	0+	12
	13	0+	0+	0+	0+	0+	0+	003	022	092	231	267	135	009	13
	14	0+	0+	0+	0+	0+	0+	0+	005	031	132	343	366	130	14
	15	0+	0+	0+	0+	0+	0+	0+	0+	005	035	206	463	860	15
16	0	851	440	185	028	003	0+	0+	0+	0+	0+	0+	0+	0+	0
	1	138	371	329	113	023	003	0+	0+	0+	0+	0+	0+	0+	1
	2	010	146	275	211	073	015	002	0+	0+	0+	0+	0+	0+	2
	3	0+	036	142	246	146	047	009	001	0+	0+	0+	0+	0+	3
	4	0+	006	051	200	204	101	028	004	0+	0+	0+	0+	0+	4
	5	0+	001	014	120	210	162	067	014	001	0+	0+	0+	0+	5
	6	0+	0+	003	055	165	198	122	039	006	0+	0+	0+	0+	6
	7	0+	0+	0+	020	101	189	175	084	019	001	0+	0+	0+	7
	8	0+	0+	0+	006	049	142	196	142	049	006	0+	0+	0+	8
	9	0+	0+	0+	001	019	084	175	189	101	020	0+	0+	0+	9
	10	0+	0+	0+	0+	006	039	122	198	165	055	003	0+	0+	10
	11	0+	0+	0+	0+	001	014	067	162	210	120	014	001	0+	11
	12	0+	0+	0+	0+	0+	004	028	101	204	200	051	006	0+	12
	13	0+	0+	0+	0+	0+	001	009	047	146	246	142	036	0+	13
	14	0+	0+	0+	0+	0+	0+	002	015	073	211	275	146	010	14
	15	0+	0+	0+	0+	0+	0+	0+	003	023	113	329	371	138	15
	16	0+	0+	0+	0+	0+	0+	0+	0+	003	028	185	440	851	16
17	0	843	418	167	023	002	0+	0+	0+	0+	0+	0+	0+	0+	0
	1	145	374	315	096	017	002	0+	0+	0+	0+	0+	0+	0+	1
	2	012	158	280	191	058	010	001	0+	0+	0+	0+	0+	0+	2
	3	001	041	156	239	125	034	005	0+	0+	0+	0+	0+	0+	3
	4	0+	008	060	209	187	080	018	002	0+	0+	0+	0+	0+	4
	5	0+	001	017	136	208	138	047	008	001	0+	0+	0+	0+	5
	6	0+	0+	004	068	178	184	094	024	003	0+	0+	0+	0+	6
	7	0+	0+	001	027	120	193	148	057	009	0+	0+	0+	0+	7
	8	0+	0+	0+	008	064	161	185	107	028	002	0+	0+	0+	8
	9	0+	0+	0+	002	028	107	185	161	064	008	0+	0+	0+	9
	10	0+	0+	0+	0+	009	057	148	193	120	027	001	0+	0+	10
	11	0+	0+	0+	0+	003	024	094	184	178	068	004	0+	0+	11
	12	0+	0+	0+	001	008	047	138	208	136	017	001	0+	0+	12
	13	0+	0+	0+	0+	0+	002	018	080	187	209	060	008	0+	13
	14	0+	0+	0+	0+	0+	0+	005	034	125	239	156	041	001	14
	15	0+	0+	0+	0+	0+	0+	001	010	058	191	280	158	012	15
	16	0+	0+	0+	0+	0+	0+	0+	002	017	096	315	374	145	16
	17	0+	0+	0+	0+	0+	0+	0+	0+	002	023	167	418	843	17

TABLE B.2 (Continued)

Binomial Probabilities

n	x	.01	.05	.10	.20	.30	.40	p .50	.60	.70	.80	.90	.95	.99	x
18	0	835	397	150	018	002	0+	0+	0+	0+	0+	0+	0+	0+	0
	1	152	376	300	081	013	001	0+	0+	0+	0+	0+	0+	0+	1
	2	013	168	284	172	046	007	001	0+	0+	0+	0+	0+	0+	2
	3	001	047	168	230	105	025	003	0+	0+	0+	0+	0+	0+	3
	4	0+	009	070	215	168	061	012	001	0+	0+	0+	0+	0+	4
	5	0+	001	022	151	202	115	033	004	0+	0+	0+	0+	0+	5
	6	0+	0+	005	082	187	166	071	015	001	0+	0+	0+	0+	6
	7	0+	0+	001	035	138	189	121	037	005	0+	0+	0+	0+	7
	8	0+	0+	0+	012	081	173	167	077	015	001	0+	0+	0+	8
	9	0+	0+	0+	003	039	128	185	128	039	003	0+	0+	0+	9
	10	0+	0+	0+	001	015	077	167	173	081	012	0+	0+	0+	10
	11	0+	0+	0+	0+	005	037	121	189	138	035	001	0+	0+	11
	12	0+	0+	0+	0+	001	015	071	166	187	082	005	0+	0+	12
	13	0+	0+	0+	0+	0+	004	033	115	202	151	022	001	0+	13
	14	0+	0+	0+	0+	0+	001	012	061	168	215	070	009	0+	14
	15	0+	0+	0+	0+	0+	0+	003	025	105	230	168	047	001	15
	16	0+	0+	0+	0+	0+	0+	001	007	046	172	284	168	013	16
	17	0+	0+	0+	0+	0+	0+	0+	001	013	081	300	376	152	17
	18	0+	0+	0+	0+	0+	0+	0+	0+	002	018	150	397	835	18
19	0	826	377	135	014	001	0+	0+	0+	0+	0+	0+	0+	0+	0
	1	159	377	285	068	009	001	0+	0+	0+	0+	0+	0+	0+	1
	2	014	179	285	154	036	005	0+	0+	0+	0+	0+	0+	0+	2
	3	001	053	180	218	087	017	002	0+	0+	0+	0+	0+	0+	3
	4	0+	011	080	218	149	047	007	001	0+	0+	0+	0+	0+	4
	5	0+	002	027	164	192	093	022	002	0+	0+	0+	0+	0+	5
	6	0+	0+	007	095	192	145	052	008	001	0+	0+	0+	0+	6
	7	0+	0+	001	044	153	180	096	024	002	0+	0+	0+	0+	7
	8	0+	0+	0+	017	098	180	144	053	008	0+	0+	0+	0+	8
	9	0+	0+	0+	005	051	146	176	098	022	001	0+	0+	0+	9
	10	0+	0+	0+	001	022	098	176	146	051	005	0+	0+	0+	10
	11	0+	0+	0+	0+	008	053	144	180	098	017	0+	0+	0+	11
	12	0+	0+	0+	0+	002	024	096	180	153	044	001	0+	0+	12
	13	0+	0+	0+	0+	001	008	052	145	192	095	007	0+	0+	13
	14	0+	0+	0+	0+	0+	002	022	093	192	164	027	002	0+	14
	15	0+	0+	0+	0+	0+	001	007	047	149	218	080	011	0+	15
	16	0+	0+	0+	0+	0+	0+	002	017	087	218	180	053	001	16
	17	0+	0+	0+	0+	0+	0+	0+	005	036	154	285	179	014	17
	18	0+	0+	0+	0+	0+	0+	0+	001	009	068	285	377	159	18
	19	0+	0+	0+	0+	0+	0+	0+	0+	001	014	135	377	826	19
20	0	818	358	122	012	001	0+	0+	0+	0+	0+	0+	0+	0+	0
	1	165	377	270	058	007	0+	0+	0+	0+	0+	0+	0+	0+	1
	2	016	189	285	137	028	003	0+	0+	0+	0+	0+	0+	0+	2
	3	001	060	190	205	072	012	001	0+	0+	0+	0+	0+	0+	3
	4	0+	013	090	218	130	035	005	0+	0+	0+	0+	0+	0+	4

TABLE B.2 (Continued)

Binomial Probabilities

n	x	.01	.05	.10	.20	.30	.40	.50	.60	.70	.80	.90	.95	.99	x
20	5	0+	002	032	175	179	075	015	001	0+	0+	0+	0+	0+	5
	6	0+	0+	009	109	192	124	037	005	0+	0+	0+	0+	0+	6
	7	0+	0+	002	055	164	166	074	015	001	0+	0+	0+	0+	7
	8	0+	0+	0+	022	114	180	120	035	004	0+	0+	0+	0+	8
	9	0+	0+	0+	007	065	160	160	071	012	0+	0+	0+	0+	9
	10	0+	0+	0+	002	031	117	176	117	031	002	0+	0+	0+	10
	11	0+	0+	0+	0+	012	071	160	160	065	007	0+	0+	0+	11
	12	0+	0+	0+	0+	004	035	120	180	114	022	0+	0+	0+	12
	13	0+	0+	0+	0+	001	015	074	166	164	055	002	0+	0+	13
	14	0+	0+	0+	0+	0+	005	037	124	192	109	009	0+	0+	14
	15	0+	0+	0+	0+	0+	001	015	075	179	175	032	002	0+	15
	16	0+	0+	0+	0+	0+	0+	005	035	130	218	090	013	0+	16
	17	0+	0+	0+	0+	0+	0+	001	012	072	205	190	060	001	17
	18	0+	0+	0+	0+	0+	0+	0+	003	028	137	285	189	016	18
	19	0+	0+	0+	0+	0+	0+	0+	0+	007	058	270	377	165	19
	20	0+	0+	0+	0+	0+	0+	0+	0+	001	012	122	358	818	20
21	0	810	341	109	009	001	0+	0+	0+	0+	0+	0+	0+	0+	0
	1	172	376	255	048	005	0+	0+	0+	0+	0+	0+	0+	0+	1
	2	017	198	284	121	022	002	0+	0+	0+	0+	0+	0+	0+	2
	3	001	066	200	192	058	009	001	0+	0+	0+	0+	0+	0+	3
	4	0+	016	100	216	113	026	003	0+	0+	0+	0+	0+	0+	4
	5	0+	003	038	183	164	059	010	001	0+	0+	0+	0+	0+	5
	6	0+	0+	011	122	188	105	026	003	0+	0+	0+	0+	0+	6
	7	0+	0+	003	065	172	149	055	009	0+	0+	0+	0+	0+	7
	8	0+	0+	001	029	129	174	097	023	002	0+	0+	0+	0+	8
	9	0+	0+	0+	010	080	168	140	050	006	0+	0+	0+	0+	9
	10	0+	0+	0+	003	041	134	168	089	018	001	0+	0+	0+	10
	11	0|	0|	0+	001	018	089	168	134	041	003	0+	0+	0+	11
	12	0+	0+	0+	0+	006	050	140	168	080	010	0+	0+	0+	12
	13	0+	0+	0+	0+	002	023	097	174	129	029	001	0+	0+	13
	14	0+	0+	0+	0+	0+	009	055	149	172	065	003	0+	0+	14
	15	0+	0+	0+	0+	0+	003	026	105	188	122	011	0+	0+	15
	16	0+	0+	0+	0+	0+	001	010	059	164	183	038	003	0+	16
	17	0+	0+	0+	0+	0+	0+	003	026	113	216	100	016	0+	17
	18	0+	0+	0+	0+	0+	0+	001	009	058	192	200	066	001	18
	19	0+	0+	0+	0+	0+	0+	0+	002	022	121	284	198	017	19
	20	0+	0+	0+	0+	0+	0+	0+	0+	005	048	255	376	172	20
	21	0+	0+	0+	0+	0+	0+	0+	0+	001	009	109	341	810	21
22	0	802	324	098	007	0+	0+	0+	0+	0+	0+	0+	0+	0+	0
	1	178	375	241	041	004	0+	0+	0+	0+	0+	0+	0+	0+	1
	2	019	207	281	107	017	001	0+	0+	0+	0+	0+	0+	0+	2
	3	001	073	208	178	047	006	0+	0+	0+	0+	0+	0+	0+	3
	4	0+	018	110	211	096	019	002	0+	0+	0+	0+	0+	0+	4

TABLE B.2 (Continued)
Binomial Probabilities

n	x	.01	.05	.10	.20	.30	.40	.50	.60	.70	.80	.90	.95	.99	x
22	5	0+	003	044	190	149	046	006	0+	0+	0+	0+	0+	0+	5
	6	0+	001	014	134	181	086	018	001	0+	0+	0+	0+	0+	6
	7	0+	0+	004	077	177	131	041	005	0+	0+	0+	0+	0+	7
	8	0+	0+	001	036	142	164	076	014	001	0+	0+	0+	0+	8
	9	0+	0+	0+	014	095	170	119	034	003	0+	0+	0+	0+	9
	10	0+	0+	0+	005	053	148	154	066	010	0+	0+	0+	0+	10
	11	0+	0+	0+	001	025	107	168	107	025	001	0+	0+	0+	11
	12	0+	0+	0+	0+	010	066	154	148	053	005	0+	0+	0+	12
	13	0+	0+	0+	0+	003	034	119	170	095	014	0+	0+	0+	13
	14	0+	0+	0+	0+	001	014	076	164	142	036	001	0+	0+	14
	15	0+	0+	0+	0+	0+	005	041	131	177	077	004	0+	0+	15
	16	0+	0+	0+	0+	0+	001	018	086	181	134	014	001	0+	16
	17	0+	0+	0+	0+	0+	0+	006	046	149	190	044	003	0+	17
	18	0+	0+	0+	0+	0+	0+	002	019	096	211	110	018	0+	18
	19	0+	0+	0+	0+	0+	0+	0+	006	047	178	208	073	001	19
	20	0+	0+	0+	0+	0+	0+	0+	001	017	107	281	207	019	20
	21	0+	0+	0+	0+	0+	0+	0+	0+	004	041	241	375	178	21
	22	0+	0+	0+	0+	0+	0+	0+	0+	0+	007	098	324	802	22
23	0	794	307	089	006	0+	0+	0+	0+	0+	0+	0+	0+	0+	0
	1	184	372	226	034	003	0+	0+	0+	0+	0+	0+	0+	0+	1
	2	020	215	277	093	013	001	0+	0+	0+	0+	0+	0+	0+	2
	3	001	079	215	163	038	004	0+	0+	0+	0+	0+	0+	0+	3
	4	0+	021	120	204	082	014	001	0+	0+	0+	0+	0+	0+	4
	5	0+	004	051	194	133	035	004	0+	0+	0+	0+	0+	0+	5
	6	0+	001	017	145	171	070	012	001	0+	0+	0+	0+	0+	6
	7	0+	0+	005	088	178	113	029	003	0+	0+	0+	0+	0+	7
	8	0+	0+	001	044	153	151	058	009	0+	0+	0+	0+	0+	8
	9	0+	0+	0+	018	109	168	097	022	002	0+	0+	0+	0+	9
	10	0+	0+	0+	006	065	157	136	046	005	0+	0+	0+	0+	10
	11	0+	0+	0+	002	033	123	161	082	014	0+	0+	0+	0+	11
	12	0+	0+	0+	0+	014	082	161	123	033	002	0+	0+	0+	12
	13	0+	0+	0+	0+	005	046	136	157	065	006	0+	0+	0+	13
	14	0+	0+	0+	0+	002	022	097	168	109	018	0+	0+	0+	14
	15	0+	0+	0+	0+	0+	009	058	151	153	044	001	0+	0+	15
	16	0+	0+	0+	0+	0+	003	029	113	178	088	005	0+	0+	16
	17	0+	0+	0+	0+	0+	001	012	070	171	145	017	001	0+	17
	18	0+	0+	0+	0+	0+	0+	004	035	133	194	051	004	0+	18
	19	0+	0+	0+	0+	0+	0+	001	014	082	204	120	021	0+	19
	20	0+	0+	0+	0+	0+	0+	0+	004	038	163	215	079	001	20
	21	0+	0+	0+	0+	0+	0+	0+	001	013	093	277	215	020	21
	22	0+	0+	0+	0+	0+	0+	0+	0+	003	034	226	372	184	22
	23	0+	0+	0+	0+	0+	0+	0+	0+	0+	006	089	307	794	23

TABLE B.2 (Continued)

Binomial Probabilities

n	x	.01	.05	.10	.20	.30	.40	p .50	.60	.70	.80	.90	.95	.99	x
24	0	786	292	080	005	0+	0+	0+	0+	0+·	0+	0+	0+	0+	0
	1	190	369	213	028	002	0+	0+	0+	0+	0+	0+	0+	0+	1
	2	022	223	272	081	010	001	0+	0+	0+	0+	0+	0+	0+	2
	3	002	086	221	149	031	003	0+	0+	0+	0+	0+	0+	0+	3
	4	0+	024	129	196	069	010	001	0+	0+	0+	0+	0+	0+	4
	5	0+	005	057	196	118	027	003	0+	0+	0+	0+	0+	0+	5
	6	0+	001	020	155	160	056	008	0+	0+	0+	0+	0+	0+	6
	7	0+	0+	006	100	176	096	021	002	0|	0|	0|	0|	0|	7
	8	0+	0+	001	053	160	136	044	005	0+	0+	0+	0+	0+	8
	9	0+	0+	0+	024	122	161	078	014	001	0+	0+	0+	0+	9
	10	0+	0+	0+	009	079	161	117	032	003	0+	0+	0+	0+	10
	11	0+	0+	0+	003	043	137	149	061	008	0+	0+	0+	0+	11
	12	0+	0+	0+	001	020	099	161	099	020	001	0+	0+	0+	12
	13	0+	0+	0+	0+	008	061	149	137	043	003	0+	0+	0+	13
	14	0+	0+	0+	0+	003	032	117	161	079	009	0+	0+	0+	14
	15	0+	0+	0+	0+	001	014	078	161	122	024	0+	0+	0+	15
	16	0+	0+	0+	0+	0+	005	044	136	160	053	001	0+	0+	16
	17	0+	0+	0+	0+	0+	002	021	096	176	100	006	0+	0+	17
	18	0+	0+	0+	0+	0+	0+	008	056	160	155	020	001	0+	18
	19	0+	0+	0+	0+	0+	0+	003	027	118	196	057	005	0+	19
	20	0+	0+	0+	0+	0+	0+	001	010	069	196	129	024	0+	20
	21	0+	0+	0+	0+	0+	0+	0+	003	031	149	221	086	002	21
	22	0+	0+	0+	0+	0+	0+	0+	001	010	081	272	223	022	22
	23	0+	0+	0+	0+	0+	0+	0+	0+	002	028	213	369	190	23
	24	0+	0+	0+	0+	0+	0+	0+	0+	0+	005	080	292	786	24
25	0	778	277	072	004	0+	0+	0+	0+	0+	0+	0+	0+	0+	0
	1	196	365	199	024	001	0+	0+	0+	0+	0+	0+	0+	0+	1
	2	024	231	266	071	007	0+	0+	0+	0+	0+	0+	0+	0+	2
	3	002	093	226	136	024	002	0+	0+	0+	0+	0+	0+	0+	3
	4	0+	027	138	187	057	007	0+	0+	0+	0+	0+	0+	0+	4
	5	0+	006	065	196	103	020	002	0+	0+	0+	0+	0+	0+	5
	6	0+	001	024	163	147	044	005	0+	0+	0+	0+	0+	0+	6
	7	0+	0+	007	111	171	080	014	001	0+	0+	0+	0+	0+	7
	8	0+	0+	002	062	165	120	032	003	0+	0+	0+	0+	0+	8
	9	0+	0+	0+	029	134	151	061	009	0+	0+	0+	0+	0+	9
	10	0+	0+	0+	012	092	161	097	021	001	0+	0+	0+	0+	10
	11	0+	0+	0+	004	054	147	133	043	004	0+	0+	0+	0+	11
	12	0+	0+	0+	001	027	114	155	076	011	0+	0+	0+	0+	12
	13	0+	0+	0+	0+	011	076	155	114	027	001	0+	0+	0+	13
	14	0+	0+	0+	0+	004	043	133	147	054	004	0+	0+	0+	14
	15	0+	0+	0+	0+	001	021	097	161	092	012	0+	0+	0+	15
	16	0+	0+	0+	0+	0+	009	061	151	134	029	0+	0+	0+	16
	17	0+	0+	0+	0+	0+	003	032	120	165	062	002	0+	0+	17
	18	0+	0+	0+	0+	0+	001	014	080	171	111	007	0+	0+	18
	19	0+	0+	0+	0+	0+	0+	005	044	147	163	024	001	0+	19

TABLE B.2 (Continued)
Binomial Probabilities

n	x	.01	.05	.10	.20	.30	.40	p .50	.60	.70	.80	.90	.95	.99	x
25	20	0+	0+	0+	0+	0+	0+	002	020	103	196	065	006	0+	20
	21	0+	0+	0+	0+	0+	0+	0+	007	057	187	138	027	0+	21
	22	0+	0+	0+	0+	0+	0+	0+	002	024	136	226	093	002	22
	23	0+	0+	0+	0+	0+	0+	0+	0+	007	071	266	231	024	23
	24	0+	0+	0+	0+	0+	0+	0+	0+	001	024	199	365	196	24
	25	0+	0+	0+	0+	0+	0+	0+	0+	0+	004	072	277	778	25

From Mosteller, F., R. E. K. Rourke, and G. B. Thomas, *Probability with Statistical Applications* (Reading, Mass.: Addison-Wesley, 1961), pp. 371–379. Reprinted with permission.

TABLE B.3

The Standard Normal Distribution
Areas under the standard normal curve from 0 to z for various values of z.

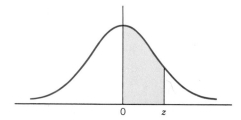

z	.00	.01	.02	.03	.04	.05	.06	.07	.08	.09
0.0	.0000	.0040	.0080	.0120	.0160	.0199	.0239	.0279	.0319	.0359
0.1	.0398	.0438	.0478	.0517	.0557	.0596	.0636	.0675	.0714	.0754
0.2	.0793	.0832	.0871	.0910	.0948	.0987	.1026	.1064	.1103	.1141
0.3	.1179	.1217	.1255	.1293	.1331	.1368	.1406	.1443	.1480	.1517
0.4	.1554	.1591	.1628	.1664	.1700	.1736	.1772	.1808	.1844	.1879
0.5	.1915	.1950	.1985	.2019.	.2054	.2088	.2123	.2157	.2190	.2224
0.6	.2258	.2291	.2324	.2357	.2389.	.2422	.2454	.2486	.2518	.2549
0.7	.2580	.2612	.2642	.2673	.2704	.2734	.2764	.2794	.2823	.2852
0.8	.2881	.2910	.2939	.2967	.2996	.3023	.3051	.3078	.3106	.3133
0.9	.3159	.3186	.3212	.3238	.3264	.3289	.3315	.3340	.3365	.3389
1.0	.3413	.3438	.3461	.3485	.3508	.3531	.3554	.3577	.3599	.3621
1.1	.3643	.3665	.3686	.3708	.3729	.3749	.3770	.3790	.3810	.3830
1.2	.3849	.3869	.3888	.3907	.3925	.3944	.3962	.3980	.3997	.4015
1.3	.4032	.4049	.4066	.4082	.4099	.4115	.4131	.4147	.4162	.4177
1.4	.4192	.4207	.4222	.4236	.4251	.4265	.4279	.4292	.4306	.4319

TABLE B.3 (Continued)

The Standard Normal Distribution
Areas under the standard normal curve from 0 to z for various values of z.

z	.00	.01	.02	.03	.04	.05	.06	.07	.08	.09
1.5	.4332	.4345	.4357	.4370	.4382	.4394	.4406	.4418	.4429	.4441
1.6	.4452	.4463	.4474	.4484	.4495	.4505	.4515	.4525	.4535	.4545
1.7	.4554	.4564	.4573	.4582	.4591	.4599	.4608	.4616	.4625	.4633
1.8	.4641	.4649	.4656	.4664	.4671	.4678	.4686	.4693	.4699	.4706
1.9	.4713	.4719	.4726	.4732	.4738	.4744	.4750	.4756	.4761	.4767
2.0	.4772	.4778	.4783	.4788	.4793	.4798	.4803	.4808	.4812	.4817
2.1	.4821	.4826	.4830	.4834	.4838	.4842	.4846	.4850	.4854	.4857
2.2	.4861	.4864	.4868	.4871	.4875	.4878	.4881	.4884	.4887	.4890
2.3	.4893	.4896	.4898	.4901	.4904	.4906	.4909	.4911	.4913	.4916
2.4	.4918	.4920	.4922	.4925	.4927	.4929	.4931	.4932	.4934	.4936
2.5	.4938	.4940	.4941	.4943	.4945	.4946	.4948	.4949	.4951	.4952
2.6	.4953	.4955	.4956	.4957	.4959	.4960	.4961	.4962	.4963	.4964
2.7	.4965	.4966	.4967	.4968	.4969	.4970	.4971	.4972	.4973	.4974
2.8	.4974	.4975	.4976	.4977	.4977	.4978	.4979	.4979	.4980	.4981
2.9	.4981	.4982	.4982	.4983	.4984	.4984	.4985	.4985	.4986	.4986
3.0	.4987	.4987	.4987	.4988	.4988	.4989	.4989	.4989	.4990	.4990
3.1	.4990	.4991	.4991	.4991	.4992	.4992	.4992	.4992	.4993	.4993
3.2	.4993	.4993	.4994	.4994	.4994	.4994	.4994	.4995	.4995	.4995
3.3	.4995	.4995	.4995	.4996	.4996	.4996	.4996	.4996	.4996	.4997
3.4	.4997	.4997	.4997	.4997	.4997	.4997	.4997	.4997	.4997	.4998
3.5	.4998	.4998	.4998	.4998	.4998	.4998	.4998	.4998	.4998	.4998
3.6	.4998	.4998	.4999	.4999	.4999	.4999	.4999	.4999	.4999	.4999
3.7	.4999	.4999	.4999	.4999	.4999	.4999	.4999	.4999	.4999	.4999
3.8	.4999	.4999	.4999	.4999	.4999	.4999	.4999	.4999	.4999	.4999
3.9	.49995	.49995	.49996	.49996	.49996	.49996	.49996	.49996	.49997	.49997
4.0	.49997									
4.5	.499997									
5.0	.4999997									

Adapted from *Standard Mathematical Tables*, 25th ed. (Boca Raton: Chemical Rubber Company Press, 1978), p. 524. Reprinted with permission.

TABLE B. 4

Student's *t* Distribution

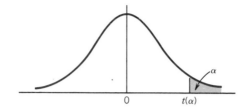

df	t(.005)	t(.01)	t(.025)	t(.05)	t(.10)	t(.25)
1	63.657	31.821	12.706	6.314	3.078	1.000
2	9.925	6.965	4.303	2.920	1.886	0.816
3	5.841	4.541	3.182	2.353	1.638	.765
4	4.604	3.747	2.776	2.132	1.533	.741
5	4.032	3.365	2.571	2.015	1.476	0.727
6	3.707	3.143	2.447	1.943	1.440	.718
7	3.499	2.998	2.365	1.895	1.415	.711
8	3.355	2.896	2.306	1.860	1.397	.706
9	3.250	2.821	2.262	1.833	1.383	.703
10	3.169	2.764	2.228	1.812	1.372	0.700
11	3.106	2.718	2.201	1.796	1.363	.697
12	3.055	2.681	2.179	1.782	1.356	.695
13	3.012	2.650	2.160	1.771	1.350	.694
14	2.977	2.624	2.145	1.761	1.345	.692
15	2.947	2.6C2	2.131	1.753	1.341	0.691
16	2.921	2.583	2.120	1.746	1.337	.690
17	2.898	2.567	2.110	1.740	1.333	.689
18	2.878	2.552	2.101	1.734	1.330	.688
19	2.861	2.539	2.093	1.729	1.328	.688
20	2.845	2.528	2.086	1.725	1.325	0.687
21	2.831	2.518	2.080	1.721	1.323	.686
22	2.819	2.508	2.074	1.717	1.321	.686
23	2.807	2.500	2.069	1.714	1.319	.685
24	2.797	2.492	2.064	1.711	1.318	.685
25	2.787	2.485	2.060	1.708	1.316	0.684
26	2.779	2.479	2.056	1.706	1.315	.684
27	2.771	2.473	2.052	1.703	1.314	.684
28	2.763	2.467	2.048	1.701	1.313	.683
29	2.756	2.462	2.045	1.699	1.311	.683
Large	2.576	2.326	1.960	1.645	1.282	.674

TABLE B.5

The Chi-Square Distribution

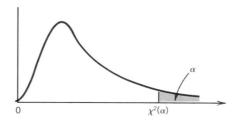

df	$\chi^2(.995)$	$\chi^2(.99)$	$\chi^2(.975)$	$\chi^2(.95)$	$\chi^2(.90)$	$\chi^2(.10)$	$\chi^2(.05)$	$\chi^2(.025)$	$\chi^2(.01)$	$\chi^2(.005)$
1	—	—	0.001	0.004	0.016	2.706	3.841	5.024	6.635	7.879
2	0.010	0.020	0.051	0.103	0.211	4.605	5.991	7.378	9.210	10.597
3	0.072	0.115	0.216	0.352	0.584	6.251	7.815	9.348	11.345	12.838
4	0.207	0.297	0.484	0.711	1.064	7.779	9.488	11.143	13.277	14.860
5	0.412	0.554	0.831	1.145	1.610	9.236	11.071	12.833	15.086	16.750
6	0.676	0.872	1.237	1.635	2.204	10.645	12.592	14.449	16.812	18.548
7	0.989	1.239	1.690	2.167	2.833	12.017	14.067	16.013	18.475	20.278
8	1.344	1.646	2.180	2.733	3.490	13.362	15.507	17.535	20.090	21.955
9	1.735	2.088	2.700	3.325	4.168	14.684	16.919	19.023	21.666	23.589
10	2.156	2.558	3.247	3.940	4.865	15.987	18.307	20.483	23.209	25.188
11	2.603	3.053	3.816	4.575	5.578	17.275	19.675	21.920	24.725	26.757
12	3.074	3.571	4.404	5.226	6.304	18.549	21.026	23.337	26.217	28.299
13	3.565	4.107	5.009	5.892	7.042	19.812	22.362	24.736	27.688	29.819
14	4.075	4.660	5.629	6.571	7.790	21.064	23.685	26.119	29.141	31.319
15	4.601	5.229	6.262	7.261	8.547	22.307	24.996	27.488	30.578	32.801
16	5.142	5.812	6.908	7.962	9.312	23.542	26.296	28.845	32.000	34.267
17	5.697	6.408	7.564	8.672	10.085	24.769	27.587	30.191	33.409	35.718
18	6.265	7.015	8.231	9.390	10.865	25.989	28.869	31.526	34.805	37.156
19	6.844	7.633	8.907	10.117	11.651	27.204	30.144	32.852	36.191	38.582
20	7.434	8.260	9.591	10.851	12.443	28.412	31.410	34.170	37.566	39.997
21	8.034	8.897	10.283	11.591	13.240	29.615	32.671	35.479	38.932	41.401
22	8.643	9.542	10.982	12.338	14.042	30.813	33.924	36.781	40.289	42.796
23	9.260	10.196	11.689	13.091	14.848	32.007	35.172	38.076	41.638	44.181
24	9.886	10.856	12.401	13.848	15.659	33.196	36.415	39.364	42.980	45.559
25	10.520	11.524	13.120	14.611	16.473	34.382	37.652	40.646	44.314	46.928
26	11.160	12.198	13.844	15.379	17.292	35.563	38.885	41.923	45.642	48.290
27	11.808	12.879	14.573	16.151	18.114	36.741	40.113	43.194	46.963	49.645
28	12.461	13.565	15.308	16.928	18.939	37.916	41.337	44.461	48.278	50.993
29	13.121	14.257	16.047	17.708	19.768	39.087	42.557	45.722	49.588	52.336
30	13.787	14.954	16.791	18.493	20.599	40.256	43.773	46.979	50.892	53.672
40	20.707	22.164	24.433	26.509	29.051	51.805	55.758	59.342	63.691	66.766

TABLE B.5 (Continued)

The Chi-Square Distribution

df	$\chi^2(.995)$	$\chi^2(.99)$	$\chi^2(.975)$	$\chi^2(.95)$	$\chi^2(.90)$	$\chi^2(.10)$	$\chi^2(.05)$	$\chi^2(.025)$	$\chi^2(.01)$	$\chi^2(.005)$
50	27.991	29.707	32.357	34.764	37.689	63.167	67.505	71.420	76.154	79.490
60	35.534	37.485	40.482	43.188	46.459	74.397	79.082	83.298	88.379	91.952
70	43.275	45.442	48.758	51.739	55.329	85.527	90.531	95.023	100.425	104.215
80	51.172	53.540	57.153	60.391	64.278	96.578	101.879	106.629	112.329	116.321
90	59.196	61.754	65.647	69.126	73.291	107.565	113.145	118.136	124.116	128.299
100	67.328	70.065	74.222	77.929	82.358	118.498	124.342	129.561	135.807	140.169

From Owen, D. B., *Handbook of Statistical Tables* (Reading, Mass.: Addison-Wesley, 1962), pp. 50–55. Reprinted with permission.

TABLE B.6
The F Distribution: Values of F (.05)

.05

F(.05)

Degrees of freedom of the numerator

df_2 \ df_1	1	2	3	4	5	6	7	8	9	10	12	15	20	24	30	40	60	120	∞
1	161.4	199.5	215.7	224.6	230.2	234.0	236.8	238.9	240.5	241.9	243.9	245.9	248.0	249.1	250.1	251.1	252.2	253.3	254.3
2	18.51	19.00	19.16	19.25	19.30	19.33	19.35	19.37	19.38	19.40	19.41	19.43	19.45	19.45	19.46	19.47	19.48	19.49	19.50
3	10.13	9.55	9.28	9.12	9.01	8.94	8.89	8.85	8.81	8.79	8.74	8.70	8.66	8.64	8.62	8.59	8.57	8.55	8.53
4	7.71	6.94	6.59	6.39	6.26	6.16	6.09	6.04	6.00	5.96	5.91	5.86	5.80	5.77	5.75	5.72	5.69	5.66	5.63
5	6.61	5.79	5.41	5.19	5.05	4.95	4.88	4.82	4.77	4.74	4.68	4.62	4.50	4.53	4.50	4.46	4.43	4.40	4.36
6	5.99	5.14	4.76	4.53	4.39	4.28	4.21	4.15	4.10	4.06	4.00	3.94	3.87	3.84	3.81	3.77	3.74	3.70	3.67
7	5.59	4.74	4.35	4.12	3.97	3.87	3.79	3.73	3.68	3.64	3.57	3.51	3.44	3.41	3.38	3.34	3.30	3.27	3.23
8	5.32	4.46	4.07	3.84	3.69	3.58	3.50	3.44	3.39	3.35	3.28	3.22	3.15	3.12	3.08	3.04	3.01	2.97	2.93
9	5.12	4.26	3.86	3.63	3.48	3.37	3.29	3.23	3.18	3.14	3.07	3.01	2.94	2.90	2.86	2.83	2.79	2.75	2.71
10	4.96	4.10	3.71	3.48	3.33	3.22	3.14	3.07	3.02	2.98	2.91	2.85	2.77	2.74	2.70	2.66	2.62	2.58	2.54
11	4.84	3.98	3.59	3.36	3.20	3.09	3.01	2.95	2.90	2.85	2.79	2.72	2.65	2.61	2.57	2.53	2.49	2.45	2.40
12	4.75	3.89	3.49	3.26	3.11	3.00	2.91	2.85	2.80	2.75	2.69	2.62	2.54	2.51	2.47	2.43	2.38	2.34	2.30
13	4.67	3.81	3.41	3.18	3.03	2.92	2.83	2.77	2.71	2.67	2.60	2.53	2.46	2.42	2.38	2.34	2.30	2.25	2.21
14	4.60	3.74	3.34	3.11	2.96	2.85	2.76	2.70	2.65	2.60	2.53	2.46	2.39	2.35	2.31	2.27	2.22	2.18	2.13
15	4.54	3.68	3.29	3.06	2.90	2.79	2.71	2.64	2.59	2.54	2.48	2.40	2.33	2.29	2.25	2.20	2.16	2.11	2.07
16	4.49	3.63	3.24	3.01	2.85	2.74	2.66	2.59	2.54	2.49	2.42	2.35	2.28	2.24	2.19	2.15	2.11	2.06	2.01
17	4.45	3.59	3.20	2.96	2.81	2.70	2.61	2.55	2.49	2.45	2.38	2.31	2.23	2.19	2.15	2.10	2.06	2.01	1.96
18	4.41	3.55	3.16	2.93	2.77	2.66	2.58	2.51	2.46	2.41	2.34	2.27	2.19	2.15	2.11	2.06	2.02	1.97	1.92
19	4.38	3.52	3.13	2.90	2.74	2.63	2.54	2.48	2.42	2.38	2.31	2.23	2.16	2.11	2.07	2.03	1.98	1.93	1.88
20	4.35	3.49	3.10	2.87	2.71	2.60	2.51	2.45	2.39	2.35	2.28	2.20	2.12	2.08	2.04	1.99	1.95	1.90	1.84
21	4.32	3.47	3.07	2.84	2.68	2.57	2.49	2.42	2.37	2.32	2.25	2.18	2.10	2.05	2.01	1.96	1.92	1.87	1.81
22	4.30	3.44	3.05	2.82	2.66	2.55	2.46	2.40	2.34	2.30	2.23	2.15	2.07	2.03	1.98	1.94	1.89	1.84	1.78
23	4.28	3.42	3.03	2.80	2.64	2.53	2.44	2.37	2.32	2.27	2.20	2.13	2.05	2.01	1.96	1.91	1.86	1.81	1.76
24	4.26	3.40	3.01	2.78	2.62	2.51	2.42	2.36	2.30	2.25	2.18	2.11	2.03	1.98	1.94	1.89	1.84	1.79	1.73
25	4.24	3.39	2.99	2.76	2.60	2.49	2.40	2.34	2.28	2.24	2.16	2.09	2.01	1.96	1.92	1.87	1.82	1.77	1.71
26	4.23	3.37	2.98	2.74	2.59	2.47	2.39	2.32	2.27	2.22	2.15	2.07	1.99	1.95	1.90	1.85	1.80	1.75	1.69
27	4.21	3.35	2.96	2.73	2.57	2.46	2.37	2.31	2.25	2.20	2.13	2.06	1.97	1.93	1.88	1.84	1.79	1.73	1.67
28	4.20	3.34	2.95	2.71	2.56	2.45	2.36	2.29	2.24	2.19	2.12	2.04	1.96	1.91	1.87	1.82	1.77	1.71	1.65
29	4.18	3.33	2.93	2.70	2.55	2.43	2.35	2.28	2.22	2.18	2.10	2.03	1.94	1.90	1.85	1.81	1.75	1.70	1.64
30	4.17	3.32	2.92	2.69	2.53	2.42	2.33	2.27	2.21	2.16	2.09	2.01	1.93	1.89	1.84	1.79	1.74	1.68	1.62
40	4.08	3.23	2.84	2.61	2.45	2.34	2.25	2.18	2.12	2.08	2.00	1.92	1.84	1.79	1.74	1.69	1.64	1.58	1.51
60	4.00	3.15	2.76	2.53	2.37	2.25	2.17	2.10	2.04	1.99	1.92	1.84	1.75	1.70	1.65	1.59	1.53	1.47	1.39
120	3.92	3.07	2.68	2.45	2.29	2.17	2.09	2.02	1.96	1.91	1.83	1.75	1.66	1.61	1.55	1.50	1.43	1.35	1.25
∞	3.84	3.00	2.60	2.37	2.21	2.10	2.01	1.94	1.88	1.83	1.75	1.67	1.57	1.52	1.46	1.39	1.32	1.22	1.00

Degrees of freedom of the denominator

TABLE B.6 (Continued)

The F Distribution: Values of $F_{(.025)}$

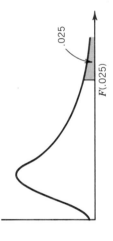

$F_{(.025)}$

Degrees of freedom of the numerator

df_1 / df_2	1	2	3	4	5	6	7	8	9	10	12	15	20	24	30	40	60	120	∞
1	647.8	799.5	864.2	899.6	921.8	937.1	948.2	956.7	963.3	968.6	976.7	984.9	993.1	997.2	1001	1006	1010	1014	1018
2	38.51	39.00	39.17	39.25	39.30	39.33	39.36	39.37	39.39	39.40	39.41	39.43	39.45	39.46	39.46	39.47	39.48	39.49	39.50
3	17.44	16.04	15.44	15.10	14.88	14.73	14.62	14.54	14.47	14.42	14.34	14.25	14.17	14.12	14.08	14.04	13.99	13.95	13.90
4	12.22	10.65	9.98	9.60	9.36	9.20	9.07	8.98	8.90	8.84	8.75	8.66	8.56	8.51	8.46	8.41	8.36	8.31	8.26
5	10.01	8.43	7.76	7.39	7.15	6.98	6.85	6.76	6.68	6.62	6.52	6.43	6.33	6.28	6.23	6.18	6.12	6.07	6.02
6	8.81	7.26	6.60	6.23	5.99	5.82	5.70	5.60	5.52	5.46	5.37	5.27	5.17	5.12	5.07	5.01	4.96	4.90	4.85
7	8.07	6.54	5.89	5.52	5.29	5.12	4.99	4.90	4.82	4.76	4.67	4.57	4.47	4.42	4.36	4.31	4.25	4.20	4.14
8	7.57	6.06	5.42	5.05	4.82	4.65	4.53	4.43	4.36	4.30	4.20	4.10	4.00	3.95	3.89	3.84	3.78	3.73	3.67
9	7.21	5.71	5.08	4.72	4.48	4.32	4.20	4.10	4.03	3.96	3.87	3.77	3.67	3.61	3.56	3.51	3.45	3.39	3.33
10	6.94	5.46	4.83	4.47	4.24	4.07	3.95	3.85	3.78	3.72	3.62	3.52	3.42	3.37	3.31	3.26	3.20	3.14	3.08
11	6.72	5.26	4.63	4.28	4.04	3.88	3.76	3.66	3.59	3.53	3.43	3.33	3.23	3.17	3.12	3.06	3.00	2.94	2.88
12	6.55	5.10	4.47	4.12	3.89	3.73	3.61	3.51	3.44	3.37	3.28	3.18	3.07	3.02	2.96	2.91	2.85	2.79	2.72
13	6.41	4.97	4.35	4.00	3.77	3.60	3.48	3.39	3.31	3.25	3.15	3.05	2.95	2.89	2.84	2.78	2.72	2.66	2.60
14	6.30	4.86	4.24	3.89	3.66	3.50	3.38	3.29	3.21	3.15	3.05	2.95	2.84	2.79	2.73	2.67	2.61	2.55	2.49
15	6.20	4.77	4.15	3.80	3.58	3.41	3.29	3.20	3.12	3.06	2.96	2.86	2.76	2.70	2.64	2.59	2.52	2.46	2.40
16	6.12	4.69	4.08	3.73	3.50	3.34	3.22	3.12	3.05	2.99	2.89	2.79	2.68	2.63	2.57	2.51	2.45	2.38	2.32
17	6.04	4.62	4.01	3.66	3.44	3.28	3.16	3.06	2.98	2.92	2.82	2.72	2.62	2.56	2.50	2.44	2.38	2.32	2.25
18	5.98	4.56	3.95	3.61	3.38	3.22	3.10	3.01	2.93	2.87	2.77	2.67	2.56	2.50	2.44	2.38	2.32	2.26	2.19
19	5.92	4.51	3.90	3.56	3.33	3.17	3.05	2.96	2.88	2.82	2.72	2.62	2.51	2.45	2.39	2.33	2.27	2.20	2.13
20	5.87	4.46	3.86	3.51	3.29	3.13	3.01	2.91	2.84	2.77	2.68	2.57	2.46	2.41	2.35	2.29	2.22	2.16	2.09
21	5.83	4.42	3.82	3.48	3.25	3.09	2.97	2.87	2.80	2.73	2.64	2.53	2.42	2.37	2.31	2.25	2.18	2.11	2.04
22	5.79	4.38	3.78	3.44	3.22	3.05	2.93	2.84	2.76	2.70	2.60	2.50	2.39	2.33	2.27	2.21	2.14	2.08	2.00
23	5.75	4.35	3.75	3.41	3.18	3.02	2.90	2.81	2.73	2.67	2.57	2.47	2.36	2.30	2.24	2.18	2.11	2.04	1.97
24	5.72	4.32	3.72	3.38	3.15	2.99	2.87	2.78	2.70	2.64	2.54	2.44	2.33	2.27	2.21	2.15	2.08	2.01	1.94
25	5.69	4.29	3.69	3.35	3.13	2.97	2.85	2.75	2.68	2.61	2.51	2.41	2.30	2.24	2.18	2.12	2.05	1.98	1.91
26	5.66	4.27	3.67	3.33	3.10	2.94	2.82	2.73	2.65	2.59	2.49	2.39	2.28	2.22	2.16	2.09	2.03	1.95	1.88
27	5.63	4.24	3.65	3.31	3.08	2.92	2.80	2.71	2.63	2.57	2.47	2.36	2.25	2.19	2.13	2.07	2.00	1.93	1.85
28	5.61	4.22	3.63	3.29	3.06	2.90	2.78	2.69	2.61	2.55	2.45	2.34	2.23	2.17	2.11	2.05	1.98	1.91	1.83
29	5.59	4.20	3.61	3.27	3.04	2.88	2.76	2.67	2.59	2.53	2.43	2.32	2.21	2.15	2.09	2.03	1.96	1.89	1.81
30	5.57	4.18	3.59	3.25	3.03	2.87	2.75	2.65	2.57	2.51	2.41	2.31	2.20	2.14	2.07	2.01	1.94	1.87	1.79
40	5.42	4.05	3.46	3.13	2.90	2.74	2.62	2.53	2.45	2.39	2.29	2.18	2.07	2.01	1.94	1.88	1.80	1.72	1.64
60	5.29	3.93	3.34	3.01	2.79	2.63	2.51	2.41	2.33	2.27	2.17	2.06	1.94	1.88	1.82	1.74	1.67	1.58	1.48
120	5.15	3.80	3.23	2.89	2.67	2.52	2.39	2.30	2.22	2.16	2.05	1.94	1.82	1.76	1.69	1.61	1.53	1.43	1.31
∞	5.02	3.69	3.12	2.79	2.57	2.41	2.29	2.19	2.11	2.05	1.94	1.83	1.71	1.64	1.57	1.48	1.39	1.27	1.00

Degrees of freedom of the denominator

TABLE B.6 (Continued)

The F distribution: Values of F (.01)

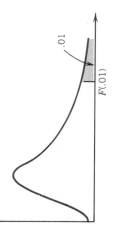

Degrees of freedom of the numerator

df_2 \ df_1	1	2	3	4	5	6	7	8	9	10	12	15	20	24	30	40	60	120	∞
1	4052	4999·5	5403	5625	5764	5859	5928	5981	6022	6056	6106	6157	6209	6235	6261	6287	6313	6339	6366
2	98·50	99·00	99·17	99·25	99·30	99·33	99·36	99·37	99·39	99·40	99·42	99·43	99·45	99·46	99·47	99·47	99·48	99·49	99·50
3	34·12	30·82	29·46	28·71	28·24	27·91	27·67	27·49	27·35	27·23	27·05	26·87	26·69	26·60	26·50	26·41	26·32	26·22	26·13
4	21·20	18·00	16·69	15·98	15·52	15·21	14·98	14·80	14·66	14·55	14·37	14·20	14·02	13·93	13·84	13·75	13·65	13·56	13·46
5	16·26	13·27	12·06	11·39	10·97	10·67	10·46	10·29	10·16	10·05	9·89	9·72	9·55	9·47	9·38	9·29	9·20	9·11	9·02
6	13·75	10·92	9·78	9·15	8·75	8·47	8·26	8·10	7·98	7·87	7·72	7·56	7·40	7·31	7·23	7·14	7·06	6·97	6·88
7	12·25	9·55	8·45	7·85	7·46	7·19	6·99	6·84	6·72	6·62	6·47	6·31	6·16	6·07	5·99	5·91	5·82	5·74	5·65
8	11·26	8·65	7·59	7·01	6·63	6·37	6·18	6·03	5·91	5·81	5·67	5·52	5·36	5·28	5·20	5·12	5·03	4·95	4·86
9	10·56	8·02	6·99	6·42	6·06	5·80	5·61	5·47	5·35	5·26	5·11	4·96	4·81	4·73	4·65	4·57	4·48	4·40	4·31
10	10·04	7·56	6·55	5·99	5·64	5·39	5·20	5·06	4·94	4·85	4·71	4·56	4·41	4·33	4·25	4·17	4·08	4·00	3·91
11	9·65	7·21	6·22	5·67	5·32	5·07	4·89	4·74	4·63	4·54	4·40	4·25	4·10	4·02	3·94	3·86	3·78	3·69	3·60
12	9·33	6·93	5·95	5·41	5·06	4·82	4·64	4·50	4·39	4·30	4·16	4·01	3·86	3·78	3·70	3·62	3·54	3·45	3·36
13	9·07	6·70	5·74	5·21	4·86	4·62	4·44	4·30	4·19	4·10	3·96	3·82	3·66	3·59	3·51	3·43	3·34	3·25	3·17
14	8·86	6·51	5·56	5·04	4·69	4·46	4·28	4·14	4·03	3·94	3·80	3·66	3·51	3·43	3·35	3·27	3·18	3·09	3·00
15	8·68	6·36	5·42	4·89	4·56	4·32	4·14	4·00	3·89	3·80	3·67	3·52	3·37	3·29	3·21	3·13	3·05	2·96	2·87
16	8·53	6·23	5·29	4·77	4·44	4·20	4·03	3·89	3·78	3·69	3·55	3·41	3·26	3·18	3·10	3·02	2·93	2·84	2·75
17	8·40	6·11	5·18	4·67	4·34	4·10	3·93	3·79	3·68	3·59	3·46	3·31	3·16	3·08	3·00	2·92	2·83	2·75	2·65
18	8·29	6·01	5·09	4·58	4·25	4·01	3·84	3·71	3·60	3·51	3·37	3·23	3·08	3·00	2·92	2·84	2·75	2·66	2·57
19	8·18	5·93	5·01	4·50	4·17	3·94	3·77	3·63	3·52	3·43	3·30	3·15	3·00	2·92	2·84	2·76	2·67	2·58	2·49
20	8·10	5·85	4·94	4·43	4·10	3·87	3·70	3·56	3·46	3·37	3·23	3·09	2·94	2·86	2·78	2·69	2·61	2·52	2·42
21	8·02	5·78	4·87	4·37	4·04	3·81	3·64	3·51	3·40	3·31	3·17	3·03	2·88	2·80	2·72	2·64	2·55	2·46	2·36
22	7·95	5·72	4·82	4·31	3·99	3·76	3·59	3·45	3·35	3·26	3·12	2·98	2·83	2·75	2·67	2·58	2·50	2·40	2·31
23	7·88	5·66	4·76	4·26	3·94	3·71	3·54	3·41	3·30	3·21	3·07	2·93	2·78	2·70	2·62	2·54	2·45	2·35	2·26
24	7·82	5·61	4·72	4·22	3·90	3·67	3·50	3·36	3·26	3·17	3·03	2·89	2·74	2·66	2·58	2·49	2·40	2·31	2·21
25	7·77	5·57	4·68	4·18	3·85	3·63	3·46	3·32	3·22	3·13	2·99	2·85	2·70	2·62	2·54	2·45	2·36	2·27	2·17
26	7·72	5·53	4·64	4·14	3·82	3·59	3·42	3·29	3·18	3·09	2·96	2·81	2·66	2·58	2·50	2·42	2·33	2·23	2·13
27	7·68	5·49	4·60	4·11	3·78	3·56	3·39	3·26	3·15	3·06	2·93	2·78	2·63	2·55	2·47	2·38	2·29	2·20	2·10
28	7·64	5·45	4·57	4·07	3·75	3·53	3·36	3·23	3·12	3·03	2·90	2·75	2·60	2·52	2·44	2·35	2·26	2·17	2·06
29	7·60	5·42	4·54	4·04	3·73	3·50	3·33	3·20	3·09	3·00	2·87	2·73	2·57	2·49	2·41	2·33	2·23	2·14	2·03
30	7·56	5·39	4·51	4·02	3·70	3·47	3·30	3·17	3·07	2·98	2·84	2·70	2·55	2·47	2·39	2·30	2·21	2·11	2·01
40	7·31	5·18	4·31	3·83	3·51	3·29	3·12	2·99	2·89	2·80	2·66	2·52	2·37	2·29	2·20	2·11	2·02	1·92	1·80
60	7·08	4·98	4·13	3·65	3·34	3·12	2·95	2·82	2·72	2·63	2·50	2·35	2·20	2·12	2·03	1·94	1·84	1·73	1·60
120	6·85	4·79	3·95	3·48	3·17	2·96	2·79	2·66	2·56	2·47	2·34	2·19	2·03	1·95	1·86	1·76	1·66	1·53	1·38
∞	6·63	4·61	3·78	3·32	3·02	2·80	2·64	2·51	2·41	2·32	2·18	2·04	1·88	1·79	1·70	1·59	1·47	1·32	1·00

Degrees of freedom of the denominator

From Pearson, E. S., and H. O. Hartley, *Biometrika Tables for Statisticians*, Vol. 1 (1958), pp. 171–173. With permission of the Biometrika Trustees.

TABLE B.7

Critical Values of r

The given values of α are for a two-tailed test.

n	$\alpha = .05$	$\alpha = .01$	n	$\alpha = .05$	$\alpha = .01$
4	.950	.999	20	.444	.561
5	.878	.959	22	.423	.537
6	.811	.917	24	.404	.515
7	.754	.875	26	.388	.496
8	.707	.834	28	.374	.479
9	.666	.798			
			30	.361	.463
10	.632	765	40	.312	.402
11	.602	.735	50	.279	.361
12	.576	.708	60	.254	.330
13	.553	.684	80	.220	.286
14	.532	.661			
			100	.196	.256
15	.514	.641	250	.124	.163
16	.497	.623	500	.088	.115
17	.482	.606	1000	.062	.081
18	.468	.590			
19	.456	.575			

Adapted from Dixon, W. J., and J. Massey, Jr., *Introduction to Statistical Analysis*, 3rd ed. (New York: McGraw-Hill, 1969), p. 569. With permission.

TABLE B.8

Critical Values for the Wilcoxon Signed-Rank Test for n = 5 to 50

One-Sided	Two-Sided	n = 5	n = 6	n = 7	n = 8	n = 9	n = 10	n = 11	n = 12	n = 13	n = 14	n = 15	n = 16
$\alpha = .05$	$\alpha = .10$	1	2	4	6	8	11	14	17	21	26	30	36
$\alpha = .025$	$\alpha = .05$		1	2	4	6	8	11	14	17	21	25	30
$\alpha = .01$	$\alpha = .02$			0	2	3	5	7	10	13	16	20	24
$\alpha = .005$	$\alpha = .01$				0	2	3	5	7	10	13	16	19

One-Sided	Two-Sided	n = 17	n = 18	n = 19	n = 20	n = 21	n = 22	n = 23	n = 24	n = 25	n = 26	n = 27	n = 28
$\alpha = .05$	$\alpha = .10$	41	47	54	60	68	75	83	92	101	110	120	130
$\alpha = .025$	$\alpha = .05$	35	40	46	52	59	66	73	81	90	98	107	117
$\alpha = .01$	$\alpha = .02$	28	33	38	43	49	56	62	69	77	85	93	102
$\alpha = .005$	$\alpha = .01$	23	28	32	37	43	49	55	61	68	76	84	92

One-Sided	Two-Sided	n = 29	n = 30	n = 31	n = 32	n = 33	n = 34	n = 35	n = 36	n = 37	n = 38	n = 39
$\alpha = .05$	$\alpha = .10$	141	152	163	175	188	201	214	228	242	256	271
$\alpha = .025$	$\alpha = .05$	127	137	148	159	171	183	195	208	222	235	250
$\alpha = .01$	$\alpha = .02$	111	120	130	141	151	162	174	186	198	211	224
$\alpha = .005$	$\alpha = .01$	100	109	118	128	138	149	160	171	183	195	208

One-Sided	Two-Sided	n = 40	n = 41	n = 42	n = 43	n = 44	n = 45	n = 46	n = 47	n = 48	n = 49	n = 50
$\alpha = .05$	$\alpha = .10$	287	303	319	336	353	371	389	408	427	446	466
$\alpha = .025$	$\alpha = .05$	264	279	295	311	327	344	361	379	397	415	434
$\alpha = .01$	$\alpha = .02$	238	252	267	281	297	313	329	345	362	380	398
$\alpha = .005$	$\alpha = .01$	221	234	248	262	277	292	307	323	339	356	373

From Wilcoxon, F., and R. A. Wilcox, *Some Rapid Approximate Statistical Procedures* (Pearl River, N.Y.: Lederle Laboratories of the American Cyanamid Company, 1964), p. 28. Reproduced with permission of the American Cyanamid Company.

TABLE B.9
Critical Values for a Mann-Whitney Test.[a]

Critical values for a one-tailed test at $\alpha = 0.01$ (roman type) and $\alpha = 0.005$ (boldface type) and for a two-tailed test at $\alpha = 0.02$ (roman type) and $\alpha = 0.01$ (boldface type).

n_2 \ n_1	1	2	3	4	5	6	7	8	9	10	11	12	13	14	15	16	17	18	19	20
1	—[b]	—	—	—	—	—	—	—	—	—	—	—	—	—	—	—	—	—	—	—
2	—	—	—	—	—	—	—	—	—	—	—	—	0	0	0	0	0	0	1	1
													—	—	—	—	—	—	**0**	**0**
3	—	—	—	—	—	—	0	0	1	1	1	2	2	2	3	3	4	4	4	5
							—	—	**0**	**0**	**0**	**1**	**1**	**1**	**2**	**2**	**2**	**2**	**3**	**3**
4	—	—	—	—	0	1	1	2	3	3	4	5	5	6	7	7	8	9	9	10
					—	**0**	**0**	**1**	**1**	**2**	**2**	**3**	**3**	**4**	**5**	**5**	**6**	**6**	**7**	**8**
5	—	—	—	0	1	2	3	4	5	6	7	8	9	10	11	12	13	14	15	16
				—	**0**	**1**	**1**	**2**	**3**	**4**	**5**	**6**	**7**	**7**	**8**	**9**	**10**	**11**	**12**	**13**
6	—	—	—	1	2	3	4	6	7	8	9	11	12	13	15	16	18	19	20	22
				0	**1**	**2**	**3**	**4**	**5**	**6**	**7**	**9**	**10**	**11**	**12**	**13**	**15**	**16**	**17**	**18**
7	—	—	0	1	3	4	6	7	9	11	12	14	16	17	19	21	23	24	26	28
			—	**0**	**1**	**3**	**4**	**6**	**7**	**9**	**10**	**12**	**13**	**15**	**16**	**18**	**19**	**21**	**22**	**24**
8	—	—	0	2	4	6	7	9	11	13	15	17	20	22	24	26	28	30	32	34
			—	**1**	**2**	**4**	**6**	**7**	**9**	**11**	**13**	**15**	**17**	**18**	**20**	**22**	**24**	**26**	**28**	**30**
9	—	—	1	3	5	7	9	11	14	16	18	21	23	26	28	31	33	36	38	40
			0	**1**	**3**	**5**	**7**	**9**	**11**	**13**	**16**	**18**	**20**	**22**	**24**	**27**	**29**	**31**	**33**	**36**
10	—	—	1	3	6	8	11	13	16	19	22	24	27	30	33	36	38	41	44	47
			0	**2**	**4**	**6**	**9**	**11**	**13**	**16**	**18**	**21**	**24**	**26**	**29**	**31**	**34**	**37**	**39**	**42**
11	—	—	1	4	7	9	12	15	18	22	25	28	31	34	37	41	44	47	50	53
			0	**2**	**5**	**7**	**10**	**13**	**16**	**18**	**21**	**24**	**27**	**30**	**33**	**36**	**39**	**42**	**45**	**48**
12	—	—	2	5	8	11	14	17	21	24	28	31	35	38	42	46	49	53	56	60
			1	**3**	**6**	**9**	**12**	**15**	**18**	**21**	**24**	**27**	**31**	**34**	**37**	**41**	**44**	**47**	**51**	**54**
13	—	0	2	5	9	12	16	20	23	27	31	35	39	43	47	51	55	59	63	67
		—	**1**	**3**	**7**	**10**	**13**	**17**	**20**	**24**	**27**	**31**	**34**	**38**	**42**	**45**	**49**	**53**	**56**	**60**
14	—	0	2	6	10	13	17	22	26	30	34	38	43	47	51	56	60	65	69	73
		—	**1**	**4**	**7**	**11**	**15**	**18**	**22**	**26**	**30**	**34**	**38**	**42**	**46**	**50**	**54**	**58**	**63**	**67**
15	—	0	3	7	11	15	19	24	28	33	37	42	47	51	56	61	66	70	75	80
		—	**2**	**5**	**8**	**12**	**16**	**20**	**24**	**29**	**33**	**37**	**42**	**46**	**51**	**55**	**60**	**64**	**69**	**73**
16	—	0	3	7	12	16	21	26	31	36	41	46	51	56	61	66	71	76	82	87
		—	**2**	**5**	**9**	**13**	**18**	**22**	**27**	**31**	**36**	**41**	**45**	**50**	**55**	**60**	**65**	**70**	**74**	**79**
17	—	0	4	8	13	18	23	28	33	38	44	49	55	60	66	71	77	82	88	93
		—	**2**	**6**	**10**	**15**	**19**	**24**	**29**	**34**	**39**	**44**	**49**	**54**	**60**	**65**	**70**	**75**	**81**	**86**
18	—	0	4	9	14	19	24	30	36	41	47	53	59	65	70	76	82	88	94	100
		—	**2**	**6**	**11**	**16**	**21**	**26**	**31**	**37**	**42**	**47**	**53**	**58**	**64**	**70**	**75**	**81**	**87**	**92**
19	—	1	4	9	15	20	26	32	38	44	50	56	63	69	75	82	88	94	101	107
		0	**3**	**7**	**12**	**17**	**22**	**28**	**33**	**39**	**45**	**51**	**56**	**63**	**69**	**74**	**81**	**87**	**93**	**99**
20	—	1	5	10	16	22	28	34	40	47	53	60	67	73	80	87	93	100	107	114
		0	**3**	**8**	**13**	**18**	**24**	**30**	**36**	**42**	**48**	**54**	**60**	**67**	**73**	**79**	**86**	**92**	**99**	**105**

[a]Discussed in Section 13.4. To be significant for any given n_1 and n_2, obtained U must be *equal to* or *less than* the value shown in the table.

[b]Dashes in the body of the table indicate that no decision is possible at the stated level of significance.

TABLE B.9 (Continued)

Critical Values for a Mann-Whitney Test

Critical values for a one-tailed test at $\alpha = 0.05$ (roman type) and $\alpha = 0.025$ (boldface type) and for a two-tailed test at $\alpha = 0.10$ (roman type) and $\alpha = 0.05$ (boldface type).

n_2 \ n_1	1	2	3	4	5	6	7	8	9	10	11	12	13	14	15	16	17	18	19	20
1	—	—	—	—	—	—	—	—	—	—	—	—	—	—	—	—	—	—	0	0
	—	—	—	—	—	—	—	—	—	—	—	—	—	—	—	—	—	—	—	—
2	—	—	—	—	0	0	0	1	1	1	1	2	2	2	3	3	3	4	4	4
	—	—	—	—	—	—	—	**0**	**0**	**0**	**0**	**1**	**1**	**1**	**1**	**1**	**2**	**2**	**2**	**2**
3	—	—	0	0	1	2	2	3	3	4	5	5	6	7	7	8	9	9	10	11
	—	—	—	—	**0**	**1**	**1**	**2**	**2**	**3**	**3**	**4**	**4**	**5**	**5**	**6**	**6**	**7**	**7**	**8**
4	—	—	0	1	2	3	4	5	6	7	8	9	10	11	12	14	15	16	17	18
	—	—	—	**0**	**1**	**2**	**3**	**4**	**4**	**5**	**6**	**7**	**8**	**9**	**10**	**11**	**11**	**12**	**13**	**13**
5	—	0	1	2	4	5	6	8	9	11	12	13	15	16	18	19	20	22	23	25
	—	—	**0**	**1**	**2**	**3**	**5**	**6**	**7**	**8**	**9**	**11**	**12**	**13**	**14**	**15**	**17**	**18**	**19**	**20**
6	—	0	2	3	5	7	8	10	12	14	16	17	19	21	23	25	26	28	30	32
	—	—	**1**	**2**	**3**	**5**	**6**	**8**	**10**	**11**	**13**	**14**	**16**	**17**	**19**	**21**	**22**	**24**	**25**	**27**
7	—	0	2	4	6	8	11	13	15	17	19	21	24	26	28	30	33	35	37	39
	—	—	**1**	**3**	**5**	**6**	**8**	**10**	**12**	**14**	**16**	**18**	**20**	**22**	**24**	**26**	**28**	**30**	**32**	**34**
8	—	1	3	5	8	10	13	15	18	20	23	26	28	31	33	36	39	41	44	47
	—	**0**	**2**	**4**	**6**	**8**	**10**	**13**	**15**	**17**	**19**	**22**	**24**	**26**	**29**	**31**	**34**	**36**	**38**	**41**
9	—	1	3	6	9	12	15	18	21	24	27	30	33	36	39	42	45	48	51	54
	—	**0**	**2**	**4**	**7**	**10**	**12**	**15**	**17**	**20**	**23**	**26**	**28**	**31**	**34**	**37**	**39**	**42**	**45**	**48**
10	—	1	4	7	11	14	17	20	24	27	31	34	37	41	44	48	51	55	58	62
	—	**0**	**3**	**5**	**8**	**11**	**14**	**17**	**20**	**23**	**26**	**29**	**33**	**36**	**39**	**42**	**45**	**48**	**52**	**55**
11	—	1	5	8	12	16	19	23	27	31	34	38	42	46	50	54	57	61	65	69
	—	**0**	**3**	**6**	**9**	**13**	**16**	**19**	**23**	**26**	**30**	**33**	**37**	**40**	**44**	**47**	**51**	**55**	**58**	**62**
12	—	2	5	9	13	17	21	26	30	34	38	42	47	51	55	60	64	68	72	77
	—	**1**	**4**	**7**	**11**	**14**	**18**	**22**	**26**	**29**	**33**	**37**	**41**	**45**	**49**	**53**	**57**	**61**	**65**	**69**
13	—	2	6	10	15	19	24	28	33	37	42	47	51	56	61	65	70	75	80	84
	—	**1**	**4**	**8**	**12**	**16**	**20**	**24**	**28**	**33**	**37**	**41**	**45**	**50**	**54**	**59**	**63**	**67**	**72**	**76**
14	—	2	7	11	16	21	26	31	36	41	46	51	56	61	66	71	77	82	87	92
	—	**1**	**5**	**9**	**13**	**17**	**22**	**26**	**31**	**36**	**40**	**45**	**50**	**55**	**59**	**64**	**69**	**74**	**78**	**83**
15	—	3	7	12	18	23	28	33	39	44	50	55	61	66	72	77	83	88	94	100
	—	**1**	**5**	**10**	**14**	**19**	**24**	**29**	**34**	**39**	**44**	**49**	**54**	**59**	**64**	**70**	**75**	**80**	**85**	**90**
16	—	3	8	14	19	25	30	36	42	48	54	60	65	71	77	83	89	95	101	107
	—	**1**	**6**	**11**	**15**	**21**	**26**	**31**	**37**	**42**	**47**	**53**	**59**	**64**	**70**	**75**	**81**	**86**	**92**	**98**
17	—	3	9	15	20	26	33	39	45	51	57	64	70	77	83	89	96	102	109	115
	—	**2**	**6**	**11**	**17**	**22**	**28**	**34**	**39**	**45**	**51**	**57**	**63**	**69**	**75**	**81**	**87**	**93**	**99**	**105**
18	—	4	9	16	22	28	35	41	48	55	61	68	75	82	88	95	102	109	116	123
	—	**2**	**7**	**12**	**18**	**24**	**30**	**36**	**42**	**48**	**55**	**61**	**67**	**74**	**80**	**86**	**93**	**99**	**106**	**112**
19	0	4	10	17	23	30	37	44	51	58	65	72	80	87	94	101	109	116	123	130
	—	**2**	**7**	**13**	**19**	**25**	**32**	**38**	**45**	**52**	**58**	**65**	**72**	**78**	**85**	**92**	**99**	**106**	**113**	**119**
20	0	4	11	18	25	32	39	47	54	62	69	77	84	92	100	107	115	123	130	138
	—	**2**	**8**	**13**	**20**	**27**	**34**	**41**	**48**	**55**	**62**	**69**	**76**	**83**	**90**	**98**	**105**	**112**	**119**	**127**

From Kirk, R., *Introductory Statistics*, pp. 423–424. Copyright © 1978 by Wadsworth, Inc. Reprinted by permission of Brooks/Cole Publishing Company, Monterey, California 93940.

TABLE B.10

Critical Values for a Two-tailed Runs Test with $\alpha = .05$

n_1 \ n_2	5	6	7	8	9	10	11	12	13	14	15
2	*	*	*	*	*	*	*	2 6	2 6	2 6	2 6
3	*	2 8	2 8	2 8	2 8	2 8	2 8	2 8	2 8	2 8	3 8
4	2 9	2 9	2 10	3 10	3 10	3 10	3 10	3 10	3 10	3 10	3 10
5	2 10	3 10	3 11	3 11	3 12	3 12	4 12	4 12	4 12	4 12	4 12
6	3 10	3 11	3 12	3 12	4 13	4 13	4 13	4 13	5 14	5 14	5 14
7	3 11	3 12	3 13	4 13	4 14	5 14	5 14	5 14	5 15	5 15	6 15
8	3 11	3 12	4 13	4 14	5 14	5 15	5 15	6 16	6 16	6 16	6 16
9	3 12	4 13	4 14	5 14	5 15	5 16	6 16	6 16	6 17	7 17	7 18
10	3 12	4 13	5 14	5 15	5 16	6 16	6 17	7 17	7 18	7 18	7 18
11	4 12	4 13	5 14	5 15	6 16	6 17	7 17	7 18	7 19	8 19	8 19
12	4 12	4 13	5 14	6 16	6 16	7 17	7 18	7 19	8 19	8 20	8 20
13	4 12	5 14	5 15	6 16	6 17	7 18	7 19	8 19	8 20	9 20	9 21
14	4 12	5 14	5 15	6 16	7 17	7 18	8 19	8 20	9 20	9 21	9 22
15	4 12	5 14	6 15	6 16	7 18	7 18	8 19	8 20	9 21	9 22	10 22

From Owen, D. B., *Handbook of Statistical Tables* (Reading, Mass.: Addison-Wesley, 1962) as adapted in Weiss, N. and M. Hassett, *Introductory Statistics* (Reading, Mass.: Addison-Wesley, 1982), p. 594. Reprinted with permission.

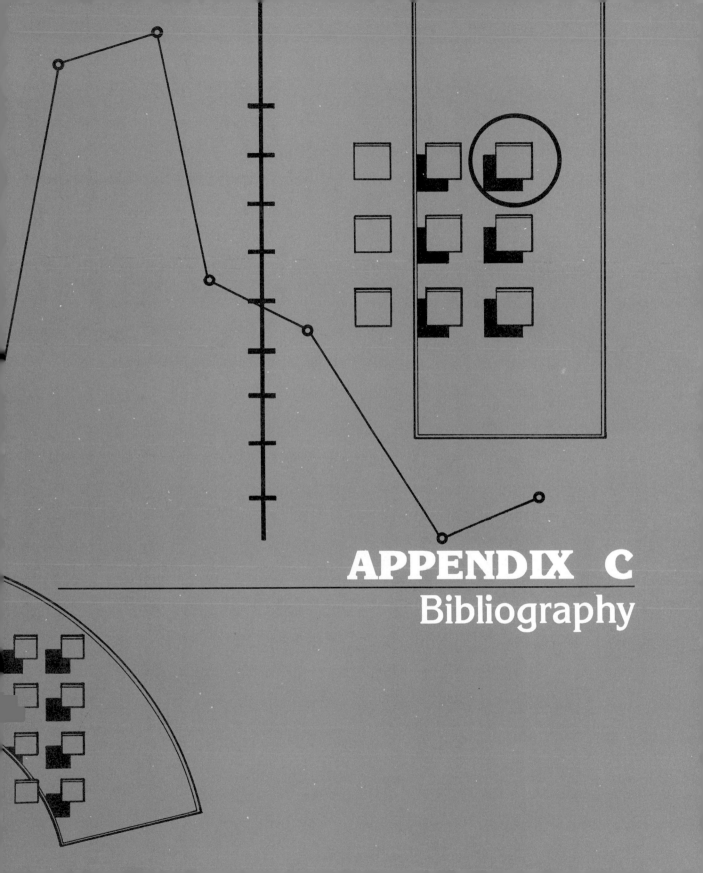

APPENDIX C
Bibliography

1. Berenson, M., D. Levine, and M. Goldstein, *Intermediate Statistical Methods and Applications* (Englewood Cliffs, N.J.: Prentice-Hall, 1983).

2. Bradley, J., *Distribution-Free Statistical Tests* (Englewood Cliffs, N.J.: Prentice-Hall, 1968).

3. Conover, W., *Practical Nonparametric Statistics* (New York: John Wiley, 1971).

4. Daniel, W., *Applied Nonparametric Statistics* (Boston: Houghton Mifflin, 1978).

5. Draper, N., and H. Smith, *Applied Regression Analysis,* 2nd ed. (New York: John Wiley, 1981).

6. Ehrenberg, A., *A Primer in Data Reduction* (New York: John Wiley, 1982).

7. Fairley, W., and F. Mosteller, *Statistics and Public Policy* (Reading, Mass.: Addison-Wesley, 1977).

8. Freedman, D., R. Pisani, and R. Purves, *Statistics* (New York: W. W. Norton, 1978).

9. Haack, D., *Statistical Literacy: A Guide to Interpretation* (North Scituate, Mass.: Duxbury, 1979).

10. Hawkins, C., and J. Weber, *Statistical Analysis* (New York: Harper & Row, 1980).

11. Hollander, M., and D. Wolfe, *Nonparametric Statistical Methods* (New York: John Wiley, 1973).

12. Huff, D., *How to Lie with Statistics* (New York: W. W. Norton, 1954).

13. Kirk, R., *Statistical Issues: A Reader for the Behavioral Sciences* (Monterey, Calif.: Brooks/Cole, 1972).

14. Kleinbaum, D., and L. Kupper, *Applied Regression Analysis and Other Multivariable Methods* (North Scituate, Mass.: Duxbury, 1978).

15. Koopmans, L., *An Introduction to Contemporary Statistics* (Boston: Duxbury, 1981).

16. Loftus, G., and E. Loftus, *Essence of Statistics* (Monterey, Calif.: Brooks/Cole, 1982).

17. Marascuilo, L., and M. McSweeney, *Nonparametric and Distribution-Free Methods for the Social Sciences* (Monterey, Calif.: Brooks/Cole, 1977).

18. Meyers, L., and N. Grossen, *Behavioral Research: Theory, Procedure, and Design* (San Francisco: Freeman, 1978).

19. Moore, D., *Statistics: Concepts and Controversies* (San Francisco: Freeman, 1979).

20. Morrison, D., *Applied Linear Statistical Methods* (Englewood Cliffs, N.J.: Prentice-Hall, 1983).

21. Mosteller, F., and R. Rourke, *Sturdy Statistics* (Reading, Mass.: Addison-Wesley, 1973).

22. Mosteller, F., R. Rourke, and G. Thomas, *Probability with Statistical Applications,* 2nd ed. (Reading, Mass.: Addison-Wesley, 1970).

23. Neter, J., W. Wasserman, and M. Kutner, *Applied Linear Statistical Models,* 2nd ed. (Homewood, Ill.: Richard D. Irwin, 1985).

24. Neter, J., W. Wasserman, and M. Kutner, *Applied Linear Regression Models* (Homewood, Ill.: Richard D. Irwin, 1983).

25. Neter, J., W. Wasserman, and G. Whitmore, *Applied Statistics* (Boston: Allyn and Bacon, 1978).

26. Phillips, J., *Statistical Thinking* (San Francisco: Freeman, 1982).

27. Ryan, B., B. Joiner, and T. Ryan, *Minitab Handbook,* 2nd ed. (Boston: Duxbury, 1985).

28. Snedecor, G., and W. Cochran, *Statistical Methods,* 7th ed. (Ames, Iowa.: The Iowa State University Press, 1980).

29. Tanur, J., ed. *Statistics: A Guide to the Unknown* (San Francisco: Holden-Day, 1972).

30. Tufte, E., *Data Analysis for Politics and Policy* (Englewood Cliffs, N.J.: Prentice-Hall, 1974).

31. Tukey, J., *Exploratory Data Analysis* (Reading, Mass.: Addison-Wesley, 1977).

32. Velleman, P., and D. Hoaglin, *Applications, Basics, and Computing of Exploratory Data Analysis* (Boston: Duxbury, 1981).

33. Weisberg, H., and B. Bowen, *An Introduction to Survey Research and Data Analysis* (San Francisco: Freeman, 1977).

34. Wheeler, M., *Lies, Damn Lies, and Statistics* (New York: Dell, 1976).

35. Wonnacott, T., and R. Wonnacott, *Introductory Statistics* (New York: John Wiley, 1977).

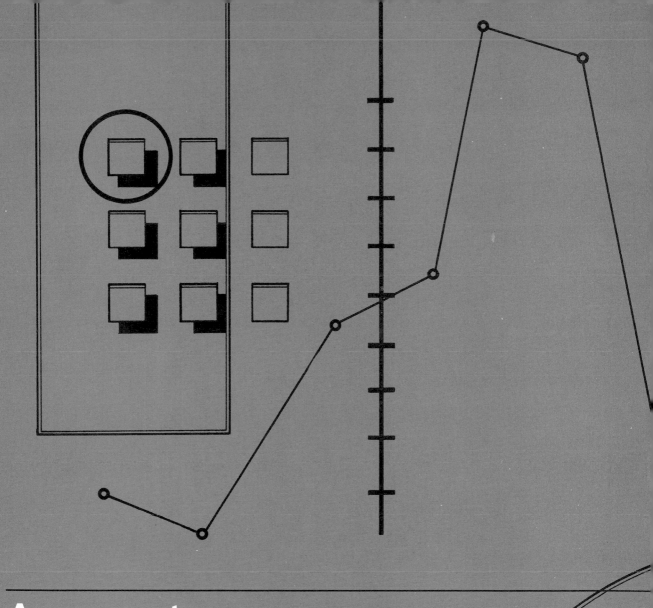

Answers to
Selected Exercises

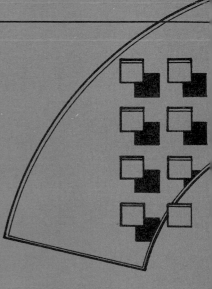

Chapter 1

1.1   A sample is a subset of a population; hence the sample and the population have common elements. However, a population may contain elements that are not part of a sample.

1.3   True   **1.5**   False   **1.7**   Inferential

1.9   Descriptive   **1.11**   Statistic   **1.13**   Parameter

1.15   **(a)**   The grade-point averages of the 750 seniors

   (b)   The average GPA of the 750 seniors   **(c)**   2.81

   (d)   No—the two samples would consist of different data values, and it would be unlikely that the average of the data values would be the same. The value of a parameter remains unchanged from sample to sample.

1.17   **(a)**   All families in the town; 1000   **(b)**   All families in the sample; 50

   (c)(i)   440   **(ii)**   430   **(d)(i)** $\frac{36}{50}$ **(ii)** $\frac{870}{1000}$

1.19   {a, b, c}   {a, b, d}   {a, c, d}   {b, c, d}

1.21   **(a)**   {00, 10, 20, 30, 40, 50, 60, 70, 80, 90}

   (b)(i)   {10, 37, 08, 99, 12, 66, 31, 85, 63, 73}

   (ii)   {08, 10, 12, 37, 31, 44, 99, 66, 85, 63}

   (iii)   {10, 12, 11, 15, 23, 19, 18, 26, 17, 20}

Chapter 2

2.1   **(a)**

x	1	4	6	9	10	12
f	5	5	3	1	2	4

(b)   Use 20 distinct data values.

2.3   **(a)**

x	20	22	24	25	26	28	30	31	32	33
f	1	3	1	2	1	1	1	1	1	1
x	34	36	39	40	41	42	48	50	53	
f	1	1	2	1	1	2	1	1	2	

(b)   36

2.5   **(a)**

x	22	32	33	36	37	39	40	41	42
f	1	6	2	1	·3	3	3	1	4
x	43	44	45	46	48	49	52	61	
f	3	3	2	1	1	4	1	1	

(b)

Class	Class Limits	Frequency
1	22–26	1
2	27–31	0
3	32–36	9
4	37–41	10
5	42–46	13
6	47–51	5
7	52–56	1
8	57–61	1

2.7 **(a)**

Class	Class Limits	Frequency
1	20–26	8
2	27–33	5
3	34–40	5
4	41–47	3
5	48–54	4

(b) 37 **(c)** 33.5 **(d)** .12

2.9

Class Limits	Class Boundaries	Class Marks
61–63	60.5–63.5	62
64–66	63.5–66.5	65
67–69	66.5–69.5	68
70–72	69.5–72.5	71
73–75	72.5–75.5	74
76–78	75.5–78.5	77
79–81	78.5–81.5	80
82–84	81.5–84.5	83

2.11 **(a)**

Class	Class Boundaries	Frequency	Cumulative Frequency
1	189.75–196.75	5	5
2	196.75–203.75	2	7
3	203.75–210.75	9	16
4	210.75–217.75	4	20
5	217.75–224.75	6	26
6	224.75–231.75	3	29
7	231.75–238.75	1	30

(b) 207.25 **(c)** 7

2.13 **(a)**

Class	Class Boundaries	Frequency	Cumulative Relative Frequency
1	152.5–171.5	1	.029
2	171.5–190.5	3	.114
3	190.5–209.5	6	.286
4	209.5–228.5	5	.429
5	228.5–247.5	7	.629
6	247.5–266.5	3	.714
7	266.5–285.5	5	.857
8	285.5–304.5	5	1.000

(b) 219 **(c)** 19 **(d)** .171 **(e)** 62.9

2.15 **(a)** **(b)**

2.17 (a)

(b)

(c)

(d) 5

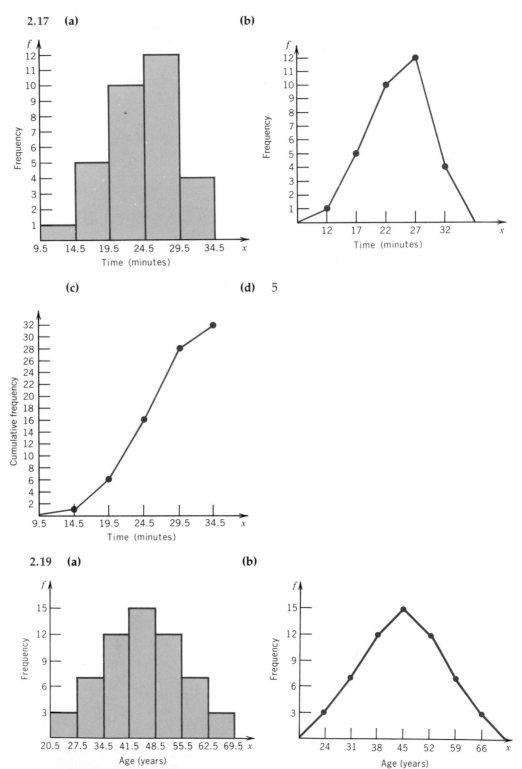

2.19 (a)

(b)

(c) **(d)** 7

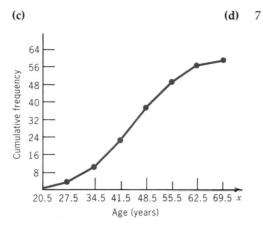

2.21 **(a)** **(b)**

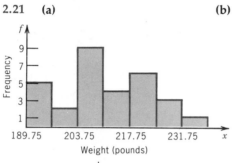

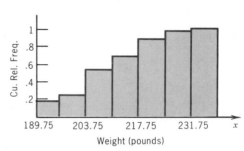

(c)

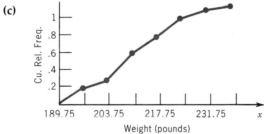

2.23 **(a)** 25 **(b)** 2–3 **(c)** Fourth **(d)** 5 **(e)** 0

2.25 **(a)** **(b)**

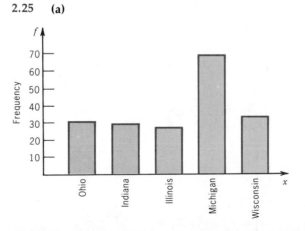

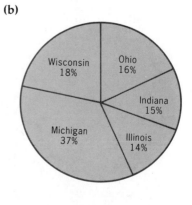

2.27   Skewed to left   Skewed to right   Bell shaped   Uniform

2.29   For example, 2|3 represents the value 23.   Skewed to the right

```
 0 | 3  6  6  7  7  8  8  9  9
 1 | 0  0  1  2  2  4  7  9  9  9
 2 | 0  3  3  3  4  5  5  6  6  7  8  8
 3 | 2  5  5  6  7
 4 | 5  6  7  8
 5 | 1  2  4
 6 |
 7 | 0  2  3
 8 | 7
 9 | 3
10 | 8
11 | 8
```

2.31   For example, 21|49 represents the value 214.9.   Roughly mound shaped

```
18 | 98
19 | 84  42  29  64  59
20 | 40  51  76  82  84  21  38  89  63
21 | 49  59  95  29  06  69
22 | 26  28  22  00  24  80  85
23 | 09  59
```

2.33   (a)

x	4.1	4.3	4.4	4.5	4.6	4.7	4.8	4.9	5.1	5.5	5.7
f	1	3	1	3	2	4	1	1	2	1	1

   (b)   80   (c)

2.35   Not worthwhile. The frequency of most data values is one.

2.37

Class	Class Limits	Frequency
1	0–6	3
2	7–13	4
3	14–20	4
4	21–27	5
5	28–34	4

2.39

Class	Class Limits	Frequency
1	19–29	4
2	30–40	6
3	41–51	2
4	52–62	2
5	63–73	4
6	74–84	2

2.41 (a)

Class Boundaries	Frequency	Relative Frequency	Cumulative Frequency
23.5–26.5	10	.05	10
26.5–29.5	30	.15	40
29.5–32.5	12	.06	52
32.5–35.5	47	.235	99
35.5–38.5	50	.25	149
38.5–41.5	28	.14	177
41.5–44.5	5	.025	182
44.5–47.5	18	.09	200

(b)

(c)(i) .505 (ii) .235

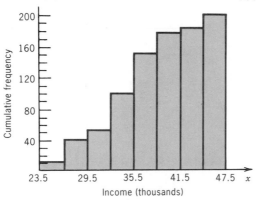

2.43 (b) $\frac{1}{10}$ Uniform

2.45 (a)

(b)

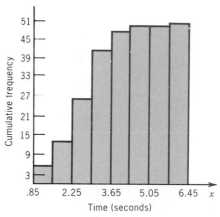

2.47 (a)

(b)

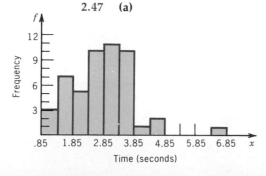

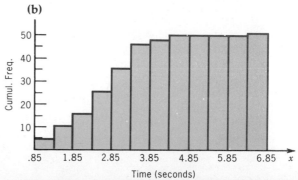

2.49   **(a)**

Class Limits	Frequency	Cumulative Frequency	Cumulative Relative Frequency
24–213	22	22	.431
214–403	18	40	.784
404–593	6	46	.902
594–783	3	49	.961
784–973	1	50	.980
974–1163	1	51	1.000

(b)

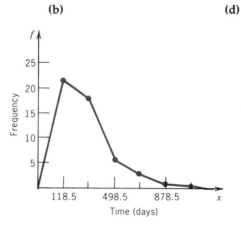

(d)

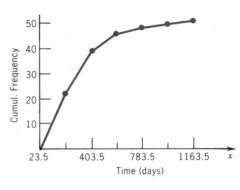

(c)   Skewed to the right   **(e)**   21.6

2.51   **(a)**   7   **(b)**   2   **(c)**   42   **(d)**   66–67   **(e)**   6   **(f)**   69
(g)   83

2.53   **(a)**   **(b)**

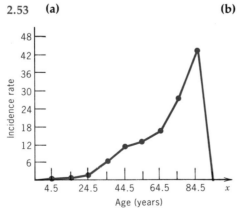

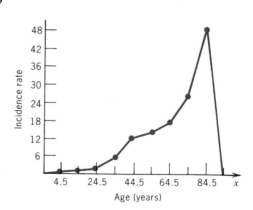

(c)   The graphs are similar. Incidence rate is a proportion. We would expect the proportion of people admitted for any particular age group to be about the same in both states.

2.57   **(a)**   376   **(b)**   321   **(c)**   389

2.59 **(a)**

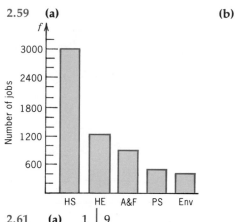

(b)

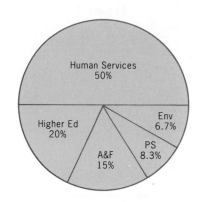

2.61 **(a)**

1	9
2	9 6 4
3	0 1 2 0 8
4	7 7 0
5	3 5
6	3 6 8
7	1 4
8	2

(c)

```
             8
             0
     4   2   0       8
     6   1   7   5   6   4
 9   9   0   7   3   3   1   2
 ─────────────────────────────
 1   2   3   4   5   6   7   8
```

Yes

(b)

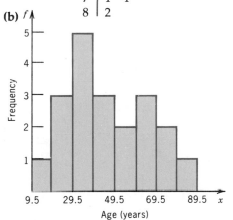

2.63   Sequence of line segments
                    form a straight line

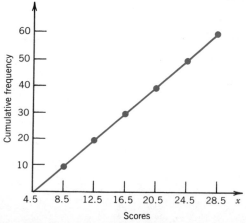

2.65 (a)

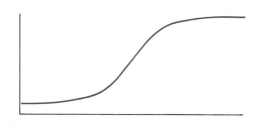

(b)

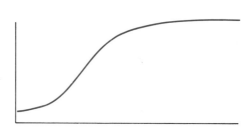

(c)

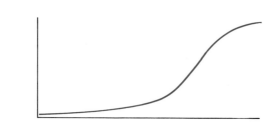

2.67 (a)   (b)

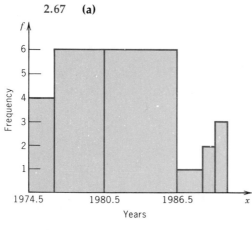

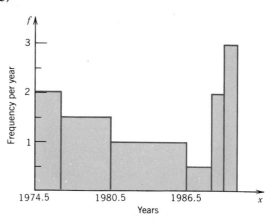

Chapter 3

3.1	(a)	8	(b)	3.5	(c)	8	(d)	12
3.3	(a)	3.2	(b)	3	(c)	4	(d)	0

3.5 **(a)** 7.89 **(b)** 5.5 **(c)** 7.6 **(d)** 7.6

3.7 **(a)** 25.5 **(b)** 3.5 **(c)** 24.95 **(d)** No mode

3.9 **(a)** 15 **(b)** 17.5

3.11 The student's reasoning is flawed. The data are not ranked.

3.13 **(a)** 42 No **(b)** 131 No

3.15 Approximately 890

3.17 **(a)** True **(b)** False **(c)** False **(d)** False

3.19 **(a)** 2.5 **(b)** 2.5 **(c)** No mode **(d)** Symmetric
    (e) Mean or median

3.21 **(a)** Mean **(b)** Median

3.23 **(a)** 20 **(b)** 63.6 **(c)** 7.97

3.25 **(a)** 232 **(b)** 5767.33 **(c)** 75.94

3.27 **(a)** No **(b)** Yes

3.29 **(a)(i)** **(ii)**

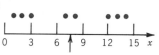 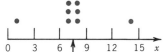

(c)(i) 5.15 **(ii)** 3.51

3.31 **(a)** 49 **(b)** 4 **(c)** 4.56

3.33 **(a)** 17.5 **(b)** 7.03

3.35 **(a)** 75.96 **(b)** 5.83

3.37 60 95 100

3.39 **(a)** .46 **(b)** 7.79 **(c)** 1.15 **(d)** 76

3.41 **(a)** 128;393 **(b)** 249 **(c)** 697 **(d)** 19.6

3.43 **(a)** 51 55 58 **(b)** 61 **(c)** 75.6

3.45 **(a)(i)** −.67 **(ii)** .67 **(iii)** −2 **(iv)** 1.67
    (b)(i) 10 **(ii)** 55 **(iii)** 77.5 **(iv)** 62.5 **(v)** 40

3.47

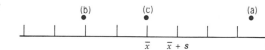

3.49 **(a)** Statistics **(b)** 87

3.51

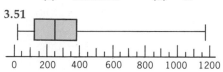

Skewed to the right

3.53

Bell-shaped

(a) 9 **(b)** 9.5 **(c)** 11 **(d)** 13 **(e)** 15.45
(f) 3.93 **(g)** 11.5 **(h)** 41.7 **(i)** −.51 **(j)** 2
(a) 18.67 **(b)** 21 **(c)** No mode **(d)** Median

3.59 **(a)**   Skewed to the right   Median   Mean   **(b)**   443,559,231
      (c)   227,030,000   **(d)**   132,130,000   **(e)**   115,522,500 No

3.61 **(a)** $\bar{x} = 2(\frac{1}{10}) + 3(\frac{6}{10}) + 5(\frac{2}{10}) + 10(\frac{1}{10}) = 4$ **(b)**   4

3.63 31

3.65 **(a)**   727   880   1124   **(b)**   545   1289
      (c)(i)   50   **(ii)**   68.06

3.67 45.68   12.80

3.69 **(a)(i)**   2   **(ii)** -2.5 **(iii)**   .75   **(iv)** -1
      (b)(i)   91   **(ii)**   87   **(iii)**   82   **(iv)**   93

3.71 Albany: 50.3   Reno: 37.3

3.73 Albany: 18.292   Reno: 13.245

3.77 **(a)**   2.016   **(b)** -1.453

3.79 **(a)**   Buffalo   **(b)**   Portland

3.81 **(a)**   25   **(b)**   75   **(c)**   25   **(d)**   72.5

3.82 **(a)**   20.4   **(b)**   20.4   **(c)**   81.6   **(d)** $s^2 \cdot k^2$

3.83 31.5

3.85 74   96   100

3.87 **(a)** 889   **(b)** 938   **(c)** 250

Chapter 4

4.1   **(a)**   {O, A, B, AB}
      (b)   {O-O, O-A, O-B, O-AB, A-O, A-A, A-B, A-AB, B-O, B-A, B-B, B-AB, AB-O, AB-A, AB-B, AB-AB}

4.3   **(a)**   {4260, 4261, 4262, 4263, 4264, 4265, 4266, 4267, 4268, 4269}
      (b)   {2468, 2648, 4268, 4628, 6248, 6428}

4.5   {xX, xY, XX, XY}

4.7 S = {0A, 0B, 0C, 0D, 0E, 1A, 1B, 1C, 1D, 1E}   **(a)**   {0D, 0E}
      (b)   {1A, 1B, 1C, 1D}
      (c)   {0A, 0B, 1A, 1B}   **(d)**   {1A, 1B, 1C, 1D, 1E}

4.9   **(a)**   {M1, M2, M3, F1, F2, F3}
      (b)   {F1, F2, F3, F4, F5, F6, M4, M5, M6}   **(c)**   {M6}   **(d)**   {M3}

4.10 **(a)** S = {M, F} with M = male and F = female   **(b)** $\frac{20}{31}$

4.11 **(a)** $\frac{6}{36}$ **(b)** $\frac{11}{36}$ **(c)** $\frac{25}{36}$ **(d)** $\frac{8}{36}$

4.13 **(a)**   {0, 1, 2, 3, 4, 5, 6, 7, 8, 9}   **(b)**   {1, 3, 5, 7, 9} $\frac{5}{10}$
      (c)   {7, 8, 9} $\frac{3}{10}$

4.15 **(a)** $\frac{6}{16}$ **(b)** $\frac{4}{16}$ **(c)** $\frac{1}{16}$ **(d)** $\frac{15}{16}$

4.17 **(a)** $\frac{582}{2809}$ **(b)** $\frac{1042}{2809}$

4.19 **(a)**   A   B   **(b)(i)** $\frac{16}{52}$ **(ii)** $\frac{15}{52}$ **(iii)** $\frac{22}{52}$

4.21 $\frac{3900}{4000}$

4.23 (a) $\frac{1100}{2000}$ (b) $\frac{1200}{2000}$ (c) $\frac{100}{2000}$ (d) $\frac{1800}{2000}$ (e) $\frac{1300}{2000}$

4.24 (a)(i) $\frac{12}{2652}$ (ii) $\frac{204}{2652}$ (iii) $\frac{396}{2652}$ (b) No

4.25 (a)(i) $\frac{16}{2704}$ (ii) $\frac{4}{52}$ (iii) $\frac{400}{2704}$ (b) $\frac{768}{140,608}$

4.27 Yes **4.29** $\frac{2}{42}$ **4.31** $\frac{30}{70}$

4.33 (a) $\frac{10}{16}$ (b) $\frac{21}{39}$ (c) $\frac{12}{19}$ (d) $\frac{21}{70}$

4.35 (a) $\frac{430}{840}$ (b) $\frac{100}{510}$ (c) No

4.37 (a) $\frac{1}{2}$ (b) $\frac{1}{3}$ **4.38** $\frac{34}{90}$

4.39 $\frac{91}{216}$ **4.41** .096 **4.43** (a) 20 (b) 160

4.44 (a) 720 (b) 1320 (d) 220

4.45 (a) 6; {a, b} {a, c} {a, d} {b, c} {b, d} {c, d}

(b) 12; ab ba ac ca ad
da bc cb bd db cd dc

4.47 17,576,000 **4.49** (a) $(10)^8 \cdot (9)^2$ (b) $(10)^6 \cdot (9)$ (c) $(10)^4$

4.51 (a) $\frac{1}{10^4}$ (b) $\frac{1}{10^4}$ (c) $\frac{10^3}{10^4}$ (d) $\frac{60}{10^4}$

4.53 (a) .0417 (b) .0033

4.55 $S = \{$R1, R2, R3, R4, B1, B2, B3, B4, G1, G2, G3, G4$\}$

4.57 (a) {cc} (b) {Cc, cC}

4.59 (a) {1ax, 2ax, 3ax, 4ax, 5ax, 1bx, 2bx, 3bx, 4bx, 5bx}

(b) {1ax, 1ay, 3ax, 3ay} (c) {1bx, 2bx, 3bx, 4bx, 5bx}

4.61 (a) $\frac{417}{1000}$ (b) $\frac{665}{1000}$ (c) $\frac{752}{1000}$

4.63 (a) $\frac{197,600}{9,366,819}$ (b) $\frac{7,786,428}{9,366,819}$ (c) 13

4.65 (a) $\frac{3}{16}$ (b) $\frac{6}{16}$ (c) $\frac{1}{16}$ (d) $\frac{7}{16}$

4.67 (a) $\frac{6}{100}$ (b) $\frac{44}{100}$ (c) $\frac{38}{100}$

4.69 (a) $\frac{1}{9}$ (b) $\frac{3}{9}$ (c) $\frac{2}{9}$

4.71 (a) $\frac{3}{10}$ (b) $\frac{63}{100}$ (c) $\frac{97}{100}$

4.73 (a) $\frac{495}{1000}$ (b) $\frac{330}{1000}$ (c) $\frac{695}{1000}$

4.75 (a) $\frac{60}{150}$ (b) $\frac{85}{150}$ (c) $\frac{40}{50}$ (d) $\frac{40}{85}$ (e) $\frac{105}{150}$ (f) No No

4.77 (a) $\frac{20}{48}$ (b) $\frac{42}{48}$

4.79 $\frac{34}{54}$ **4.81** 72 **4.83** 20,358,520

4.85 (a) $\frac{1}{5040}$ (b) $\frac{24}{5040}$ (c) $\frac{120}{5040}$ (d) $\frac{720}{5040}$

Chapter 5

5.1 (a) Discrete (b) Continuous (c) Continuous

(d) Discrete (e) Continuous

5.3 (a) 3 4 5 6 7 8 9 10 11 12 13 14 15 16 17 18

(b) 1 2 3 4 5 6 (c) 2 4 6 8 10 12

5.5 (a) Nonnegative numbers (b) Continuous

5.7 (a) 0, 1, 2, . . . , 500 (b) Discrete

5.9

x	25,000	25,700	26,600	27,200	27,300	27,800	28,400	29,000
P(x)	1/15	2/15	1/15	1/15	4/15	2/15	2/15	2/15

5.11

x	70	72	73	74	75	76	77	78	79	80
P(x)	1/25	1/25	2/25	4/25	4/25	3/25	2/25	3/25	3/25	2/25

5.13 **(a)**

x	0	1
P(x)	.466	.534

(b)

x	0	1
P(x)	.583	.417

5.15 No: $P(x)$ cannot be negative   **5.17** Yes   **5.19** No: $\sum P(x) \neq 1$

5.21

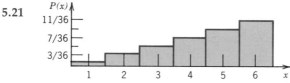

5.23

x	0	1	2	3
P(x)	17/50	19/50	6/50	8/50

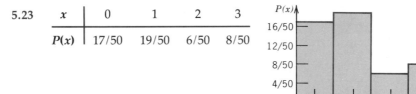

5.25

x	0	1	2	3
P(x)	1/6	2/6	2/6	1/6

5.27 **(a)** 2   **(b)** 1   **(c)** 1   **(d)** Skewed to the left

5.29 **(a)** $\frac{3}{2}$ **(b)** $\frac{3}{4}$ **(c)** $\dfrac{\sqrt{3}}{2}$ **(d)** Symmetric

5.31 **(a)** 7   **(b)** 5.83   **(c)** 2.42

5.33 **(a)** 2.65   **(b)** 1.03   **(c)** 1.01

5.35 **(a)(i)** 4.5   **(ii)** 2.87

5.37 **(a)** 56.3 cents   **(b)** $5.63

5.39 −86 cents

5.41 **(a)** Yes
(b) No The number of trials is not determined in advance.
(c) No   The trials are dependent.   **(d)** Yes
(e) No   The trials are dependent.

5.43

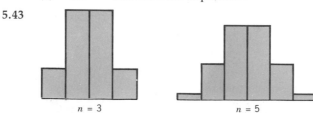

5.45 (a) .026 (b) .154 (c) .346

5.47 (a) .031 (b) .217 (c) .461 (d) .147 (e) .996

5.49 .264

5.51 (a) .117 (b) .249 (c) .127

5.53 (a) .177 (b) .090 (c) .973

5.55 (a) 4 (b) $\frac{16}{5}$ (c) 1.79

5.57 (a) 4 (b) $\frac{8}{3}$ (c) 1.63

5.59 (a) 3.475 (b) 3.354 (c) 1.832

5.61 (a) 62.37 (b) 7.26 (c) 6.52 (d) 23.09 4.67 1.73

(e) Probably not

5.63 (a) $\frac{20}{6}$ (b) $\frac{20}{36}$ (c) .75

5.65 400 $\frac{1}{4}$ **5.69** Yes **5.71** No, because $P(4)$ is less than zero

5.73 (a) 4 (b) 16.33 (c) 4.04

5.75 (a) $\frac{3}{12}$ (b) 2.75 (c) 2.52 (d) 1.59

5.77 $\frac{1}{3}$

5.79 (a)

x	0	1	2	3
$P(x)$	.729	.243	.027	.001

(b) .3 (c) .27 (d) .52

5.81 (a) 1 (b) 1.0004 (c) 1.0002 (d)

(e) .6321

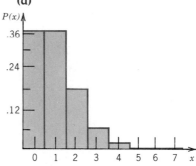

5.83 (a) .8 (b) .48 (c) .69 (d)

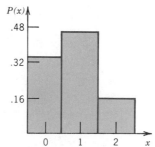

5.85 −2.7 cents **5.87** $2

5.89 (a) .312 (b) .016 (c) .032 (d) .984

5.91 (a) .082 (b) .004 (c) .029

5.93 (a) .444 (b) .027

| 5.95 | **(a)** | 4.5 | **(b)** | 3.15 | **(c)** | 1.77 |
| 5.97 | **(a)** | 4.25 | **(b)** | .64 | **(c)** | .80 |

Chapter 6

| 6.1 | **(a)** $\frac{1}{4}$ | **(b)** $\frac{3}{4}$ | **(c)** $\frac{7}{10}$ | **(d)** $\frac{9}{10}$ | **(e)** 1 | **(f)** 0 |

6.3   **(a)** .0456   **(b)** .5   **(c)** .1359   **(d)** .8413   **(e)** .4772
      (f) .6826   **(g)** .9544

6.5   **(a)** 5   **(b)** $\frac{1}{16}$ **(c)** $\frac{1}{2}$ **(d)** $\frac{55}{64}$

6.7   **(a)** .2486   **(b)** .0316   **(c)** .9375   **(d)** .9686
      (e) .0456   **(f)** .9544

6.9   **(a)** 1.72   **(b)** -1.24 **(c)** 1.41
      (d) -2.45 **(e)** 2.05   **(f)** $-.67$

6.10   **(a)** 2.33   **(b)** -1.65 **(c)** -1.28
      (d) 2.58   **(e)** .84

6.11   A = .0287   B = .6628   C = .2417   D = .0668

6.13   **(a)** 50   **(b)** 15.87   **(c)** 93.32   **(d)** 2.28
      (e) 77.34   **(f)** 37.21   **(g)** 13.59   **(h)** 50.62
      (i) 86.64   **(j)** 2.14

6.15   **(a)** 216.75   **(b)** 190.25   **(c)** 67   **(d)** 2.28

6.17   a = 75   b = 93

6.19   **(a)** .1056   **(b)** .3372   **(c)** .7492

6.21   **(a)** 2.28   **(b)** 54.06

6.23   **(a)** 789.6   **(b)** 59.52

6.25   **(a)** .047   **(b)** .0485

6.27   **(a)** .192   **(b)** .1896

6.29   **(a)** .203   **(b)** .2059

6.31   .0104

6.33   **(a)** .0013   **(b)** The data do not support the board's claim.

6.35   If the chances of a boy or girl being born were equally likely, only about 8 in 10,000 times would 120 or more girls be born out of 195 births. The administrator's claim appears justified.

6.37   **(a)**

$\bar{x}$	0	1	2	3	4
$P(\bar{x})$	1/9	2/9	3/9	2/9	1/9

      (b)(i) 2   **(ii)** 4/3   **(c)(i)** 2   **(ii)** 8/3

6.39   **(a)** $\mu_{\bar{x}} = 325$ $\sigma_{\bar{x}} = 2$ **(b)** .1587   **(c)** .0606
      (d) .0668   **(e)** .8185

6.41   **(a)** 200   **(b)** 5   **(c)** .6826   **(d)** .0228

6.43   .015

6.45   **(a)** .8413   **(b)** .0062   **(c)** 5.767

6.47 **(a)** .2835 **(b)** .1354 **(c)** .3486 **(d)** .7165
(e) .8545

6.49 **(a)** $\frac{3}{4}$ **(b)** $\frac{5}{16}$ **(c)** $\frac{1}{16}$ **(d)** $\frac{7}{16}$

6.51 **(a)** 1 **(b)** .0497 **(c)** .2325 **(d)** .6065
(e) .6322

6.53 **(a)** 1.96 **(b)** .67 **(c)** 0 **(d)** −1.04

6.55 **(a)** 154.4 **(b)** 213 **(c)** 69.15 **(d)** 2.28

6.57 **(a)** 78.81 **(b)** 201.25 minutes

6.59 **(a)** .0062 **(b)** .4938 **(c)** .8944

6.61 **(a)** .166 **(b)** .1635

6.63 **(a)** .196 **(b)** .1985

6.65 **(a)** .387 **(b)** .3859

6.67 **(a)** .1711 **(b)** .9596 **(c)** .9978 **(d)** .0606
(e) .0398 **(f)** .0242

6.69 .6772

6.71 **(a)** .8413 **(b)** .8164 **(c)** .9544 **(d)** .0228

6.73 **(a)** 4.5 8.25
(b)(i) 4.5 .23 **(ii)** 4.5 .17 **(iii)** 4.5 .08
(c)(i) .7016 **(ii)** .7776 **(iii)** .9182
(d)(i) .3015 **(ii)** .2709 **(iii)** .1922

Chapter 7

7.1 **(a)(i)** 3.92 **(ii)** $121.08 < \mu < 128.92$
(b)(i) 6.45 **(ii)** $199.55 < \mu < 212.45$
(c)(i) .55 **(ii)** $153.45 < \mu < 154.55$
(d)(i) 6.53 **(ii)** $302.47 < \mu < 315.53$
(e)(i) 2.58 **(ii)** $37.42 < \mu < 42.58$
(f)(i) .83 **(ii)** $77.17 < \mu < 78.83$

7.3 **(a)** $25.5 < \mu < 26.5$ **(b)** 198 **7.5** $.63 < \mu < .81$

7.7 **(a)** $63.43 < \mu < 66.57$ **(b)** 1.57 **(c)** Yes

7.9 **(a)** $\$17.88 < \mu < \$18.72$ **(b)** $\$17.66 < \mu < \$18.94$
(c) Longer Shorter

7.11 Match 10 ± 3, 10 ± 4.5, 10 ± 5, 10 ± 8 with 80%, 90%, 95%, and 99%, respectively.

7.13 **(a)** $39.86\% < \mu < 43.74\%$ **(b)** 1.94%

7.15 34 **7.17** **(a)** 69 **(b)** Larger **(c)** Larger

7.19 **(a)** 22.5 **(b)** 3.92 **(c)** 225

7.21 **(a)** $99.245 < \mu < 114.517$ **(b)** Yes

7.22 The administrator concludes that birthweights at CMH are smaller, on average, than those nationwide.

7.23   **(a)** $49.56 < \mu < 55.38$ **(b)**   No

7.25   **(a)** $\hat{p} = .50$ $.45 < p < .55$ **(b)** $x = 225$ $.23 < p < .27$

       (c) $\hat{p} = .71$ $.63 < p < .79$ **(d)** $x = 1025$ $.80 < p < .84$

       (e) $\hat{p} = .37$ $.35 < p < .39$ **(f)** $x = 610$ $.59 < p < .63$

7.27 $.83 < p < .97$

7.29   **(a)** $.08 < p < .18$ **(b)** $.24 < p < .32$

7.31   **(a)** $.02 < p < .10$ **(b)**   No

7.33   **(a)**   .6   **(b)** $.49 < p < .71$ **(c)**   .11

7.35   **(a)** $.48 < p < .56$ **(b)**   .04   **(c)**   No

7.37   757   **7.39**   **(a)**   1702   **(b)**   4161

7.41   **(a)(i)** $H_0: \mu = 72$; $H_a: \mu > 72$ **(ii)**   Right-tailed

       (iii)   Type I: The data suggest that the mean amount of rainfall per year is more than 72, when it's not.

               Type II: The data do not suggest that the mean amount of rainfall per year is more than 72, when it is.

       (b)(i) $H_0: \mu = 250$; $H_a: \mu \neq 250$ **(ii)**   Two-tailed

       (iii)   Type I: The data suggest that the mean number of books borrowed per day is not 250, when it is.

               Type II: The data suggest that the mean number of books borrowed per day is 250, when it's not

       (c)(i) $H_0: \mu = 78$; $H_a: \mu < 78$ **(ii)**   Left-tailed

       (iii)   Type I: The data suggest that the mean July temperature is less than 78, when it's not.

               Type II: The data do not suggest that the mean July temperature is less than 78, when it is.

       (d)(i) $H_0: \mu = 36$; $H_a: \mu > 36$ **(ii)**   Right-tailed

       (iii)   Type I: The data suggest that the mean age is more than 36, when it's not.

               Type II: The data do not suggest that the mean age is more than 36, when it is.

       (e)(i) $H_0: \mu = \$29,500$; $H_a: \mu < \$29,500$ **(ii)**   Left-tailed

       (iii)   Type I: The data suggest that the mean salary is less than $29,500, when it's not.

               Type II: The data do not suggest that the mean salary is less than $29,500, when it is.

       (f)(i) $H_0: \mu = 18,000$; $H_a: \mu \neq 18,000$ **(ii)**   Two-tailed

       (iii)   Type I: The data suggest that the mean number of families is not 18,000, when it is.

               Type II: The data suggest that the mean number of families is 18,000, when it's not.

       (g)(i) $H_0: \mu = 2.3$; $H_a: \mu > 2.3$ **(ii)**   Right-tailed

       (iii)   Type I: The data suggest that the mean grade-point average is more than 2.3, when it's not.

               Type II: The data do not suggest that the mean grade-point average is more than 2.3, when it is.

7.43 **(a)** $H_0: \mu = 90$; $H_a: \mu \neq 90$ **(b)(i)** $z \leq -2.58$ or $z \geq 2.58$
(b)(ii) $z \leq -1.96$ or $z \geq 1.96$ **(b)(iii)** $z \leq -1.65$ or $z \geq 1.65$

7.45 **(a)(i)** No **(ii)** No **(iii)** Yes **(b)(i)** No **(ii)** Yes
(iii) Yes
(c)(i) Yes **(ii)** Yes **(iii)** Yes

7.47 $H_0: \mu = 150$; $H_a: \mu > 150$ $\alpha = .05$ Observed value: $z = 2.53$
Critical value: $z = 1.65$ Reject H_0 Type I = The data suggest that the mean
time is more than 150 minutes, when it's not.

7.49 $H_0: \mu = 15$; $H_a: \mu > 15$ $\alpha = .05$ Observed value: $z = 3.01$
Critical value: $z = 1.65$ Reject H_0
The data suggest that the mean depression score is more than 15.

7.51 $H_0: \mu = 500$; $H_a: \mu < 500$ $\alpha = .01$ Observed value: $z = -1.49$
Critical value: $z = -2.33$ Fail to reject H_0
The data do not suggest that the mean depth is less than 500 feet.

7.52 $H_0: \mu = 27$; $H_a: \mu > 27$ $\alpha = .01$ Observed value: $z = 3.16$
Critical value: $z = 2.33$ Reject H_0
The data suggest that the mean breaking strength is more than 27 pounds per
square inch.

7.53 $H_0: \mu = 27$; $H_a: \mu > 27$ $\alpha = .01$ Observed value: $z = 1.26$
Critical value: $z = 2.33$ Fail to reject H_0
The data do not suggest that the mean breaking strength is more than 27 pounds
per square inch.

7.55 **(a)** $H_0: \mu = 700$; $H_a: \mu \neq 700$ $\alpha = .05$ Observed value: $z = -2.25$
Critical values: $z = \pm1.96$ Reject H_0 The data suggest that the mean
number of hours is not 700.
(b) Type I **(c)** .05

7.57 .0401 Reject H_0

7.59 .2033 Fail to reject H_0

7.61 .0802 Fail to reject H_0

7.63 $H_0: \mu = 30$; $H_a: \mu > 30$ **(a)** .1587 **(b)** .0808 **(c)** .0359
(d) $H_0: \mu = 30$; $H_a: \mu \neq 30$ .3174 .1616 .0718

7.65 $H_0: \mu = 12$; $H_a: \mu < 12$ Observed value: $z = -1.5$ **(a)** .0668
(b)(i) Yes **(ii)** No **(iii)** No **(c)(i)** Type I **(ii)** Type II
(iii) Type II

7.67 $H_0: \mu = 15$; $H_a: \mu > 15$ $\alpha = .05$ Observed value: $z = 3.01$
P value $= .0013$ Reject H_0
The data suggest that the mean depression score is more than 15.

7.69 $H_0: \mu = 500$; $H_a: \mu < 500$ $\alpha = .01$ Observed value: $z = -1.49$
P value $= .0681$ Fail to reject H_0
The data do not suggest that the mean depth is less than 500 feet.

7.71 **(a)** P value $= .0174$ **(b)** Larger than or equal to .0174

7.73 **(a)(i)** $H_0: p = .53$; $H_a: p \neq .53$ **(ii)** Two-tailed
(iii) Type I: The data suggest that p is not .53, when it is.
Type II: The data suggest that p is .53, when it's not.

(b)(i) H_0: $p = .06$; H_a: $p < .06$ **(ii)** Left-tailed
(iii) Type I: The data suggest that p is less than .06, when it's not.
Type II: The data do not suggest that p is less than .06, when it is.

(c)(i) H_0: $p = .90$; H_a: $p < .90$ **(ii)** Left-tailed
(iii) Type I: The data suggest that p is less than .90, when it's not.
Type II: The data do not suggest that p is less than .90, when it is.

(d)(i) H_0: $p = .07$; H_a: $p > .07$ **(ii)** Right-tailed
(iii) Type I: The data suggest that p is more than .07, when it's not.
Type II: The data do not suggest that p is more than .07, when it is.

(e)(i) H_0: $p = .40$; H_a: $p \neq .40$ **(ii)** Two-tailed
(iii) Type I: The data suggest that p is not .40, when it is.
Type II: The data suggest that p is .40, when it's not.

(f)(i) H_0: $p = .05$; H_a: $p < .05$ **(ii)** Left-tailed
(iii) Type I: The data suggest that p is less than .05, when it's not.
Type II: The data do not suggest that p is less than .05, when it is.

(g)(i) H_0: $p = .30$; H_a: $p > .30$ **(ii)** Right-tailed
(iii) Type I: The data suggest that p is more than .30, when it's not.
Type II: The data do not suggest that p is more than .30, when it is.

7.75 H_0: $p = .70$; H_a: $p < .70$
(a)(i) $z \leq -1.28$ **(ii)** Observed value: $z = -1.09$ Fail to reject H_0
(P value $= .1379$) **(iii)** Type II

(b)(i) $z \leq -1.65$ **(ii)** Observed value: $z = -2.18$ Reject H_0
(P value $= .0146$) **(iii)** Type I

(c)(i) $z \leq -2.33$ **(ii)** Observed value: $z = -2.18$ Fail to reject H_0
(P value $= .0146$) **(iii)** Type II

(d)(i) $z \leq -2.33$ **(ii)** Observed value: $z = -4.36$ Reject H_0
(P value $= 0$) **(iii)** Type I

7.77 H_0: $p = .5$; H: $p > .5$ $\alpha = .10$ Observed value: $z = .71$
Critical value: $z = 1.28$ Fail to reject H_0 (P value $= .2388$)
The data do not suggest that the proportion p is more than .5.

7.79 H_0: $p = .20$; H_a: $p > .20$ $\alpha = .01$ Observed value: $z = 3$
P value $= .0013$ Reject H_0
The data suggest that the proportion p is more than .20.

7.81 **(a)** H_0: $p = .20$; H_a: $p \neq .20$ $\alpha = .05$ Observed value: $z = 1.19$
Critical values: $z = \pm 1.96$ Fail to reject H_0

(b) P value $= .2340$
The data do not suggest that the proportion p is unequal to .20.

7.83 H_0: $p = .90$; H_a: $p < .90$ $\alpha = .01$ Observed value: $z = -1.89$
Critical value: $z = -2.33$ Fail to reject H_0 (P value $= .0294$)
The data do not suggest that the proportion p is less than .90.

7.85 **(a)** H_0: $p = .154$; H_a: $p < .154$ $\alpha = .025$ Observed value: $z = -6.52$
Critical value: $z = -1.96$ Reject H_0 (P value $= 0$)
The data suggest that the proportion p is less than .154.

(b) H_0: $p = .057$; H_a: $p < .057$ $\alpha = .025$ Observed value: $z = -1.73$
Critical value: $z = -1.96$ Fail to reject H_0 (P value $= .0418$)
The data do not suggest that the proportion p is less than .057.

7.87 **(a)** $208.48 < \mu < 211.52$ **(b)** 1.52

7.89 $\$26,819.82 < \mu < \$31,980.18$ Yes

7.91 $70.19 < \mu < 72.57$

7.93 **(a)** $\bar{x} = \$27,500$ **(b)** 90%

7.95 289 **7.97** 196

7.98 **(a)** 69 **(b)** 167 **7.99** **(b)** 97

7.101 **(a)** $x = 50$ $.40 < p < .60$ **(b)** $\hat{p} = .40$ $.33 < p < .47$

(c) $x = 288$ $.66 < p < .78$ **(d)** $\hat{p} = .20$ $.15 < p < .25$

(e) $x = 36$ $.50 < p < .70$ **(f)** $\hat{p} = .60$ $.42 < p < .78$

7.103 **(a)** $.37 < p < .47$ **(b)** .05

7.105 **(a)** 385 **(b)** 1068 **(c)** 9604

7.107 **(a)** $H_0: \mu = 450; H_a: \mu \neq 450$ $\alpha = .01$ Observed value: $z = 2.76$
Critical values: $z = \pm 2.58$ Reject H_0

(b) P value $= .0058$
The data suggest that the mean length is not 450 inches.

7.109 $H_0: \mu = 12; H_a: \mu < 12$ $\alpha = .05$ Observed value: $z = -1.42$
Critical value: $z = -1.65$ Fail to reject H_0 (P value $= .0778$)
The data do not suggest that the mean length is less than 12 inches.

7.111 $H_0: \mu = 16; H_a: \mu \neq 16$ Observed value: $z = -3$
(a) P value $= .0026$ **(b)** Reject H_0 in both cases

7.113 $H_0: \mu = 60; H_a: \mu < 60$ $\alpha = .05$ Observed value: $z = -4.38$
Critical value: $z - -1.65$ Reject II_0 (P value $= 0$)
The data suggest that the mean down time per week is less than 60 minutes.

7.115 $H_0: p = .20; H_a: p > .20$ $\alpha = .05$ Observed value: $z = 1.77$
Critical value: $z = 1.65$ Reject H_0 (P value $= .0384$)
The data suggest that the proportion p is more than .20.

7.117 $H_0: p = .25; H_a: p > .25$ $\alpha = .05$ Observed value: $z = 1.58$
Critical value: $z = 1.65$ Fail to reject H_0 (P value $= .0571$)
The data do not suggest that the proportion p is more than .25.

7.119 **(a)** .3632 **(b)** .2358 **(d)** .6293 **(e)** .0475

7.120 **(a)** .087 **(b)** .471

7.121 **(a)** .317 **(b)** .162

Chapter 8

8.1 **(a)(i)** 3.499 **(ii)** 2.998 **(iii)** 2.365 **(iv)** 1.895 **(v)** 1.415

(b)(i) 3.055 **(ii)** 2.681 **(iii)** 2.179 **(iv)** 1.782 **(v)** 1.356

(c)(i) 2.787 **(ii)** 2.485 **(iii)** 2.060 **(iv)** 1.708 **(v)** 1.316

8.2 **(a)** 9 **(b)** .01 **(c)** 1.323 **(f)** -4.604 **(h)** .90

8.3 **(a)** .95 **(b)** .05 **(c)** .99 **(d)** .90 **(e)** .075 **(f)** .94

8.5 **(a)** $H_0: \mu = 16; H_a: \mu > 16$ $\alpha = .10$ Observed value: $t = 1.87$
Critical value: $t = 1.350$ Reject H_0
$(.025 < P$ value $< .05)$ Type I

(b) H_0: $\mu = 27$; H_a: $\mu < 27$ $\alpha = .05$ Observed value: $t = -1.71$
Critical value: $t = -1.860$ Fail to reject H_0
$(.05 < P$ value $< .10)$ Type II

(c) H_0: $\mu = 30$; H_a: $\mu \neq 30$ $\alpha = .01$ Observed value: $t = -3.06$
Critical values: $t = \pm4.032$ Fail to reject H_0
$(.02 < P$ value $< .05)$ Type II

(d) H_0: $\mu = 125$; H_a: $\mu > 125$ $\alpha = .05$ Observed value: $z = 1.05$
Critical value: $z = 1.65$ Fail to reject H_0
$(P$ value $= .1469)$ Type II

(e) H_0: $\mu = 50$; H_a: $\mu < 50$ $\alpha = .025$ Observed value: $t = -2.24$
Critical value: $t = -2.093$ Reject H_0
$(.01 < P$ value $< .025)$ Type I

(f) H_0: $\mu = 60$; H_a: $\mu \neq 60$ $\alpha = .10$ Observed value: $t = 1.77$
Critical values: $t = \pm1.895$ Fail to reject H_0
$(.10 < P$ value $< .20)$ Type II

8.7   (a) H_0: $\mu = \$30{,}750$; H_a: $\mu > \$30{,}750$ Observed value: $t = 2.50$
Critical value: $t = 1.860$ for $\alpha = .05$, and $t = 2.896$ for $\alpha = .01$
Reject H_0 for $\alpha = .05$; Fail to reject H_0 for $\alpha = .01$

(b) $.01 < P$ value $< .025$

8.9 H_0: $\mu = \$6000$; H_a: $\mu \neq \$6000$ $\alpha = .05$ Observed value: $t = 2.98$
Critical values: $t = \pm2.093$ Reject H_0 $(P$ value $< .01)$
The data suggest that the mean contribution is not $6000.

8.11 H_0: $\mu = 160$; H_a: $\mu < 160$ Observed value: $t = -3.57$
P value $< .005$ $\alpha \geq .005$

8.13 H_0: $\mu = 10$; H_a: $\mu > 10$ Observed value: $t = 1.70$
$.05 < P$ value $< .10$ $\alpha \geq .10$

8.15 H_0: $\mu = 36$; H_a: $\mu \neq 36$ $\alpha = .05$ Observed value: $t = -2.42$
Critical values: $t = \pm2.056$ Reject H_0 $(.02 < P$ value $< .05)$
The data suggest that the mean trait anxiety level is not 36.

8.17 H_0: $\mu = 15$; H_a: $\mu \neq 15$ $\alpha = .05$ Observed value: $t = 1.71$
Critical values: $t = \pm2.074$ Fail to reject H_0 $(.10 < P$ value $< .20)$
The data do not suggest that the mean time is unequal to 15.

8.19 H_0: $\mu = 13$; H_a: $\mu > 13$ $\alpha = .05$ Observed value: $t = 1.24$
Critical value: $t = 1.796$ Fail to reject H_0 $(.10 < P$ value $< .25)$
The data do not suggest that the mean time is more than 13.

8.21   (a) $21.55 < \mu < 29.05$ (b)   No   (c)   3.75

8.23   (a) $64.56 < \mu < 67.04$ (b)   1.24

8.25 $12.82 < \mu < 13.96$

8.26   (a)(i)   21.666   (ii)   19.023   (iii)   16.919   (iv)   3.325   (v)   2.088
(b)(i)   30.578   (ii)   27.488   (iii)   24.996   (iv)   7.261   (v)   5.229
(c)(i)   44.314   (ii)   40.646   (iii)   37.652   (iv)   14.611   (v)   11.524

8.27   (a)   16.812   (b)   .872   (c)   12.592
(d)   10.645   (e)   .872   (f)   1.635

8.28   (a)   11   (b)   .95   (c)   3.94   (d)   25
(e)   .10   (f)   27.587   (g)   12.443   (h)   3.053

8.29 **(a)** .10 **(b)** .01 **(c)** .04 **(d)** .85

8.31 **(a)** $H_0: \sigma^2 = 100$; $H_a: \sigma^2 > 100$ $\alpha = .05$ Observed value: $\chi^2 = 25.95$
Critical value: $\chi^2 = 24.996$ Reject H_0 $(.025 < P$ value $< .05)$ Type I

(b) $H_0: \sigma^2 = 70$; $H_a: \sigma^2 < 70$ $\alpha = .01$ Observed value: $\chi^2 = 2.00$
Critical value: $\chi^2 = 3.053$ Reject H_0 $(P$ value $< .005)$ Type I

(c) $H_0: \sigma^2 = 50$; $H_a: \sigma^2 \neq 50$ $\alpha = .10$ Observed value: $\chi^2 = 11.99$
Critical values: $\chi^2 = 9.39, 28.869$ Fail to reject H_0
$(P$ value $> .20)$ Type II

(d) $H_0: \sigma^2 = 30$; $H_a: \sigma^2 > 30$ $\alpha = .10$ Observed value: $\chi^2 = 12$
Critical value: $\chi^2 = 13.362$ Fail to reject H_0 $(P$ value $> .10)$ Type II

(e) $H_0: \sigma^2 = 65$; $H_a: \sigma^2 < 65$ $\alpha = .05$ Observed value: $\chi^2 = 12.99$
Critical value: $\chi^2 = 11.591$ Fail to reject H_0
$(.05 < P$ value $< .10)$ Type II

(f) $H_0: \sigma^2 = 20$; $H_a: \sigma^2 \neq 20$ $\alpha = .01$ Observed value: $\chi^2 = 48$
Critical values: $\chi^2 = 10.52, 46.928$ Reject H_0 $(P$ value $< .01)$ Type I

8.33 **(a)** $H_0: \sigma^2 = .0001$; $H_a: \sigma^2 > .0001$ $\alpha = .01$ Observed value: $\chi^2 = 28$
Critical value: $\chi^2 = 29.141$ Fail to reject H_0

(b) $.01 < P$ value $< .025$
The data do not suggest that the variance is more than .0001.

8.35 **(a)** $H_0: \sigma^2 = 4$; $H_a: \sigma^2 \neq 4$ $\alpha = .05$ Observed value: $\chi^2 = 28.5$
Critical values: $\chi^2 = 8.907, 32.852$ Fail to reject H_0

(b) $.10 < P$ value $< .20$
The data do not suggest that the variance is unequal to 4.

8.37 $H_0: \sigma = 20$; $H_a: \sigma > 20$ $\alpha = .01$ Observed value: $\chi^2 = 48.64$
Critical value: $\chi^2 = 36.191$ Reject H_0 $(P$ value $< .005)$
The data suggest that the standard deviation is more than 20.

8.39 $H_0: \sigma^2 = 156$; $H_a: \sigma^2 < 156$ $\alpha = .01$ Observed value: $\chi^2 = 4.40$
Critical value: $\chi^2 = 1.646$ Fail to reject H_0 $(P$ value $> .10)$
The data do not suggest that the variance is less than 156.

8.41 **(a)** $.0001 < \sigma^2 < .0006$ **(b)** $.56 < \sigma^2 < 1.66$ **(c)** $9.46 < \sigma^2 < 97.2$

8.42 **(a)** 2.718 **(b)** 1.714 **(c)** 1.345 **(d)** -2.080

8.43 **(a)** .80 **(b)** .04 **(c)** .975 **(d)** .85

8.45 **(a)** $H_0: \mu = 150$; $H_a: \mu > 150$ $\alpha = .05$ Observed value: $t = 1.87$
Critical value: $t = 1.771$ Reject H_0

(b) $.025 < P$ value $< .05$
The data suggest that the mean number of calories is more than 150.

8.47 $H_0: \mu = .12$; $H_a: \mu \neq .12$ $\alpha = .10$ Observed value: $z = -1.70$
Critical values: $z = \pm 1.65$ Reject H_0 $(P$ value $= .0892)$
The data suggest that the mean diameter is not .12.
(*Note*: We did a z-test because σ is known.)

8.49 **(a)** $H_0: \mu = \$600$; $H_a: \mu < \$600$ $\alpha = .05$ Observed value: $t = -1.94$
Critical value: $t = -1.761$ Reject H_0

(b) $.025 < P$ value $< .05$
The data suggest that the mean tax deduction for charities is less than $600.

8.51 $H_0: \mu = 13; H_a: \mu \neq 13$ $\alpha = .05$ Observed value: $t = -2.37$
Critical values: $t = \pm 2.056$ Reject H_0 $(.02 < P$ value $< .05)$
The data suggest that the mean tension score is not 13.

8.53 $H_0: \mu = 130; H_a: \mu > 130$ $\alpha = .05$ Observed value: $t = 2.96$
Critical value: $t = 1.761$ Reject H_0 $(.005 < P$ value $< .01)$
The data suggest that the mean lactate level is more than 130.

8.55   **(a)** $150.06 < \mu < 152.34$ **(b)** $10.65 < \mu < 12.35$
(c) $.105 < \mu < .123$

8.56   **(a)**   .99   **(c)**   .09

8.57   **(a)**   26.217   **(b)**   30.144   **(c)**   16.791   **(d)**   69.126

8.59   **(a)** $H_0: \sigma^2 = 9; H_a: \sigma^2 < 9$ $\alpha = .01$ Observed value: $\chi^2 = 8.32$
Critical value: $\chi^2 = 4.107$ Fail to reject H_0

(b) P value $> .10$
The data do not suggest that the variance is less than 9.

8.61   **(a)** $H_0: \mu = 10; H_a: \mu \neq 10$ $\alpha = .05$ Observed value: $t = 1.81$
Critical values: $t = \pm 2.064$ Fail to reject H_0 $(.05 < P$ value $< .10)$
The data do not suggest that the mean fat content is unequal to 10%.

(b) $H_0: \sigma = .75; H_a: \sigma \neq .75$ $\alpha = .05$ Observed value: $\chi^2 = 29.39$
Critical values: $\chi^2 = 12.401, 39.364$ Fail to reject H_0
$(P$ value $> .20)$ The data do not suggest that the standard deviation is unequal
to .75%.

8.63 $H_0: \mu = 450; H_a: \mu < 450$ Observed value: $t = -2.10$

(a) $\alpha = .05$ Critical value: $t = -1.895$ Reject H_0

(b) $\alpha = .01$ Critical value: $t = -2.998$ Fail to reject H_0
$(.025 < P$ value $< .05)$

8.65   **(b)** $H_0: \mu = 450; H_a: \mu < 450$ Observed value: $t = -3.14$
(i) $\alpha = .05$ Critical value: $t = -1.895$ Reject H_0
(ii) $\alpha = .01$ Critical value: $t = -2.998$ Reject H_0
$(.005 < P$ value $< .01)$

8.67   **(b)** $H_0: \mu = 450; H_a: \mu < 450$ Observed value: $t = -1.26$
(i) $\alpha = .05$ Critical value: $t = -1.895$ Fail to reject H_0
(ii) $\alpha = .01$ Critical value: $t = -2.998$ Fail to reject H_0
$(.10 < P$ value $< .25)$

8.69   10   15   20   25

Chapter 9

9.1

		(i) $F(.01)$	(ii) $F(.025)$	(iii) $F(.05)$
	(a)	6.63	4.82	3.69
	(b)	10.29	6.76	4.82
	(c)	4.41	3.42	2.77
	(d)	12.25	8.07	5.59

9.2   **(a)**   4.01   **(b)**   4.77   **(c)**   3.13   **(d)**   3.07

9.3   **(a)(i)** $H_0: \sigma_A^2 = \sigma_B^2; H_a: \sigma_A^2 \neq \sigma_B^2$ $\alpha = .05$ Observed value: $F = 4.3$
Critical value: $F = 4.76$ Fail to reject H_0 $(.05 < P$ value $< .10)$

(ii)   Type II

(b)(i) $H_0: \sigma_A^2 = \sigma_B^2$; $H_a: \sigma_A^2 \neq \sigma_B^2$ $\alpha = .05$ Observed value: $F = 4.3$
Critical value: $F = 3.95$ Reject H_0 $(.02 < P \text{ value} < .05)$

  (ii) Type I

(c)(i) $H_0: \sigma_B^2 = \sigma_A^2$; $H_a: \sigma_B^2 > \sigma_A^2$ $\alpha = .01$ Observed value: $F = 4$
Critical value: $F = 2.78$ Reject H_0 $(P \text{ value} < .01)$

  (ii) Type I

(d)(i) $H_0: \sigma_A^2 = \sigma_B^2$; $H_a: \sigma_A^2 > \sigma_B^2$ $\alpha = .05$ Observed value: $F = 2.5$
Critical value: $F = 2.04$ Reject H_0 $(.01 < P \text{ value} < .025)$

  (ii) Type I

(e)(i) $H_0: \sigma_A^2 = \sigma_B^2$; $H_a: \sigma_A^2 \neq \sigma_B^2$ $\alpha = .10$ Observed value: $F = 10$
Critical value: $F = 3.63$ Reject H_0 $(P \text{ value} < .02)$

  (ii) Type I

(f)(i) $H_0: \sigma_A^2 = \sigma_B^2$; $H_a: \sigma_A^2 > \sigma_B^2$ $\alpha = .01$ Observed value: $F = 10$
Critical value: $F = 14.66$ Fail to reject H_0
$(.01 < P \text{ value} < .025)$

  (ii) Type II

9.4 **(a)** $.90 < \dfrac{\sigma_A^2}{\sigma_B^2} < 16.99$ **(c)** $.10 < \dfrac{\sigma_A^2}{\sigma_B^2} < .70$ **(e)** $2.75 < \dfrac{\sigma_A^2}{\sigma_B^2} < 60.00$

9.5 $H_0: \sigma_B^2 = \sigma_A^2$; $H_a: \sigma_B^2 > \sigma_A^2$ $\alpha = .05$ Observed value: $F = 2.2$
Critical value: $F = 1.98$ Reject H_0 $(.025 < P \text{ value} < .05)$
The data suggest that the variance of brand A is less than the variance of brand B.

9.7 $H_0: \sigma_A^2 = \sigma_B^2$; $H_a: \sigma_A^2 > \sigma_B^2$ $\alpha - .01$ Observed value: $F = 2$
Critical value: $F = 6.03$ Fail to reject H_0 $(P \text{ value} > .05)$
The data do not suggest that the variance of clerk A is more than the variance of clerk B.

9.9 $H_0: \sigma_A^2 = \sigma_B^2$; $H_a: \sigma_A^2 > \sigma_B^2$ $\alpha = .05$ Observed value: $F = 2$
  (b) First table; Critical value: $F = 1.69$
Second table; Critical value: $F = 2.98$
First table, reject H_0 $(.01 < P \text{ value} < .025)$
Second table, fail to reject H_0 $(P \text{ value} > .05)$

9.10 **(a)(i)** $H_0: \mu_d = 0$; $H_a: \mu_d \neq 0$ $\alpha = .05$ Observed value: $t = 1.41$
Critical values: $t = \pm 2.365$ Fail to reject H_0 $(.20 < P \text{ value} < .50)$

  (ii) Type II

  (b)(i) $H_0: \mu_d = 0$; $H_a: \mu_d > 0$ $\alpha = .05$ Observed value: $t = 2.84$
Critical value: $t = 2.015$ Reject H_0 $(.01 < P \text{ value} < .025)$

  (ii) Type I

9.11 $H_0: \mu_A - \mu_B = 0$; $H_a: \mu_A - \mu_B \neq 0$ $\alpha = .05$ Observed value: $t = 2.47$
Critical values: $t = \pm 2.201$ Reject H_0 $(.02 < P \text{ value} < .05)$
The data suggest that the mean systolic blood pressure is different between smokers and nonsmokers.

9.13 $H_0: \mu_A - \mu_B = 0$; $H_a: \mu_A - \mu_B > 0$ Observed value: $t = 2.31$
  (a) $.025 < P \text{ value} < .05$

  (b) Reject H_0 at the 5% and 10% levels, but fail to reject H_0 at the 1% level of significance.

9.15 $H_0: \mu_d = 0; H_a: \mu_d > 0$ $\alpha = .01$ Observed value: $z = 9.72$
Critical value: $z = 2.33$ Reject H_0 (P value $= 0$)
The data suggest that the mean thickness of cake A is larger.

9.17 **(a)** $-1.62 < \mu_d < 6.38$ **(b)** $1.02 < \mu_d < 5.98$
(c) $-4.59 < \mu_d < .87$ **(d)** $10.34 < \mu_d < 16.32$

9.19 **(b)** $H_0: \mu_A - \mu_B = 0; H_a: \mu_A - \mu_B > 0$ $\alpha = .05$
Observed value: (First data set): $t = 11.01$ (Second data set): $t = 1.32$
Critical value: $t = 2.015$
(First data set): Reject H_0 (P value $< .005$)
(Second data set): Fail to reject H_0 ($.10 < P$ value $< .25$)

9.21 **(a)** $H_0: \mu_A - \mu_B = 0; H_a: \mu_A - \mu_B > 0$ $\alpha = .05$ Observed value:
$z = 1.33$ Critical value: $z = 1.65$ Fail to reject H_0

(b)(i) No   **(ii)** Yes **(iii)** Yes

(c) $\alpha = .10$ Critical value: $z = 1.28$ Reject H_0
(P value $= .0918$)

9.23 **(a)** $H_0: \mu_B - \mu_A = 0; H_a: \mu_B - \mu_A < 0$ $\alpha = .05$ Observed value:
$z = -1.42$ Critical values: $z = -1.65$ Fail to reject H_0

(b) P value $= .0778$
The data do not suggest that μ_B is less than μ_A.

9.25 $H_0: \mu_A - \mu_B = 0; H_a: \mu_A - \mu_B \neq 0$ $\alpha = .01$ Observed value: $z = .81$
Critical values: $z = \pm 2.58$ Fail to reject H_0 (P value $= .4180$)
The data do not suggest that there is a difference in mean elapsed time for the
two types of vehicles.

9.27 **(a)** $-.32 < \mu_A - \mu_B < -.04$
(b) $.70 < \mu_A - \mu_B < 15.30$
(c) $-64.65 < \mu_A - \mu_B < 4.65$

9.29 **(a)** $H_0: \mu_B - \mu_A = 10; H_a: \mu_B - \mu_A \neq 10$ $\alpha = .05$ Observed value: $t = 1.03$
Critical values: $t = \pm 2.145$ Fail to reject H_0 ($.20 < P$ value $< .50$)
The data do not suggest that the mean of population B is 10 more than the
mean of population A.

(b) $H_0: \mu_A - \mu_B = 0; H_a: \mu_A - \mu_B < 0$ $\alpha = .10$ Observed value:
$t = -2.68$
Critical value: $t = -1.440$ Reject H_0 ($.01 < P$ value $< .025$)
The data suggest that the mean of population A is less than the mean of
population B.

9.31 $H_0: \sigma_A^2 = \sigma_B^2; H_a: \sigma_A^2 \neq \sigma_B^2$ $\alpha = .10$ Observed value: $F = 1.17$
Critical value: $F = 3.58$ Fail to reject H_0 (P value $> .10$) Assume $\sigma_A^2 = \sigma_B^2$
$H_0: \mu_A - \mu_B = 0; H_a: \mu_A - \mu_B > 0$ $\alpha = .10$ Observed value: $t = 2.14$
Critical value: $t = 1.345$ Reject H_0 ($.025 < P$ value $< .05$)
The data suggest that test A has a longer mean completion time.

9.33 **(a)** $H_0: \mu_A - \mu_B = 0; H_a: \mu_A - \mu_B < 0$ $\alpha = .05$ Observed value:
$t = -2.07$ Critical value: $t = -1.753$ Reject H_0

(b) ($.025 < P$ value $< .05$)
The data suggest that rotated workers have lower mean depression levels
than non-rotated workers.

9.35 $H_0: \mu_A - \mu_B = 0$; $H_a: \mu_A - \mu_B \neq 0$ $\alpha = .05$
Observed value: $t = -2.19$ Critical values: $t = \pm 2.228$
Fail to reject H_0 (.05 < P value < .10)
The data do not suggest a difference in the population means.

9.37 $H_0: \sigma_A^2 = \sigma_B^2$; $H_a: \sigma_A^2 \neq \sigma_B^2$ $\alpha = .02$ Observed value: $F = 1.68$
Critical value: $F = 5.35$ Fail to reject H_0 (P value > .10)   Assume $\sigma_A^2 = \sigma_B^2$
$H_0: \mu_A - \mu_B = 8$; $H_a: \mu_A - \mu_B \neq 8$ $\alpha = .02$ Observed value: $t = 2.09$
Critical values: $t = \pm 2.552$ Fail to reject H_0 (.05 < P value < .10)
The data do not suggest that the mean height of male runners is 8 centimeters more than the mean height of female runners.

9.39 **(a)** $-38.24 < \mu_A - \mu_B < -1.76$ **(b)** $2.84 < \mu_A - \mu_B < 29.16$
(c) $6.89 < \mu_A - \mu_B < 19.31$

9.40 **(a)** $\hat{p}_1 = .700$ $\hat{p}_2 = .771$ $\hat{p} = .729$
$H_0: p_1 - p_2 = 0$; $H_a: p_1 - p_2 \neq 0$ $\alpha = .05$ Observed value: $z = -1.62$
Critical values: $z = \pm 1.96$ Fail to reject H_0 (P value = .1052)
The data do not suggest that the proportions p_1 and p_2 for populations A and B are different.

9.41 **(a)** $-.155 < p_1 - p_2 < .013$ **(b)** $.010 < p_1 - p_2 < .106$
(c) $.044 < p_1 - p_2 < .158$

9.43 **(a)** $\hat{p} = .567$
(b) Let $\hat{p}_1$ and $\hat{p}_2$ be the sample proportions from the eastern and western parts of the state, respectively.
$\hat{p}_1 = .6$ $\hat{p}_2 = .533$
$H_0: p_1 - p_2 = 0$; $H_a: p_1 - p_2 \neq 0$ $\alpha = .05$ Observed value: $z = 1.17$
Critical values: $z = \pm 1.96$ Fail to reject H_0 (P value = .2420)
The data do not suggest a difference in her support between the eastern and western parts of the state.

9.45 **(a)** $H_0: p_B - p_W = 0$; $H_a: p_B - p_W < 0$ $\alpha = .025$ Observed value: $z = -3.75$ P value = .0001   Reject H_0
The data suggest that the population proportion of blacks selected is less than that of whites.
(b) Yes

9.47 **(a)** $-.277 < p_A - p_B < -.007$ **(b)** $-.045 < p_1 - p_2 < .179$ No

9.49

	(i)	(ii)	(iii)
(a)	6.42	4.72	3.63
(b)	14.66	8.90	6.00
(c)	2.20	1.94	1.74

9.51 $H_0: \sigma_A^2 = \sigma_B^2$; $H_a: \sigma_A^2 > \sigma_B^2$ $\alpha = .05$ Observed value: $F = 1.06$
Critical value: $F = 1.69$ Fail to reject H_0 (P value > .05)
The data do not suggest that the variability in males' diastolic blood pressure is larger than in females' diastolic blood pressure.

9.53 **(a)** $H_0: \mu_A - \mu_B = 0$; $H_a: \mu_A - \mu_B < 0$ $\alpha = .05$ Observed value:
$t = -2.32$ Critical value: $t = -1.833$ Reject H_0
(b) .01 < P value < .025
The data suggest that, on average, it's quicker by car.

9.55 **(a)** $-2.489 < \mu_A - \mu_B < -.291$ **(b)** $-1.68 < \mu_A - \mu_B < 4.82$

9.57 H_0: $\mu_A - \mu_B = 0$; H_a: $\mu_A - \mu_B \neq 0$ $\quad \alpha = .01$ $\quad$ Observed value: $z = -3.29$
Critical values: $z = \pm 2.58$ $\quad$ Reject H_0 $\quad$ (P value $= .0010$)
The data suggest a difference in mean percent calories from fats.

9.59 (a) H_0: $\sigma_A^2 = \sigma_B^2$; H_a: $\sigma_A^2 \neq \sigma_B^2$ $\quad \alpha = .05$ $\quad$ Observed value: $F = 1.5$
Critical value: $F = 3.44$ Fail to reject H_0 (P value $> .10$) Assume $\sigma_A^2 = \sigma_B^2$
H_0: $\mu_A - \mu_B = 0$; H_a: $\mu_A - \mu_B < 0$ $\quad \alpha = .05$ $\quad$ Observed value: $t = -2.39$
Critical value: $t = -1.721$ $\quad$ Reject H_0 $\quad$ ($.01 < P$ value $< .025$)
The data suggest that the mean of A is smaller than the mean of B.

(b) H_0: $\sigma_A^2 = \sigma_B^2$; H_a: $\sigma_A^2 \neq \sigma_B^2$ $\quad \alpha = .10$ $\quad$ Observed value: $F = 5$
Critical value: $F = 2.94$ $\quad$ Reject H_0 $\quad$ (P value $< .02$) $\quad$ Assume $\sigma_A^2 \neq \sigma_B^2$
H_0: $\mu_A - \mu_B = 0$; H_a: $\mu_A - \mu_B > 0$ $\quad \alpha = .10$ $\quad$ Observed value: $t = 2.22$
Critical value: $t = 1.383$ $\quad$ Reject H_0 $\quad$ ($.025 < P$ value $< .05$)
The data suggest that the mean of A is larger than the mean of B.

9.61 (a) H_0: $\mu_A - \mu_B = 0$; H_a: $\mu_A - \mu_B < 0$ $\quad \alpha = .05$ $\quad$ Observed value:
$t = -1.55$ $\quad$ Critical value: $t = -1.734$ $\quad$ Fail to reject H_0

(b) $.05 < P$ value $< .10$
The data do not suggest that the drug reduces the mean time for rats to complete the maze

9.63 (a) $-16.84 < \mu_A - \mu_B < -1.16$ $\qquad$ (b) $-14.38 < \mu_A - \mu_B < .38$

(c) $1.74 < \mu_A - \mu_B < 18.26$

9.65 (a) Let A and B represent men and women, respectively.
H_0: $p_A - p_B = 0$; H_a: $p_A - p_B \neq 0$ $\quad \alpha = .01$ $\quad$ Observed value: $z = 2.72$
Critical values: $z = \pm 2.58$ $\quad$ Reject H_0

(b) P value $= .0066$
The data suggest a difference in population proportion of attitudes between men and women on the capital punishment question.

9.67 (a) H_0: $p_B - p_W = 0$; H_a: $p_B - p_W < 0$ $\quad \alpha = .025$ $\quad$ Observed value:
$z = -2.40$
Critical value: $z = -1.96$ $\quad$ Reject H_0 $\quad$ (P value $= .0082$)
The data suggest that the population proportion of blacks who pass is less than that of whites.

(b) No

9.69 (a) $-.218 < p_A - p_B < .018$ $\qquad$ (b) $.007 < p_A - p_B < .233$

(c) $-.199 < p_A - p_B < .019$

9.71

df for F	df for t	$F(.05)$	$t(.025)$	$t^2(.025)$
(1, 3)	3	10.13	3.182	10.13
(1, 6)	6	5.99	2.447	5.99
(1, 8)	8	5.32	2.306	5.32
(1, 10)	10	4.96	2.228	4.96
(1, 14)	14	4.60	2.145	4.60
(1, 20)	20	4.35	2.086	4.35

The F distribution with $(1, k)$ degrees of freedom is the same as a t^2 distribution, where t has k degrees of freedom.

Chapter 10

10.1 **(a)** $b_0 = 4$ and $b_1 = 7$· **(b)** $b_0 = 13$ and $b_1 = 5$

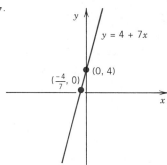

(c) $b_0 = 8$ and $b_1 = 0$ **(d)** $b_0 = 9/2$ and $b_1 = -2$ **(e)** $b_0 = 4$ and $b_1 = 3$

10.3 **(b)** $\hat{y} = 4 + 2x$ **(c)**

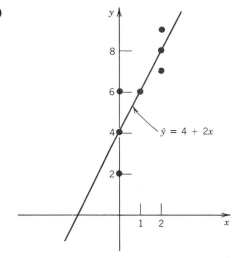

10.5 **(b)** $\hat{y} = 2.8 - .6x$

10.7 $\hat{y} = 4 + 2x$

10.9 $\hat{y} = 3.175 + .654x$

10.11 **(a)** $\hat{y} = 3.816 - .072x$ **(b)** 1.728

10.13 **(a)** $\hat{y} = 10.044 - .221x$ **(b)** 7.834

10.15 **(a)** $\hat{y} = 80.54 - 1.95x$ **(b)** 14.24

10.17 **(a)** $\hat{y} = 56.90 - .31x$ **(b)** 29.62

10.19 **(a)** .84 **(b)** $(.84)^2 = .7056$ About 71% of variation in y is explained by the relationship with x.

10.21 **(a)** $-.95$ **(b)** $(-.95)^2 = .9025$ About 90% of variation in y is explained by the relationship with x.

10.23 .173

10.25 $-.91$ $r^2 = .83$

10.27 $r^2 = .60$; about 60% of the variation in miles per gallon is explained by horsepower.

10.29 **(b)** About 8% of the variation in approval rating is explained by unemployment.

(c) .283

10.31 **(a)(i)** 89.2 **(ii)** 84.1 **(b)** $r^2 = .943$

10.33 **(a)** $s_x = 2.121320$ $s_y = 2.915476$ **(b)** .687184
(c) $\hat{y} = 1.17 + .94x$

10.35 $H_0: \rho = 0; H_a: \rho \neq 0$ $\alpha = .01$ Observed value: $r = .84$
Critical values: $r = \pm.875$ Fail to reject H_0
The data do not suggest a linear relationship.

10.37 $H_0: \rho = 0; H_a: \rho \neq 0$ $\alpha = .05$ Observed value: $r = -.95$
Critical values: $r = \pm.878$ Reject H_0
The data suggest a linear relationship.

10.39 $H_0: \rho = 0; H_a: \rho \neq 0$ $\alpha = .05$ Observed value: $r = 1$
Critical values: $r = \pm.95$ Reject H_0
The data suggest a linear relationship.

10.41 **(a)** $H_0: \rho = 0; H_a: \rho \neq 0$ $\alpha = .05$ Observed value: $r = .96$
Critical values: $r = \pm.754$ Reject H_0
The data suggest a linear relationship.

(b) $\hat{y} = 5.727 + 25.837x$ **(c)** 70.320

10.43 $H_0: \rho = 0; H_a: \rho \neq 0$ $\alpha = .01$ Observed value: $r = .77$
Critical values: $r = \pm.834$ Fail to reject H_0
The data do not suggest a linear relationship.

10.45 $H_0: \rho = 0; H_a: \rho \neq 0$ $\alpha = .05$ Observed value: $r = -.98$
Critical values: $r = \pm.878$ Reject H_0
The data suggest a linear relationship.

10.47 **(a)** $\hat{y} = 14 - 1.2x$ **(b)** 5.6 **(c)** 3.23 to 7.97
(d) 4.33 to 6.87

10.49 **(a)** $H_0: \rho = 0; H_a: \rho \neq 0$ $\alpha = .05$ Observed value: $r = -.78$
Critical values: $r = \pm.361$ Reject H_0

(b) 21.98 to 36.02 **(c)** 27.50 to 30.50

10.51 **(a)** 10.97 to 22.17 **(b)** 14.74 to 18.40

10.55 **(a)** 100 .5 **(b)** 300 .25 **(c)** 10 .2
(d) 60 150 **(e)** 12 40 **(f)** 32 8
(g) 0 50 **(h)** 75 1

10.57 **(b)** $H_0: \rho = 0; H_a: \rho \neq 0$ $\alpha = .01$ Observed value: $r = .97$
Critical values: $r = \pm.917$ Reject H_0
The data suggest a linear relationship.

(c) 31.6 (per 100,000)

10.59 **(b)** $H_0: \rho = 0; H_a: \rho \neq 0$ $\alpha = .05$ Observed value: $r = .59$
Critical values: $r = \pm.754$ Fail to reject H_0
The data do not suggest a linear relationship.

10.61 **(b)** $H_0: \rho = 0$; $H_a: \rho \neq 0$ $\alpha = .05$ Observed value: $r = -.90$
Critical values: $r = \pm.811$ Reject H_0
The data suggest a linear relationship.

(c) $\hat{y} = 66.29 - .6x$ **(d)** 24.29

10.63 **(b)** $H_0: \rho = 0$; $H_a: \rho \neq 0$ $\alpha = .05$ Observed value: $r = -.88$
Critical values: $r = \pm.707$ Reject H_0
The data suggest a linear relationship.

(c) $\hat{y} = 5.314 - .543x$ **(d)** $450

10.65 **(a)** 8.15 to 40.43 **(b)** 16.84 to 31.74
10.67 **(a)** 3.781 to 5.219 **(b)** 4.260 to 4.740

Chapter 11

11.1 **(a)** 7.5 (2, 12) 3.89 Reject H_0

(b) 3 (2, 12) 3.89 Fail to reject H_0

(c) 6 (1, 10) 10.04 Fail to reject H_0

(d) 12 (1, 10) 10.04 Reject H_0

(e) 4 (4, 15) 3.06 Reject H_0

(f) 2 (4, 15) 3.06 Fail to reject H_0

11.3 $H_0: \mu_A = \mu_B = \mu_C$; H_a: Not all the means are equal $\alpha = .05$
Observed value: $F = 4.75$ Critical value: $F = 3.89$ Reject H_0
$(.025 < P$ value $< .05)$ The data suggest that not all the means are the same.

11.5 $H_0: \mu_M = \mu_F$; $H_a: \mu_M \neq \mu_F$ $\alpha = .01$
Observed value: $F = 6.63$ Critical value: $F = 8.29$ Fail to reject H_0
$(.01 < P$ value $< .025)$ The data do not suggest that the means are different.

11.7 **(a)(i)** (4, 25) **(ii)** $F = 2.76$ **(iii)** $F = 2.17$ **(iv)** No (P value $> .05$)

(b)(i) Remains the same **(ii)** Larger **(iii)** Larger **(iv)** $F = 4.57$
Yes (P value $< .01$)

11.9

Source	SS	df	MS	F statistic
Between samples	120	1	120	9.47
Within samples	76	6	12.667	
Total	196	7		

$H_0: \mu_A = \mu_B$; $H_a: \mu_A \neq \mu_B$ $\alpha = .05$
Observed value: $F = 9.47$ Critical value: $F = 5.99$ Reject H_0
$(.01 < P$ value $< .025)$ The data suggest that the means are different.

11.11

Source	SS	df	MS	F statistic
Between samples	63.6	2	31.8	3.59
Within samples	62	7	8.86	
Total	125.6	9		

H_0: $\mu_A = \mu_B = \mu_C$; H_a: Not all the means are equal $\qquad \alpha = .05$
Observed value: $F = 3.59$ $\qquad$ Critical value: $F = 4.74$ $\qquad$ Fail to reject H_0
(P value $> .05$) $\qquad$ The data do not suggest that there is a difference in means.

11.13 **(a)**

Source	SS	df	MS	F statistic
Between samples	250	5	50	8.33
Within samples	150	25	6	
Total	400	30		

(b) H_0: $\mu_A = \mu_B = \mu_C = \mu_D = \mu_E = \mu_F$; H_a: Not all the means are equal
$\alpha = .05$ $\qquad$ Observed value: $F = 8.33$ $\qquad$ Critical value: $F = 2.60$
Reject H_0 (P value $< .01$) $\qquad$ The data suggest that not all the means are
the same.

11.15 **(a)**

Source	SS	df	MS	F statistic
Between samples	128	4	32	2
Within samples	160	10	16	
Total	288	14		

(b) H_0: $\mu_A = \mu_B = \mu_C = \mu_D = \mu_E$; H_a: Not all the means are equal
$\alpha = .05$ $\qquad$ Observed value: $F = 2$ $\qquad$ Critical value: $F = 3.48$
Fail to reject H_0 (P value $> .05$) The data do not suggest that there is a
difference in means.

11.17 H_0: $\mu_A = \mu_B = \mu_C$; H_a: Not all the means are equal $\qquad \alpha = .05$
Observed value: $F = 4.03$ $\qquad$ Critical value: $F = 3.63$ $\qquad$ Reject H_0
($.025 < P$ value $< .05$) $\qquad$ The data suggest that not all the means are the same.

11.19 H_0: $\mu_A = \mu_B = \mu_C = \mu_D$; H_a: Not all the means are equal $\qquad \alpha = .05$
Observed value: $F = 4.97$ $\qquad$ Critical value: $F = 3.10$ $\qquad$ Reject H_0
(P value $< .01$) $\qquad$ The data suggest that not all the means are the same.

11.21 H_0: $\mu_{CR} = \mu_M = \mu_{CO}$; H_a: Not all the means are equal $\qquad \alpha = .01$
Observed value: $F = 13.49$ $\qquad$ Critical value: $F = 4.98$ $\qquad$ Reject H_0
(P value $< .01$) $\qquad$ The data suggest that not all the means are the same.

11.23 **(a)** SSB $= 120$ $\qquad$ SSW $= 76$ $\qquad$ **(b)** SSB $= 98$ $\qquad$ SSW $= 196$

$\qquad\qquad$ **(c)** SSB $= 63.6$ $\qquad$ SSW $= 62$ $\qquad$ **(d)** SSB $= 275.71$ $\qquad$ SSW $= 190$

11.25

Source	SS	df	MS	F statistic
Between samples	322.5	2	161.25	5.64
Within samples	200	7	28.571	
Total	522.5	9		

$H_0: \mu_A = \mu_B = \mu_C$; H_a: Not all the means are equal $\alpha = .05$
Observed value: $F = 5.64$ Critical value: $F = 4.74$ Reject H_0
$(.025 < P$ value $< .05)$ The data suggest that not all the means are the same.

11.27

Source	SS	df	MS	F statistic
Between samples	132	4	33	3.3
Within samples	400	40	10	
Total	532	44		

$H_0: \mu_A = \mu_B = \mu_C = \mu_D = \mu_E$; H_a: Not all the means are equal
$\alpha = .05$ Observed value: $F = 3.3$ Critical value: $F = 2.61$
Reject H_0 $(.01 < P$ value $< .025)$
The data suggest that not all the means are the same.

11.29 $H_0: \mu_A = \mu_B = \mu_C$; H_a: Not all the means are equal $\alpha = .05$
Observed value: $F = 5.82$ Critical value: $F = 4.46$ Reject H_0
$(.025 < P$ value $< .05)$ The data suggest that not all the means are the same.

11.31 $H_0: \mu_A = \mu_B = \mu_C$; H_a: Not all the means are equal $\alpha = .05$
Observed value: $F = 6.01$ Critical value: $F = 3.74$ Reject H_0
$(.01 < P$ value $< .025)$ The data suggest that not all the means are the same.

11.33 $H_0: \mu_A = \mu_B = \mu_C$; H_a: Not all the means are equal $\alpha = .05$
Observed value: $F = 4.30$ Critical value: $F = 3.89$ Reject H_0
$(.025 < P$ value $< .05)$ The data suggest that not all the means are the same.

11.35 $H_0: \mu_E = \mu_G = \mu_N$; H_a: Not all the means are equal $\alpha = .05$
Observed value: $F = 1.81$ Critical value: $F = 3.15$ Fail to reject H_0
$(P$ value $> .05)$ The data do not suggest that there is a difference in means.

11.37   **(b)** $H_0: \mu_A = \mu_B = \mu_C = \mu_D = \mu_E = \mu_F$; H_a: Not all the means are equal
    $\alpha = .05$ Observed value: $F = 0.22$ Critical value: $F = 3.11$
    Fail to reject H_0 $(P$ value $> .05)$
    The data do not suggest that there is a difference in means.

11.39   **(a)**   The ANOVA table is identical with that obtained in Exercise 11.38.

Chapter 12

12.1 $H_0: p_1 = p_2 = p_3 = p_4 = \frac{1}{4}$; H_a: H_0 is not true $\alpha = .05$
Observed value: $\chi^2 = 5.04$ Critical value: $\chi^2 = 7.815$ Fail to reject H_0
$(P$ value $> .10)$

12.3 $H_0: p_1 = p_2 = \frac{1}{10}, p_3 = p_4 = \frac{1}{5}, p_5 = \frac{2}{5}$; H_a: H_0 is not true $\alpha = .10$
Observed value: $\chi^2 = 9.58$ Critical value: $\chi^2 = 7.779$ Reject H_0
$(.025 < P$ value $< .05)$

12.5 $H_0: p_1 = .30, p_2 = .05, p_3 = .50, p_4 = .15; H_a: H_0$ is not true $\alpha = .05$
Observed value: $\chi^2 = 10.89$ Critical value: $\chi^2 = 7.815$ Reject H_0
($.01 < P$ value $< .025$)   The data suggest that the official's claim is not true.

12.7 $H_0: p_1 = \frac{1}{10}, p_2 = p_3 = p_4 = \frac{3}{10}; H_a: H_0$ is not true $\alpha = .10$
Observed value: $\chi^2 = 3.08$ Critical value: $\chi^2 = 6.251$ Fail to reject H_0
(P value $> .10$)   The data do not dispute the owner's assertion.

12.9 $H_0: p_1 = p_2 = p_3 = p_4 = p_5 = p_6 = p_7 = p_8 = p_9 = p_{10} = \frac{1}{10}; H_a: H_0$ is not true
$\alpha = .05$ Observed value: $\chi^2 = 6.40$ Critical value: $\chi^2 = 16.919$
Fail to reject H_0 (P value $> .10$)   The data do not dispute the claim.

12.11 H_0: The characteristics of rating and age are independent.
H_a: The characteristics of rating and age are related.
$\alpha = .05$ Observed value: $\chi^2 = 2.92$ Critical value: $\chi^2 = 5.991$
Fail to reject H_0 (P value $> .10$)
The data do not suggest that the characteristics of rating and age are related.

12.13 H_0: The characteristics of gender and CHD are independent.
H_a: The characteristics of gender and CHD are related.
$\alpha = .05$ Observed value: $\chi^2 = 0.48$ Critical value: $\chi^2 = 3.841$
Fail to reject H_0 (P value $> .10$)
The data do not suggest that the characteristics of gender and CHD are related.

12.15 H_0: The characteristics of sex and salary are independent.
H_a: The characteristics of sex and salary are related.
$\alpha = .10$ Observed value: $\chi^2 = 7.40$ Critical value: $\chi^2 = 6.251$
Reject H_0 ($.05 < P$ value $< .10$)
The data suggest that the characteristics of sex and salary are related.

12.17   **(a)(i)**   0   **(ii)**   0   **(iii)**   0

(c)(i)

	B1	B2	B3
A1	20	10	30
A2	60	30	90
A3	180	90	270

(ii)

	B1	B2	B3
A1	20	10	30
A2	60	30	90
A3	30	15	45

12.19 H_0: The dice are homogeneous with respect to number of dots showing
$H_a: H_0$ is not true
$\alpha = .05$ Observed value: $\chi^2 = 2.07$ Critical value: $\chi^2 = 11.071$
Fail to reject H_0 (P value $> .10$)
There is insufficient evidence to conclude that the dice are not homogeneous.

12.21 H_0: The appropriate populations of males and females are homogeneous
with respect to belief of discrimination in salaries
$H_a: H_0$ is not true
$\alpha = .01$ Observed value: $\chi^2 = 7.34$ Critical value: $\chi^2 = 6.635$
Reject H_0 ($.005 < P$ value $< .01$)
The data suggest that the appropriate populations are not homogeneous.

12.25 $H_0: p_1 = .10, p_2 = .20, p_3 = .30, p_4 = .25, p_5 = .15; H_a: H_0$ is not true
$\alpha = .05$ Observed value: $\chi^2 = 6.30$ Critical value: $\chi^2 = 9.488$
Fail to reject H_0 (P value $> .10$)
The data do not suggest the new teacher's grading policy is different from that
of the department.

12.27 H_0: $p_1 = .70$, $p_2 = .20$, $p_3 = .08$, $p_4 = .02$; H_a: H_0 is not true
$\alpha = .05$ Observed value: $\chi^2 = 14.7$ Critical value: $\chi^2 = 7.815$
Reject H_0 (P value $< .005$)
The data suggest that the modes of transportation to work have changed.

12.29 H_0: The characteristics of gender and CHD are independent.
H_a: The characteristics of gender and CHD are related.
$\alpha = .10$ Observed value: $\chi^2 = 0.63$ Critical value: $\chi^2 = 2.706$
Fail to reject H_0 (P value $> .10$)
The data do not suggest that the characteristics of gender and CHD are related.

12.31 H_0: The male and female populations are homogeneous with respect to age.
H_a: H_0 is not true.
$\alpha = .05$ Observed value: $\chi^2 = 6.86$ Critical value: $\chi^2 = 5.991$
Reject H_0 ($.025 < P$ value $< .05$)
The data suggest that the male and female populations are not homogeneous with respect to age.

12.33 **(a)** H_0: The attitude concerning racial quotas in hiring is homogeneous with respect to appropriate populations of Democrats, Republicans, and Independents.
H_a: H_0 is not true.

(b) Same

Chapter 13

13.1 **(a)** $\{0, 1, 2, 10, 11, 12\}$ **(b)** 0.063 **(c)** $\{0, 1, 2\}$ **(d)** 0.07

(e) $\{11, 12, 13, 14, 15\}$ **(f)** 0.047

(g) $\{0, 1, 2, 3, 4, 5\}$ **(h)** 0.032

(i) 0.124 **(j)** 0.031 **(k)** $\{0, 1, 2, 3, 4, 12, 13, 14, 15, 16\}$

(l) $\{14, 15, 16, 17, 18, 19\}$

13.3 H_0: Md $= 2.8$; H_a: Md > 2.8 $\alpha = .05$
Observed value: $x = 8$ Critical region: $\{8, 9, 10\}$ Reject H_0
The data support the manager's claim.

13.5 H_0: Md $= 0$; H_a: Md > 0 $\alpha = .05$
Let x be the number of differences (weight before $-$ weight after) that are larger than zero.
Observed value: $x = 6$ Critical region: $\{7, 8\}$ Fail to reject H_0
The data do not suggest that the diet is effective.

13.7 **(a)**

Difference $D = x - 10$	Magnitude $|D|$	Signed Rank
8	8	6
2	2	1
-6	6	-4
-9	9	-7
5	5	3
7	7	5
16	16	9
-10	10	-8
4	4	2

(b) H_0: Md $= 10$; H_a: Md > 10 $\alpha = .05$
Observed value: $W^- = 19$ Critical value: $c = 8$ Fail to reject H_0

13.9   **(a)**

Difference $D = x - 50$	Magnitude $\|D\|$	Signed Rank
-8	8	-4
-11	11	-5
-13	13	-6
5	5	3
-4	4	-2
2	2	1
-17	17	-7

(b) H_0: Md $= 50$; H_a: Md < 50 $\alpha = .05$
Observed value: $W^+ = 4$ Critical value: $c = 4$ Reject H_0

13.11 H_0: Md $= 10$; H_a: Md > 10 $\alpha = .01$
Observed value: $W^- = 19$ Critical value: $c = 20$ Reject H_0
The data suggest that the lotion does not perform as advertised.

13.13 H_0: Md $= 520$; H_a: Md > 520 $\alpha = .01$
Observed value: $W^- = 39.5$ Critical value: $c = 43$ Reject H_0
The data suggest that the advertising campaign was successful.

13.15 H_0: Md$_{x_1-x_2} = 0$; H_a: Md$_{x_1-x_2} < 0$ $\alpha = .05$

(a)   Observed value: $W^+ = 101$ Critical value: $c = 110$ Reject H_0

(b)   Observed value: $z = -1.89$ Critical value: $z = -1.65$
Reject H_0
The data suggest that judge A tended to give lower performance ratings
than judge B.

13.16

	Critical Value	Decision		Critical Value	Decision
(a)	11	Fail to reject H_0	**(b)**	21	Reject H_0
(c)	90	Reject H_0	**(d)**	31	Fail to reject H_0
(e)	45	Reject H_0	**(f)**	44	Fail to reject H_0
(g)	1	Fail to reject H_0	**(h)**	60	Reject H_0
(i)	7	Reject H_0			

13.17   **(a)**

Rank (x)	Rank (y)
1	4
2	6
3	8
5	10
7	
9	

(b) H_0: Md$_x =$ Md$_y$; H_a: Md$_x \neq$ Md$_y$ $\alpha = .10$
Observed value: $U = 6$ Critical value: $c = 3$ Fail to reject H_0

13.19 H_0: $\text{Md}_x = \text{Md}_y$; H_a: $\text{Md}_x < \text{Md}_y$ $\alpha = .05$
Observed value: $U_x = 10$ Critical value: $c = 13$ Reject H_0
The data suggest that the median endurance level of soccer players is larger than that for football players.

13.21 H_0: $\text{Md}_x = \text{Md}_y$; H_a: $\text{Md}_x > \text{Md}_y$ $\alpha = .05$
Observed value: $U_y = 26$ Critical value: $c = 27$ Reject H_0
The data support the spokesman's claim.

13.23

	c_1	c_2	R	Decision
(a)	2	9	2	Reject H_0
(b)	3	11	12	Reject H_0
(c)	3	12	4	Fail to reject H_0
(d)	3	11	8	Fail to reject H_0
(e)	3	12	7	Fail to reject H_0
(f)	4	13	9	Fail to reject H_0

13.25 H_0: The process is random; H_a: It is not random $\alpha = .05$
Observed value: $R = 3$ Critical values: $c_1 = 2$, $c_2 = 8$
Fail to reject H_0
There is insufficient evidence to conclude that the process is not random.

13.27 H_0: $\text{Md} = 625$; H_a: $\text{Md} > 625$ $\alpha = .05$
(a) Observed value: $x = 13$ Critical region: $\{14, 15, 16, 17, 18, 19, 20\}$
Fail to reject H_0
(b) Observed value: $W^- = 40$ Critical value: $c = 60$ Reject H_0

13.29 H_0: $\text{Md} = 200$; H_a: $\text{Md} < 200$ $\alpha = .05$
(a) Observed value: $x = 2$ Critical region: $\{0, 1\}$ Fail to reject H_0
(b) Observed value: $W^+ = 5$ Critical value: $c = 6$ Reject H_0

13.31 H_0: $\text{Md}_x = \text{Md}_y$; H_a: $\text{Md}_x < \text{Md}_y$ $\alpha = .05$
Observed value: $U_x = 11$ Critical value: $c = 15$ Reject H_0
The data suggest that girls tend to read with more comprehension than boys upon entering the fourth grade.

13.33 H_0: $\text{Md}_A = \text{Md}_B$; H_a: $\text{Md}_A < \text{Md}_B$ $\alpha = .01$
Observed value: $U_A - 8$ Critical value: $c = 3$ Fail to reject H_0
The data do not support the claim that the median price of building lots was less in resort area A than in resort area B.

13.35 H_0: The process is random; H_a: It is not random.
(a) Observed value: $R = 8$ Critical values: $c_1 = 8$, $c_2 = 19$
Reject H_0
(b) Observed value: $z = -2.25$ Critical values: $z = \pm 1.96$
Reject H_0
The data suggest a lack of randomness.

Subject Index